T0213950

Lecture Notes in Artificial Intelligence   12197

Subseries of Lecture Notes in Computer Science

More information about this series at http://www.springer.com/series/1244

Dylan D. Schmorrow · Cali M. Fidopiastis (Eds.)

Augmented Cognition

Human Cognition and Behavior

14th International Conference, AC 2020
Held as Part of the 22nd HCI International Conference, HCII 2020
Copenhagen, Denmark, July 19–24, 2020
Proceedings, Part II

 Springer

Editors
Dylan D. Schmorrow
Soar Technology Inc.
Orlando, FL, USA

Cali M. Fidopiastis
Design Interactive, Inc.
Orlando, FL, USA

ISSN 0302-9743   ISSN 1611-3349 (electronic)
Lecture Notes in Artificial Intelligence
ISBN 978-3-030-50438-0   ISBN 978-3-030-50439-7 (eBook)
https://doi.org/10.1007/978-3-030-50439-7

LNCS Sublibrary: SL7 – Artificial Intelligence

This Springer imprint is published by the registered company Springer Nature Switzerland AG
The registered company address is: Gewerbestrasse 11, 6330 Cham, Switzerland

Foreword

The 22nd International Conference on Human-Computer Interaction, HCI International 2020 (HCII 2020), was planned to be held at the AC Bella Sky Hotel and Bella Center, Copenhagen, Denmark, during July 19–24, 2020. Due to the COVID-19 coronavirus pandemic and the resolution of the Danish government not to allow events larger than 500 people to be hosted until September 1, 2020, HCII 2020 had to be held virtually. It incorporated the 21 thematic areas and affiliated conferences listed on the following page.

A total of 6,326 individuals from academia, research institutes, industry, and governmental agencies from 97 countries submitted contributions, and 1,439 papers and 238 posters were included in the conference proceedings. These contributions address the latest research and development efforts and highlight the human aspects of design and use of computing systems. The contributions thoroughly cover the entire field of human-computer interaction, addressing major advances in knowledge and effective use of computers in a variety of application areas. The volumes constituting the full set of the conference proceedings are listed in the following pages.

The HCI International (HCII) conference also offers the option of "late-breaking work" which applies both for papers and posters and the corresponding volume(s) of the proceedings will be published just after the conference. Full papers will be included in the "HCII 2020 - Late Breaking Papers" volume of the proceedings to be published in the Springer LNCS series, while poster extended abstracts will be included as short papers in the "HCII 2020 - Late Breaking Posters" volume to be published in the Springer CCIS series.

I would like to thank the program board chairs and the members of the program boards of all thematic areas and affiliated conferences for their contribution to the highest scientific quality and the overall success of the HCI International 2020 conference.

This conference would not have been possible without the continuous and unwavering support and advice of the founder, Conference General Chair Emeritus and Conference Scientific Advisor Prof. Gavriel Salvendy. For his outstanding efforts, I would like to express my appreciation to the communications chair and editor of HCI International News, Dr. Abbas Moallem.

July 2020   Constantine Stephanidis

HCI International 2020 Thematic Areas
and Affiliated Conferences

Thematic areas:

- HCI 2020: Human-Computer Interaction
- HIMI 2020: Human Interface and the Management of Information

Affiliated conferences:

- EPCE: 17th International Conference on Engineering Psychology and Cognitive Ergonomics
- UAHCI: 14th International Conference on Universal Access in Human-Computer Interaction
- VAMR: 12th International Conference on Virtual, Augmented and Mixed Reality
- CCD: 12th International Conference on Cross-Cultural Design
- SCSM: 12th International Conference on Social Computing and Social Media
- AC: 14th International Conference on Augmented Cognition
- DHM: 11th International Conference on Digital Human Modeling and Applications in Health, Safety, Ergonomics and Risk Management
- DUXU: 9th International Conference on Design, User Experience and Usability
- DAPI: 8th International Conference on Distributed, Ambient and Pervasive Interactions
- HCIBGO: 7th International Conference on HCI in Business, Government and Organizations
- LCT: 7th International Conference on Learning and Collaboration Technologies
- ITAP: 6th International Conference on Human Aspects of IT for the Aged Population
- HCI-CPT: Second International Conference on HCI for Cybersecurity, Privacy and Trust
- HCI-Games: Second International Conference on HCI in Games
- MobiTAS: Second International Conference on HCI in Mobility, Transport and Automotive Systems
- AIS: Second International Conference on Adaptive Instructional Systems
- C&C: 8th International Conference on Culture and Computing
- MOBILE: First International Conference on Design, Operation and Evaluation of Mobile Communications
- AI-HCI: First International Conference on Artificial Intelligence in HCI

Conference Proceedings Volumes Full List

http://2020.hci.international/proceedings

14th International Conference on Augmented Cognition (AC 2020)

Program Board Chairs: **Dylan D. Schmorrow, Soar Technology Inc., USA, and Cali M. Fidopiastis, Design Interactive, Inc., USA**

- Amy Bolton, USA
- Martha E. Crosby, USA
- Fausto De Carvalho, Portugal
- Daniel Dolgin, USA
- Sven Fuchs, Germany
- Rodolphe Gentili, USA
- Monte Hancock, USA
- Frank Hannigan, USA
- Kurtulus Izzetoglu, USA
- Ion Juvina, USA
- Chang S. Nam, USA
- Sarah Ostadabbas, USA
- Mannes Poel, The Netherlands
- Stefan Sütterlin, Norway
- Suraj Sood, USA
- Ayoung Suh, Hong Kong
- Georgios Triantafyllidis, Denmark
- Melissa Walwanis, USA

The full list with the Program Board Chairs and the members of the Program Boards of all thematic areas and affiliated conferences is available online at:

http://www.hci.international/board-members-2020.php

HCI International 2021

The 23rd International Conference on Human-Computer Interaction, HCI International 2021 (HCII 2021), will be held jointly with the affiliated conferences in Washington DC, USA, at the Washington Hilton Hotel, July 24–29, 2021. It will cover a broad spectrum of themes related to Human-Computer Interaction (HCI), including theoretical issues, methods, tools, processes, and case studies in HCI design, as well as novel interaction techniques, interfaces, and applications. The proceedings will be published by Springer. More information will be available on the conference website: http://2021.hci.international/.

General Chair
Prof. Constantine Stephanidis
University of Crete and ICS-FORTH
Heraklion, Crete, Greece
Email: general_chair@hcii2021.org

http://2021.hci.international/

Contents – Part II

Contents – Part I

Augmented Cognition in Learning

Assessing Intravenous Catheterization Simulation Training of Nursing Students Using Functional Near-Infrared Spectroscopy (fNIRs)

Mehmet Emin Aksoy[1,2], Kurtulus Izzetoglu[3], Atahan Agrali[1(✉)], Esra Ugur[4],
Vildan Kocatepe[4], Dilek Kitapcioglu[2], Engin Baysoy[1], and Ukke Karabacak[4]

[1] Biomedical Device Technology Department,
Acibadem Mehmet Ali Aydinlar University, Istanbul, Turkey
atahanagrali@gmail.com
[2] Center of Advanced Simulation and Education (CASE),
Acibadem Mehmet Ali Aydinlar University, Istanbul, Turkey
[3] School of Biomedical Engineering, Science and Health Systems, Drexel University,
Philadelphia, USA
[4] Department of Nursing, Faculty of Health Sciences,
Acibadem Mehmet Ali Aydinlar University, Istanbul, Turkey

Abstract. Training of healthcare providers, throughout their undergraduate, graduate and postgraduate includes a wide range of simulation based training modalities. Serious gaming has become an important training modality besides other simulation systems and serious gaming based educational modules have been implemented into the educational curriculum of nursing training. The aim of this study was to investigate whether functional near-infrared spectroscopy (fNIRS) for monitoring hemodynamic response of the prefrontal cortex can be used as an additional tool for evaluating performances of nursing students in the intravenous catheterization training. The twenty three participants were recruited for this study. The novice group consisted of fifteen untrained nursing students and the expert group had eight senior nurses. Training performances of students evaluated over, pre and posttest scores and fNIRS measurements from task trainer application. Novice group completed the protocol on four sessions on four different days. On day 1 and day 30, they performed on task trainer procedure. On day 2 and day 7, participants of the novice group were trained only by using the Virtual Intravenous Simulator Module. The expert group only performed on the task trainer on days 1 and 30. Participants' performance scores and fNIRS results of the 1st and 30th day training sessions were compared. fNIRS measurements revealed significant changes in prefrontal cortex (PFC) region oxygenation of novice group, while their post-test scores increased. Our findings suggests that, fNIRS could be used as an objective, supportive assessment techniques for effective monitor of expertise development of the nursing trainees.

Keywords: Nursing education · I.V. catheterization · Optical brain imaging

© Springer Nature Switzerland AG 2020
D. D. Schmorrow and C. M. Fidopiastis (Eds.): HCII 2020, LNAI 12197, pp. 3–13, 2020.
https://doi.org/10.1007/978-3-030-50439-7_1

1 Introduction

Over the last two decades, simulation based medical trainings have been widely used in nursing trainings in order to improve patient safety and self confidence levels of nursing students and to meet today's demands in the healthcare services [1–3].

Serious gaming is becoming an important modality among other simulation modalities used for undergraduate and post graduate levels of nursing trainings [4–6]. Serious gaming provides interactivity and multiple gaming environments thus actively engaging and motivating the learners to in the process of learning [6]. Another advantage of serious gaming is that it provides a location and time independent learning opportunity for learners [5]. Serious gaming modules with various learning contents like advanced life support, intravenous (I.V.) catheterization, basic life support or patient encounter scenarios are nowadays commonly used for undergraduate and post graduate level of healthcare training [7].

Integrated scoring systems of simulators or observational analysis based scoring criteria prepared by senior nursing educators are nowadays the most commonly tools to evaluate the outcome of the simulation trainings.

Due to the limitations of the traditional assessment techniques in terms of development of expertise during practice of complex cognitive and visuomotor tasks, educators are looking for new supportive assessment techniques [8]. As learning is associated with complex processes of the human brain, measuring cognitive and mental workload of trainees is crucial while performing the critical tasks [8, 9].

It is known that specific regions of the human brain are activated during motor execution, verbalization and observation [10–12]. In parallel to this, oxygen consumption of these specific areas of the brain varies due to the hemodynamic changes which are characterized by the alterations of oxygenated hemoglobin (HbO2) and cerebral hemoglobin (HHb) chromophores [13].

Performing psychomotor skills affect the amount of hemodynamic response in different parts of brain. Measuring the alterations of hemodynamic changes in prefrontal cortex (PFC) of the human brain provides valuable data about novel motor skill acquisition, attention, working memory and learning [14, 15]. Electroencephalography (EEG), functional Magnetic Resonance Imaging (fMRI), Magnetoencephalography (MEG), Positron Emission Tomography (PET) and functional Near-Infrared Spectroscopy (fNIRS) are neuroimaging techniques commonly used for monitoring human brain activation in response to a given specific task. fNIRS system, which measures neuronal activities in human brain by using near-infrared light, has been chosen as the neuroimaging technique for this study as it is a non-invasive, affordable and practical cerebral hemodynamic monitoring system [15–20].

fNIRS imaging technique has already been used to measure cognitive and mental workload in various studies about task specific training procedures such as open surgery, laparoscopic surgery, and robot assisted surgery trainings [14, 15, 19, 22–24].

The aim of this study is to investigate whether fNIRS technology can be used as an additional assessment tool for I.V. catheterization training of nursing students by measuring hemodynamic responses induced by the task-specific cognitive workload in the PFC regions of the participants.

2 Materials and Methods

2.1 Participants

Twenty three participants were recruited for this study. All twenty three participants were provided written informed consent, which has been reviewed and approved by the Ethical Committee of Acıbadem Mehmet Ali Aydinlar University. Participants were divided into two groups as novices and experts. The novice group consisted of fifteen untrained nursing students and the expert group had eight senior nurses (Table 1).

Table 1. Distribution of the participants

Group		N	Average age (SD)	Female: Male
Novice	First year nursing students	15	18.8($\pm$0.7)	13:2
Expert	Senior nurses	8	33.5($\pm$6.05)	8:0
Total		23	23.9($\pm$7.9)	21:2

2.2 Experimental Protocol

The protocol was run for 30 days and data acquisition occurred on 1st, 2nd, 7th and 30th days. The novice group attended all 4 days of the data acquisition sessions. Initial stage of the protocol, novice participants took a pretest on the theoretical knowledge level about the I.V. catheterization procedure. Following the pretest, novice group continued the experimental protocol by applying intravenous catheterization on the task trainer module and as the last task of the day 1, novice group participants completed the serious gaming intravenous catheterization training module. The Virtual I.V. Simulator (Laerdal; Norway; www.laerdal.com) was used as the I.V. training module. On the second and third sessions, participants of the novice group completed only the Virtual I.V. Simulator module. On the 30th day of the experiment, novice group started the session with the Virtual I.V. simulator training module and the task trainer module followed. After the final session of the protocol, a posttest was given to novice group in order to measure the differences of theoretical knowledge level between day 1 and day 30. The expert group completed the protocol in two sessions, on day 1 and on day 30, only by using the task trainer (see Fig. 1).

Three types of measurements, theoretical test scores, I.V. Virtual Simulator scores and cortical oxygenation changes from the prefrontal cortex via fNIRS, were captured during the experiment (see Fig. 2). In addition to the scores, participants were camera recorded during performing on the task trainer modules and screen recordings of the computer running the I.V. Virtual Simulator module were also gathered.

Pretest and Posttest. A theoretical test were given to the participants of the novice group to compare their knowledge levels about the I.V. catheterization process at the beginning and at the end of the study. The theoretical test was prepared by the faculty of School

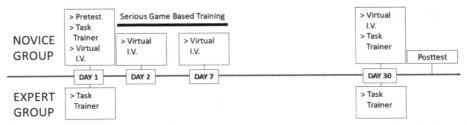

Fig. 1. Timeline of the experimental protocol and data acquisition sessions of 1st, 2nd, 7th and 30th days for both groups.

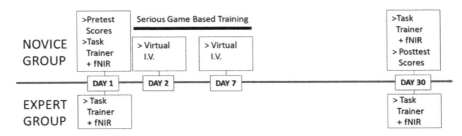

Fig. 2. Types of data acquired on sessions.

of Nursing of Acıbadem Mehmet Ali Aydinlar University and it is a part of the standard curriculum for the evaluation of I.V. catheterization training classes. In total there are 17 questions in the test and it consists of 2 multiple choice scenario analysis questions and 15 true/false survey questions (See Appendix for the short version example of the test).

Task Trainer Module. A manikin arm designed for intravenous catheterization practice. Task trainers, utilized in most clinical courses of the nursing schools, has been described as the traditional method for teaching peripheral vascular access [25, 26]. During the task trainer sessions of the protocol, participants of both groups wore the fNIR headband, which recorded the hemodynamic responses (see Fig. 3.a).

Virtual I.V. Simulator Module were utilized in this study as the intravenous catheterization training simulator module. It presents a fully interactive platform for self-directed learning of trainees during performing of intravenous catheterization skills [27]. The Virtual I.V. simulator module has haptic features and feedback mechanism that provides feeling of skin stretch, palpation, size, and insertion forces alongside an assessment tool that stores performance scores of trainees (see Fig. 3.b). Data acquired from this modality were not included in the analysis. The aim of using this modality was to train the participants and to increase their knowledge on I.V. catheterization procedure.

Functional Near-Infrared Spectroscopy. fNIRS is safe, portable and a non-invasive optical brain imaging modality. Continuous wave fNIRS Imager 1200 system (fNIRS Devices LLC, Potomac, MD) was used to record the hemodynamic response from the prefrontal cortex. The system has three components: a headband that carries the 16 channel sensor pad, a control box and a computer for acquiring and recording fNIRS data.

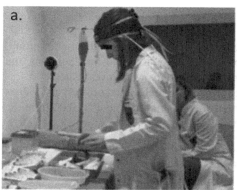 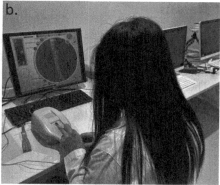

Fig. 3. a. A participant is performing I.V. catheterization on the task trainer. b. A participant using Virtual I.V. Simulation Module

The sensor pad has 4 light emitting diodes (LED) as light sources and 10 detectors. The 2.5 cm of light source and detector separation was designed to monitor anterior PFC underlying the forehead [18].

2.3 Data Analysis

fNIRS Data. The raw (light intensity) data recorded through fNIRs were filtered using a low pass finite impulse hamming filter. Sliding window motion artifact rejection filter was applied in order to exclude noise from the data, which were caused by the motion artifacts [28]. Oxygenation values for each of the 16 channels were calculated by applying The Modified Beer Lambert Law [8]. We have selected Oxy marker- difference between the oxyHb and deoxyHb – for the statistical analysis and channels of the left hemisphere as the regions of interest, which is an area known to be associated with working memory during learning and training studies [8, 22, 29].

Statistical Analysis. The normality of continuous variables were investigated by Shapiro-Wilk's test. Descriptive statistics were presented using mean and standard deviation for normally distributed variables and median (and minimum-maximum) for the non-normally distributed variables. For comparison of two non-normally distributed dependent groups Wilcoxon Signed Rank test was used. For comparison of two non-normally distributed independent groups Mann Whitney U test was used. Non-parametric statistical methods were utilized for values with skewed distribution. Statistical significance was accepted when two-sided p value was lower than 0.05. Statistical analysis was performed using the MedCalc Statistical Software version 12.7.7 (MedCalc Software, Ostend, Belgium, www.medcalc.org). Virtual I.V. Simulator scores excluded from statistical analysis of this study.

3   Results

In this study, the differences between the pretest and posttest scores of the novice group, changes of oxy values of the novice group between the initial and final sessions, the differences of oxy values between novice and expert groups were investigated.

There is a statistically significant difference between pretest and posttest test values (Wilcoxon test p < 0.05). Novice group scored significantly higher (P = 0.007) on the posttest when compared with the pretest (see Table 2).

Table 2. Pretest and Posttest Scores of the Novice Group

	Day 1 – Pretest Mean + SD	Day 30 – Posttest Mean + SD	P Value
Test scores	78.82 ± 6.6	87.95 ± 8.31	0.007

Wilcoxon test

The average oxygen consumption of the novice group on the left PFC were significantly higher during the first task trainer session when compared to the final (day 30) task trainer session (Wilcoxon test p < 0.05, p = 0.002). Average oxygen consumption of the novice group were significantly higher during the first task trainer session when comparing to the expert group (Mann-Whitney U test p < 0.05, p = 0.004). There are no significant differences of the left PFC oxygen consumption between groups during the day 30 task trainer session (see Table 3).

Table 3. Novice vs. Expert, Day 1 vs. Day 30, Task Trainer- Oxy Comparison

	Day 1 – OXY Mean + SD Med.(Min − Max)	Day 30 – OXY Mean + SD Med.(Min − Max)	P Value*
Novice	3.8 ± 1.69 3.52 (1.48 − 7.17)	1.52 ± 1.5 2.13 (−1.84 − 3.64)	***0.002**
Expert	1.42 ± 1.37 1.33 (−0.17 − 4.14)	1.17 ± 1.27 1.76 (−0.67 − 2.43)	*0.735
P Value	**0.004**	0.368	

Mann Whitney U test, *Wilcoxon test

These results indicate that within 30 days of the protocol; oxygen consumption of the novice group significantly decreased while performing I.V. catheterization on a task trainer. Between the beginning and end of the protocol, novice group managed to significantly increase their scores on the theoretical test (See Fig. 4).

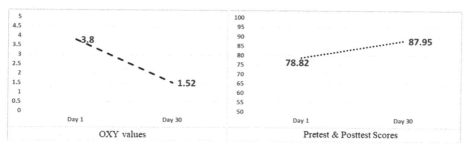

Fig. 4. Oxy values and test scores comparison of the novice group, first day vs. last day of the experimental protocol.

Oxygen consumption of the expert group slightly dropped, as expected there is no statistical significance of this minor shift. On the first day of the experimental protocol, novice groups' oxy values were significantly higher than the experts. However, on day 30, novice group oxy values arrived at a similar level of the oxy values expert group (see Fig. 5).

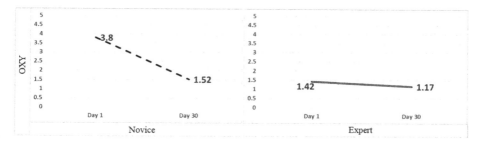

Fig. 5. Comparison of novice and expert groups' oxy values on day 1 and day 30

4 Discussion and Conclusion

Regarding the patient safety, simulation based training has a crucial role for education of nursing students before they work on the real patient in clinical setting. In the field of simulation, it was shown that repetitive simulation sessions increase the performance and self-confidence levels of nursing students in several clinical practices [30]. Simulation based education has been integrated into the curriculum of our university's nursing school as a learning and assessment tool in the last seven years. Task trainers and different simulation modalities like serious game based modalities and mannequin based simulators are used for this purpose at our university. In the last 18 months, we have been focusing on revealing the efficiency of using fNIRS measurements for monitoring the improvement of learners' cognitive and mental workloads during simulation based trainings [31].

As it was revealed in the prior similar studies, using fNIRS technology in the brain-based learning approach may serve as an objective and complementary assessment tool to generate indicators or benchmarks to assess the outcome of the simulation based sessions [14, 15, 19, 22–24]. In this study, the evaluation of the oxygenation (OXY values) at the left PFC regions of the novice group revealed statistically significant changes between first and last sessions of the study indicated less oxygen use on the left PFC regions during the last session of the study. The OXY levels of the expert group were statistically similar on day 1 and day 30. Similar OXY values were obtained from the experts and the novices in their left PFC regions during the last (30th day) task trainer session. Pre and Posttest scores of novice group have a correlation with fNIRs measurements associated with hemodynamic responses of the PFC. These data clearly indicate that novice group had almost reached similar levels of oxygen consumption with expert group in their left PFC at the end of the training protocol.

Further studies are required in-depth analysis for the fNIRS measurements can be used in combination with the existing scoring systems currently utilized for simulation based trainings. With the help of the new generation fNIRS devices, potential to portable and wireless, the measurements will be much easier and ecologically-valid for the educators and learners.

Disclosure. fNIR Devices, LLC manufactures the optical brain imaging instrument and licensed IP and know-how from Drexel University. Dr. K. Izzetoglu was involved in the technology development and thus offered a minor share in the startup firm, fNIR Devices, LLC.

Appendix

A short sample of the theoretical test novice group took on Day 1 and Day 30.

A 22-year-old patient with acute appendicitis was admitted to your clinic. Vascular access is need to apply I.V. fluid therapy to the patient, your task is to apply I.V. catheter to patient. Please answer the following questions related to the case above.

- Please identify and select the option with the correct pairings for the patient in the case above.

I. Hand	a. Basilic Vein
II. Arm	b. Femoral Vein
III. Leg	c. Cephalic and median vein
IV. Foot	d. dorsal metacarpal vein
	e. dorsal metatarsal arch veins

a) I-c, II-a, III-e, IV-b-d
b) I-d, II-a-c, III-b, IV-e
c) I-e, II-a-b, III-d, IV-e
d) I-a, II-b, III-c, IV, d-e
e) I-b, II-d, III-a-e, IV-c

Determine whether the statements below are **True** or **False** by marking the appropriate box next to the statement.

- When applying into visible veins, gauge is inserted with 15° angle.   T   F

- After applying the drug, you need to untie the tourniquet.   T   F

- 20-22 gauge catheters are often preferred for catheter administration.   T   F

- The tourniquet is placed 30 cm above the aplication area.   T   F

References

1. Benner, P.: Educating nurses: a call for radical transformation-how far have we come? J. Nurs. Educ. **51**(4), 183–184 (2012)
2. Berndt, J.: Patient safety and simulation in prelicensure nursing education: an integrative review. Teach. Learn. Nurs. **9**(1), 16–22 (2014)
3. Cummings, C.L., Connelly, L.K.: Can nursing students' confidence levels increase with repeated simulation activities? Nurse Educ. Today **36**, 419–421 (2016)
4. Johnsen, H., Fossum, M., Vivekananda-Schmidt, P., Fruhling, A., Slettebø, Å.: A serious game for teaching nursing students clinical reasoning and decision-making skills. Studies in health technology and informatics. In: Sermeus, W., Procter, P.M., Weber, P. (eds.) E-health for All: Every Level Collaboration - from Project to Realization, vol. 225, pp. 905–906. IOS Press, Amsterdam (2016)
5. Aksoy, M., Sayali, E.: Serious gaming as an additional learning tool for medical education. Int. J. Educ. Technol. Learn. **5**, 52–59 (2019)

6. Elaachak, L., Belahbib, A., Bouhorma, M.: A digital revolution in Nursing Education-the serious games. In: 2016 5th International Conference on Multimedia Computing and Systems, pp. 705–709. IEEE (2016)
7. Keleekai, N.L., et al.: Improving nurses' peripheral intravenous catheter insertion knowledge, confidence, and skills using a simulation-based blended learning program: a randomized trial. Simul. Healthcare **11**(6), 376 (2016)
8. Ayaz, H., Shewokis, P.A., Bunce, S., Izzetoglu, K., Willems, B., Onaral, B.: Optical brain monitoring for operator training and mental workload assessment. Neuroimage **59**(1), 36–47 (2012)
9. Pool, C.R.: Brain-based learning and students. Educ. Digest **63**(3), 10–16 (1997)
10. Grezes, J., Decety, J.: Functional anatomy of execution, mental simulation, observation, and verb generation of actions: a meta-analysis. Hum. Brain Map. **12**(1), 1–19 (2001)
11. Gallese, V., Goldman, A.: Mirror neurons and the simulation theory of mind-reading. Trends Cogn. Sci. **2**(12), 493–501 (1998)
12. Wulf, G., Shea, C., Lewthwaite, A.: Motor skill learning and performance: a review of influential factors. Med. Educ. **44**(1), 75–84 (2010)
13. Modi, H.N., Singh, H., Yang, G.Z., Darzi, A., Leff, D.R.: A decade of imaging surgeons' brain function (part I): terminology, techniques, and clinical translation. Surgery **162**, 1121–1130 (2017)
14. Shetty, K., Leff, D.R., Orihuela-Espina, F., Yang, G.Z., Darzi, A.: Persistent prefrontal engagement despite improvements in laparoscopic technical skill. JAMA Surg. **151**(7), 682–684 (2016)
15. Ohuchida, K., et al.: The frontal cortex is activated during learning of endoscopic procedures. Surg. Endosc. **23**(10), 2296–2301 (2009)
16. Rolfe, P.: In vivo near-infrared spectroscopy. Annu. Rev. Biomed. Eng. **2**, 715–754 (2000)
17. Izzetoglu, M., Bunce, S.C., Izzetoglu, K., Onaral, B., Pourrezaei, K.: Functional brain imaging using near-infrared technology. IEEE Eng. Med. Biol. Mag. **26**, 38–46 (2007)
18. Izzetoglu, M., et al.: Functional near-infrared neuroimaging. IEEE Trans. Neural Syst. Rehabil. Eng. **13**(2), 153–159 (2005)
19. Modi, H.N., Singh, H., Yang, G.Z., Darzi, A., Leff, D.R.: A decade of imaging surgeons' brain function (part II): a systematic review of applications for technical and nontechnical skills assessment. Surgery **162**, 1130–1139 (2017)
20. Bunce, S.C., et al.: Implementation of fNIRS for monitoring levels of expertise and mental workload. In: Schmorrow, D.D., Fidopiastis, C.M. (eds.) FAC 2011. LNCS (LNAI), vol. 6780, pp. 13–22. Springer, Heidelberg (2011). https://doi.org/10.1007/978-3-642-21852-1_2
21. Tian, F., et al.: Quantification of functional near infrared spectroscopy to assess cortical reorganization in children with cerebral palsy. Opt. Express **18**(25), 25973–25986 (2010)
22. Leff, D.R., et al.: Functional prefrontal reorganization accompanies learning-associated refinements in surgery: a manifold embedding approach. Comput. Aided Surg. **13**, 325–339 (2008)
23. Leff, D.R., Orihuela-Espina, F., Atallah, L., Darzi, A., Yang, G.-Z.: Functional near infrared spectroscopy in novice and expert surgeons – a manifold embedding approach. In: Ayache, N., Ourselin, S., Maeder, A. (eds.) MICCAI 2007. LNCS, vol. 4792, pp. 270–277. Springer, Heidelberg (2007). https://doi.org/10.1007/978-3-540-75759-7_33
24. Mylonas, G.P., Kwok, K.-W., Darzi, A., Yang, G.-Z.: Gaze-contingent motor channelling and haptic constraints for minimally invasive robotic surgery. In: Metaxas, D., Axel, L., Fichtinger, G., Székely, G. (eds.) MICCAI 2008. LNCS, vol. 5242, pp. 676–683. Springer, Heidelberg (2008). https://doi.org/10.1007/978-3-540-85990-1_81
25. Hayden, J.: Use of simulation in nursing education: national survey results. J. Nurs. Regul. **1**(3), 52–57 (2010)

26. Alexandrou, E., Ramjan, L., Murphy, J., Hunt, L., Betihavas, V., Frost, S.A.: Training of undergraduate clinicians in vascular access: an integrative review. J. Assoc. Vasc. Access **17**(3), 146–158 (2012)
27. Laerdal Medical, Virtual I.V.: Directions for Use (2009)
28. Ayaz, H., Izzetoglu, M., Shewokis, P.A., Onaral, B.: Sliding-window motion artifact rejection for Functional Near-Infrared Spectroscopy. In: Annual International Conference of the IEEE Engineering in Medicine and Biology, pp. 6567–6570. IEEE, Buenos Aires (2010)
29. Kelly, A.M., Garavan, H.: Human functional neuroimaging of brain changes associated with practice. Cereb. Cortex (2005)
30. Labrague, L.J., McEnroe-Petitte, D.M., Bowling, A.M., Nwafor, C.E., Tsaras, K.: High-fidelity simulation and nursing students' anxiety and self-confidence: a systematic review. Nurs. Forum **54**(3), 358–368 (2019)
31. Aksoy, E., Izzetoglu, K., Baysoy, E., Agrali, A., Kitapcioglu, D., Onaral, B.: Performance monitoring via functional near infrared spectroscopy for virtual reality based basic life support training. Front. Neurosci. **13**, 1336 (2019)

Using fMRI to Predict Training Effectiveness in Visual Scene Analysis

Joseph D. Borders$^{(\boxtimes)}$, Bethany Dennis, Birken Noesen, and Assaf Harel

Department of Psychology, Wright State University, Dayton, OH, USA
borders.9@wright.edu

Abstract. Visual analysis of complex real-world scenes (e.g. overhead imagery) is a skill essential to many professional domains. However, little is currently known about how this skill is formed and develops with experience. The present work adopts a neuroergonomic approach to uncover the underlying mechanisms associated with the acquisition of scene expertise, and establish neurobehavioral markers for the effectiveness of training in scene imagery analysis. We conducted an intensive six-session behavioral training study combined with multiple functional MRI scans using a large set of high-resolution color images of real-world scenes varying in their viewpoint (aerial/terrestrial) and naturalness (manmade/natural). Participants were trained to categorize the scenes at a specific-subordinate level (e.g. suspension bridge). Participants categorized the same stimuli for five sessions; the sixth session consisted of a novel set of scenes. Following training, participants categorized the scenes faster and more accurately, reflecting memory-based improvement. Learning also generalized to novel scene images, demonstrating learning transfer, a hallmark of perceptual expertise. Critically, brain activity in scene-selective cortex across all sessions significantly correlated with learning transfer effects. Moreover, baseline activity (pre-training) was highly predictive of subsequent perceptual performance. Whole-brain activity following training indicated changes to scene- and object-selective cortex, as well as posterior-parietal cortex, suggesting potential involvement of top-down visuospatial-attentional networks. We conclude that scene-selective activity can be used to predict enhancement in perceptual performance following training in scene categorization and ultimately be used to reveal the point when trainees transition to an expert-user level, reducing costs and enhancing existing training paradigms.

Keywords: Expertise · Training · Neuroergonomics

1 Introduction

Visual analysis of complex real-world scenes is essential to a variety of professional contexts, primarily in the defense sector, in fields such as overhead imagery analysis, remote sensing, and target detection, as well as in several civilian contexts (e.g. land surveys, weather prediction). And while automated scene recognition systems are continuing to improve their ability to detect, track, and classify scenes [1], there are two unique human properties that cannot be easily emulated. The first is the ability to rapidly

© Springer Nature Switzerland AG 2020
D. D. Schmorrow and C. M. Fidopiastis (Eds.): HCII 2020, LNAI 12197, pp. 14–26, 2020.
https://doi.org/10.1007/978-3-030-50439-7_2

and flexibly categorize scenes along a number of dimensions and across various behavioral contexts [2]. The second is the ability to detect the "unknown unknowns"; the anomalies that are rarely, if ever encountered and therefore, cannot be modeled easily [3]. This suggests a need for deeper understanding of how humans develop expertise in scene analysis, key to which is determining the neural circuits and mechanisms subserving scene expertise. Uncovering the neural substrates of scene expertise has both theoretical and applied advantages. First, understanding how the brain gives rise to proficiency in complex scene understanding can provide critical insights into the nature of mechanisms involved in it and provide constraints on theories of expertise acquisition [4, 5]. Second, neural markers of scene analysis can be used as implicit measures of learning benchmarks (i.e., novice, proficient, expert) [5, 6]. Third, neuroimaging tools, such as functional magnetic resonance imaging (fMRI) can provide diagnostic metrics to determine operational readiness as well as measures of selection and assessment [7]. The current paper describes how such a neuroergonomic approach can be utilized to establish neuromarkers of expertise acquisition to inform individualized training paradigms with the ultimate goal of optimizing performance in imagery analysis.

The main challenge of studying the development of scene expertise is that, as noted above, humans are already extremely adept at recognizing complex visual scenes, which leaves very little room for improvement (or corresponding neural changes). One solution to this challenge is studying one type of scene imagery that does require intensive training, namely, aerial scenes. Aerial scene recognition can be extremely demanding. Through evolution, our scene recognition system developed to extract diagnostic features reflecting statistical regularities of the world from an earth-bound perspective (e.g. gravitational frame; [8]). As such, most people are simply not familiar with these analogous statistical regularities in aerial imagery. Moreover, aerial scenes are much more homogeneous than terrestrial scenes and thus, more difficult to discriminate [9]. To overcome this challenge, intensive training is usually required. However, there is currently very little research on how people recognize aerial scenes [10], and specifically, on how expertise in aerial scene recognition is developed [11]. Neuroimaging studies report that aerial scene recognition is supported by some of the same areas that are engaged in everyday (terrestrial) scene recognition [12, 13]. These areas, such as the PPA (parahippocampal place area [14]), OPA (occipital place area [15]), and RSC/MPA (retrosplenial complex/medial place area [16]) comprise a network of dedicated cortical areas in posterior-parietal, occipital, and medio-temporal cortex, which interacts with additional large-scale cortical networks to support spatial memory and navigation [17]. However, no study to date (to the best of our knowledge), has explored how increased experience in aerial scene recognition modulates the scene-selective network.

We suggest that we can harness existing theoretical neural frameworks of visual expertise to overcome this gap in knowledge, and ultimately predict what will change in people who undergo intensive training in aerial scene recognition. Specifically, the interactive framework of expertise [4, 18] suggests that visual expertise is a controlled, interactive process, rather than a purely perceptual skill. Visual expertise, according to this view, develops from the reciprocal interactions between the visual system and multiple top-down factors, including semantic knowledge, top-down attentional control, and task relevance, and it is these interactions that enable the ability to flexibly access domain-specific information at multiple scales and levels guided by multiple recognition goals [4, 18–20]. Accordingly, extensive experience within a given visual domain should manifest in the recruitment of multiple systems, reflected in widespread neural activity beyond visual cortex, primarily in areas involved in higher-level cognitive processes. To test these ideas in the domain of aerial scene recognition, we conducted the present study. We trained naïve participants to categorize aerial scenes, their terrestrial counterparts (see below), and measured neural activity before, during and after training using fMRI. Our objective was to two-fold: (1) uncover how training in aerial scene categorization impacts memory and perception, and (2) assess the extent to which these behavioral changes are associated with changes in neural activity in dedicated scene-selective regions, and higher-level cognitive brain regions. Our central hypothesis is that relating specific behavioral changes associated with the scene expertise acquisition to specific changes in neural activity will determine the importance of changes in neural scene representations as markers of scene expertise.

2 Methods

Subjects. 26 subjects (13 female, age range: 19–34) volunteered to participate in the experiment for monetary compensation. Subjects came from the Wright State University community, had normal or corrected-to-normal visual acuity, and no history of psychiatric or neurological disorders. Subjects provided their written informed consent, as well as signed a HIPAA authorization form, all approved by the Wright State University Institutional Review Board (IRB). Subjects were randomly assigned to either the experimental (n = 14, 5 female, age range: 19–33) or control group (n = 12, 8 female, age range: 20–34) (see details below).

Stimuli. We used a large set of high-resolution color scene images, consisting of 480 individual images spanning 6 dimensions: Naturalness (manmade/natural), Category (e.g. "airport"), Sub-category (e.g. "military airbase"), and Exemplar (a specific place, e.g., "Edwards AFB") and Viewpoint (terrestrial/aerial). Each exemplar presented a single place, both in a terrestrial and an aerial viewpoint. We used two viewpoints of the same place to determine the extent to which scene expertise involves an increase in the observer's ability to extract invariant information from one viewpoint and apply it to another. Terrestrial images were collected from the Internet. Aerial images were collected using Bing maps (bing.com/maps), and controlled for altitude (approximately 400–600 ft.). Each subset of aerial and terrestrial images consisted of 240 images; 120 manmade and 120 natural scenes, 4 categories for each dimension: airports, bridges, stadiums, settlements and mountains, deserts, land types, and bodies of water. Each of these categories contained three sub-categories, and ten exemplars within each sub-category. For example, the category "Deserts" contained three types of deserts: Sandy, Shrub and Rocky, and each of these types of desert contained ten individual images of specific desert landscapes (Fig. 1).

Procedures. A full training regimen consisted of behavioral training sessions interleaved with MRI scanning sessions. Behavioral training regimens for both groups comprised of six sessions. Half of the stimuli were used within the first five sessions (each

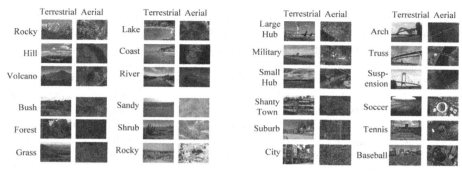

Fig. 1. Examples of scene stimuli used in the study and their category labels. Each image depicted the same place from a terrestrial and aerial viewpoint (depicted here are one of ten exemplars for each category). The images span both natural and manmade categories, four categories within each (Natural: Mountains, Land Types, Bodies of Water, Deserts; Manmade: Airports, Settlements, Bridges, Stadiums). Each subcategory contained three subordinate categories as labeled here (labels were not presented during the neuroimaging sessions).

participant's stimulus set was pseudorandomly selected to conform to the stimulus hierarchy), and the sixth session consisted of the other half of the stimulus list and were previously unseen. In addition to the behavioral training, participants were scanned four times: pre-training (baseline), post-training, and two scans during the course of training. Both the experimental and control groups saw the same scenes they were trained with in the scanner.

Behavioral Training Paradigm. Participants in the experimental group were trained to categorize the scenes at the sub-category level using a category verification paradigm. Each trial began with the presentation of a sub-category label (e.g. "sandy desert"), followed by a scene image. Participants task was to determine whether the image fits the category scene label or not (feedback was provided at the end of each trial). Every image was presented three times during the session, culminating in a 720 individual trials per training session. The control group was exposed to the same stimuli for the same number of times, but performed an orthogonal fixation cross task, which did not require active categorization. Participants in both groups viewed the stimuli on a 24″ LCD monitor at a viewing distance of approximately 57 cm. Each image was 600×400 pixels in size (subtending $15.88° \times 10.52°$ visual angle). All stimuli were displayed at the center of the screen on top of a gray background.

Behavioral Analysis. Mean reaction times and accuracy scores were calculated for each condition. Any outliers were replaced with the participants' overall mean across conditions within-session. Inverse Efficiency Scores (IES) were calculated to accommodate for any speed-accuracy trade-off. IES was calculated by taking mean response time divided by the proportion correct [21]. The individual IES means were then submitted to ANOVAs assessing two metrics of performance: within-set learning (memory), (2) generalization to new exemplars (perceptual expertise).

fMRI Procedures. In the scanner, subjects performed a 1-back task, which required only minimal knowledge of/or engagement with the scene images, and thus allowed us to compare the two groups irrespective of the subjects' overall engagement and experience. We employed a typical block-design fMRI experiment with Naturalness and Viewpoint as independent variables, yielding 4 conditions: manmade-terrestrial, manmade-natural, natural-terrestrial, and natural-aerial. A single scan session consisted of 4 experimental runs: all 240 images were presented within a single run. A run comprised of 20 blocks, each of the 4 conditions presented in 5 blocks, pseudorandomized within a session across subjects. Each block consisted of 12 individual scene images (+2 repetitions for the 1-back task), spanning the 60 individual scenes/condition. In addition, each subject completed an independent block-design localizer scan to identify scene - (PPA, RSC/MPA, OPA), object - (LOC: Lateral Occipital Complex) and face-selective regions (FFA: Fusiform Face Area) and non-selective, early visual cortex (EVC) (for full detail, see [19]).

Magnetic Resonance Imaging Parameters. Participants were scanned in a 3T GE MRI (Discovery 750w, GE Healthcare, Madison, WI) using a 24-channel head coil. We obtained the BOLD contrast with gradient-echo echo-planar imaging (EPI) sequence: time repetition (TR) = 2 ms, time echo (TE) = 24 ms, flip angle = 90°, field of view (FOV) 24 $\times$ 24 cm^2, and matrix size 96 $\times$ 96 (in-plane resolution of 3 $\times$ 3 mm^2). Each scan volume consisted of 27 axial slices of 3 mm thickness and 0.5 mm gap covering the majority of the cortex. High-resolution (1.1 $\times$ 1.1 mm^2) T1-weighted anatomic images of the same orientation and thickness as the EPI slices were also acquired in each scan session to facilitate the incorporation of the functional data into the 3D Talairach space.

fMRI Preprocessing. We used BrainVoyager software package (Brain Innovation, Maastricht, The Netherlands) to analyze the fMRI data. The first 3 images of each functional scan were removed. The functional images were superimposed on 2D anatomic images and incorporated into the 3D normalized Talairach space through trilinear-sinc interpolation. Preprocessing of functional scans included 3D motion correction, slice scan time correction, spatial smoothing (FWHM = 5 mm), linear trend removal, and low-frequency filtering.

fMRI Analyses. For each participant a general linear model was applied to the time course of activity in each voxel for each one of the experimental conditions. To obtain the multi-subject group activation maps, for each experiment, the time courses of subjects from the 2 groups were z-normalized. This was achieved using a random effect (RE) procedure. We created separate activation maps for each of the 4 conditions (manmade-aerial, manmade-terrestrial, natural-aerial, and natural-terrestrial), for each group for each session. To establish the specific effects of training, we derived activation maps depicting the interaction between Group (experimental/control), Session (pre-training baseline/post-training), and Viewpoint (aerial/terrestrial) for the manmade and natural scenes, separately ($p < 0.01$, RE, corrected). In addition, for each subject we sampled the time courses of activation in the scene experiment separately in each independently-defined ROI (FFA, PPA, LOC, OPA, EVC) for each scan session. These response magnitudes were averaged across hemispheres and used for the neurobehavioral correlations presented below.

3 Results

3.1 Behavioral Findings

To assess the efficiency of the training regimen, we employed two behavioral metrics: *(1) Recognition memory:* within-set learning (learning across the first five sessions), and *(2) perceptual expertise:* defined as subjects' ability to generalize their learning to new exemplars (transfer of learning). We found robust training effects across both measures, specified next.

Memory: Learning Within-Session. Significant learning occurred within the five sessions (main effect of Session; $F(4, 52) = 12.86, p < .001, \eta p^2 = .50$). A significant linear decrease in IES was noted, with better performance with increasing levels of practice ($F(1, 13) = 16.66, p < .01, \eta p^2 = .56$) (Fig. 2). The most substantial learning occurred between the first and second session, improving by approximately 12% ($p < .001$). Performance improved with session (evident in the linear trend), and post-hoc comparisons between consecutive levels did not reach significance (p's $> .05$), with the exception of sessions 3 and 4 ($p = .046$). Focusing on the effects of Naturalness and Viewpoint, we found that natural scenes were categorized more efficiently than manmade scenes ($F(1, 13) = 36.10, p < .001, \eta p^2 = .74$). This gap was minimized as learning progressed, as evidenced by a Session-by-Naturalness interaction ($F(4, 52) = 9.73, p < .001, \eta p^2 = .43$). Terrestrial scenes were categorized more efficiently than aerial scenes ($F(1, 13) = 129.69, p < .001, \eta p^2 = .91$), and critically, this gap also decreased over the course of training ($F(4, 52) = 14.74, p < .001, \eta p^2 = .53$). Naturalness and Viewpoint interacted ($F(1, 13) = 44.70, p < .001, \eta p^2 = .76$); the difference between aerial and terrestrial scenes was more pronounced in the manmade compared to natural scenes. Confirming that the above training effects do not simply reflect the effect of repeated exposure to the stimuli, the control group did not show any significant interaction effects with Session ($F(4, 44) = .04, p = .99, \eta p^2 = .001$), nor a Session main effect ($F(4, 44) = 1.38, p = .26, \eta p^2 = .11$). This supports the notion that the present training effects stem from the specific categorization paradigm.

Perceptual Expertise: Learning Transfer. One of the hallmarks of perceptual expertise is learning transfer to novel exemplars from the same category [22, 23]. Such signatures of expertise were also found in the present study, as supported by the following lines of evidence. First, categorization of novel images after the completion of the training regimen was markedly better than categorization at the onset of training ($t(13) = 2.55$, $p = .02$; Fig. 2). Second, average performance in subsequent sessions (second through fifth) was equivalent to performance in the sixth session ($F(1, 13) = .86, p = .37, \eta p^2 = .06$). In other words, subjects categorized new scene images as well as they did with the repeated, familiar ones. Third, resonating the Viewpoint by Naturalness interaction reported above, the minimization of the gap between aerial and terrestrial performance evident during training was also manifest with the novel images, suggesting that subjects learned to extract viewpoint-invariant features that could have been applied to a novel set of scenes (Naturalness × Session (1 vs 6) × Viewpoint: $F(1, 25) = 11.02, p < .01$, $\eta p^2 = .31$). In comparison, the control subjects showed no improvement between the

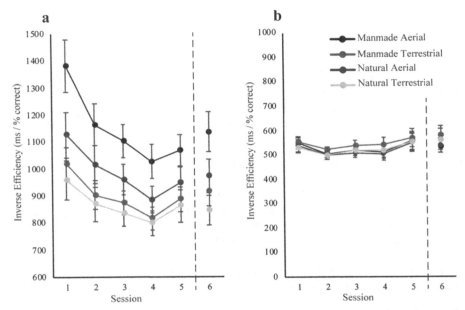

Fig. 2. Learning trajectories across training sessions as a function of scene naturalness (manmade, natural) and viewpoint (aerial, terrestrial) for the experimental group (a) and the controls (b). Efficiency of learning (in milliseconds/percent correct) increases with training with same scenes (Sessions 1–5) and with novel scenes (Session 6).

first and sixth session ($t(11) = -.31, p = .77$), ruling out the possibility that the effects of the sixth session stem from overall faster response times as a result of motor learning.

3.2 Neuroimaging Findings

Subjects were scanned before, during, and after their training sessions. Our goal was to determine (1) how brain activity supporting the perception of terrestrial and aerial scenes changes with training, and (2) the extent to which such brain activity predicts individual differences in training effectiveness. The following analyses address these two goals, respectively.

Whole-Brain Findings. To establish the training effects in recognizing aerial scenes imagery, we contrasted the neural activity associated with perceiving aerial scenes (compared to terrestrial) in the experimental group compared to the controls at pre- vs. post-training session. We applied this contrast to both Naturalness conditions (manmade and natural). Training with aerial compared to terrestrial scenes resulted in changes to occipito-temporal cortex, including areas involved in visual scene recognition (parahippocampal cortex and transverse occipital sulcus), and object recognition (e.g. lateral ventral-occipitotemporal cortex) (Fig. 3). Notably, we also found training-specific activations in areas outside of visual cortex, primarily parietal cortex (Fig. 3), suggesting potential involvement of top-down attention or feedback from visuospatial areas dedicated to the mapping of visual space. This supports our hypothesis that the acquisition of visual expertise involves changes not only to dedicated visual areas, but also the recruitment of top-down attentional control networks.

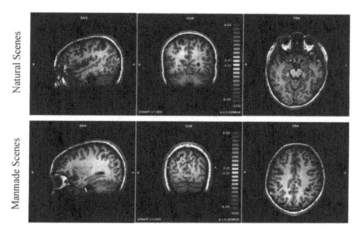

Fig. 3. Group-averaged whole brain statistical activation maps depicting the neural responses sensitive to Viewpoint (aerial vs. terrestrial) as a function of training (the interaction between Group, Session, Viewpoint: [(aerial - terrestrial)$_{\text{session 4}}$ - (aerial - terrestrial)$_{\text{session 1}}$]$_{\text{experimentals}}$ > [(aerial - terrestrial)$_{\text{session 4}}$ - (aerial - terrestrial)$_{\text{session 1}}$]$_{\text{controls}}$ for the natural and manmade scenes (top and bottom row, respectively).

Neural-Behavioral Correlations. To identify potential neural markers of perceptual expertise, we identified functionally-specialized visual regions of interest (ROIs) with the objective of utilizing changes in neural signal over the course of training to predict training efficiency. Our primary ROI was PPA, as this region is specialized for scene perception and has been previously shown to be involved in aerial recognition [12, 13]. We found a significantly positive correlation ($r = .75, p = .001$) between PPA response magnitude across the four scanning sessions with the behavioral effect of perceptual expertise (perceptual expertise index = IES performance in session 6 vs. performance in session 2) (Fig. 4a). In contrast, the correlation between average response magnitude in a control region like FFA (a high-level visual region implicated in visual object expertise) did not significantly correlate with perceptual expertise effects ($r = -.34, p = .12$). We also examined the extent to which it is possible to predict not only learning transfer, but also within-set learning (i.e. performance in session 1 vs. session 5) based on PPA average response magnitude (Fig. 4b). Here, we did not observe a significant correlation ($r = .31, p < .15$), suggesting that PPA plays a specific role in the acquisition of expertise in scene recognition, namely, perceptual generalization rather than associative memory per se.

Lastly, to examine how early during the course of training the correlation between PPA activity and perceptual expertise emerges, we correlated PPA response magnitude

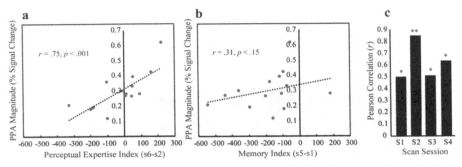

Fig. 4. (a) Correlation between average PPA response magnitude and perceptual expertise index (s6-s2; IES). (b) Correlation between average PPA magnitude and memory index/learning (s5-s1; IES). (c) Pearson correlations for PPA response magnitude and perceptual expertise index across the four scanning sessions. *p < .05, **p < .01

in each scan session with the participant's perceptual expertise index. Strikingly, we observed a robust correlation in the first session ($r = .47, p < .05$) with subsequent perceptual expertise effects. That is, before any training sessions occurred the participants' PPA response correlated with their perceptual expertise index. Further, this correlation was strengthened after only two training sessions ($r = .79, p < .001$), and remained significant across scan three and four ($r = .50, p < .05; r = .54, p < .05$, respectively) (Fig. 4c). The correlation between neural activity and behavior was specific to PPA, as no additional visual ROIs demonstrated similar correlations (all p's > .05), indicating that the current results do not reflect a general effect of attention.

4   Discussion

The current study was conducted to demonstrate the feasibility of using neuroimaging as a tool for assessment of the effectiveness of training in recognition of aerial and terrestrial scene imagery. To address that, we trained a group of healthy volunteers to categorize scenes from a variety of real-world scene categories. We found that a relatively brief period of training, not more than ten hours is sufficient to produce substantial learning effects, both in terms of long-term memory and more remarkably, in terms of perceptual performance, with successful generalization to novel images. Complementing these findings, we found substantial changes in neural activity with increasing levels of experience. These changes included regions in the occipito-temporal cortex related to high-level visual processing as well as in posterior parietal cortex, a region dedicated to higher-level functions, primarily attention and visuospatial processing. Notably, these effects were most pronounced when contrasting training in the trained group relative to the controls on the difference between aerial and terrestrial viewpoints following the completion of the full training regimen. Lastly, we found that the magnitude of activity in dedicated scene-selective cortex (area PPA) was highly related to perceptual expertise effects in scene recognition, to the extent that activity in pre-training baseline could successfully predict subsequent training effectiveness. Moreover, the effects we observed were highly specific, both region- and process-wise (PPA was the only visual area to show a correlation with perceptual expertise) suggesting that neural activity in PPA can be used as a neuromarker of training in aerial and terrestrial scene recognition.

The current study is among a few studies to date to study the neural substrates of aerial scene recognition. Several works have examined how the brain integrates information from a route perspective to a survey perspective (terrestrial and aerial, respectively) [12, 13], but none have specifically assessed the impact of *accruing experience* in recognizing aerial imagery. In that respect, the current study fills an important gap in the literature, especially given the recent need to understand the human factors related to ISR, largely due to potential limits with automated image analysis solutions [24]. By studying the dynamics of expertise acquisition in scene recognition and its underlying neural mechanisms, we can gain both practical and theoretical insight about the nature of training, which would ultimately translate to enhanced scene recognition performance. In line with this rationale, our findings provide the following lessons:

First, training in visual scene analysis recruits both memory and perceptual processes with the former enhancing the latter. Implementing a category verification paradigm for training allowed us to assess changes to associative memory and visual discrimination. As participants improved in their memory skills, we noted a subsequent improvement in visual discrimination in the form of learning transfer. This is in line with Ericsson's prominent theory of expertise [25] suggesting that real-world experts acquire memory skills to encode relevant information to their long-term working memory supported by highly developed semantic knowledge structures, which can be automatically retrieved to facilitate perceptual processes (e.g. scene recognition). One intriguing possibility raised by the current findings, however, is that individual differences in perception might underlie the successful perceptual performance at the completion of training. This conjecture is based on our finding that neural activity in a dedicated scene-selective region, the PPA, was highly predictive of improvement in perceptual abilities already at the outset of training, even before any actual learning took place. This suggests a richer interactive process between perceptual and memory systems supporting visual expertise [14, 18]. Future research on training should take this dynamic into account, including accommodating for individual differences in memory and perception.

Second, combining neuroimaging and behavioral data can uncover the dynamics of learning. The current study demonstrates that training can be rigorously quantified based on a neurobehavioral metric, which combines and is sensitive to changes along behavioral and neural dimensions. We suggest that such a neurobehavioral metric can be used to (1) determine which skills are acquired at which times during training (in our case, perception, memory, top-down attention - as revealed in the whole-brain analysis), (2) to establish the point in time in which learning-specific neural activity plateaus, even prior to behavioral observations, potentially indicating a transition from a novice level to an experienced-user level. Significantly, this can ultimately shorten the training period (potentially reducing training costs) and be used to provide ongoing feedback about the learner's current expertise state, and even enable adaptive individualized training regimens.

Third, neuroimaging can be used as an effective tool for selection and assessment. Given the previous two points, it is clear that scene-selective activity can be used to predict enhancement in perceptual performance following training in scene categorization and thus serve as a potential neuromarker of expertise acquisition. This is a proof of concept that neuroimaging, as well as other neurophysiological measures, can be utilized as means for implicitly assessing the effectiveness of a training program, including its scope, length, and impact. Such a combined neurobehavioral approach would not only provide implicit means to determine field readiness, but will also serve as a diagnostic screening tool - either prior to or during training. One obvious concern, of course, is the costs associated with imaging, but this can be quickly offset given the expected time reduction and increased efficiency that will be provided by individualized training programs.

Lastly, from a more theoretical perspective, our findings shed light on the nature of expertise in visual analysis of complex real-world scenes, particularly of aerial scene recognition. First, recognizing places from an aerial perspective utilizes the same scene-selective areas involved in the recognition of terrestrial perspectives, in line with previous studies [12, 13]. However, over the course of training additional areas (e.g. object-related areas in occipitotemporal cortex and posterior parietal cortex) are recruited, indicating that additional processes are being used to facilitate the task of generalizing across viewpoints. Second, training in semantic scene categorization impacted the neural representations of scenes presented in an altogether different task context, arguably suggesting that learning to recognize scenes is achieved in a task-independent fashion (see also [26]). It still remains to be seen to what extent behavioral relevance, task-context, and level of categorization impact the development of scene expertise. Third, it should be noted that while we used the term 'expertise', we do not by any means consider our trained participants, with their about ten hours of training, to be full-fledged experts with years of experience. Nevertheless, the current results demonstrate the impact of training in scene categorization and suggest potential mechanisms underlying the acquisition of expertise in aerial scene recognition. Future research is needed to assess the extent to which our training approximates real-world expertise, and by extension address the question of what form prior experience has on visual processing. One possibility is that experience develops incrementally, suggesting a quantitative difference between novice and expert [27]. Alternatively, expertise might entail qualitative shifts in representations, manifesting in differential neural activations with varying degrees of experience [28].

In summary, our study demonstrates that naïve participants can be trained to develop expertise that goes beyond their initial capabilities in scene analysis. We showed that we can track changes in performance using behavioral measures and neural signatures of scene processing. These signatures are task- and region-specific and can be used as future neuro-metrics for training effectiveness and acquisition of expertise in scene analysis.

References

1. Cheng, G., Han, J., Lu, X.: Remote sensing image scene classification: benchmark and state of the art. Proc. IEEE **105**(10), 1865–1883 (2017)
2. Malcolm, G.L., Groen, I.I., Baker, C.I.: Making sense of real-world scenes. Trends Cogn. Sci. **20**(11), 843–856 (2016)
3. Drew, T., Evans, K., Võ, M.L.H., Jacobson, F.L., Wolfe, J.M.: Informatics in radiology: what can you see in a single glance and how might this guide visual search in medical images? Radiographics **33**(1), 263–274 (2013)

4. Harel, A., Kravitz, D., Baker, C.I.: Beyond perceptual expertise: revisiting the neural substrates of expert object recognition. Front. Hum. Neurosci. **7**, 885 (2013)
5. Sestito, M., Flach, J., Harel, A.: Grasping the world from a cockpit: perspectives on embodied neural mechanisms underlying human performance and ergonomics in aviation context. Theor. Issues Ergon. Sci. **19**(6), 692–711 (2018)
6. Sestito, M., Harel, A., Nador, J., Flach, J.: Investigating neural sensorimotor mechanisms underlying flight expertise in pilots: preliminary data from an EEG study. Front. Hum. Neurosci. **12**, 489 (2018)
7. Tracey, I., Flower, R.: The warrior in the machine: neuroscience goes to war. Nat. Rev. Neurosci. **15**(12), 825–834 (2014)
8. Ringer, R.V., Loschky, L.C.: Head in the clouds, feet on the ground: applying our terrestrial minds to satellite perspectives. In: Remote Sensing and Cognition. pp. 77–100. CRC Press (2018)
9. Loschky, L.C., Ringer, R.V., Ellis, K., Hansen, B.C.: Comparing rapid scene categorization of aerial and terrestrial views: a new perspective on scene gist. J. Vis. **15**(6), 11 (2015)
10. Lloyd, R., Hodgson, M.E., Stokes, A.: Visual categorization with aerial photographs. Ann. Assoc. Am. Geogr. **92**(2), 241–266 (2002)
11. Šikl, R., Svatoňová, H., Děchtěrenko, F., Urbánek, T.: Visual recognition memory for scenes in aerial photographs: exploring the role of expertise. Acta Psychologica **197**, 23–31 (2019)
12. Shelton, A.L., Gabrieli, J.D.E.: Neural correlates of encoding space from route and survey perspectives. J. Neurosci. **22**, 2711–2717 (2002)
13. Barra, J., Laou, L., Poline, J.B., Lebihan, D., Berthoz, A.: Does an oblique/slanted perspective during virtual navigation engage both egocentric and allocentric brain strategies? PLoS One **7**(11), e49537 (2012)
14. Epstein, R., Kanwisher, N.: A cortical representation of the local visual environment. Nature **392**(6676), 598–601 (1998)
15. Dilks, D.D., Julian, J.B., Paunov, A.M., Kanwisher, N.: The occipital place area is causally and selectively involved in scene perception. J. Neurosci. **33**(4), 1331–1336 (2013)
16. Silson, E.H., Steel, A.D., Baker, C.I.: Scene-selectivity and retinotopy in medial parietal cortex. Front. Hum. Neurosci. **10**, 412 (2016)
17. Epstein, R.A., Baker, C.I.: Scene perception in the human brain. Ann. Rev. Vis. Sci. **5**(1), 373–397 (2019)
18. Harel, A.: What is special about expertise? Visual expertise reveals the interactive nature of real-world object recognition. Neuropsychologia **83**, 88–99 (2016)
19. Harel, A., Gilaie-Dotan, S., Malach, R., Bentin, S.: Top-down engagement modulates the neural expressions of visual expertise. Cereb. Cortex **20**(10), 2304–2318 (2010)
20. Harel, A., Ullman, S., Harari, D., Bentin, S.: Basic-level categorization of intermediate complexity fragments reveals top-down effects of expertise in visual perception. J. Vis. **11**(8), 1–13 (2011)
21. Bruyer, R., Brysbaert, M.: Combining speed and accuracy in cognitive psychology: is the inverse efficiency score (IES) a better dependent variable than the mean reaction time (RT) and the percentage of errors (PE)? Psychologica Belgica **51**(1), 5–13 (2011)
22. Gopher, D., Well, M., Bareket, T.: Transfer of skill from a computer game trainer to flight. Hum. Factors **36**(3), 387–405 (1994)
23. Tanaka, J.W., Curran, T., Sheinberg, D.L.: The training and transfer of real-world perceptual expertise. Psychol. Sci. **16**(2), 145–151 (2005)
24. Kim, J., Zeng, H., Ghadiyaram, D., Lee, S., Zhang, L., Bovik, A.C.: Deep convolutional neural models for picture-quality prediction: challenges and solutions to data-driven image quality assessment. IEEE Signal Process. Mag. **34**(6), 130–141 (2017)
25. Ericsson, K.A., Kintsch, W.: Long-term working memory. Psychol. Rev. **102**(2), 211–245 (1995)

26. Hansen, N.E., Noesen, B.T., Nador, J.D., Harel, A.: The influence of behavioral relevance on the processing of global scene properties: an ERP study. Neuropsychologia **114**, 168–180 (2018)
27. Gauthier, I.: Domain-specific and domain-general individual differences in visual object recognition. Curr. Dir. Psychol. Sci. **27**(2), 97–102 (2018)
28. Harley, E.M., et al.: Engagement of fusiform cortex and disengagement of lateral occipital cortex in the acquisition of radiological expertise. Cereb. Cortex **19**(11), 2746–2754 (2009)

Synthetic Expertise

Ron Fulbright[(⊠)] and Grover Walters[(⊠)]

University of South Carolina Upstate, 800 University Way, Spartanburg, SC 29303, USA
{rfulbright,gwalters}@uscupstate.edu

Abstract. We will soon be surrounded by artificial systems capable of cognitive performance rivaling or exceeding a human expert in specific domains of discourse. However, these "cogs" need not be capable of full general artificial intelligence nor able to function in a stand-alone manner. Instead, cogs and humans will work together in collaboration each compensating for the weaknesses of the other and together achieve synthetic expertise as an ensemble. This paper reviews the nature of expertise, the Expertise Level to describe the skills required of an expert, and knowledge stores required by an expert. By collaboration, cogs augment human cognitive ability in a human/cog ensemble. This paper introduces six Levels of Cognitive Augmentation to describe the balance of cognitive processing in the human/cog ensemble. Because these cogs will be available to the mass market via common devices and inexpensive applications, they will lead to the Democratization of Expertise and a new cognitive systems era promising to change how we live, work, and play. The future will belong to those best able to communicate, coordinate, and collaborate with cognitive systems.

Keywords: Cognitive systems · Cognitive augmentation · Synthetic expertise · Cognitive architecture · Knowledge level · Expertise level

1 Introduction

The idea of enhancing human cognitive ability with artificial systems is not new. In the 1640s, mathematician Blaise Pascal created a mechanical calculator [1]. Yet, the Pascaline, and the abacus thousands of years before that, were just mechanical aids executing basic arithmetic operations. The human does all the real thinking. In the 1840s, Ada Lovelace envisioned artificial systems based on Babbage's machines assisting humans in musical composition [2, 3]. In the 1940s, Vannevar Bush envisioned the Memex and discussed how employing associative linking could enhance a human's ability to store and retrieve information [4]. The Memex made the human more efficient but did not actually do any of the thinking on its own. In the 1950s, Ross Ashby coined the term *intelligence amplification* maintaining human intelligence could be synthetically enhanced by increasing the human's ability to make appropriate selections on a persistent basis [5]. But again, the human does all of the thinking. The synthetic aids just make the human more efficient. In the 1960s, J.C.R. Licklider and Douglas Engelbart envisioned human/computer symbiosis—humans and artificial systems working together in

© Springer Nature Switzerland AG 2020
D. D. Schmorrow and C. M. Fidopiastis (Eds.): HCII 2020, LNAI 12197, pp. 27–48, 2020.
https://doi.org/10.1007/978-3-030-50439-7_3

co-dependent fashion to achieve performance greater than either could by working alone [17, 18].

Over thirty years ago, Apple, Inc. envisioned an intelligent assistant called the Knowledge Navigator [6]. The Knowledge Navigator was an artificial executive assistant capable of natural language understanding, independent knowledge gathering and processing, and high-level reasoning and task execution. The Knowledge Navigator concept was well ahead of its time and not taken seriously. However, some of its features are seen in current voice-controlled "digital assistants" such as Siri, Cortana, and the Amazon Echo (Alexa). Interestingly, the Knowledge Navigator was envisioned as a collaborator rather than a stand-alone artificial intelligence, a feature we argue is critical.

In 2011, a cognitive computing system built by IBM, called Watson, defeated two of the most successful human Jeopardy champions of all time [7]. Watson received clues in natural language and gave answers in natural spoken language. Watson's answers were the result of searching and deeply reasoning about millions of pieces of information and aggregation of partial results with confidence ratios. Watson was programmed to *learn* how to play Jeopardy, which it did in many training games with live human players before the match [8, 9]. Watson *practiced* and achieved expert-level performance within the narrow domain of playing Jeopardy. Watson represents a new kind of computer system called *cognitive systems* [3, 10]. IBM has been commercializing Watson technology ever since.

In 2016, Google's AlphaGo defeated the reigning world champion in Go, a game vastly more complex than Chess [13, 14]. In 2017, a version called AlphaGo Zero learned how to play Go by playing games with itself and not relying on any data from human games [15]. AlphaGo Zero exceeded the capabilities of AlphaGo in only three days. Also in 2017, a generalized version of the learning algorithm called AlphaZero was developed capable of learning any game. While Watson required many person-years of engineering effort to program and teach the craft of Jeopardy, AlphaZero achieved expert-level performance in the games of Chess, Go, and Shogi after only a few hours of unsupervised self-training [16].

These recent achievements herald a new type of artificial entity, one able to achieve, in a short amount of time, expert-level performance in a domain without special knowledge engineering or human input. Beyond playing games, artificial systems are now better at predicting mortality than human doctors [40], detecting signs of child depression through speech [41], detecting lung cancer in X-Rays [42, 43]. Systems can even find discoveries in old scientific papers missed by humans [44].

In 2014, IBM released a video showing two humans interacting with and collaborating with an artificial assistant based on Watson technology [11]. IBM envisions systems acting as partners and collaborators with humans. The similarity between the IBM Watson video and the Knowledge Navigator is striking. John Kelly, Senior Vice President and Director of Research at IBM describes the coming revolution in cognitive augmentation as follows [12]:

"The goal isn't to replace human thinking with machine thinking. Rather humans and machines will collaborate to produce better results—each bringing their own superior skills to the partnership."

We believe the ability for systems to learn on their own how to achieve expert-level performance combined with cognitive system technology will lead to a multitude of mass-market apps and intelligent devices able to perform high-level cognitive processing. Millions of humans around the world will work daily with and collaborate with these systems we call *cogs*. This future will belong to those of us better able to collaborate with these systems to achieve expert-level performance in multiple domains—*synthetic expertise*.

2   Literature

2.1   Human/Computer Symbiosis

Engelbart and Licklider envisioned human/computer symbiosis in the 1960s. Licklider imagined humans and computers becoming mutually interdependent, each complementing the other [17]. However, Licklider envisioned the artificial aids merely assisting with the preparation work leading up to the actual thinking which would be done by the human. In 1962, Engelbart developed the famous H-LAM/T framework modeling an augmented human as part of a system consisting of: the human, language (concepts, symbols, representations), artifacts (physical objects), methodologies (procedures, know-how), and training [18]. As shown in Fig. 1, Engelbart's framework envisions a human interacting, and working together on a task, with an artificial entity To perform the task, the system executes a series of processes, some performed by the human (explicit-human processes), others performed by artificial means (explicit-artifact processes), and still others performed by a combination of human and machine (composite processes). Engelbart's artifacts were never envisioned to do any of the high-level thinking. We feel as though the *artifacts* themselves are about to change. Recent advances in machine learning and artificial intelligence research indicates the artifacts are quickly becoming able to perform human-like cognitive processing.

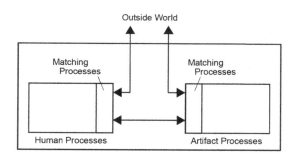

Fig. 1. Engelbart's H-LAM/T framework

2.2   The Nature of Expertise

What it means to be an expert has been debated for decades. Traditional definitions of expertise rely on *knowledge* (the know-that's) and *skills* (the know-how's). Gobet offers a more general definition maintaining an expert *"obtains results vastly superior to those obtained by the majority of the population"* [34]. De Groot established the importance of *perception* in expertise in that an expert perceives the most important aspects of a situation faster than novices [45]. Formally, experts are goal-driven intelligent agents where set of goals, *G,* and a set of utility values, *U,* drive the expert's perceiving, reasoning and acting over time [23, 24]. Intelligent agents perceive a subset, *T,* of possible states, *S,* of the environment and perform actions from set of actions, *A,* to effect changes on the environment.

In their influential study of experts, Chase and Simon found experts compile a large amount of domain-specific knowledge from years of experience—on the order of 50,000 pieces [21]. Steels later describes this as deep domain knowledge and identified: problem-solving methods, and task models as needed by an expert [22]. Problem-solving methods are to solve a problem and a task model is knowledge about how to do something. An expert must know both generic and domain-specific problem-solving methods and tasks. In a new situation, an expert can perceive the most important features quicker than a novice. Experts then match the current situation to their enormous store of deep domain knowledge and efficiently extract knowledge and potential solutions from memory. Furthermore, an expert applies this greater knowledge to the situation more efficiently and quicker than a novice making experts superior problem solvers.

Fulbright extended the description of experts by defining the fundamental skills of an expert based on the skills identified in Bloom's Taxonomy [46]. These skills, described at the Expertise Level, and the knowledge stores, described at the Knowledge Level, form Fulbright's Model of Expertise shown in Fig. 2.

Cognitive scientists have studied and modeled human cognition for decades. The most successful cognitive architecture to date, begun by pioneer Allen Newell and now led by John Laird, is the Soar architecture [25]. Fulbright recently applied the Model of Expertise shown in Fig. 2 to the Soar architecture as shown in Fig. 3 [46].

2.3　Human Cognitive Augmentation

Fulbright described a human/cog ensemble, much like Engelbart's HLAM/T framework, where the human does some cognitive processing and the cog does some cognitive processing [28]. With *W* being a measure of cognitive processing, comparing the amount of cognitive processing done by each component yields a metric called the *augmentation factor, A^+*

$$A^+ = \frac{W_{cog}}{W_{hum}} \tag{1}$$

If the human does more cognitive processing than the cog, $A^+ < 1$; but when the cog starts performing more cognitive processing than the human, $A^+ > 1$ and increases without bound as the capability of cogs grows. At some point in the future, human cognitive processing may become vanishingly small relative to the cog. When cogs develop to the point of being truly *artificial experts,* able to perform at the expert level without human contribution, the human component in Eq. (1) falls to zero and the idea of being cognitively augmented will be a senseless quantity to measure. Until that time

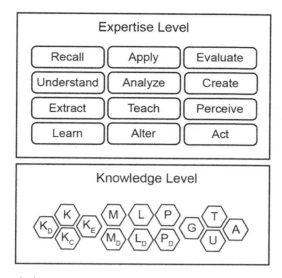

Knowledge
K   declarative knowledge statements
K_D domain-specific knowledge
K_C common-sense knowledge
K_E episodic knowledge   **G** goals to achieve
M/M$_D$ world models   **U** utility values
L/L$_D$ task models   **T** perceivable states
P/P$_D$ problem-solving models   **A** actions

Skills
Perceive   sense/interpret the environment
Act   perform action affecting environment
Recall   remember; store/retrieve knowledge
Understand classify, categorize, discuss, explain, identify
Apply   implement, solve, use knowledge
Analyze   compare, contrast, experiment
Evaluate   appraise, judge, value, critique
Create   design, construct, develop, synthesize
Extract   match/retrieve deep knowledge
Learn   modify existing knowledge
Teach   convey knowlege/skills to others
Alter   modify goals

Fig. 2. Model of expertise

though, the world will belong to those humans best able to partner with and collaborate with cogs.

These humans will be cognitively augmented. A human working alone performs W_{hum} amount of cognitive processing. However, a human working with one or more

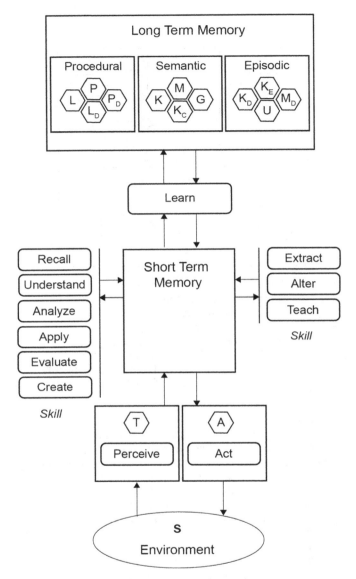

Fig. 3. Soar model of expertise

cogs performs an increased amount of cognitive processing [29, 30]

$$W^* = W_{hum} + \sum W_{cog}^i \tag{2}$$

where i is the number of cogs in the collaboration and W^* is the total cognitive processing done by the ensemble. To an observer outside of the ensemble, it appears the human is performing at a much higher level than expected. When W^* reaches or exceeds the level of an expert in the domain of discourse, the human/cog ensemble will have achieved *synthetic expertise* as shown in Eq. (3).

$$W^* \geq W_{expert} \tag{3}$$

Fulbright described the measurement of the augmentation of two specific cognitive capabilities, *cognitive accuracy* and *cognitive precision* [30]. Given a problem to solve, cognitive accuracy involves the ability to synthesize the best solution and cognitive precision involves being able to synthesize nothing but the best solution. In a case study, with the cog component being simulated by expert "suggestions," cognitive accuracy of the human was improved by 74% and cognitive precision was improved by 28%.

3 Synthetic Expertise

Biological systems capable of performing all skills and acquiring/possessing all knowledge in the Model of Expertise shown in Fig. 2 are *human experts*. Non-biological systems capable of the same are called *artificial experts*. We envision a future in which artificial experts achieve or exceed the performance of human experts in virtually every domain. However, we think it will be some time before fully autonomous artificial experts exist. In the immediate future, humans and artificial systems will work together to achieve expertise as an ensemble—*synthetic expertise*. We choose the word "synthetic" rather than "artificial" because the word artificial carries a connotation of not being real. We feel as though the cognitive processing performed by cogs and the ensemble is real even though it may be very different from human cognitive processing.

We call the artificial collaborators *cogs*. Cogs are intelligent agents—entities able to rationally act toward achieving a goal [24]. However, the term *intelligent agent* refers to a wide range of systems, from very simple systems such as a thermostat in your home to very complex systems, such as artificially intelligent experts. In the study of synthetic expertise, the term *cog* is defined as:

> *cog:* an intelligent agent, device, or algorithm able to perform, mimic, or replace one or more cognitive processes performed by a human or a cognitive process needed to achieve a goal.

It is important to note cogs can be artificially intelligent but do not necessarily have to be. However, cogs are expected to be relatively complex because, for synthetic expertise, cogs should perform part of, or all of, at least one of the fundamental skills identified in the Model of Expertise shown in Fig. 2: *recall, apply, understand, evaluate, analyze, extract, alter, learn, teach, perceive, act,* and *create*. Cogs also need not be terribly broad in scope nor deep in performance. They can be narrow and shallow agents. Cogs also need not fully implement a cognitive process, but instead may perform only a portion of a cognitive process with the human performing the remainder of that process. Computers have tremendous advantage over humans in some endeavors such as number crunching, speed of operations, and data storage. Some cogs will leverage their advantage in these kinds of functions in support of a jointly-performed cognitive process. Cognitive processing in. a human/cog ensemble will therefore be a combination of biological cognitive processing and non-biological cognitive processing.

Indeed, we see this already beginning to happen both at the professional level and at the personal level. Every day, millions of people issue voice commands to virtual assistants like Apple's Siri, Microsoft's Cortana, Google Assistant, and Amazon's Alexa and a host of applications on computers and handheld electronic devices. These assistants can understand spoken natural language commands and reply by spoken natural language. In the professional world, professionals such as doctors are using cognitive systems to diagnose malignant tumors and bankers are using cognitive systems to analyze risk profiles. In some domains, the cogs already outperform humans.

At the most basic level we envision a human interacting with a cog as shown in Fig. 4. The human and cog ensemble form an Engelbart-style system with the human component performing some of the cognitive work, W_H and the cog performing some of the cognitive work, W_C. Information flows into the ensemble and out of the ensemble, S_{in} and S_{out} respectively, and the total cognitive work performed by the ensemble is W^*. The difference between Fig. 4 and Engelbart's HLAM/T framework is the cogs are capable of high-level human-like cognitive processing and act as peer collaborators working with humans rather than mere tools. As cogs become more advanced, the human/cog collaboration will become more collegial in nature.

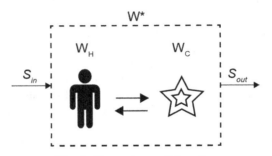

Fig. 4. Human/cog collaboration

Recalling Eq. (3), when the cognitive performance of the ensemble reaches or exceeds that of an expert, the ensemble has achieved *synthetic expertise*. While one day, a single cog may be developed capable of expert performance in any domain, for the foreseeable future, the domain of discourse of the human/cog ensemble will be limited. A cog may help raise a person's performance in only one domain of discourse, or even a subset of a particular domain. Therefore, we expect the near future to see humans employing a number of different cogs, each with different capabilities. In fact, we see this today as well. People employ a number of different apps and devices throughout the day. The difference in the cognitive systems future will be the apps and devices will be capable of high-level cognitive processing. Furthermore, we certainly expect more than one human to be involved in come collaborations forming a virtual team where the cogs will be seen as just another team member.

In this future, communication and collaboration within the ensemble/virtual team is key and is a fertile area for future research. Humans will certainly converse with other humans (human/human). Likewise, cogs will converse with other cogs (cog/cog) and humans will converse with cogs (human/cog). The dynamics of each of these three realms of communication and collaboration are quite different and should be explored in future research. In fact, research is already underway. The fields of human/computer interaction (HCI), human/autonomy teaming (HAT), and augmented cognition (AugCog) are currently quite active. The fields of distributed artificial intelligence (DAI), multiagent systems, negotiation, planning, and communication, and computer-supported cooperative work (CSCW) are older fields of study but are quite relevant. Fields such as human-centered design, augmented reality (AR), virtual reality (VR), enhanced reality (ER), and brain-computer interfaces (BCI) lead in promising directions. Information design (ID), user experience design (UX), and information architecture (IA) have important contributions.

3.1 The Human/Cog Ensemble

In the cog future, the *collaborate* skill is critical for both humans and cogs. Because the human and the cog are physically independent agents, both must *perceive, act,* and *collaborate.* Adding *collaborate* to the skills and knowledge stores identified in the Model of Expertise (Fig. 2) yields the depiction of synthetic expertise shown in Fig. 5. The human/cog ensemble must perform all skills and maintain all knowledge stores. Skills are performed solely by the human (corresponding to Engelbart's human-explicit processes), solely by the cog (corresponding to Engelbart's artifact-explicit processes), or by a combination of human and cog effort (Engelbart's composite processes). To the outside world, it does not matter which entity performs a skill as long as the skills are performed by the ensemble.

For the immediate future, cogs will perform lower-order skills and humans will perform higher-order skills. As an example, consider the situation with today's virtual assistants, like Siri, as shown in Fig. 6. Assume a person asks Siri what time it is while performing a task. The human performs an action (*act*) by speaking the command "Siri, what time is it?" Through the smartphone's microphone, Siri *perceives* the spoken command, and *analyzes* it to *understand* the user is asking for the time (even though this is a rudimentary form of understanding). Siri then *recalls* the time from the internal clock on smartphone, formulates a spoken response, and articulates the reply back to the user (*act*). In this situation, the cog is not doing a large amount of cognitive processing. The human is doing most of the thinking as represented by more of the fundamental skills being shown on the human side. $A^+ < 1$ in this situation. However, as cognitive systems evolve, they will be able to perform more of the higher-order fundamental skills themselves.

3.2 Cognitive Augmentation

In the coming cognitive systems era, cognitive processing in human/cog ensembles will be a mixture of human and cog processing resulting in the augmentation of the human's cognitive processing. It will be many years before fully artificial intelligences become available to the mass market. In the meantime, there will be human/cog ensembles achieving varying amounts of cognitive augmentation. Here, we define the Levels of

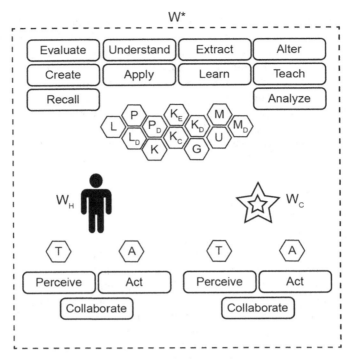

Fig. 5. Synthetic expertise

Cognitive Augmentation ranging from no augmentation at all (all human thinking) to fully artificial intelligence (no human thinking) as shown in Fig. 7.

Until now, computers and software humans use represent Level 1 cognitive augmentation (assistive tools). Recent advances in deep learning and unsupervised learning have produced Level 2 cognitive augmentation. But as the abilities of cogs improves, we will see Level 3 and Level 4 cognitive augmentation leading eventually to fully artificial intelligence, Level 5, in which no human cognitive processing will be required.

The promise of cogs, intelligent agents, cognitive systems, and artificial intelligence in general, is superior performance. If P is a measure of human performance working alone and P^* is a measure of human/cog performance, then we expect

$$P^* > P \tag{4}$$

so we can calculate the percentage change realized by the human working with a cog

$$\Delta P = \frac{P^* - P}{P}. \tag{5}$$

How does one measure performance in a particular domain of discourse? This may vary widely from domain to domain but in general, we may seek to reduce commonly measured quantities such as *time*, *effort*, and *cost*. Or we may seek to increase quantities such as:

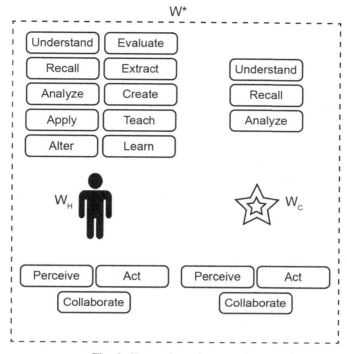

Fig. 6. Human-intensive expertise

Level 0: No Augmentation
   human performs all cogntiive processing

Level 1: Assistive Tools
   abacus, calculators, software, etc.

Level 2: Low-Level Cognition
   pattern recognition, classification, speech
   human makes all high-level decisions

Level 3: High-Level Cognition
   concept understanding, critique,
   conversational natural language

Level 4: Creative Autonomy
   human-inspired, unsupervised synthesis

Level 5: Artificial Intelligence
   no human cognitive processing

Fig. 7. Levels of cognitive augmentation

- Quality
- Revenue
- Efficiency
- Number of Transactions
- Number of Actions Completed
- Number of Customers Serviced
- Level of Cognition Achieved

As an example, consider a deep-learning algorithm able to detect lung cancers better than human doctors [42]. The rate of false positives and false negatives by human evaluation of low-dose computed tomography (LDCT) scans delay treatment of lung cancers until the cancer has reached an advanced stage. However, the algorithm outperforms humans in recognizing problem areas reducing false positives by 11% and false negatives by 5%. Therefore, the human/cog ensemble achieves better performance by a measurable extent. Another way of putting this is by working with the cog, the doctor's performance in enhanced.

In dermatology, Google's Inception v4 (a convolutional neural network) was trained and validated using dermoscopic images and corresponding diagnoses of melanoma [47]. Performance of this cog against 58 human dermatologists was measured using a 100-image testbed. Measured was the sensitivity (the proportion of people with the disease with a positive result), the specificity (the proportion of people without the disease with a negative result), and the ROC AUC (a performance measurement for classification problem at various thresholds settings). Results are shown in Fig. 8. The cog outperformed the group of human dermatologists by significant percentages suggesting in the future, the human dermatologists would improve their performance by working with this cog.

	Human	Cog	Improvement
Sensitivity:	86.6%	95.0%	+9.7%
Specificity:	71.3%	82.5%	+15.7%
ROC AUC:	0.79	0.86	+8.9%

Fig. 8. Human vs. cog in lesion classification

In the field of diabetic retinopathy, a study evaluated the diagnostic performance of an autonomous artificial intelligence system, a cog, for the automated detection of diabetic retinopathy (DR) and Diabetic Macular Edema (DME) [48]. The cog exceeded all pre-specified superiority goals as shown in Fig. 9.

This begs an important question. Doctors use other artificial devices to perform their craft. Thermometers and stethoscopes enhance a doctor's performance. Why are cogs different? The answer is yes, these tools enhance human performance. Humans have been making and using tools for millennia and indeed this is one differentiating characteristic of humans. Engelbart and Licklider's vision of "human augmentation" in the 1960s was for computers to be tools making humans better and more efficient

	Goal	Cog	Improvement
Sensitivity:	>85.0%	87.2%	+2.6%
Specificity:	>82.5%	90.7%	+9.9%

Fig. 9. Cog performance in diabetic retinopathy

at thinking and problem solving. Yet, they envisioned the human as doing most of the thinking. We are now beginning to see cognitive systems able to do more than a mere tool, they are able to perform some of the high-level thinking on their own. Today, some of the highest-level skills identified in the Model of Expertise (Fig. 2) such as *understand* and *evaluate* are beyond current cog technology, but the ability of cognitive systems is gaining rapidly.

Research areas such as task learning, problem-solving method learning, goal assessment, strategic planning, common sense knowledge learning, generalization and specification are all critical areas of future research.

3.3   Knowledge of the Ensemble

In a human/cog ensemble, the human will possess some of the knowledge stores identified in the Model of Expertise and the cog will possess some of the knowledge stores. The knowledge of the ensemble should be viewed as a combination of human-maintained knowledge and cog-maintained knowledge. In most cases, we expect both the human and cog to possess and maintain their own versions of the knowledge stores and communicate contents to each other when necessary. However, we recognize existing and future technology able to combine these stores by connecting the human mind directly with a computer. With technology like this, a knowledge store could be shared directly by human and cog without needing communication.

Until such technology becomes available, cogs have a unique and important advantage over humans. Cogs can simply download knowledge from an external source—even another cog. Currently, it is not possible to simply dump information directly into a human brain. However, cogs can simply transmit knowledge directly from an external source, such as the Internet. With global communication via the Internet, cogs will have near instantaneous access to knowledge far beyond its own and be able to obtain this remote knowledge with minimal effort.

Figure 10 depicts a cog, involved in a local human/cog ensemble, sending and receiving knowledge to and from remote stores via the Internet. The figure shows domain-specific domain knowledge (K_D) and domain-specific problem-solving knowledge (P_D) being obtained from two different remote sources. As far as the local ensemble is concerned, once downloaded, the knowledge stores obtained remotely are no different from locally-produced knowledge. It is as if the ensemble had always been in possession of this knowledge. Any knowledge store can be imported partially or entirely from a remote source. We have described the human/cog ensemble as a local entity but in reality, with pervasive Internet connectivity, a human/cog ensemble is a combination of local knowledge and all other available knowledge. In the cog era, instead of benefitting from one cog, humans will actually be benefitting from millions of cogs. The local/remote line will tend to blur and this vast artificial knowledge will just be assumed to be available "in the cloud" anytime we want it, much how we view Internet-based services today.

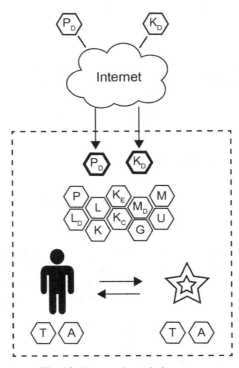

Fig. 10. Remote knowledge stores

This makes possible one of the most exciting features of the coming cog era. Once a cog learns or synthesizes something, any other cog in existence then or in the future can obtain the knowledge. Millions of cogs interacting with users on a daily basis, continually learning and communicating knowledge to all other cogs, will lead to exponential knowledge creation and evolution of cog capabilities. We expect cogs and cog-based knowledge to evolve very quickly. Fields such as knowledge representation, knowledge management, semantic Web, software agents, distributed artificial intelligence, and ontologies are all important future research areas along with automated knowledge discovery.

3.4 Composite Skills

The Model of Expertise shown in Fig. 2, identifies fundamental skills necessary for expertise: *recall, apply, understand, evaluate, analyze, extract, alter, learn, teach, perceive, act,* and *create.* However, one quickly recognizes this is not a complete list of everything an expert does. Experts are certainly expected to do things like: summarize, conceptualize, theorize, classify, categorize, visualize, predict, forecast, define, explain, suggest, compare, assess, prioritize, organize, analyze, recommend, motivate, inspire, etc.

Many of these are included within the scope of the fundamental skills identified and defined in Fig. 2. For example, classify and categorize fall under the *understand* fundamental skill. Others, are higher-order composite skills made up by a combination of fundamental skills. For example, predicting and forecasting involve a combination of the *understand, apply, create,* and *evaluate* fundamentals. In another example, explaining is a combination of *understanding* and *analyzing.*

A promising area for future research is to define constructions of fundamental skills for higher-order skills. Doing so would make it possible to compare and contrast future implementations of the higher-order skills and possibly lead to establishing effective metrics for the skills. Another area of future research is to focus on the fundamental skills themselves. The Model of Expertise identifies 12 fundamental skills and including *collaborate* as we suggest here makes 13. Are these the only fundamental skills experts need to possess? Is it possible to identify additional fundamental skills?

3.5 Grammar of Action

We envision a cognitive systems future in which humans will interact with and collaborate with multiple cogs. Today, it is easy to imagine this interaction to be based on natural language, traditional computer displays, keyboards, mice, and touchscreens. However, we believe there will be and should be a greater bandwidth for communication between human and cog. An exciting area of future research is to explore these possibilities which might include: gesture recognition, body language, gait analysis, facial expressions, emotional awareness, activity recognition, and behavioral recognition among others.

Harper described the importance of considering the grammars of action of new technologies [49]. Whenever new technologies are adopted, human behavior changes to accommodate and integrate the new technology into our daily lives. As a result, a set of nouns, verbs, adjectives, and phrases arises because of the new technology. This is called a *grammar of action.*

For example, consider the computer mouse. Words like *click, double click, scroll wheel,* and *mouse pointer* describe how to perform tasks such as *select, drag, drop,* and *rubber band* items on the computer screen. These words existed before the computer mouse was invented but new meanings were adopted as a result of their use in the context of human/computer interaction. Similarly, *pinch, spread, tap, double tap, swipe left,* and *swipe right* are included in the grammar of action for touchscreen interfaces.

In the cog era, we expect a new grammar of action to arise as the result of mass-market adoption and use of cognitive systems. An interesting area of future research is to predict what new nouns, verbs, and adjectives will be invented. However, we expect more than words will be invented. What new gestures will be invented to facilitate easy and effective human/cog interaction? How will the cog use the human's facial expressions, body language, and emotional state to enhance and guide its interaction and collaboration with humans? We anticipate elements of the "cog grammar of action" to connect to the higher-order skills described in the last section. For example, one can imagine a certain hand gesture to become the normal way for a human to express to the cog "please explain that to me."

4 The Cog Era

We are at the very beginning of the cog era and its evolution will play out over the next few decades. The cog era will, for the first time, give humans artificial systems able to perform some of the high-level thinking. This will create a new industry and a new market sure to change things for us culturally, legally, socially, and legally.

The Cog Market. We expect the emergence of personal cogs intended for the mass-market. These cogs will be bought and sold by average people through existing Internet sales channels much in the same way apps, music, and other items are sold now. We will both be able to purchase or rent our own cogs and also be able to subscribe to cloud-based cog services and information. This will give every person access to professional-level expertise in any domain. This *democratization of expertise* will lead to changes similar in scope to the way the democratization of computing and information has changed us over the last few decades.

Expertise as a Commodity. Because cogs learn from humans, we expect the need for experts in a field to work with cogs and develop their own unique store of knowledge. Entities in industries such as financial services, investment services, legal, medical, news, politics, and technology will compete in offering access to their "superior" store of knowledge created through the interaction of their experts and their cogs. In the cog era, knowledge will become an economic commodity.

Teacher Cogs. We expect cogs to become intelligent tutoring systems. Through customized and personalized interaction with a person, teacher cogs will impart this knowledge to the student in ways similar to the master/apprentice model of education. The best teacher cogs will be personal cogs able to remember every interaction with a person over an extended period of time, even years or decades. Imagine an algebra cog able to answer a question by a 35-year old who it has been working with since grade school. We anticipate teacher cogs to evolve for every subject taught in schools and beyond. We think students of future generations will start using cogs all throughout their education and then retain the cogs, and years of interaction, through the rest of their lives. Again, we foresee vigorous competition arising from different teacher cog providers attempting to bring to the market the best teacher cog for a particular subject matter.

Advisor, Coach, Self-help, and Pet Cogs. It is natural for humans to form emotional relationships with anything, biological or artificial, they can interact with. Indeed, people form emotional relationships with animals and technology today. We foresee cog technology giving personalities to artificial systems. Since cogs will be able to give expert-level advice in any domain, we predict the evolution of a host of self-help cogs ranging from relationship advice to life/work balance, grief counseling, faith-based counseling and beyond. People will confide intimate details to these cogs and receive advice of great personal value and satisfaction. People will spend hours conversing with their personal self-help/companionship cogs. We can easily envision the development of virtual pets with cog-based personalities and communication abilities. In the cog era, we will love our cog pets. Indeed, we have already witnessed the beginning of these kind of applications. Today, hundreds of millions of people have used "synthetic frend" cogs Xiaoice in China, Rinna in Japan, and Ruuh in India.

Productivity Cogs. We predict every productivity application in use today will become enhanced by cog technology in the future. Indeed, applications like word processors, spreadsheets, presentation editors, Web browsers, entertainment apps, games, graphics editing, etc. may become a primary interface point for humans and cogs. Cog capabilities will both be built into the applications themselves and provide expert-level collaboration to the user and also evolve into stand-alone cogs for a particular task. For example, we can imagine a future version of Microsoft Word coming complete with embedded creative writing cog services. We can also imagine purchasing a creative writing cog from an app store operating independently of a specific word processor.

Personal productivity cogs will understand our recent context in a deep manner and use that to customize their assistance and interaction with us. Imagine, for example, a word processing cog that understands you are writing about the future of cognitive processing but also knows that you have communicated with several others via email on that and related topics and can also take into consideration every article or Web page you have accessed in recent months while researching the paper. Such a cog knows a lot about you personally and can combine that knowledge with its own searching and reasoning about the millions of documents it has searched on the Internet. Personal productivity cogs will become our intelligent assistants.

Collaborative Cognition. In addition to enhancing current productivity applications, we expect an entirely new genre of cog-based productivity app to arise, *collaborative cognition*. We envision new kinds of problem solving, brainstorming, business/competitive/market analysis, and big data analysis. We foresee multi-cog "collaborative virtual team" applications being created. Collaborative cogs will become our artificial intelligent team members. Again, we see a vigorous and dynamic competitive market arising around the idea of collaborative cogs. By partnering with humans, cogs achieve ever-increasing levels of knowledge in a particular area. Therefore, considerable market value will be attached to collaborative cogs that have worked with the best experts in the field. The cog era will bring forth a new kind of virtual consultant.

Research Cogs. We foresee future graduate students, entrepreneurs, scientists and any of us creative and inquisitive people conducting research by conversing with their research cog(s) instead of searching and reading scores of journal articles and technical papers. Today, we tell graduate students the first step in their research is to go out and read as many articles, books, and papers as they can find about their topic. Future research students' first action will be to sit down with his research cog and ask "So, what is the current state of the art in <insert domain here>."

Cogs will far exceed the ability of any human in consuming billions of articles, papers, books, Web pages, emails, text messages, and videos. Even if a person spent their entire professional life learning and researching a particular subject is not able to read and understand everything available about that subject. Yet, future researchers will be able to start their education from that vantage point by the use of research cogs. In the cog era, the best new insights and discoveries will come from the interaction between researchers and their research cogs.

Here again we see evidence of knowledge becoming a commodity. Today, we may be able to learn a great deal from the notebooks of great inventors like Tesla, Edison, and DaVinci. In fact, notebooks of inventors like these are worth millions of dollars. But imagine how valuable it would be if we had access to Einstein's personal research cog he used for years while he was synthesizing the theory of relativity. In the cog era, not only will cogs assist us in coming up with great discoveries, they will also record and preserve that interaction for future generations. Such cogs will be enormously valuable both economically and socially.

Discovery Engines. Even though we envision cogs partnering with humans, we expect cogs to evolve to be able to perform on their own. We fully expect cogs working semi-autonomously to discover significant new theories, laws, proofs, associations, correlations, etc. In the cog era, the cumulative knowledge of the human race will increase by the combined effort of millions of cogs all over the world. In fact, we foresee an explosion of knowledge, an exponential growth, when cogs begin working with the knowledge generated by other cogs. This kind of cognitive work can proceed without the intervention of a human and therefore proceed at a dramatically accelerated rate. We can easily foresee the point in time where production of new knowledge by cogs exceeds, forever, the production of new knowledge by humans.

In fact, we anticipate a class of discovery engine cogs whose sole purpose is to reason about enormous stores of knowledge and continuously generate new knowledge of ever-increasing value resulting ultimately in new discoveries that would have never been discovered by humans or, at the very least, taken humans hundreds if not thousands of years to discover.

Cognitive Property Rights. The cog era will bring forth new questions, challenges, and opportunities in intellectual property rights. For example, if a discovery cog makes an important new discovery, who owns the intellectual property rights to that discovery? An easy answer might be "whoever owned the cog." But, as we have described, we anticipate cogs conferring with other cogs and using knowledge generated by other cogs. So a cog's work and results are far from being in isolation. We predict existing patent, copyright, trademark, and service mark laws will have to be extended to accommodate the explosion of knowledge in the cog era.

5   Democratization of Expertise

As described earlier, non-experts will be able to achieve or exceed expert-level performance in virtually any domain by working with cogs achieving Level 3 or Level 4 cognitive augmentation. When a large number of these cogs become available to the masses via the cog market as described above, we will be in a future where any average person could be a *synthetic expert*. We call this the *democratization of expertise*.

When expertise becomes available to the masses, changes will occur. Democratization of expertise will disrupt many professions. While we certainly do not anticipate the demise of doctors, lawyers, and accountants, their professions may change when their former customers have access to expert-level information and services possibly superior to what they could have supplied. What are the consequences and possibilities when millions of us can achieve expert-level performance in virtually any domain of discourse?

6   Conclusion

We have introduced the concept of *synthetic expertise* and have defined it as the ability of an average person to achieve expert-level performance by virtue of working with and collaborating with artificial entities (cogs) capable of high-level cognitive processing. Humans working in collaboration with cogs in a human/cog ensemble are cognitively augmented as a result of the collaboration. Over time, as the capabilities of cogs improve, humans will perform less and less of the thinking. We have used the balance of cognitive effort between human and cog to formulate the Levels of Cognitive Augmentation to describe the phenomenon. The Levels of Cognitive Augmentation can be used in the future to compare and contrast different systems and approaches.

To describe *expertise*, we have combined previous work in cognitive science, cognitive architectures, and artificial intelligence with the notion of expertise from education pedagogy to formulate the Model of Expertise. The Model of Expertise includes a Knowledge Level and an Expertise Level description of the fundamental knowledge and skills required by an expert.

Cogs will continue to improve and take on more of the skills defined in the Model of Expertise. These cogs will also become available to everyone via mass-market apps, services, and devices. Expertise becoming available to the masses is something we call the *democratization of expertise* and will usher in many social, cultural, societal, and legal changes.

We have identified several interesting areas of future research relating to synthetic expertise:

- Human/cog communication (HCI, AR, VR, ER, human/brain interfaces)
- Human/cog teaming and collaboration (HAT, CSCW)
- Cog/cog communication
- Fundamental skills of an expert
- Composite skills of an expert
- Grammar of action associated with cogs
- Human-centered design
- Task learning
- Problem-Solving method learning
- Goal assessment
- Strategic planning
- Common sense knowledge learning
- Intelligent agent theory

- Software agents
- Intellectual property
- Automated knowledge discovery

Some characteristics and challenges of the coming cognitive systems era have been described. Besides mass-market adoption of cog technology, we see expertise and knowledge becoming commodities leading us to interesting futures involving artificially-generated knowledge and legal battles over ownership of knowledge. We also describe a future in which people routinely collaborate with, learn from, and commiserate with cogs. In much the same way computer and Internet technology has woven itself into every fiber of life, we expect cognitive system technology to do the same.

References

1. Chapman, S.: Blaise Pascal (1623–1662) Tercentenary of the calculating machine. Nature **150**, 508–509 (1942)
2. Hooper, R.: Ada Lovelace: My brain is more than merely mortal. New Scientist (2015). https://www.newscientist.com/article/dn22385-ada-lovelace-my-brain-is-more-than-merely-mortal. Accessed Nov 2015
3. Isaacson, W.: The Innovators: How a Group of Hackers, Geniuses, and Geeks Created the Digital Revolution. Simon & Schuster, New York (2014)
4. Bush, V.: As We May Think. The Atlantic, July 1945
5. Ashby, W.R.: An Introduction to Cybernetics. Chapman and Hall, London (1956)
6. Apple: Knowledge Navigator. You Tube (1987). https://www.youtube.com/watch?v=JIE8xk6Rl1w. Accessed Apr 2016
7. Jackson, J.: IBM Watson Vanquishes Human Jeopardy Foes. PC World (2015). http://www.pcworld.com/article/219893/ibm_watson_vanquishes_human_jeopardy_foes.html. Accessed May 2015
8. Ferrucci, D.A.: Introduction to this is Watson. IBM J. Res. Dev. **56**(3/4) (2012)
9. Ferrucci, D., et al.: Building Watson: an overview of the DeepQA project. AI Mag. **31**(3), 59–79 (2010)
10. Wladawsky-Berger, I.: The Era of Augmented Cognition. The Wall Street Journal: CIO Report (2015). http://blogs.wsj.com/cio/2013/06/28/the-era-of-augmented-cognition/. Accessed May
11. Gil, D.: Cognitive systems and the future of expertise. YouTube (2019). https://www.youtube.com/watch?v=0heqP8d6vtQ. Accessed May 2019
12. Kelly, J.E., Hamm, S.: Smart Machines: IBMs Watson and the Era of Cognitive Computing. Columbia Business School Publishing, Columbia University Press, New York (2013)
13. Silver, D., et al.: Mastering the game of Go with deep neural networks and tree search. Nature **529**, 484 (2016)
14. DeepMind: The story of AlphaGo so far. DeepMind (2018). https://deepmind.com/research/alphago/. Accessed Feb 2018
15. DeepMind: AlphaGo Zero: learning from scratch. DeepMind (2018). https://deepmind.com/blog/alphago-zero-learning-scratch/. Accessed Feb 2018
16. ChessBase: AlphaZero: Comparing Orangutans and Apples. ChessBase (2018). https://en.chessbase.com/post/alpha-zero-comparing-orang-utans-and-apples. Accessed Feb 2018
17. Licklider, J.C.R.: Man-computer symbiosis. IRE Trans. Hum. Factors Electron. **HFE-1**, 4–11 (1960)

18. Engelbart, D.C.: Augmenting Human Intellect: A Conceptual Framework. Summary Report AFOSR-3233. Stanford Research Institute, Menlo Park, CA, October 1962
19. Newell, A.: Unified Theories of Cognition. Harvard University Press, Cambridge (1990)
20. Newell, A.: The knowledge level. Artif. Intell. **18**(1), 87–127 (1982)
21. Chase, W., Simon, H.: Perception in chess. Cogn. Psychol. **4**, 55–81 (1973)
22. Steels, L.: Components of expertise. AI Mag. **11**(2), 28 (1990)
23. Genesereth, M., Nilsson, N.: Logical Foundations of Artificial Intelligence. Morgan Kaufmann, Burlington (1987)
24. Russell, S., Norvig, P.: Artificial Intelligence: A Modern Approach, 3rd edn. Pearson, London (2009)
25. Laird, J.E.: The SOAR Cognitive Architecture. The MIT Press, Cambridge (2012)
26. Bloom, B.S., Engelhart, M.D., Furst, E.J., Hill, W.J., Krathwohl, D.R.: Taxonomy of Educational Objectives: The Classification of Educational Goals. Handbook I: Cognitive Domain. David McKay Company, New York (1956)
27. Anderson, L.W., Krathwohl, D.R. (eds): A Taxonomy for Learning, Teaching, and Assessing: A Revision of Bloom's Taxonomy of Educational Objectives (2001)
28. Fulbright, R.: Cognitive augmentation metrics using representational information theory. In: Schmorrow, D.D., Fidopiastis, C.M. (eds.) AC 2017. LNCS (LNAI), vol. 10285, pp. 36–55. Springer, Cham (2017). https://doi.org/10.1007/978-3-319-58625-0_3
29. Fulbright, R.: On measuring cognition and cognitive augmentation. In: Yamamoto, S., Mori, H. (eds.) HIMI 2018. LNCS, vol. 10905, pp. 494–507. Springer, Cham (2018). https://doi.org/10.1007/978-3-319-92046-7_41
30. Fulbright, R.: Calculating cognitive augmentation – a case study. In: Schmorrow, D.D., Fidopiastis, C.M. (eds.) HCII 2019. LNCS (LNAI), vol. 11580, pp. 533–545. Springer, Cham (2019). https://doi.org/10.1007/978-3-030-22419-6_38
31. Fulbright, R.: How personal cognitive augmentation will lead to the democratization of expertise. In: Fourth Annual Conference on Advances in Cognitive Systems, Evanston, IL, June 2016. http://www.cogsys.org/posters/2016. Accessed Jan 2017
32. Fulbright, R.: The cogs are coming: the coming revolution of cognitive computing. In: Proceedings of the 2016 Association of Small Computer Users in Education (ASCUE) Conference, June 2016
33. Fulbright, R.: ASCUE 2067: how we will attend posthumously. In: Proceedings of the 2017 Association of Small Computer Users in Education (ASCUE) Conference, June 2017
34. Gobet, F.: Understanding Expertise: A Multidisciplinary Approach. Palgrave, London (2016)
35. Gobet, F., Simon, H.: Five seconds or sixty? Presentation time in expert memory. Cogn. Sci. **24**(4), 651–682 (2000)
36. Gobet, F., Chassy, P.: Expertise and intuition: a tale of three theories. Minds Mach. **19**, 151–180 (2009). https://doi.org/10.1007/s11023-008-9131-5
37. Dreyfus, H.L.: What Computers Can't Do: A Crtique of Artificial Reason. The MIT Press, Cambridge (1972)
38. Dreyfus, H.L., Dreyfus, S.E.: Mind Over Machine: The Power of Human Intuition and Expertise in the Era of the Computer. Free Press, New York (1988)
39. Minsky, M.: Frame-system theory. In: Johnson-Laird, P.N., Watson, P.C. (eds.) Thinking, Readings in Cognitive Science. Cambridge University Press, Cambridge (1977)
40. Wehner, M.: AI is now better at predicting mortality than human doctors. New York Post, 14 May 2019. https://nypost.com/2019/05/14/ai-is-now-better-at-predicting-mortality-than-human-doctors/?utm_campaign=partnerfeed&utm_medium=syndicated&utm_source=flipboard. Accessed June 2019
41. Lavars, N.: Machine learning algorithm detects signals of child depression through speech. New Atlas, 7 May 2019. https://newatlas.com/machine-learning-algorithm-depression/59573/. Accessed June 2019

42. Sandoiu, A.: Artificial intelligence better than humans at spotting lung cancer. Medical News Today Newsletter, 20 May 2019. https://www.medicalnewstoday.com/articles/325223.php#1. Accessed Nov 2019
43. Towers-Clark, C.: The Cutting-Edge of AI Cancer Detection. Forbes, 30 April 2019. https://www.forbes.com/sites/charlestowersclark/2019/04/30/the-cutting-edge-of-ai-cancer-detection/#45235ee7733. Accessed June 2019
44. Gregory, M.: AI Trained on Old Scientific Papers Makes Discoveries Humans Missed. Vice (2019). https://www.vice.com/en_in/article/neagpb/ai-trained-on-old-
45. De Groot, A.D.: Thought and Choice in Chess. Mouton, The Hague (1965)
46. Fulbright, R.: The expertise level. In: Schmorrow, D.D., Fidopiastis, C.M. (eds.) HCI 2020. LNAI, vol. 12197, pp. 49–68. Springer, Cham (2020)
47. Haenssle, H.A., et al.: Man against machine: diagnostic performance of a deep learning convolutional neural network for dermoscopic melanoma recognition in comparison to 58 dermatologists. Ann. Oncol. 29(8), 1836–1842 (2018). https://academic.oup.com/annonc/article/29/8/1836/5004443. Accessed Nov 2019
48. Abràmoff, M.D., Lavin, P.T., Birch, M., Shah, N., Folk, J.C.: Pivotal trial of an autonomous AI-based diagnostic system for detection of diabetic retinopathy in primary care offices. Digit. Med. 1, 39 (2018)
49. Harper, R.: The role of HCI in the age of AI. Int. J. Hum.-Comput. Interact. 35 (2019). https://www.tandfonline.com/doi/abs/10.1080/10447318.2019.1631527. Accessed Jan 2020

The Expertise Level

Ron Fulbright[(⊠)]

University of South Carolina Upstate, 800 University Way, Spartanburg, SC 29303, USA
rfulbright@uscupstate.edu

Abstract. Computers are quickly gaining on us. Artificial systems are now exceeding the performance of human experts in several domains. However, we do not yet have a deep definition of expertise. This paper examines the nature of expertise and presents an abstract knowledge-level and skill-level description of expertise. A new level lying above the Knowledge Level, called the Expertise Level, is introduced to describe the skills of an expert without having to worry about details of the knowledge required. The Model of Expertise is introduced combining the knowledge-level and expertise-level descriptions. Application of the model to the fields of cognitive architectures and human cognitive augmentation is demonstrated and several famous intelligent systems are analyzed with the model.

Keywords: Cognitive systems · Cognitive augmentation · Synthetic expertise · Cognitive architecture · Knowledge Level · Expertise Level

1 Introduction

Artificial systems are gaining on us! Powered by new machine learning and reasoning methods, artificial systems are beginning to exceed expert human performance in many domains. IBM's Deep Blue, defeated the reigning human chess champion in 1997 [1]. In 2011, a cognitive system built by IBM, called Watson, defeated the two most successful human champions of all time in the game of Jeopardy! [2, 3]. In 2016, Google's AlphaGo defeated the reigning world champion in Go, a game vastly more complex than Chess [4, 5]. In 2017, a version called AlphaGo Zero learned how to play Go by playing games with itself not relying on any data from human games [6]. AlphaGo Zero exceeded the capabilities of AlphaGo in only three days. Also in 2017, a generalized version of the learning algorithm called AlphaZero was developed capable of learning any game. After only a few hours of self-training, AlphaZero achieved expert-level performance in the games of Chess, Go, and Shogi [7].

This technology goes far beyond playing games. Computers are now better at predicting mortality than human doctors [8], detecting early signs of heart failure [9], detecting signs of child depression through speech [10], and can even find discoveries in old scientific papers missed by humans [11]. Many other examples of artificial systems achieving expert-level performance exist.

© Springer Nature Switzerland AG 2020
D. D. Schmorrow and C. M. Fidopiastis (Eds.): HCII 2020, LNAI 12197, pp. 49–68, 2020.
https://doi.org/10.1007/978-3-030-50439-7_4

2 Literature

2.1 What is an Expert?

What does it mean to be an expert? What is expertise? Are these systems really artificial experts? To answer these kinds of questions, one needs a model of expertise to compare them to. The nature of intelligence and expertise has been debated for decades. To motivate the Model of Expertise presented in this paper, we draw from research in artificial intelligence, cognitive science, intelligent agents, and educational pedagogy.

As Gobet points out, traditional definitions of expertise rely on knowledge (what an expert knows) and skills (what an expert knows how to do) [12]. Some definitions say an expert knows more than a novice while other definitions say an expert can do more than a novice. While an expert is certainly expected to know about their topic and be able to perform skills related to that topic, not everyone who is knowledgeable and skillful in a domain is an expert in that domain. Simply knowing more and being able to do more is not enough. Therefore, Gobet gives a results-based definition of expertise:

> "…an expert is somebody who obtains results vastly superior to those obtained by the majority of the population."

This definition immediately runs into the venerable debates involving Searle's Chinese room [13] and the Turing test [14]. Is a machine yielding results like an expert really an expert? Answering these kinds of questions is difficult because we lack a deep model of expertise. In their influential study of experts, Chase and Simon state: [15]

> "…a major component of expertise is the ability to recognize a very large number of specific relevant cues when they are present in any situation, and then to retrieve from memory information about what to do when those particular cues are noticed. Because of this knowledge and recognition capability, experts can respond to new situations very rapidly and usually with considerable accuracy."

Experts look at a current situation and match it to an enormous store of domain-specific knowledge. Steels later describes this as deep domain knowledge [16]. Experts acquire this enormous amount of domain knowledge from experience (something we now call episodic memory). It is estimated experts possesses at least 50,000 pieces of domain-specific knowledge requiring on the order of 10,000 h of experience. Even though these estimates have been debated, it is generally agreed experts possess vast domain knowledge and experience. Experts extract from memory much more knowledge, both implicit and explicit, than novices. Furthermore, an expert applies this greater knowledge to the situation at hand more efficiently and quickly than a novice. Therefore, experts are better and more efficient problem solvers than novices.

The ability of an expert to quickly jump to the correct solution has been called intuition. A great deal of effort has gone into defining and studying intuition. Dreyfus and Dreyfus (Dreyfus 1972) and (Dreyfus and Dreyfus 1988) argue intuition is a holistic human property not able to be captured by a computer program [17, 18]. However, Simon et al. argue intuition is just the ultra-efficient matching and retrieval of "chunks" of knowledge and know-how.

The idea of a "chunk" of information has been associated with artificial intelligence research and cognitive science for decades dating back to pioneers Newell and Simon. Gobet and Chassy argue the traditional notion of a "chunk" is too simple and instead, introduce the notion of a "template" as a chunk with static components and variable, or dynamic components, resembling a complex data structure [19]. Templates are similar to other knowledge representation mechanisms in artificial intelligence such as Minsky frames [20] and models in intelligent agent theory [21]. As Gobet and Simon contend, templates allow an expert to quickly process information at different levels of abstraction yielding the extreme performance consistent with intuition [22].

DeGroot experimentally established the importance of perception in expertise [23]. When perceiving a situation in the environment, an expert is able to see the most important things quicker than a novice. Being able to perceive the most important cues and retrieve knowledge form one's experience and quickly apply it are hallmarks of expertise.

Steels identified the following as needed by experts: deep domain knowledge, problem-solving methods, and task models [16]. Problem-solving methods are how one goes about solving a problem. There are generic problem solutions applicable to almost every domain of discourse such as "how to average a list of numbers." However, there are also domain-specific problem-solving methods applicable to only a specific domain or very small collection of domains or even just one domain. A task model is knowledge about how to do something. For example, how to remove a faucet is a task an expert plumber would know. As with problem solutions, there are generic tasks and domain-specific tasks. Summarizing, the basic requirements for an expert are:

- the ability to experience and learn domain knowledge
- learn task knowledge
- learn problem-solving knowledge
- perceive a current situation
- match the current situation with known domain knowledge
- retrieve knowledge relevant to the situation
- apply knowledge and know-how
- achieve superior results

2.2 The Knowledge Level

Cognitive scientists have studied human cognition for many decades with the hope of being able to create artificial entities able to perform human-level cognition. Newell recognized computer systems are described at many different levels and defined the Knowledge Level as a means to analyze the knowledge of intelligent agents at an abstract level as shown in Fig. 1 [24]. The lower levels represent physical elements from which the system is constructed (electronic devices and circuits). The higher levels represent the logical elements of the system (logic circuits, registers, symbols in programs).

In general, a level "abstracts away" the details of the lower level. For example, consider a computer programmer writing a line of code storing a value into a variable. The programmer is operating at the Program/Symbol Level and never thinks about how the value is stored in registers and ultimately is physically realized as voltage potentials in electronic circuits. Details of the physical levels are abstracted away at

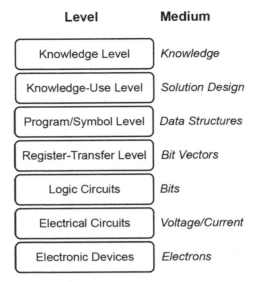

Fig. 1. The Knowledge Level

the Program Level. Likewise, at the Knowledge Level, implementation details of how knowledge is represented in computer programs is abstracted away. This allows us to talk about knowledge in implementation-independent terms facilitating generic analysis about intelligent agents.

Steels added the Knowledge-Use Level between the Knowledge Level and the Program Level to address issues like task decomposition, execution, scheduling, software architecture, and data structure design [16]. This level is geared toward implementation and is quite dependent on implementation details but is necessary to bridge the gap between the Knowledge Level and the Program/Symbol Level.

2.3 Cognitive Architecture

Cognitive scientists have spent much effort analyzing and modeling human intelligence and cognition. One of the most successful models of human cognition is the Soar model shown in Fig. 2 began by Newell and evolved by his students and others for over thirty years [25]. Although not explicitly a part of the Soar model, the figure shows the Soar model situated in an environment with perceive and act functions. The agent can learn new procedural knowledge (how to do things), semantic knowledge (knowledge about things), and episodic knowledge (knowledge about its experiences). New knowledge can be learned by reinforcement learning, semantic learning, or episodic learning.

Pieces of this knowledge from long term memory, as well as perceptions, are brought into a working area of memory, short term memory, where they are processed by the appraisal and decision functions. Soar is a model of human cognition but is not necessarily a model of expertise. We will later update the Soar architecture to include elements to support expertise.

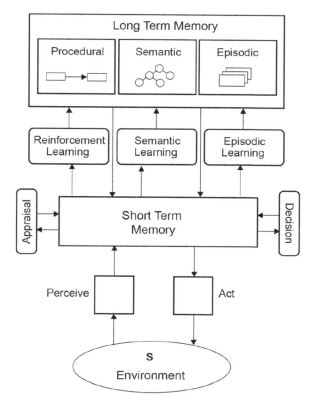

Fig. 2. The Soar model of cognition

2.4 Formal Intelligent Agent Models

Researchers in artificial intelligence have defined several architectural models of intelligent agents. These models are of interest to us in this paper because an expert is certainly an intelligent agent. Figure 3 shows a formal model of a goal-driven, utility-based learning/evolving intelligent agent [21, 26].

Situated in an environment, intelligent agents repeatedly execute the perceive-reason-act-learn cycle. Through various sensors, the see function allows agents to perceive the environment, S, as best they can. Agents can perceive only a subset, T, of the environment. Every agent has a set of actions, A, they can perform via the do function. The agent selects the appropriate action through the action function. Every action causes the environment to change state.

Models, M, are internal descriptions of objects, situations, and the real world. The model function matches incomplete data from the agent's perceptions with its models to classify what it is currently encountering. For example, a "danger" model would allow an agent to recognize a hazardous situation.

Intelligent agents continually work to achieve one or more goals, *G*. The *alter* function allows the agent to change its goals over time. Agents can change their sensitivity

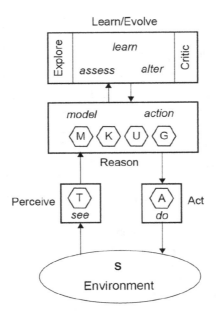

{**K, M, U, G, S, T, A**, see, do, action, model, learn, alter, assess}

K   Set of general knowledge statements
M   Set of internal states/models
U   Set of utility values
G   Set of goals the agent is trying to achieve
S   Set of states for the environment
T   Partitions of **S** distinguishable by the agent
A   Set of actions the agent can perform

see $S \rightarrow T$
do $A \times S \rightarrow S$
action $M \times G \times K \times U \times T \rightarrow A$
model $M \times G \times K \times U \times T \rightarrow M$
learn $M \times G \times K \times U \times T \rightarrow K$
alter $M \times G \times K \times U \times T \rightarrow G$
assess $M \times G \times K \times U \times T \rightarrow U$

Fig. 3. Formal model of a learning/evolving intelligent agent

to things using a set of utility values, **U**. The *assess* function allows the agent to adjust its utility model. This is important because the urgency of certain actions and goals changes over time and with changes in a dynamic environment. For example, the goal "recharge batteries" might have a low utility value at the beginning of a journey but as the battery charge level gets lower, the utility value of the "recharge batteries" goal rises and eventually becomes the most important goal. An intelligent agent maintains a set of general knowledge, **K**, and can learn new knowledge.

2.5 Bloom's Taxonomy

There are other notions of "expertise" from outside the fields of artificial intelligence and cognitive science. Originally published in the 1950s, and revised in the 1990s, Bloom's Taxonomy was developed in the educational field as a common language about learning and educational goals to aid in the development of courses and assessments [27, 28].

As shown in Fig. 4, Bloom's taxonomy consists of a set of verbs describing fundamental cognitive processes at different levels of difficulty. The idea behind Bloom's Taxonomy is when learning a new subject, a student able to perform these processes, has demonstrated proficiency in the subject matter. The processes are listed in order from the simplest (remember) to the most complex (create). Here, we propose these processes also describe expert performance in a particular domain. The verbs in Bloom's Taxonomy identify skills we expect an expert to possess.

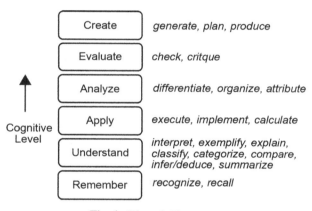

Fig. 4. Bloom's Taxonomy

We note two things as we look at historical notions of intelligence and expertise. First, we encounter both knowledge and skills as requirements of an expert. Therefore, a model of expertise must be able to represent both. Second, while there are many models of intelligence and human cognition, there are no comprehensive models of expertise. While it may be true all experts are intelligent agents, it is not true all intelligent agents are experts. Therefore, we seek a comprehensive model of expertise and wish for this model to be implementation independent and apply to human and the artificial.

In this paper, we first introduce a new abstract level called the Expertise Level. Lying above the Knowledge Level, the Expertise Level describes the skills needed by an expert. We use the skills identified by Simon, Steels, and Gobet augmented with the skills from Bloom's Taxonomy to form this description. We also develop a Knowledge Level description of expertise describing the knowledge required of an expert. For this description, we combine the knowledge identified in various cognitive architectures discussed earlier. We then show how to apply the Model of Expertise by showing how it can be incorporated into the Soar architecture and how it can be used in the field of human cognitive augmentation. We then use the Model of Expertise to discuss and characterize current systems.

3   The Model of Expertise

3.1   The Expertise Level

As described earlier, Simon, Steels, and Gobet identify kinds of knowledge an expert must have and kinds of functions or actions an expert must perform. Newell's Knowledge Level is suitable for holding a description of an expert's knowledge. However, a full model of expertise must accommodate both knowledge and skills.

As shown in Fig. 5, we extend Newell's levels and create a new level called the Expertise Level above the Knowledge Level to represent skills an expert must possess. At the Expertise Level, we talk about what an expert does—the skills—and not worry about the details of the knowledge required to perform these skills. Therefore, the medium of the Expertise Level is skills.

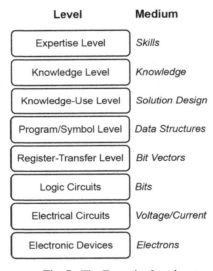

Fig. 5. The Expertise Level

3.2   The Expertise Level Description of Experts

What skills does an expert need to have? We start with the six skills identified in Bloom's Taxonomy: *recall, understand, apply, analyze, evaluate,* and *create.* These skills were identified because in the education field, a student demonstrating ability in all six skills is considered to have achieved a mastery of the subject matter. An expert certainly has mastery of subject matter, so any model of expertise should include these six skills.

An expert certainly is considered to be an intelligent agent operating in an environment. As such, following the example set by researchers from the intelligent agent theory field, the expert must sense the environment (perceive) and perform actions (act) to effect changes on the environment. Common in the intelligent agent and cognitive science fields is the notion of learning. An expert acquires knowledge and know-how via experience through the learn skill. Future refinement of the model may identify several different kinds of learning and therefore add skills, but here we represent all types of learning with the single learn skill. From the work of Simon, Steels, and Gobet, the extract skill represents the expert's ability to match perceptions to stored deep domain knowledge and knowledge, procedures, and tasks relevant to the situation.

Experts are goal-driven intelligent agents with the ability to change goals over time. The *alter* skill allows the expert to change its goals as it evolves. We also believe experts must be utility-driven intelligent agents and must have the ability to adjust its utility values. We do not create a new skill for this because this ability is subsumed by the *evaluate* skill already in our list of skills. We feel an expert should be able to *teach* about their domain of discourse, so include this skill in our list.

Therefore, as shown in Fig. 6, twelve skills are identified at the Expertise Level description of an expert: *recall, apply, evaluate, understand, analyze, create, extract, teach, perceive, learn, alter, and act.*

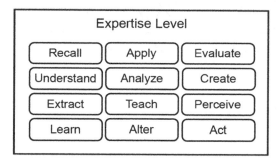

Fig. 6. The Expertise Level Description of Experts

3.3 The Knowledge Level Description of Experts

To create the Knowledge Level description of an expert, we combine ideas from intelligent agent theory, cognitive architectures, and cognitive science. An expert is an agent able to perceive the environment. Because of limitations in its sensory systems, an expert perceives only a subset of the possible states of the environment, T. The expert also has a set of actions, A, it can perform to change the environment. Because experts are goal-driven and utility-driven evolving agents, G represents the set of goals and U represents the set of utility values.

In addition to deep domain knowledge K_D (knowledge about the domain) experts possess general background knowledge K (generic knowledge about things), common-sense knowledge K_C, and episodic knowledge K_E (knowledge from and about experiences). A model, similar to Gobet's templates and Minksy's frames, is an internal representation allowing the expert to classify its perceptions and recognize or differentiate situations it encounters. For example, an expert plumber would have an idea of what a leaky faucet looks, sounds, and acts like based on experience allowing the plumber to quickly recognize a leaky faucet. Some models are domain-specific, M_D, and other models are generic, M. In humans, the collection and depth of models is attained from years of experience. As models are learned from experience, creating and maintaining M_D requires K_E and K_D as a minimum but may also involve other knowledge stores. Following Simon and Steels, an expert must know how to solve problems in a generic sense, P, and how to solve problems with domain-specific methods, P_D. In addition, experts must also know how to perform generic tasks, L, and domain-specific tasks, L_D.

Therefore, we have identified 14 knowledge stores an expert maintains as shown in Fig. 7.

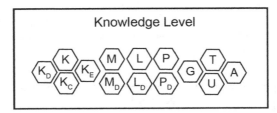

Fig. 7. The Knowledge Level Description of Experts

3.4   The Model of Expertise

Combining the Expertise Level description of experts shown in Fig. 6 with the Knowledge Level description shown in Fig. 7 yields the full Model of Expertise shown in Fig. 8. Experts require 12 fundamental skills: *recall, apply, evaluate, understand, analyze, create, extract, teach, perceive, learn, alter,* and *act.* When working with another entity, experts need the *collaborate* skill as well.

Experts also require the maintenance of 14 knowledge stores: 4 general types of knowledge (generic, domain-specific, common-sense, and episodic), generic and domain-specific models, generic and domain-specific task models, generic and domain-specific problem-solving models, a set of goals, utility values, a set of actions, and a set of perceived environmental states.

Unlike other cognitive models, the Model of Expertise introduced here can be applied to biological experts (humans) and to artificial experts (cognitive systems and artificial intelligence). Because the Model of Expertise is based on abstract Expertise Level and Knowledge Level descriptions, implementation details of how an entity carries out a skill or implements a knowledge store is not specified. This leaves implementation details up to the entities themselves. Humans implement the skills and knowledge stores quite differently than computers do and different cognitive systems developed in the cog era will implement them differently from each other.

The Model of Expertise introduced here can be applied to other cognitive architectures and cognitive models as well as is demonstrated in the next section. Our hope is

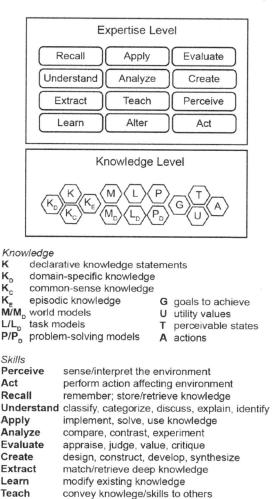

Knowledge
K   declarative knowledge statements
K_D domain-specific knowledge
K_C common-sense knowledge
K_E episodic knowledge   G goals to achieve
M/M_D world models   U utility values
L/L_D task models   T perceivable states
P/P_D problem-solving models   A actions

Skills
Perceive   sense/interpret the environment
Act   perform action affecting environment
Recall   remember; store/retrieve knowledge
Understand classify, categorize, discuss, explain, identify
Apply   implement, solve, use knowledge
Analyze   compare, contrast, experiment
Evaluate   appraise, judge, value, critique
Create   design, construct, develop, synthesize
Extract   match/retrieve deep knowledge
Learn   modify existing knowledge
Teach   convey knowlege/skills to others
Alter   modify goals

Fig. 8. The Model of Expertise

the Model of Expertise can serve as a common ground and common language facilitating comparison, contrast, and analysis of different systems, models, architectures, and designs.

This version of the Model of Expertise contains 12 fundamental skills (13 including the *collaborate* skill) and 14 knowledge stores. Future research may identify additional skills and additional knowledge stores and we invite collaborative efforts along these lines of thought.

3.5 Composite Processes/Activities

An expert certainly performs more actions than the skills identified in the Model of Expertise. It is important to note the skills listed in the Model of Expertise are fundamental in nature. Other, higher-level processes and activities are composites combining one or more fundamental skills. Examples are the *justify* activity and the *predict* process. Whenever a human or an artificial system arrives at a decision, it is common for someone to ask for justification as to why that decision was made. To justify a decision or action, the expert would exercise a combination of the *recall, analyze,* and *understand* skills. To *predict* an outcome, the expert would exercise a combination of *recall, analyze, evaluate,* and *apply* skills. Future research may very well identify additional fundamental skills. However, in doing so, care should be taken to identify skills which cannot be composed of combination of the fundamental skills in the model. Other examples of composite processes/activities include:

Check me	Conjecture
Clarify	Conceptualize
Define	Debate
Emphasize	Exemplify
Explain how	Explain when
Explain where	Explain why
Expand the scope	Expound/Elucidate
Gather evidence of	Gather information on
Give me alternatives	Give me analogies
How do you feel about	Illustrate/Depict
Inspire me	Make a case for
Make a case against	Make me feel better about
Monitor and notify me	Motivate me
Narrow the scope	Organize
Predict	Prioritize
Show me	Show me associations
Simplify	Summarize
Theorize	Think differently than me
Visualize	What if
What is best for me	What is the cost of
What is most important	What is this least like
What is this most like	

4 Applications

4.1 The Soar Model of Expertise

Cognitive scientists have studied and modeled human cognition for decades. The most successful cognitive architecture to date, begun by pioneer Allen Newell and now led by John Laird, is the Soar architecture shown in Fig. 2. The Model of Expertise can be applied to and incorporated into the Soar architecture as shown in Fig. 9.

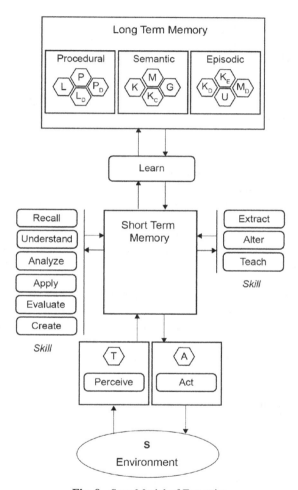

Fig. 9. Soar Model of Expertise

The knowledge stores are located in long-term memory (T and A are shown associated with the *perceive* and *act* functions) and are brought into short-term working memory by one or more of the skills. Reinforcement, semantic, and episodic learning continually updates the knowledge stores in long term memory. Higher-order processes defined in Soar are composite processes resulting from a combination of the fundamental skills in the Model of Expertise. Future research should document and analyze these processes and ground them to the skills and knowledge stores identified in the Model of Expertise introduced here.

4.2 Cognitive Augmentation/Synthetic Expertise

We will soon be surrounded by artificial systems designed for the mass market capable of cognitive performance rivaling or exceeding a human expert in specific domains of discourse. Indeed, we see the beginning of this era now with voice-activated assistants and applications on our smartphones and other devices. John Kelly, Senior Vice President and Director of Research at IBM describes the coming revolution in cognitive augmentation as follows [29]:

> "The goal isn't to replace human thinking with machine thinking. Rather humans and machines will collaborate to produce better results—each bringing their own superior skills to the partnership."

The future lies in humans collaborating with artificial systems capable of high-level cognition. Engelbart was one of the first who thought of a human interacting with an artificial entity as a system [30]. While working together on a task, some processes are performed by the human (explicit-human processes), others are performed by artificial means (explicit-artifact processes), and others are performed by a combination of human and machine (composite processes). In the cognitive systems future, cognition will be a mixture of biological and artificial thinking. The human component in this ensemble will be said to have been cognitively augmented. We can represent cognitive augmentation using our Model of Expertise as shown in Fig. 10.

The figure depicts a human working in collaboration with an artificial entity, a cognitive system, called a cog. As in Engelbart's framework, some of the skills are performed by the human and some are performed by the cog. In some cases, portions of a skill are performed by both the human and the cog. The human performs an amount of the *cognitive work* (W_H) and the cog performs an amount of the cognitive work (W_C). The cognitive work performed by the entire ensemble is W^*. The most important thing is all skills identified in the Model of Expertise are performed by the human/cog ensemble. It does not matter to the outside world whether or not a biological or artificial entity performs a skill.

Because the human and the artificial system are physically independent entities, we have drawn each with perceive and act skills and an additional skill, *collaborate*, has been added. This is necessary because in order to work together the human and the artificial entity must collaborate. In the figure, the knowledge stores are represented as not belonging solely to either entity. Physically, the human will have its own version of all knowledge stores in the formal model and the cog will have its own knowledge stores. However, logically, the knowledge of the ensemble will be a combination of the human and artificial knowledge sources. In fact, the entire cognitive performance of the human/cog ensemble is the emergent result of human/artificial collaboration and cognition. When the ensemble can achieve results exceeding most in the population it

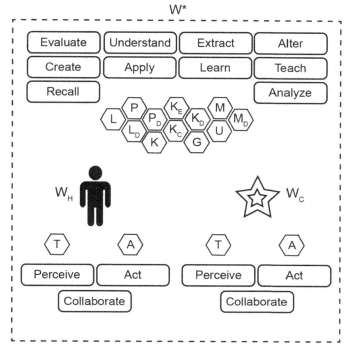

Fig. 10. Synthetic Expertise

will have achieved the Gobet definition of expertise described earlier of achieving results superior to most. The ensemble will have achieved *synthetic expertise.*

An average human and non-expert in a field of study acting alone is able to perform to a certain level. The same human working in collaboration with a cog will be able to perform at a higher level, even to the level of an expert. Therefore, to the world outside of the human/cog ensemble, the human will appear to be cognitively augmented.

4.3 Cognitive System Architectures

The Model of Expertise introduced in this paper can be used to design future cognitive systems. An example is Lois, an artificial companion and caretaker for the elderly shown in Fig. 11.

Lois is composed of several different models allowing it to monitor the status and well-being of the elder. Episodic memory allows Lois to remember every interaction with the elder and use the knowledge to learn new knowledge, tasks, and activities. We invite researchers to develop more such models based on the Model of Expertise.

5   Discussion

In the introduction, several recent and impressive success stories were discussed featuring an artificial system performing better than human experts in a particular domain. Using

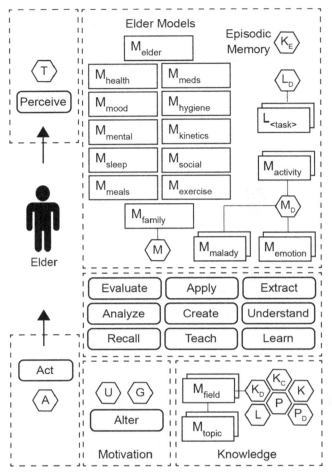

Fig. 11. Synthetic Elderly Companion (Lois)

the Model of Expertise introduced in Fig. 8, we can now discuss these and other systems in new ways. The Model of Expertise identifies fundamental skills and types of knowledge required of an expert. How do real systems stack up against the model?

The systems discussed in the introduction, DeepBlue, Watson, AlphaGo, AlphaGo Zero, and AlphaZero, all perform some form of *perceive, analyze, evaluate,* and *act* skills. However, the scope of these functions is limited to the domain of discourse—the state of the gameboard and scoring the effect of a move. The problem here is the limited scope of the games themselves. Even though these systems have achieved expert-level performance, the scope of the expertise is small. Most systems in existence today share this characteristic of being narrowly focused. We call this *narrow expertise.* Systems able to learn general principles applicable to other domains is a very active area of research. This line of thought leads to another question. Should we expect an artificial expert to perform well in other domains? We do not necessarily expect this from human experts.

For example, if an expert plumber is a terrible carpenter, we do not think less of him or her as a plumber.

No system discussed in the introduction can be said to *create*. If a Chess or Go program plays a sequence of moves never before played in the history of the game, did the system create it or just stumble upon it? This leads to an interesting argument involving understanding and this is a debate that has raged throughout the history of artificial intelligence research. Few would agree DeepBlue, Watson, AlphaGo, AlphaGo Zero, and AlphaZero actually understand the games they play and many would argue true creation cannot happen without understanding.

However, recently, systems have been developed to create faces (thisperso ndoesnotexist.com), music (www.aiva.ai), art (obviousart.com), and news stories (openai.com/blog/better-language-models). These systems do not understand what they are doing and their sole purpose is to create. Do these systems possess a true *create* skill? The images of the fake people created are indistinguishable from photographs of real people. Likewise, the music, art, and language are indistinguishable from that generated by humans and therein lies the problem. Recalling the Gobet definition of expertise, experts are expected to achieve results superior than most others. Therefore, an argument against these systems is the generated faces, music, portraits, and news briefs are not better, they are simply the same as those created by humans. This might pass the Turing test criterion involving artificial behavior a human cannot distinguish from human behavior, but the criterion for expertise is higher according to Gobet's definition. With experts, we are looking for exemplar behavior.

Unfortunately, whether not an object is an exemplar is highly subjective in most cases. However, we would not be surprised if, soon, an artificial system creates a hit song or an award-winning bit of poetry or prose. Even if systems achieve something like this, the question still remains: do systems understand what they have done? That brings us back to the above question. Strong AI proponents maintain systems must know and understand what they are doing to be considered intelligent whereas weak AI proponents require only results. Therefore, at best, all known systems at this time are examples of weak expertise.

None of the systems discussed so far exhibit the *teach* skill. The field of intelligent tutoring systems has been an active area of research for several decades and several tutoring systems exist, especially in mathematics. However, these systems are designed and built solely for the purpose of teaching specific knowledge/skills in a particular domain, not to be an expert in the domain. These systems are not expert mathematicians. Therefore, today we have systems achieving expertise but are not teachers and we have systems that teach but are not experts.

So far we have focused discussion on the skills in the Model of Expertise. What about the knowledge of an expert? All game-playing systems have some form of goals and utility value mechanism to execute strategy. However, a system's ability to *alter* and assess/*evaluate* new goals is limited to the game play context. There are only a few goals of interest in game play. For example, AlphaGo would never synthesize a goal of "acquire a ham sandwich." Artificial expertise has not yet been achieved in domains where complex goals and utility values are necessary. This again speaks to the narrow expertise and weak expertise nature of systems today as discussed earlier.

One of the great achievements of systems like AlphaGo, AlphaGo Zero, and AlphaZero has been the knowledge acquired via semi-supervised or unsupervised machine deep learning (K_D). Because of their narrowness, these systems do not learn generic knowledge like common-sense knowledge nor general knowledge (K, K_C). Again, no known system requiring extensive common-sense or generic knowledge has achieved expert-level performance.

Neither do these systems acquire episodic knowledge (K_E). For example, a human Go player would be able to recount a particularly interesting game against an opponent in which they learned an effective bit of strategy. A human would be able to tell you when the game was, who the opponent was, what they were feeling at the time, what was happening in the world about that time, etc. This is deep episodic memory—capturing the entire experience. Researchers are working on experience and context capture, common-sense learning, and general knowledge acquisition, but as of yet, these systems appear to be separate efforts no yet included into an artificial expert.

The degree of problem-solving knowledge (L, L_D) in these systems is difficult to assess without detailed analysis of how they work. One can certainly make the case these systems can learn and recall solutions and even strategies. But again, the narrowness of the game playing domain means these systems are not learning generic problem-solving knowledge (L) and any domain-specific problem-solving knowledge (L_D) is limited in scope and breadth.

Task knowledge is similarly restricted in game-playing systems. Since the set of actions, (A) is limited to game-related movement and placement of pieces, these systems do not have to perform an array of different kinds of tasks. Consider the difference between AlphaGo and an expert human carpenter. An expert knows how to perform hundreds if not thousands, of tasks related to carpentry (e.g. ways to make different kinds of joins), knows how to use dozens, if not hundreds, of different kinds of tools, and knows dozens of ways to apply paint, sealer, and varnish. The narrowness of existing systems limits task and problem-solving knowledge.

Game-playing systems certainly recognize situations and respond accordingly. For example, a Chess program will recognize its king is in check. It is unclear, but probably not the case, these systems use templates, frames, or other kinds of dynamic symbolic models. Medical diagnosis systems involve mostly pattern recognition and classification. For example, when detecting cancer in a radiograph, an artificial system will detect a pattern (part of its *perceive* and *understand* skillset) and then compare it (*match, analyze, evaluate*) to examples of known cancerous radiographs. If deploying an artificial neural network, the network's response to a stimulus (the radiograph) is compared to responses to known cancerous radiographs. Therefore, modeling seems to be built into the system itself rather than exist as a standalone knowledge store.

In the Model of Expertise, the extract skill represents the ability of an expert to *match* perceptions to stored knowledge and then retrieve a quantity of deep domain, procedural, and task knowledge (P, P_D, L, L_D). The systems discussed do not perform in this manner primarily because they do not seek to implement an entire expert (a doctor or game-playing person). Instead, today's systems should be considered to be low-level cogs better suited to be employed in a human/cog ensemble to achieve synthetic expertise.

6   Conclusion

A new abstract level, the Expertise Level, has been introduced to lie above Newell's Knowledge Level. Skills are the medium at the Expertise Level. An abstract Knowledge Level and Expertise-Level description of expertise has been introduced drawing from previous research in cognitive science, artificial intelligence, cognitive architectures, and cognitive augmentation. The Knowledge Level description describes the kinds of knowledge stores an expert must possess and the Expertise Level description describes the skills an expert must possess. Together, the Knowledge Level and Expertise Level description of expertise forms the Model of Expertise. The Model of Expertise can be used in all fields involving the study of intelligence and cognition including: cognitive science, cognitive architectures, artificial intelligence, cognitive systems, and cognitive augmentation. Specifically, we would like to see the Model of Expertise and the Expertise Level used to guide and assess future systems capable of artificial expertise and synthetic expertise, particularly in the cognitive systems area.

Integration of our model of expertise into the Soar cognitive architecture has been demonstrated. This could facilitate development of artificial expertise systems and exploration of human cognition based on the Soar architecture. Also demonstrated was using the Model of Expertise in the field of human cognitive augmentation to describe synthetic expertise whereby cognitive output of the ensemble is a mixture of biological and artificial cognition.

Finally, we discussed and an analyzed current game-playing systems using the Model of Expertise identifying them as examples of narrow expertise. Using the Model of Expertise, we may now distinguish narrow expertise from broad expertise and weak expertise from strong expertise by the degree to which systems utilize the knowledge and all skills identified at the Expertise Level of the Model of Expertise. Future cognitive systems will be developed for domains requiring extensive use of all skills and knowledge stores defined in the Model of Expertise.

References

1. IBM: Deep Blue. IBM (2018). http://www03.ibm.com/ibm/history/ibm100/us/en/icons/deepblue/. Accessed Feb 2018
2. Jackson, J.: IBM Watson Vanquishes Human Jeopardy Foes. PC World (2011). http://www.pcworld.com/article/219893/ibm_watson_vanquishes_human_jeopardy_foes.html. Accessed May 2015
3. Ferrucci, D.A.: Introduction to this is Watson. IBM J. Res. Dev. **56**(3/4) (2012)
4. Silver, D., et al.: Mastering the game of Go with deep neural networks and tree search. Nature **529**, 484 (2016)
5. DeepMind: The story of AlphaGo so far. DeepMind (2018). https://deepmind.com/research/alphago/. Accessed Feb 2018
6. DeepMind: AlphaGo Zero: learning from scratch. Deep-Mind (2018). https://deepmind.com/blog/alphago-zero-learning-scratch/. Accessed Feb 2018
7. ChessBase: AlphaZero: Comparing Orangutans and Apples. ChessBase (2018). https://en.chessbase.com/post/alpha-zero-comparing-orang-utans-and-apples. Accessed Feb 2018
8. Wehner, M.: AI is now better at predicting mortality than human doctors. New York Post, 14 May 2019

9. O'Hara, J.: Artificial intelligence can help detect early signs of heart failure. Mayo Clinic News Network, 16 May 2019. https://newsnetwork.mayoclinic.org/discussion/artificial-int elligence-can-help-detect-early-signs-of-heart-failure/. Accessed Aug 2019

10. Lavars, N.: Machine learning algorithm detects signals of child depression through speech. New Atlas (2019). https://newatlas.com/machine-learning-algorithm-depression/ 59573/. Accessed Aug 2019

11. Gregory, M.: AI Trained on Old Scientific Papers Makes Discoveries Humans Missed. Vice (2019). https://www.vice.com/en_in/article/neagpb/ai-trained-on-old-scientific-papers-makes-discoveries-humans-missed. Accessed Aug 2019

12. Gobet, F.: Understanding Expertise: A Multidisciplinary Approach. Palgrave, London (2016)

13. Searle, J.: Minds, brains and programs. Behav. Brain Sci. **3**(3), 417–457 (1980)

14. Turing, A.: Computing machinery and intelligence. Mind **236**, 433–460 (1950)

15. Chase, W., Simon, H.: Perception in Chess. Cogn. Psychol. **4**, 55–81 (1973)

16. Steels, L.: Components of expertise. AI Mag. **11**(2), 28 (1990)

17. Dreyfus, H.L.: What Computers Can't Do: A Critique of Artificial Reason. The MIT Press, Cambridge (1972)

18. Dreyfus, H.L., Dreyfus, S.E.: Mind Over Machine: The Power of Human Intuition and Expertise in the Era of the Computer. Free Press, New York (1988)

19. Gobet, F., Chassy, P.: Expertise and intuition: a tale of three theories. Minds Mach. **19**, 151–180 (2009). https://doi.org/10.1007/s11023-008-9131-5

20. Minsky, M.: Frame-system theory. In: Johnson-Laird, P.N., Watson, P.C. (eds.) Thinking, Readings in Cognitive Science. Cambridge University Press, Cambridge (1977)

21. Russell, S., Norvig, P.: Artificial Intelligence: A Modern Approach, 3rd edn. Pearson, London (2009)

22. Gobet, F., Simon, H.: Five seconds or sixty? Presentation time in expert memory. Cogn. Sci. **24**(4), 651–682 (2000)

23. De Groot, A.D.: Thought and Choice in Chess. Mouton, The Hague (1965)

24. Newell, A.: The knowledge level. Artif. Intell. **18**(1), 87–127 (1982)

25. Laird, J.E.: The Soar Cognitive Architecture. The MIT Press, Cambridge (2012)

26. Genesereth, M., Nilsson, N.: Logical Foundations of Artificial Intelligence. Morgan Kaufmann, Burlington (1987)

27. Bloom, B.S., Engelhart, M.D., Furst, E.J., Hill, W.H., Krath-wohl, D.R.: Taxonomy of Educational Objectives: The Classification of Educational Goals. Handbook I: Cognitive Domain. David McKay Company, New York (1956)

28. Anderson, L.W., Krathwohl, D.R., Airasian, P.W., Cruikshank, K.A., Mayer, R.E.: A Taxonomy for Learning, Teaching, and Assessing: A Revision of Bloom's Taxonomy of Educational Objectives. Pearson, London (2001)

29. Kelly, J.E., Hamm, S.: Smart Machines: IBMs Watson and the Era of Cognitive Computing. Columbia Business School Publishing, Columbia University Press, New York (2013)

30. Engelbart, D.C.: Augmenting Human Intellect: A Conceptual Framework, Summary Report AFOSR-3233. Stanford Research Institute, Menlo Park (1962)

The Dark Sides of Technology - Barriers to Work-Integrated Learning

Camilla Gjellebæk[1]([envelope]) [iD], Ann Svensson[2] [iD], and Catharina Bjørkquist[1] [iD]

[1] Østfold University College, Halden, Norway
camilla.gjellebak@hiof.no
[2] University West, Trollhättan, Sweden

Abstract. Digitalization and technology are interventions seen as a solution to the increasing demand for healthcare services, but the associated changes of the services are characterized by multiple challenges. A work-integrated learning approach implies that the learning outcome is related to the learning environment and the learning affordances available at the actual workplace. To shape workplace affordances it is of great importance to get a deeper understanding of the social practices. This paper will explore a wide range of managers' and professionals' emotions, moods and feelings related to digitalization and new ways of providing healthcare services, as well as the professionals' knowledge and experiences. Zhang's affective response model (ARM) will be used as a systematic approach and framework to gain knowledge of how professionals and managers experience and experience digitization of municipal health services. The research question is: **How can knowledge about dark sides of technology reduce barriers to work-integrated learning?**

This paper is based on a longitudinal study with a qualitative approach. Focus group discussions were used as method for collecting data. The findings and themes crystallized through the content analysis were then applied to the Affective Response Model as a systematic approach to gain more knowledge about professionals and managers' experiences and how that knowledge can reduce the barriers to work-integrated learning. Understanding of, and consciousness about the dark sides of technology and the professionals' affective responses may support the digitalization of the sector and the development of the new ways of providing healthcare services.

Keywords: Dark sides of technology · Focus groups · Work-integrated learning

1 Introduction

1.1 Background and Challenges in Healthcare Services

Society is facing vast challenges regarding the compliance between people in need for healthcare and resources available to meet these needs [1–4]. Representatives from public as well as private sector have pointed at digitalization and technological interventions as solutions to the "care crisis" [5–7], but implementation of new technologies and

© Springer Nature Switzerland AG 2020
D. D. Schmorrow and C. M. Fidopiastis (Eds.): HCII 2020, LNAI 12197, pp. 69–85, 2020.
https://doi.org/10.1007/978-3-030-50439-7_5

the associated changes of the services are characterized by multiple challenges [8]. Contextual factors such as organizational issues, technological infrastructure and human actions represent such challenges [5, 9, 10]. Human actions are often a result of the individuals' affective and emotional responses to the surrounding circumstances. The dark side of technology is a concept used to describe "a broad collection of 'negative' phenomena that are associated with the use of IT, and that have the potential to infringe the well-being of individuals, organizations and societies" [11]. It seems to be a duality between uncovering the dark side of technology and the individuals' responses on the one hand and the ability to facilitate for learning and development of skills on the other hand. The increased demand for healthcare and new ways of providing healthcare services are factors that entails consequences for working life, workplace practice and necessary occupational requirements [12, 13]. Even if there are high expectations to the impact of digitalization and the introduction of technology, the success rate has until now been referred to as low [14]. According to Barakat [15], the absence of needed knowledge and skills among healthcare professionals regarding the use of technology represents a barrier to the digitalization of the sector and to the opportunity of reclaiming the expected impact. The value of technical competencies is underscored [13], and Lapão [16] states that the "development of digital skills by health workers is critical". In an era characterized by an increasing amount of workplaces demanding complex skills and mastery of new technologies, here is a need for organizations to integrate learning as part of practice and to make the workplace the locus of learning [17].

The emphasize on learning among healthcare professionals as an issue beyond traditional education, creates a need for integrating work and learning [18, 19]. Hattinger, Eriksson, Malmskiöld and Svensson [20] describe work-integrated learning as "a combination of education and practice in the workplace". Work-integrated learning is summarized to concern the development of skills needed when improving the way work is provided, as well as developing the work organization itself [21]. Work-integrated learning is seen as situated [22]. Situated learning is implying that learning outcome and the opportunity for learning offered the employees are affected by the learning environment and the learning affordances at the actual workplace [21]. Affordances provided by the workplace are situational factors like activities, interpersonal dynamics, rules and norms of practice that provide opportunities and invite the employees to engage as learners [23–26]. Examples of affordances include providing the employees with access to video instructions demonstrating the use of a new device or utility, or by providing a work schedule with integrated time for professional reflection. The managers are responsible for facilitating an upskilling of the professionals, supporting and motivating them to engage in learning situations at the workplace [13]. To be able to facilitate the affordances best suited for the professionals to learn, it is decisive to understand how individuals learn and what might be a barrier to learning [27].

Each individuals' acceptance of technology is reported as an important predictor to whether or not to start and to continue to use technology as part of work [28]. In many social contexts, affect and emotional responses are critical factors in human decisions, behavior and action [29, 30]. The awareness of professionals' affective and emotional responses to digitalization and technology can enable work-integrated learning through affordances that match the individuals' responses. Being unaware of affective responses can result in dark sides of technology. In addition to the potential of infringing the well-being of individuals, organizations and societies, the dark side of technology may lead to dissatisfaction, low morale, decreased work quality and resistance [31], poor job engagement and information overload [32]. Consequently, it may also represent a barrier to the development of digital skills and work-integrated learning. When studying "Two Decades of the Dark Side in the Information System Basket", Pirrkalainen and Salo [32] identified a need for further studies that can contribute to the identification of aspects decisive when planning for workplace affordances and work-integrated learning.

The interest in "dark side-research" increases and will be an important contribution when developing interventions and workplace affordances aiming at reducing the barriers to work-integrated learning as well as the challenges related to the digitalization of the healthcare services [32]. The Technology Acceptance Model (TAM) has been used in various studies when analyzing implementation of IT and personnel's acceptance of technology (e.g. [33]). The model are focusing on cognitive reactions and factors influencing an individual's intention to use new technology [34]. It has mainly been used in studies with a quantitative approach, and some have called attention to its limitation due to impact factors involved in the acceptance of health information technology [33]. In this paper we will explore a range of managers' and professionals' emotions, moods and feelings related to digitalization and new ways of providing healthcare services, as well as the professionals' knowledge and experiences. According to Zhang [29], "a robust understanding of affect may also have practical implications for design, acceptance, use, and management of ICTs". Zhang's affective response model (ARM) will be used as a systematic approach and framework to gain knowledge of how professionals and managers experience and experience digitization of municipal health services [29]. Using ARM will contribute to an identification of aspects decisive when planning for workplace affordances and work-integrated learning. The research question is: **How can knowledge about dark sides of technology reduce barriers to work-integrated learning?**

In Sect. 2 we present a short explanation of The Affective Response Model (ARM) with its taxonomy of five dimensions describing different perspectives on stimulus to affective responses. The methodological approach for the study is presented in Sect. 3. Section 4 consists of results from the focus groups, before discussing how knowledge about dark sides of technology can reduce barriers to work-integrated learning in Sect. 5. In Sect. 6 we will present concluding reflections.

2   The Affective Response Model (ARM)

2.1   Need for Knowledge

The way of providing healthcare services is within an ongoing translation according to digitalization and introduction of technology, with the demand for knowledge and work-integrated learning for healthcare professionals as one of the barriers. Digitalization and changes in the healthcare services creates a demand for new knowledge, and new knowledge represents a push factor for development of the services. Learning and new knowledge are decisive when changing work practices in healthcare organizations.

2.2   Conceptualizing Affective Responses Through the Taxonomy of ARM

The ARM-model consists of two parts; 1) a taxonomy of five dimensions for conceptualizing various types of affective responses and 2) a set of propositions that describe the relationships among various types of affective concepts, resulting in a nomological network [29]. According to Agogo & Hess [30] the five dimensions of the ARM-model provide a taxonomy which can be helpful when differentiating affective concepts. In this paper the analysis will concentrate on the first part of ARM with the taxonomy aiming at conceptualizing the affective responses professionals and managers in healthcare services experience when digitalizing workplaces and developing the ways of providing the services. According to Zhang [29], emotions, moods and feelings are included in the description of the umbrella term affect. Affect "can explain a significant amount of variance in one's cognition and behavior…" [29], and that is why it is important to obtain an understanding of affect. The taxonomy of the five dimensions provides a theoretically bounded framework, and that framework can be a help when trying to understand the meaning of affective concepts related to different kinds of human interaction with information and communication technology (ICT). When applying the taxonomy of affective concepts in an ICT context, the main interest is related to the human interaction with technology and ICT. The technology and ICT may act like an object or it may be the interaction as a behavior representing the stimulus. The definition of stimulus applied by Zhang [29] is "as something or some event in one's environment that person reacts or respond to" (Fig. 1).

Fig. 1. Illustration of ARM-model (based on Zhang [29])

Residing Dimension

In the first dimension of the taxonomy, Zhang [29] describes three categories of referent objects (residing dimension). In the case when the person itself represents the object, it implies that the person gets in a mood or has a negative attitude against technology without being in contact with any contact with or awareness of any stimulus. When ICT provokes a response "regardless of who perceives or interprets it" [29], it often means that the design or attribute of the technology causes the response and that the stimulus is within the ICT/technology. The last example of referent object in the residing dimension is when the response appears to be residing between a person and an ICT stimulus. When the affective concept is situated between a person and ICT it means that the affective response of one person could vary according to which the ICT representing the stimulus.

On the other hand, the same ICT stimulus could provoke different affective responses for different persons.

Temporal Dimension
Agogo and Hess [30] refers to affective responses that could be both state-like and trait-like. The terms state-like and trait-like are connected to time and the value of the affective concepts are constrained by time [29]. A trait-like concept can be exemplified by a personality trait as a stable value that appears in the same way in similar situations consistently. Concepts characterized as state-like are temporary values that relates to a certain condition or environment where a person experience an affect related to ICT.

Stimulus Specificity – Object vs. Behavior Stimulus
In the third dimension of ARM, the stimulus is divided into the two aspects object and behavior. According to Agogo & Hess [30], the object stimulus refer to situations where the response is related directly to the technology. Thrift [35] uses toys as an example of an object stimulus. A toy represents an object stimulus when advertisement is presenting it as an object in itself, not giving any focus on how it is possible to interact with the toy (Ibid.) The aspect of behavior as the specificity for the stimulus appears when a person experiences a response due to a "behavior on objects". While a person getting nervous by seeing a computer exemplifies the object stimulus, a person experiencing the stimulus in relation to action is an example of the behavior stimulus.

Stimulus Specificity – Particular vs. General Stimulus
The difference between general and particular stimuli imply that a person can experience one kind of affective response when exposed to a particular ICT like Google Docs, but confronted with an opinion of technology for collaboration online in general, the same person may experience a negative affective response.

Dimension of Process vs. Outcome
The last of the dimensions in the ARM-model concerns the time of evaluation, and as a consequence, the basis for the affective response. Process-based affective responses or evaluations often appears like an immediate response to an encounter with ICT. When the interaction with the technology endures over some time, the response will be based on the outcome of the experience.

Practical Implications for the Use of the ARM-Model
The ARM-model and the taxonomy of five dimensions can be used to better understand the reasons for, and the sources of certain affective responses.

3 Methodological Approach

3.1 Qualitative Approach and Participants

The study of professionals' and managers' knowledge and experiences, emotions, moods and feelings related to digitalization and new ways of providing healthcare services, is based on data collected through a preliminary project and the initial phase of a longitudinal study called eTeam. The eTeam project is an Interreg Sweden-Norway project aiming at promoting collaboration and exchange of knowledge related to digitalization of the healthcare sector and new ways of providing services between representatives from municipalities, academia and private sector. When exploring professionals' affective responses and the dark sides of technology, the study emphasizes on interpreting and seeing through the eyes of the research participants. This approach, focusing on the participants' description of their experiences and perceptions in the context of digitalization and new ways of providing healthcare services, motivates for a qualitative approach [36, 37].

The collection of data took place within the framework of the eTeam project, and interviews in focus groups were used as the method. Collaboration and the exchange of experiences are essential factors in the eTeam project, and by using the focus group method, we were able to include several participants representing different municipalities from both sides of the southern border between Norway and Sweden. The participants had a variety of professional backgrounds, different positions in the organizations and varying degree of experience with digitalization and development of new ways to provide healthcare services. Independent of different backgrounds, the focus groups facilitated interaction in the different groups when they were questioned on topics related to their experiences [37]. Participating in focus groups may result in a feeling of group pressure to agree, and may therefore limit the participants from presenting controversies [38]. On the other hand, it is a suitable method for revealing what the participants may agree upon and their common experiences, but different viewpoints and experiences may also emerge. Conducting focus groups facilitate the mobilization and activation of the participants in a way that is not possible in individual interviews [39]. Interaction in the groups contributed to new insight that we most probably not have had access to through other methods. Furthermore, it was important to get knowledge about the different participants' experiences and affective responses at an individual level, but at the same time that knowledge was meant to be the starting point when planning for workplace affordances and work-integrated learning on an organizational level.

The municipalities were included based on that they were in process of digitalizing services, implementing technology and developing new ways of providing healthcare. With one exception, these are rural municipalities with more densely populated centers. Scandinavian healthcare services are predominantly public, organized by the municipalities and funded by taxes. Three rounds of focus groups were conducted. From two to ten persons participated in each group, but the average number of participants were eight to ten. In the first round, the groups were organized nationally on each side of the border. Some of the groups were composed of managers within healthcare services, whilst professionals like assistant nurses, nurses and physical and occupational therapists working close to the patients and users, participated in the others. The composition of the groups with the professionals aimed at including different occupations in order to identify and discuss different professional perspectives. Moreover, one interview was conducted with two participants from an IT department. In the next two rounds, the participants were organized in mixed groups with Norwegian and Swedish participants, and the groups were composed of both managers and professionals. In order to access the participants, contact persons in the municipalities recruited the participants. The criterion was that the participants should be involved in digitalization processes or that they were to be involved at a later stage. They all signed an informed consent.

3.2 Data Analysis

The interviews with the different focus groups were transcribed and data were analyzed using content analysis. Conventional content analysis describes a phenomenon by developing codes through multiple readings of the interviews. Such an inductive approach helps discover meaningful underlying patterns [40]. To reduce the risk of misinterpretations of the data due to the researchers' preunderstanding, all researchers discussed in detail the findings and their systematization during the analysis process [41]. The findings and themes crystallized through the content analysis were then applied to the Affective Response Model as a systematic approach to understand how to predict and understand the participants' responses to technology and ICT. Applying the main themes from the analysis of the findings into the taxonomy of the five dimensions of the ARM-model, made it possible to differentiate affective concepts from the participants' knowledge and experience.

4 Results

4.1 Summarized Analysis of Results

The analysis of the focus group interviews identifies various barriers for both work-integrated learning as well as digitalization of the healthcare services. The participants' points at challenges and dark sides of technology originated from the changing needs in the work practice when digitalizing the services. They have experienced many small-scale implementation projects with varied outcome, and the situation is characterized by a constantly introduction to new technologies. Some of the municipalities' healthcare organizations are aware of the demand for knowledge to facilitate the digitalization and development of the services, but at the same time, they identify shortcomings related to strategies necessary if the healthcare sector should concentrate and push development of services fitted for the future. Through exploring the professionals' and managers' knowledge and experiences related to digitalization and new ways of providing healthcare services, different responses to ICT were identified. Lack of understanding of ICT and technology, limited possibility for collaboration and teamwork and lack of opportunity for training and learning were all elements that were crystallized from the content analysis.

Understanding of and Attitudes to ICT and Technology

There are differences in how the professionals perceive the ongoing digitalization with introduction of ICT and technology. Some refer to technologies as technical artifacts they have to handle, and have to learn to handle. Others refer to them more according to their functions, for example as "smart house", providing the opportunity to be connected to things and people and to be able to monitor, raise alarms and communicate. One of the participants gave an example of a patient having the opportunity to call and speak to a nurse out of the ordinary visits. In all three rounds and in all groups there were participants sharing negative experiences with ICT, but at the same time, they expressed to have an all over positive attitude to digitalization and the technology introduced as part of that process. Several of the managers had experiences with female professionals, age of 50–60 years old, having a negative attitude to ICT and technology in general. According to the same managers, the digital development and lack of technological knowledge and experience in this group were the explanation for the negativity. Both managers and professionals perceived the patients as quite positive to the technology, and it was referred to several patients above 80 years in age paying their invoice on the internet as well as using Skype for communication.

Some professionals are aware of that they have to change the routines of how to perform their work, related to the digitalization and development of new ways of providing healthcare services. The professionals' talk about how time they earlier used on providing care directly to the patients, they now have to spend on providing technical support to the patients and their relatives who are dealing with security alarms and monitoring their disease at a distance. Some of the professionals express a dilemma related to technologies providing opportunities for better control of the patients and their diseases and the reliability gets better, when simultaneously the same technology requires an increasingly amount of resources from the professionals and the organizations. One of the nurses expressed her frustration and said that if more patients are given security alarms that they use to call for help and need support to administrate, the professionals have to change the way they work to be able to respond all the calls and provide both support on technology on top of the ordinary healthcare service. Based on their experiences from work, the participants talked about a need for developing competence related to ICT and technology. Not all professionals are accustomed to use ICT and other technologies such as digital devices or computers in their work. In order to be able to digitalize the healthcare services they expressed a need to learn and understand more about the functionality of the specific technology introduced and how the new technology could communicate with systems already in place. At the same time, several of the participants expressed to have competence related to technology, and they told about professionals in their workplace handling ICT and technology in a competent way. Therefore, the knowledge as well as the need for learning among the professionals seems to be extremely varied.

Because of the introduction of ICT and other technologies, both managers and professionals participating in the groups knew colleagues experiencing the new conditions for work as a threat to themselves and to their own work. Being afraid of what is unknown was a statement that were repeated in various groups. The participants that had been working with digitalization for a while told about how they had experienced colleagues, by learning and gaining more understanding, beginning to perceive ICT and other technologies as positive opportunities for their work and for the patients. Frequent change of technologies resulted in frustration for many professionals. They talked about how the frequent change of devices made it difficult for the professionals to be updated and to have knowledge and understanding of all existing technologies. One of the participants said that in situations where the professionals do not know the technologies; it is difficult for them to provide support to the patient and to facilitate the intended use in the patient's home. Some of the participants also expressed their frustration related to the professionals' being responsible for the technology and its functionality when placed in the patients' homes. They were also engaged by discussions related to technology from a private marked, and shared experiences of having to help patients with technology that were nor provided by the healthcare services. As one of the participants expressed: "... every single person over 65 years of age is soon better equipped to use ICT and other technology than professionals in the healthcare services".

Collaboration and Teamwork

Through the discussions, it appeared that the majority of the participants experience a lack of collaboration and arenas for practicing teamwork. When the participants were talking about teamwork, it was a common view that teamwork should involve different stakeholders like different departments in the healthcare services, the department responsible for IT and digitalization, as well as stakeholders from private sector, patients and relatives. Both managers and professionals pointed at a "common language" as a prerequisite for collaboration, and they concluded that it is often a lack of such a vocabulary. The Norwegian Labour and Welfare Administrations (NAV) and their technical aids center was one of the examples of a very important collaborator for the municipal healthcare services. Even if the technical aids center were seen as an important partner, the participants mentioned various barriers to the interaction with them. The technical aids center are providing different utilities and tools to patients and people living with disabilities, but several of the professionals expressed to be uncertain about what utilities the center could offer and what rules and laws that regulated the rights of different patients and different diagnosis. The experienced lack of collaboration was not only tangible related to different sectors like healthcare and the technical aids center, but as well within the healthcare services. The participants talked about how different professional roles are authorized to prescribe different kinds of technical aid and services. A physiotherapist explained it as "there are different prescribers for different utilities and tools, depending on who is responsible for the expenses and budget". In all the groups, they had focus on the necessity of more collaboration and they talked about how different professions and different roles could complement each other with knowledge and experiences. Participants from one municipality shared their experiences with groups that were built up by different professions representing different departments. These groups have a mandate to work with digitalization and development of services together, independent of the traditional working lines. A nurse expressed that "we need a sort of care technician, who are able to supervise and support the caring staff". In lack of the role "care technician", the same participant said that if the workplace provide arenas where interdisciplinary teams may collaborate, it would facilitate sharing of knowledge for both managers and professionals.

As described in the previous section, many participants felt frustration related to all the resources they had to spend on supporting and guiding of both colleagues, patients and relatives when introducing technology. They called for more collaboration with instances like public libraries and schools since they meant that they had a defined responsibility for public education.

Learning, Training and Competence Development

Some of the managers talked about the need for a new type of changes in the healthcare services. According to those managers, both managers and professionals are used to constantly changes, but they felt the actual demands for developing healthcare services for the future without having the answer of the concrete needs of the future, much more challenging. Several participants point at the importance of developing more competence among the managers. Professionals in different groups agreed that the professionals' ability to learn and to increase their knowledge about digitalization and development of services for the future depends on the knowledge of the managers.

The overall experience among the participants is that the persons, who work in departments where technology is introduced, receive specific training on how to use a given technology. The professionals' perceived outcome of specific training varied, but many of them referred to the way training was organized as a challenge. It was a widespread experience that one or two, more or less random chosen colleagues received training and were given the responsibility for sharing knowledge with the rest of the professionals. Professionals required more quality assurance regarding the internal training related to specific technologies. Thus, the individual learning processes were characterized by trial and error. As a response to the demand for learning and training, some of the managers explained that they organized visits to departments that already had experience with technology in addition to the specific training. The same managers added that their department had established routines to dedicate 1–2 days a year focusing on new technology and related training needs. Other managers referred to positive experiences with video-recorded instructions and manuals going along with the introduction of a new technological device. According to the managers, professionals were satisfied with the opportunity to use video instructions. The positive experience with video instructions were related to the possibility to watch the video one or several times according to their individual needs. Both managers and professionals shared stories about the significant differences in learning capabilities among the persons working in the healthcare services. Some managers had experiences with professionals having problems reading traditional manuals and instructions. They pointed at language skills and age as additional challenges. Some of the participants had experiences with eLearning afforded as a tool for learning and increasing of knowledge, but both the managers and the professionals argued that eLearning depended on a lot off initiative from each individual that were going to attend the eLearning course. It appeared to be a widespread understanding that the best effect of learning could be achieved through learning and training organized and structured in line with issues the professionals themselves experience as particularly challenging. One of the participating nurses expressed that "it is a difference between the general knowledge you need to provide a professional work and the specific knowledge you need at the actual workplace".

5   Discussion

5.1   Summarizing of Results

The aim of this paper is to explore knowledge and experiences with digitalization and development of new ways of providing healthcare services among managers and professionals in municipal healthcare services. Through the participant's discussions and by analyzing the results emotions, moods and other responses to ICY and technology are identified. The results show that the professionals lack understanding of innovative technologies and together with their differences in attitudes to the technologies, challenges will arise when digitalizing the services and the processes were the healthcare services for the future are to be developed. The results also imply that both professionals and managers are in lack of possibilities and arenas were they could collaborate and work in teams. According to our findings, there are an articulated need for opportunities for learning and development of knowledge. The demand for learning and knowledge is not answered, and the result may be an experience of the dark sides of technology and negative responses towards digitalization and development of new ways of providing healthcare services. Knowledge, experiences, emotions and moods are identified through the analysis of the focus groups, but also by the participants in the groups underway in the process with the three rounds of focus groups. In the following section, the results will be discussed in light of the ARM-model and its taxonomy of five dimensions.

5.2   Affective Responses to Digitalization and Introduction of Technology

The understanding of and attitude to digitalization and introduction of ICT and other technologies may be related to different dimensions in the ARM-model depending on each individual. Some participants referred to the introduction of technology at their workplace as "something" they just had to take into use, and that associates with the 4th dimension when the stimulus and the effective response is connected to technology in general. Little attention on specific technologies can be related to what the ARM-model describes as a general dimension, which means that the affective response to technology "is applicable to a general class of technologies" [30]. When the technology is referred to in general terms, it may be more difficult to make plans for learning affordances and it may indicate a need for learning on a more general, superior level. Organizing for work-integrated learning at the workplace demand on the one side suitable affordances provided by the workplace, and on the other side it is dependent on the engagement of the individuals working there [24]. The workplace affordances are decisive for the individual engagement, and the "right" affordances may motive to engagement. Female professionals at the age over 50 years old were reported to be a group with several examples of negative attitude to ICT and technology. Since the digitalization of the healthcare sector is a relative new phenomenon, the negative attitude is likely to be explained as a response that "is just there". A response "just being there" indicates that it is residing within the person independent of a concrete technology. A person with a response connected to the residing dimension where the person itself represents the referent object, it is also likely that the person refers to technology in general terms as in the 4th dimension. Many participants talked about being afraid of the unknown. That type of affective response may be connected to the concept were the referent object of the residing dimension lays in the intersection of person and technology. The person feels afraid, not because of his or her personality or stable mood and the fear is not a result of a technological stimulus. The fear lies in sort of an intersection between the technology and the person, meaning that another technology may or may not provoke the same response. And the same technology may or may not result in the same affective response for another person [29].

Planning for workplace affordances will be challenging when there are persons with the described attitude to technology. Looking into the temporal dimension will be off interest to identify if the attitude and the affective responses are trait-like or state-like. If the person itself represents the referent object and they assume a distance to the technology, it is probably a permanent, trait-like response. Facing affective responses that are trait-like, means that the managers or the persons facilitating the affordance will have to prepare for a more systematic approach than if the response were only fleeting connected to a certain situation. Affective responses can be both state-like and trait-like, and according to Agogo and Hess [30] a trait-like response can change to a state-like when introducing systematic programs to reduce for example affective responses like computer anxiety. That mechanism were also discussed in the groups when the participants claimed learning and more knowledge would reduce the feeling of fear and uncertainty. Even if a trait-like response may represent a more challenging attitude to change, the advantage is the predictability. Having good knowledge about the concept of a persons' or a groups' affective response may be helpful when planning for affordances and for work-integrated learning. Through the focus groups it appeared that all the participants had experiences of the dark sides of technology, but in spite of negative experiences the majority expressed to have an all over positive attitude to digitalization and technology. In light of the process- and outcome-based dimension, the "duality" of the all over experience may indicate that they have some negative experiences related to the process-based evaluation of the digitalization, whilst their all over satisfaction is related to an outcome-based evaluation.

The dark side of technology is reflected through the frustration they feel when having to prioritize support on ICT and other technologies to colleagues, patients and relatives at the expense of the time traditionally used to more typical caring tasks together with the patient. The experience of "loosing" the possibility to spend time with the patients aligns with a study of Mort, Finch and May [42], concluding that introduction of technology is likely to lead to a bigger distance between patient and those providing the healthcare services and to a objectification of the patient [42]. The ones that feel frustration just thinking about the technology, directs their responses to the technology as a device representing something that makes it difficult to carry out their work as they are used to. The other expression of affective response in the 3rd dimension can appear when the person are to use a computer, and the stimulus will be "using computers" [29]. When the behavior and interaction with the technology are focused, it may be easier for the professionals to see value in the process of digitalization and introduction of technology. Interacting with a technology may provoke positive affective response by developing a new form of patient-provider relationship on the basis of sharing data, prescribing therapies, asking for advice through ICT [43].

Among the participants, there were a unison view that lack of knowledge represent a barrier for the digitalization of the healthcare services, and an experienced shortage of knowledge will result in a negative response when evaluating the use of technology in a process-based manner, seldom or never reaching an measurable outcome. A parallel problematic may be related to the experienced lack of arenas practicing teamwork and collaboration. Struggle and difficulties along the process, makes it challenging to reach an outcome, that may lead to new insight and a further development of both knowledge and collaboration opportunities. When professionals have negative experiences with technology as a medium for collaboration, it will create a negative affective response to the intersection of person and technology stimulus [30]. The negative affective response formed by the person or professionals' response to the stimulus might have a negative effect on the collaboration per se, even if ICT and information systems for healthcare expected to support collaboration and teamwork [44]. In alignment with the experienced lack of collaboration, the participant's points at a need for a common language. Related to the group "other stakeholders", the participants from the municipalities wishes to increase their collaboration with the technical aids center. The participants' experiencing not having a common language or other qualifications necessary for collaborating, will consequently experience the dark sides of technology. That may easy become a situation where the person finds him or herself on the wane with no positive responses to technology, and the challenges related to engage in learning and knowledge development.

An average of the participants were pointing at the importance of learning more and developing more knowledge and generic competencies that are useful across different clinical context [45]. In addition to the experienced need for generic competence, both managers and professionals expressed a need for a new, more specific type of knowledge. In accordance to Barakat [15] and Lapão [16], the new type of knowledge can be characterized as state-like because the response were actuated due to a fleeting situation dominated by lack of knowledge related to ICT, technology and the ability to contribute to continuously development of the services. In the focus groups, the participants presented some examples of affordances aiming at facilitating learning at the workplace. In the example with the eLearning courses, the affective response were related to a specific technology and it is easier to evaluate both the process and the outcome when it relates to a specific technology. When developing knowledge and especially in relation with video-recorded instructions and eLearning, the response will be connected to a specific behavior with the technology [30]. The overall affective responses to the eLearning courses were negative. The technology, both the process- and the outcome-based evaluation, as well as technology as an object and the behavior with the technology are easy to follow, and still the participants do not find any motivation to engage in eLearning. The organizational training, were the instructors are randomly pointed out and the professionals often felt that they were supposed to learn by trial and error, leads to affective responses as the dark sides of technology. The process-based evaluation will be negative because the professionals experience too much problems in their interaction and "behavior" with the technology. In those two examples, the workplace affordances are not fitted with the needs of the professionals.

6 Conclusive Reflections

Challenges related to learning and knowledge are identified and analyzed by using the ARM-model. The challenges originate from the changing needs in the work practice when digitalizing the healthcare services. Small scale projects were technology is introduces are often conducted to try out different technologies, whereas the experiences are very varied. The importance of workplace affordances that are in line with the professionals needs are decisive. The affective responses and the consequences those responses will have on each individual's motivation to engage in learning at the workplace, as well as the well-being of individuals, organizations and societies, will depend highly on how managers and professionals experience the quality of the learning environment facilitated by the workplace affordances and enhancement of competence. Powerlessness and avoidance of responsibility may be examples of affective responses in situations where there is lack of a common language that enables collaboration, clarification of which department and which roles who are responsible for decisions related to use and prescription of technology as part of the healthcare service.

A learning environment, knowledge and competence is decisive when it comes to the process of digitalization the healthcare services and introducing technology and new ways of providing the services. Knowledge about the dark side of technology and use of ICT is important in a society where managers and professionals in healthcare services have to relate to new technologies and new ways of providing services constantly.

This paper is a contribution to the identification of a systematic way to increase knowledge about the affective responses of managers and professionals in municipal healthcare services. That knowledge are decisive when planning and organizing for workplace affordances and work-integrated learning to be able to shape the affordances in a way that engage the different individuals. A limitation for in the study was that many of the participants had limited experience with ICT and other technologies, and that made the foundation of their contribution equally restricted. Further studies should focus on the connection between the use of ARM as a systematic approach to knowledge about affective responses and how it should be used in practice to increase the effective output when organizing for work-integrated learning.

Acknowledgements. The project has received support from the European Regional Development Fund through the Interreg Sverige-Norge. The authors wish to acknowledge the contributions of managers and professionals as participants in the preliminary project as well as longitudinal study reported above. As part of the project group Nina Fladeby, Kerstin Grundén and Lena Larsson being part of collecting and analysing of data.

References

1. Blix, M.: Framtidens välfärd och den åldrande befolkningen. Delutredning från Framtidskommissionen. Regeringskansliet, Stockholm (2013)
2. Chidzambwa, L.: The social considerations for moving health services into the home: a telecare perspective. Health Policy Technol. **2**, 10–25 (2013). https://doi.org/10.1016/j.hlpt.2012.12.003

3. Hofmann, B.: Ethical challenges with welfare technology: a review of the literature. Sci. Eng. Ethics **19**, 389–406 (2013). https://doi.org/10.1007/s11948-011-9348-1
4. Nilsen, E.R., et al.: Exploring resistance to implementation of welfare technology in municipal healthcare services - a longitudinal case study. BMC Health Serv. Res. **16** (2016). https://doi.org/10.1186/s12913-016-1913-5
5. Cresswell, K., Sheikh, A.: Organizational issues in the implementation and adoption of health information technology innovations: an interpretative review. Int. J. Med. Inform. **82**, e73–e86 (2013). https://doi.org/10.1016/j.ijmedinf.2012.10.007
6. Lindberg, I., Lindberg, B., Söderberg, S.: Patients' and healthcare personnel's experiences of health coaching with online self-management in the renewing health project. Int. J. Telemed. Appl. **2017** (2017). https://doi.org/10.1155/2017/9306192
7. Marasinghe, K.M.: Assistive technologies in reducing caregiver burden among informal caregivers of older adults: a systematic review. Disabil. Rehabil. Assistive Technol. **11**, 353–360 (2016). https://doi.org/10.3109/17483107.2015.1087061
8. Xyrichis, A., et al.: Healthcare stakeholders' perceptions and experiences of factors affecting the implementation of critical care telemedicine (CCT): qualitative evidence synthesis. Cochrane Database Syst. Rev. **2017** (2017). https://doi.org/10.1002/14651858.CD012876
9. Abbott, C., et al.: Emerging issues and current trends in assistive technology use 2007–2010: practising, assisting and enabling learning for all. Disabil. Rehabil. Assistive Technol. **9**, 453–462 (2014). https://doi.org/10.3109/17483107.2013.840862
10. Black, A.D., et al.: The impact of eHealth on the quality and safety of health care: a systematic overview. PLoS Med. **8**, e1000387 (2011). https://doi.org/10.1371/journal.pmed.1000387
11. Tarafdar, M., Cooper, C.L., Stich, J.-F.: The technostress trifecta - techno eustress, techno distress and design: theoretical directions and an agenda for research. **29**, 6–42 (2019). https://doi.org/10.1111/isj.12169
12. Billett, S.: The clinical teacher's toolbox. Readiness and learning in health care education. Clin. Teach. **12**, 367–372 (2015). https://doi.org/10.1111/tct.12477
13. Cortellazzo, L., Bruni, E., Zampieri, R.: The role of leadership in a digitalized world: a review. Front. Psychol. **10** (2019). https://doi.org/10.3389/fpsyg.2019.01938
14. Gjestsen, M.T., Wiig, S., Testad, I.: What are the key contextual factors when preparing for successful implementation of assistive living technology in primary elderly care? A case study from Norway. BMJ Open **2017**, e015455 (2017). https://doi.org/10.1136/bmjopen-2016-01545
15. Barakat, A., et al.: eHealth technology competencies for health professionals working in home care to support older adults to age in place: outcomes of a two-day collaborative workshop. Medicine 2.0 **2**, e10 (2013). https://doi.org/10.2196/med20.2711
16. Lapão, L.: Digitalization of healthcare: where is the evidence of the impact on healthcare workforce' performance? In: Ugon, A., et al. (eds.) Building Continents of Knowledge in Oceans of Data. The Future of Co-created eHealth, pp. 646–650. IOS Press (2018)
17. Olsen, D.S., Tikkanen, T.: The developing field of workplace learning and the contribution of PIAAC. Int. J. Lifelong Educ. **37**, 546–559 (2018). https://doi.org/10.1080/02601370.2018.1497720
18. Skule, S.: Learning conditions at work: a framework to understand and assess informal learning in the workplace. Int. J. Train. Dev. **8**, 8–20 (2004). https://doi.org/10.1111/j.1360-3736.2004.00192.x
19. Ellström, P.-E.: Integrating learning and work: problems and prospects. Hum. Resour. Dev. Q. **12**, 421–435 (2001). https://doi.org/10.1002/hrdq.1006
20. Hattinger, M., et al.: Work-integrated learning and co-creation of knowledge: design of collaborative technology enhanced learning activities. In: Ghazawneh, A., Nørbjerg, J., Pries-Heje, J. (eds.) Proceedings of the 37th Information Systems Research Seminar in Scandinavia (IRIS 37), pp. 1–15 (2014)

21. Pennbrant, S., Nunstedt, H.: The work-integrated learning combined with the portfolio method-a pedagogical strategy and tool in nursing education for developing professional competence. J. Nurs. Educ. Pract. **8**, 8–15 (2017). https://doi.org/10.5430/jnep.v8n2p8

22. Pennbrant, S., Svensson, L.: Nursing and learning – healthcare pedagogics and work-integrated learning. High. Educ. Skills Work-Based Learn. **8**, 179–194 (2018). https://doi.org/10.1108/HESWBL-08-2017-0048

23. Billett, S.: Learning through work: workplace affordances and individual engagement. J. Workplace Learn. **13**, 209–214 (2001). https://doi.org/10.1108/EUM0000000005548

24. Billett, S.: Workplace participatory practices: conceptualising workplaces as learning environments. J. Workplace Learn. **16**, 312–424 (2004). https://doi.org/10.1108/13665620410550295

25. Chen, H.C., et al.: Legitimate workplace roles and activities for early learners. Med. Educ. **48**, 136–145 (2014). https://doi.org/10.1111/medu.12316

26. Chen, H.C., Teherani, A.: Workplace affordances to increase learner engagement in the clinical workplace. Med. Educ. **49**, 1184–1186 (2015). https://doi.org/10.1111/medu.12888

27. Billett, S., Smith, R., Barker, M.: Understanding work, learning and the remaking of cultural practices. Stud. Continuing Educ. **27**, 219–237 (2005). https://doi.org/10.1080/01580370500376564

28. Rienties, B., et al.: Why some teachers easily learn to use a new virtual learning environment: a technology acceptance perspective. Interact. Learn. Environ. **24**, 539–552 (2016). https://doi.org/10.1080/10494820.2014.881394

29. Zhang, P.: The affective response model: a theoretical framework of affective concepts and their relationships in the ICT context. Manag. Inf. Syst. Q. (MISQ) **37**, 247–274 (2013). https://doi.org/10.25300/MISQ/2013/37.1.11

30. Agogo, D., Hess, T.J.: "How does tech make you feel?" A review and examination of negative affective responses to technology use. Eur. J. Inf. Syst. **27**, 570–599 (2018). https://doi.org/10.1080/0960085X.2018.1435230

31. Bhattacherjee, A., et al.: User response to mandatory IT use: a coping theory perspective. Eur. J. Inf. Syst. **27**, 395–414 (2018). https://doi.org/10.1057/s41303-017-0047-0

32. Pirkkalainen, H., Salo, M.: Two decades of the dark side in the information systems basket: suggesting five areas for future research. In: Proceedings of the 24th European Conference on Information Systems, ECIS 2016, p. 101 (2016)

33. Shubber, M., et al.: Acceptance of video conferencing in healthcare planning in hospitals. In: Proceedings of the Association for Information Systems, AMCIS 2018 (2018)

34. Charness, N., Boot, W.R.: Technology, gaming, and social networking. In: Schaie, K.W., Willis, S.L. (eds.) Handbook of the Psychology of Aging, pp. 389–407. Academic Press, San Diego (2016)

35. Thrift, N.: Knowing Capitalism. SAGE Publications Ltd., Thousand Oaks (2005)

36. Myers, M.D.: Qualitative Research in Business and Management, 3rd edn. SAGE Publications Limited, Thousand Oaks (2019)

37. Bryman, A.: Social Research Methods, 5th edn. Oxford University Press, Oxford (2016)

38. Brandt, B.: Gruppeintervju: perspektiv, relasjoner og kontekst [Group interviews: perspectives, relations and context]. In: Kalleberg, R., Holter, H. (eds.) Kvalitative metoder i samfunnsforskning [Qualitative methods in social science research], pp. 145–165. Universitetsforlaget, Oslo (1996)

39. Andvig, B.: Förändran i hermeneutisk forskning – ett exempel från fokusgruppsamtal [Wondering in hermeneutic research: An example from focus group interviews]. In: Lassenius, E., Severinsson, E. (eds.) Hermeneutik i vårdpraxis: det nära, det flyktiga, det dolda [Hermeneutics in care practice], pp. 197–209. Gleerups Utbildning AB, Malmö (2014)

40. Graneheim, U.H., Lindgren, B.-M., Lundman, B.: Methodological challenges in qualitative content analysis: a discussion paper. Nurse Educ. Today **56**, 29–34 (2017). https://doi.org/10.1016/j.nedt.2017.06.002

41. Malterud, K.: The art and science of clinical knowledge: evidence beyond measures and numbers. Lancet **358**, 397–400 (2001). https://doi.org/10.1016/S0140-6736(01)05548-9

42. Mort, M., Finch, T., May, C.: Making and unmaking telepatients: identity and governance in new health technologies. Sci. Technol. Hum. Values **34**, 9–33 (2009). https://doi.org/10.1177/0162243907311274

43. Piras, E.M., Miele, F.: On digital intimacy: redefining provider–patient relationships in remote monitoring. Sociol. Health Illness **41**, 116–131 (2019). https://doi.org/10.1111/1467-9566.12947

44. Lluch, M.: Healthcare professionals' organisational barriers to health information technologies-a literature review. Int. J. Med. Inform. **80**, 849–862 (2011). https://doi.org/10.1016/j.ijmedinf.2011.09.005

45. Pijl-Zieber, E.M., et al.: Competence and competency-based nursing education: finding our way through the issues. Nurse Educ. Today **34**, 676–678 (2014). https://doi.org/10.1016/j.nedt.2013.09.007

Exploring the Effects of Immersive Virtual Reality on Learning Outcomes: A Two-Path Model

Yongqian Lin$^{(\boxtimes)}$ ⓘ, Guan Wang, and Ayoung Suh ⓘ

School of Creative Media, City University of Hong Kong, Kowloon, Hong Kong, China
Yongqian.Lin@my.cityu.edu.hk,
{guan.wang,Ayoung.Suh}@cityu.edu.hk

Abstract. Immersive virtual reality (VR) has attracted widespread attention and has been increasingly adopted in education. However, the influence of immersive VR on learning outcomes is still unclear, with conflicting results emerging from the research. The contradictory research can be attributed to the fact that little research has systematically analyzed the effects of immersive VR on learning processes and ultimately on learning outcomes. Therefore, this study focuses on analyzing how immersive VR affects learning outcomes and explaining the reasons behind the contradictory research about immersive VR's effects. A survey was conducted in a laboratory setting in which the participants were asked to play an immersive VR application for learning. The results show that immersive VR affects learning outcomes through affective and cognitive paths. In the affective path, immersive VR features influence learning outcomes through the mediation of immersion and enjoyment. In the cognitive path, immersive VR features influence learning outcomes through the mediation of usefulness, control and active learning, and cognitive benefits. By providing a nuanced understanding of the effects of immersive VR on learning outcomes, this study contributes to the VR literature for learning.

Keywords: Immersive VR · Learning outcomes · VR features · Affective path · Cognitive path · Immersion

1 Introduction

Virtual reality (VR) is a technology that generates a three-dimensional (3-D) virtual environment in which users interact with virtual objects through sensing devices [1]. Immersive VR allows users to fully immerse themselves in a virtual environment through stereoscopic displays, such as the head-mounted display (HMD) and cave automatic virtual environment (CAVE) [2]. Immersive VR is currently applied in many fields, including education [3], aerospace [4], and entertainment [2]. Educational institutions are inclined to progressively adopt educational immersive VR applications and displays to facilitate students' learning due to the potential benefits of immersive VR (e.g., immersion, interaction, imagination) [3, 5]. Therefore, research on the impact of immersive VR on learning outcomes has become increasingly important [5, 6].

© Springer Nature Switzerland AG 2020
D. D. Schmorrow and C. M. Fidopiastis (Eds.): HCII 2020, LNAI 12197, pp. 86–105, 2020.
https://doi.org/10.1007/978-3-030-50439-7_6

In recent decades, researchers have empirically investigated how immersive VR enhances learning outcomes. While some have proven that immersive VR has positive effects on various dimensions of learning outcomes [7–9], others have failed to find significant effects [10, 11]. These conflicting results have puzzled educators and learners alike, leaving them wondering whether they should use immersive VR, and if so, in which situations immersive VR enhances learning outcomes. Although some studies have provided a few perspectives, proposing that immersive VR's impact will be influenced by cognitive load and summary habit [3, 10], there is still a lack of research systematically exploring the influencing process of immersive VR to provide possible explanations for the conflicting results. Therefore, this research intends to fill this gap.

This research builds on the two-path model proposed by Makransky and Petersen [12] to explore how immersive VR influences learning outcomes through both affective and cognitive paths. By extending our understanding of the underlying mechanisms by which immersive VR influences learning outcomes, this study provides useful insights into design strategies that will harness the benefits of immersive VR in education.

2 Literature Review

2.1 Immersive VR

Immersive VR is a system with special hardware that provides users with the psychophysical experiences of being completely isolated from the physical world outside by providing a sense of being fully immersed in a virtual environment [13]. It allows the users to interact with a virtual environment via an HMD or CAVE, which blocks the users' visual access to the physical world and projects a stereoscopic virtual environment [14–16]. Immersive VR techniques, such as 3-D images, high-quality audiovisual effects, and accurate motion tracking functions, are used to enhance the users' feelings of immersion [17].

To gain a deeper understanding of immersive VR, it is necessary to explore its technological features. According to Lee et al. [18] and Whitelock et al. [19], the most prominent dimensions of VR features are vision and movement. Regarding immersive VR, the vision features include representational fidelity and aesthetic quality. Representational fidelity refers to the realistic degree of the virtual objects and virtual world, which is enhanced by 3-D images, scene content, and smooth motion of the virtual objects [20]. The aesthetic quality, which describes the visual aspects of the virtual environment [21], is also essential because it influences the users' motivations and interests [22, 23]. Due to its representational fidelity and aesthetic quality, immersive VR has great potential to attract and retain users.

The movement features of immersive VR include immediacy of control and interactivity. Immediacy of control describes the ability to smoothly change the viewpoint and to manipulate objects with expected continuity in the virtual world [18]. In addition, immersive VR has another important feature: interactivity, which is defined as the degree to which users influence the form or content of the immersive virtual environment [24]. Highly interactive VR not only allows users to navigate the virtual world but also to explore, control, and even modify this virtual environment [24, 25]. Interactivity, a core advantage of immersive VR, distinguishes immersive VR from other technologies because it provides its users with effectiveness and motivation through multiple interaction possibilities, which most other technologies fail to offer [24]. For instance, immersive VR attracts users by detecting their head and finger movements, while a great number of technologies only respond to users' clicking a mouse.

2.2 Immersive VR in Education

Immersive VR has been adopted in the education and training industry, including science and mathematics education [26–28], firefighting training [29], and mine-rescuing training [30]. Researchers have examined the effects of immersive VR on learning outcomes [6, 9, 31]; however, there is still much doubt about whether it truly increases learning outcomes [14, 16].

Inconsistent Results. Considerable research has proven the positive influence of immersive VR on learning outcomes, including learning motivation [8, 32, 33], perceived learning effectiveness [6, 34, 35], and objective learning performance [36, 37]. It was reported that immersive VR might increase learners' motivation through providing interactive experiences [3], immersion [38], and realism [39]. As for perceived learning effectiveness, it would be stimulated by the learners' perceived presence [32] and their openness to immersive VR [35]. Evidence has also shown that immersive VR increased objective learning performance, which was even better than other traditional teaching methods [36, 37].

However, there are several studies showing that the impact of immersive VR on learning outcomes was insignificant [12, 40, 41]. Immersive VR did not effectively improve the performance of knowledge transfer, although the students felt a strong sense of immersion [11] or presence [42]. One possible explanation was that immersive VR devices were so sophisticated that the learners felt cognitively overloaded in the learning process [10, 42]. In conclusion, although immersive VR provides learners with a strong sense of immersion, it cannot guarantee the improvement of learning outcomes.

The Mediating Role of Learning Processes. To investigate the effects of immersive VR on learning outcomes, it is important to further identify when and how to use VR's features to support the learning process, which ultimately enhance learning outcomes. Lee et al. [18] proposed that desktop VR features affected learning through interaction experience (i.e., usability) and psychological factors (i.e., presence, motivation, cognitive benefits, control and active learning, and reflective thinking). Based on the discussion above, this study focuses on analyzing immersive VR features and explaining the process of how immersive VR affects learning outcomes.

3   Theoretical Background and Model

3.1   Technology-Mediated Learning

Technology-mediated learning (TML) describes a situation in which learners interact with learning materials, peers, or lecturers through the intermediary role of advanced information technologies [43]. TML theory suggests that information technology and institutional strategy ultimately influence the learning outcomes through psychological learning process [43]. Studies focusing on how learning is mediated by the intervention of VR have gradually developed TML theory in the VR context [12, 18]. In the work of Makransky and Petersen [12], two paths (i.e., the affective and cognitive paths) are proposed to explain how desktop VR influences learning (see Fig. 1). The affective learning path is concerned with the internalization of learners' positive emotions and attitudes toward learning [44, 45]. With positive emotions, learners are driven to learn more and apply their knowledge after learning [46]; therefore, affective factors play a crucial role in influencing learning motivation [47] and in the quantity as well as quality of knowledge learned [48]. The cognitive learning path is concerned with the process of improving cognitive abilities and acquiring cognitive objectives [49]. According to Bloom [50], cognitive learning includes memorizing, comprehending, applying, analyzing, synthesizing, and evaluating knowledge, which eventually enhances learners' acquisition of knowledge, skills, and abilities [49].

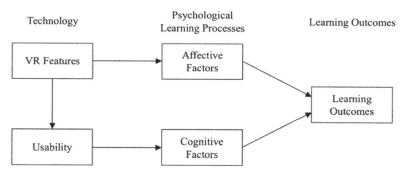

Fig. 1. A TML framework in the VR context

Based on Makransky and Petersen's work [12], this study proposes an adapted model of TML theory in the immersive VR context, as shown in Fig. 2 and Fig. 3. There are several differences between our model and Makransky and Petersen's model [12]. First, this model updates *immersive VR features* (i.e., representational fidelity, immediacy of control, interactivity, and aesthetic quality) to describe immersive VR.

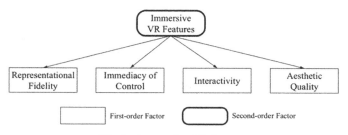

Fig. 2. Immersive VR features

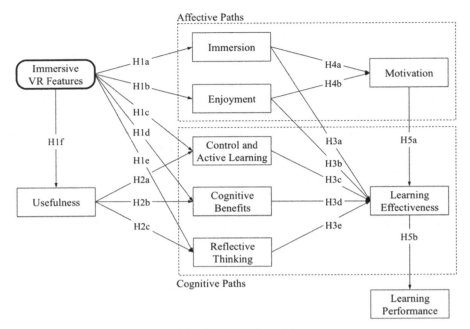

Fig. 3. Research model

Second, this study proposes immersion and enjoyment as the affective factors, and control and active learning, cognitive benefits, and reflective thinking as the cognitive factors. *Immersion* describes the sensation in which learners are fully involved in the artificial world and get confused about the virtual and real world [44, 51]. *Enjoyment* pertains to the extent to which people find the learning activities enjoyable [52]. Immersion and enjoyment as affective factors indicate the positive emotional responses during the learning process. Three cognitive factors are included in the model. *Control and active learning* pertains to the instructional design, which allows learners to actively make their own decisions and eventually feel competent as well as self-determined in learning [18]. As for *cognitive benefits*, it indicates that the learners have improved memorization, understanding, and application of knowledge [18]. *Reflective thinking* is defined as a state of mind whereby learners inquire about their doubts [12].

Third, combining the work of Lee et al. [18] and Makransky and Petersen [12], this research adopts motivation, learning effectiveness, and learning performance as learning outcomes. Motivation and learning effectiveness are subjective constructs that reflect learners' viewpoints toward learning. *Motivation* refers to an internal psychological state that stimulates learners' behaviors and indicates a direction [53]. *Learning effectiveness* measures the extent to which learners believe that they have gained knowledge [54]. As an objective construct, *learning performance* refers to the learners' overall knowledge and skills acquisition after the VR intervention [55].

3.2 Research Hypotheses

According to TML theory, immersive VR features enhance both immersion and enjoyment. Through increasing the realistic degree of the virtual environment, immersive VR allows learners to easily feel as though they are in the virtual world, enhancing their feeling of immersion [56, 57]. Moreover, with diverse interaction methods, smooth movements, and instant feedback, immersive VR enables learners to directly interact with the 3-D virtual world as they would in the real world, increasing their feeling of immersion [58, 59]. Furthermore, immersive VR enables learners to navigate the virtual world and control the virtual objects in different ways, bringing them interesting experiences and increasing their enjoyable feelings. There is extensive evidence proving that immersive VR features are significantly relative to learners' enjoyment [6, 12]. Based on the discussion, we propose the following hypotheses.

Hypotheses 1a–b: Immersive VR features positively influence a) immersion and b) enjoyment.

TML theory suggests that immersive VR has a positive impact on cognitive factors, including control and active learning, cognitive benefits, and reflective thinking. With vision and movement features, immersive VR not only visualizes knowledge; it also allows learners to actively control the learning objects, which enhances their understanding, memorization, and reflective thinking of the learning content with less cognitive effort than traditional learning [27, 60–62]. Based on the discussion, we propose the following hypotheses.

Hypotheses 1c–e: Immersive VR features positively influence c) control and active learning, d) cognitive benefits, and e) reflective thinking.

Based on TML theory, the effects of the technology features on cognitive learning processes are mediated by the interaction experience, which is represented by technology usability, including ease of use and usefulness [12, 18]. In this hypothesized model, ease of use is excluded because immersive VR is not always easy for learners to use. Learners tend to believe that learning with immersive VR is useful because immersive VR provides diverse learning experiences and timely feedback on their behaviors [63, 64]. When learners feel that immersive VR is useful, they might also think it is relevant, important, and valuable to their learning, resulting in their willingness to actively control learning activities [65]. Moreover, because perceived usefulness increases learning initiative, it offers the possibility of promoting conceptual understanding [66], and significantly influence reflective thinking [12, 18]. Based on the discussion, we propose the following hypotheses.

Hypothesis 1f: Immersive VR features positively influence usefulness.
Hypotheses 2a–c: Usefulness positively influences a) control and active learning, b) cognitive benefits, and c) reflective thinking.

TML theory proposes that all psychological factors have positive effects on learning effectiveness. Immersion and enjoyment are proven to promote the positive impact of immersive VR on learning [18, 67, 68]. When immersed in the virtual environment, learners are not disturbed by the outside world; they are focused on learning the content, making their learning effective. When learning with immersive VR is enjoyable, the learners will gain high interests and improve their learning efficiency, which enhances their perceived learning effectiveness [68, 69].

When learners have high control and perform active learning, they make decisions about their learning path, learning pace, and instruction methods [70]. As a result, the learners are able to adjust the learning process according to their personal situations and discover the most suitable learning methods, which allows them to feel their learning effectiveness [18, 71]. With cognitive benefits, learners gain better understanding, memorization, and application of knowledge in immersive VR [18]. After improving their cognitive abilities, learners can comprehend, recall, and apply knowledge in a shorter time, which leads to their perceived learning effectiveness. Reflecting thinking enhances learning effectiveness by enabling learners to critically reflect on what they have learned and their doubts [61, 72, 73]. By gaining new knowledge after reflection, learners tend to believe that their learning is effective. Based on the discussion, we propose the following hypotheses.

Hypotheses 3a–e: a) Immersion, b) enjoyment, c) control and active learning, d) cognitive benefits, and e) reflective thinking positively influence learning effectiveness.

Affective factors are proven to have a positive impact on learning motivation [12]. Immersion provided by immersive VR allows learners to fully engage in virtual learning activities, providing them with a different learning environment and unique learning experiences. These novel learning experiences enhanced by immersion stimulate the learners' motivation and curiosity to explore the virtual world and learn more [74]. As for enjoyment, it is believed to positively affect the learners' motivation because enjoyment is one of the internal needs that enhances intrinsic motivation in learning [75]. When the learners' inherent needs for happiness are fulfilled, they will be motivated to learn. Based on the discussion, we propose the following hypotheses.

Hypotheses 4a–b: a) Immersion and b) enjoyment positively influence motivation.

Motivated learners are likely to make a positive evaluation of their learning effectiveness [43, 64, 76]. If learners are motivated, they will actively devote more time and energy to learn and gain more knowledge; therefore, they tend to feel that the learning is effective [77]. The relationship between learning effectiveness and learning performance has also been widely discussed. Learners usually think learning is effective because they believe that they gain more new knowledge, which will simultaneously improve their learning performance. Moreover, with positive perception of the learning effectiveness, the learners are more willing to learn, resulting in the improvement of learning performance [76, 77]. Based on the discussion, we propose the following hypotheses.

Hypothesis 5a: Motivation positively influences learning effectiveness.

Hypothesis 5b: Learning effectiveness positively influences learning performance.

4   Methods

4.1   Research Subjects and Procedures

To test the hypotheses we propose, this study employed a survey method. Sixty undergraduate and postgraduate students were recruited from a university in Hong Kong. First, the participants were asked to answer demographic questions and biology questions related to the immersive VR learning content, which took about 10 min. Next, the participants played an immersive VR game with HMD for about 10 min. Last, the participants were asked to answer a list of biology questions in the first session again and fill in a questionnaire containing all constructs in the research model.

4.2   Software and Hardware

Participants were asked to play an educational VR application, *The Body VR*. This VR application applied immersive animations and narration to instructing learners about the knowledge of cells and the human body. A simulative and enlarged human body system was designed to allow the learners to shuttle through the human body. The learners could look around the virtual environment at 360 degrees. They were also allowed to touch, rotate, and send out virtual objects, such as red blood cells and white blood cells. Figure 4 shows a screenshot of this game. A 2017 HTC Vive® as the HMD was used to offer the fully immersive VR experiences. Two handheld motion controllers were provided for interaction during the experiences.

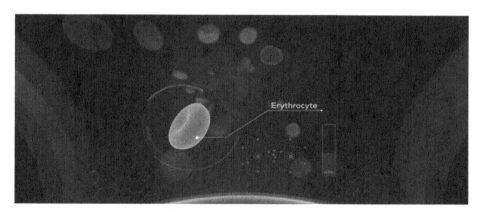

Fig. 4. The screenshot of the Body VR

4.3   Questionnaires

The questionnaires consisted of demographic questions, a biology test, and measurement items. The demographic questions included age, gender, nationality, major, and educational background. The biology test questions were multiple-choice questions about the information introduced in the learning application. These questions were used to test the prior knowledge as well as the learning performance of the participants. Two pilot tests with 10 students were conducted to ensure the test performances were differentiated. The sequence of the questions used before and after playing was counterbalanced. The measurement items of the other constructs were adapted based on the previous research, and they were measured with a 5-point Likert scale, as shown in the appendix.

5 Results

Among the sixty participants, 41.7% (n = 25) of them were male, and 58.3% (n = 35) were female. Their average age was 20. SmartPLS was used to test the model. The next session shows the results of the measurement model and the structural model.

5.1 Measurement Model

We assessed the measurement model through an estimated coefficient or loading, convergent validity, and discriminant validity. All items load significantly on their latent constructs. We assessed the convergent validity through composite reliability (CR) [47], cronbach's alpha (CA), and average variance extracted (AVE). The lowest value of CR and CA should be 0.7, while the lowest value of AVE should be 0.5 [78]. All constructs meet the requirement (see Table 1).

Table 1. CR, CA, and AVE

Constructs	CR	CA	AVE
Immersive VR features	0.855	0.773	0.597
Usefulness	0.876	0.788	0.703
Immersion	0.879	0.799	0.709
Enjoyment	0.967	0.949	0.907
Control and active learning	0.864	0.803	0.562
Cognitive benefits	0.877	0.823	0.589
Reflective thinking	0.880	0.821	0.647
Motivation	0.919	0.889	0.694
Learning effectiveness	0.859	0.798	0.551

The correlational method was applied to evaluate the discriminant validity. The correlations between the constructs should be lower than the squared root of AVE. Discriminant validity is achieved in this model (see Table 2).

Table 2. Correlation between the constructs

	1	2	3	4	5	6	7	8	9	10
1	**0.768**									
2	0.748	**0.750**								
3	0.726	0.719	**0.742**							
4	0.505	0.511	0.405	**0.953**						
5	0.289	0.451	0.310	0.468	**0.842**					
6	0.720	0.675	0.707	0.674	0.501	**0.833**				
7	0.297	0.255	0.415	0.262	0.061	0.339	1			
8	0.717	0.694	0.586	0.435	0.388	0.666	0.179	**0.804**		
9	0.757	0.741	0.661	0.474	0.349	0.693	0.325	0.720	**0.838**	
10	0.563	0.557	0.510	0.628	0.549	0.719	0.184	0.434	0.536	**0.772**

Notes: Squared root of AVE for each latent construct is given in diagonals.
1 = Cognitive Benefits; 2 = Control and Active Learning; 3 = Learning Effectiveness; 4 = Enjoyment; 5 = Immersion; 6 = Motivation; 7 = Learning Performance; 8 = Reflective Thinking; 9 = Usefulness; 10 = Immersive VR Features.

5.2 Structural Model

Figure 5 and Fig. 6 display the results of the structural model. The parameters of the model include the path coefficients and squared multiple correlation (R^2). The path coefficients illustrate the effects of a variable as a cause of another variable, indicating the effective connectivity between the constructs. R^2 explains to what extent the variance of a construct is explained by the independent constructs. In general, the results show that most hypotheses have been supported, except H1d, H1e, H3a, H3b, and H3e.

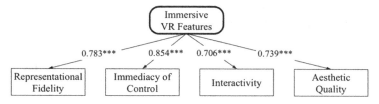

Fig. 5. Loadings of immersive VR features

As shown in Fig. 6, the model explained 49.8% of the varience in motivation, 65.6% of the varience in learning effectiveness, and 42.0% of the varience in learning performance. Moreover, the immersive VR features are significant antecedents to usefulness (beta = 0.536, t = 5.543), immersion (beta = 0.549, t = 7.386), enjoyment (beta = 0.628, t = 8.206), and control and active learning (beta = 0.225, t = 2.022). Usefulness is a significant antecedent to control and active learning (beta = 0.620, t = 7.055), cognitive benefits (beta = 0.638, t = 6.359), and reflective thinking (beta = 0.683, t =

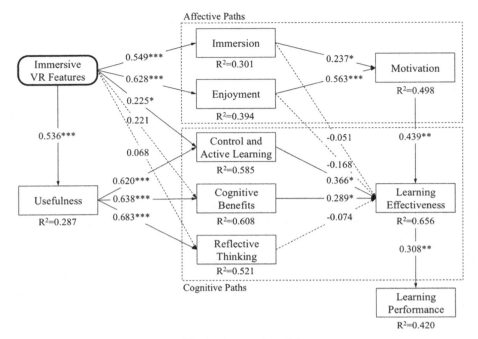

Fig. 6. Structural model

7.380). Immersion (beta = 0.237, t = 2.036) and enjoyment (beta = 0.563, t = 3.606) are significant antecedents to motivation. Control and active learning (beta = 0.366, t = 2.267), cognitive benefits (beta = 0.289, t = 1.992), and motivation (beta = 0.439, t = 2.952) are significant antecedents to learning effectiveness. Learning effectiveness (beta = 0.308, t = 2.813) is a significant antecedent to learning performance. However, immersive VR features are not significant antecedents to cognitive benefits (beta = 0.221, t = 1.832) and reflective thinking (beta = 0.068, t = 0.661). Immersion (beta = −0.051, t = 0.501) and enjoyment (beta = −0.168, t = 1.301) are not significant antecedents to learning effectiveness. The control variables of this model include age (beta = 0.240, t = 1.910), gender (beta = 0.041, t = 0.334), degree (beta = −0.199, t = 1.641), interest in technology (beta = 0.037, t = 0.356), previous VR playing experiences (beta = 0.131, t = 1.214), and prior knowledge (beta = 0.395, t = 3.238).

6 Discussion

This research was set to explore how immersive VR influences learning outcomes. To do so (1) we conceptualized and operationalized new technological features to describe immersive VR and (2) we unpacked the paths from the immersive VR features to the learning outcomes, focusing on the affective and cognitive factors.

First, we proposed four subdimensions of immersive VR features, and we empirically tested the constructs. The convergent and discriminant validity satisfied the criteria, indicating that two new VR features (i.e., interactivity and aesthetic quality) are valid factors that describe immersive VR's characteristics.

Second, we tested how immersive VR features influenced learning performance through the affective path. The results indicate that immersive VR features have a direct impact on the affective factors (i.e., immersion and enjoyment). We did not find the direct effects of the affective factors on learning effectiveness and performance. Instead, the affective factors significantly increase motivation, which further influences learning effectiveness and performance.

Third, we also tested how the immersive VR features influenced learning performance through the cognitive path. The results indicate that the effects of immersive VR features on the cognitive factors (i.e., control and active learning, cognitive benefits, and reflective thinking) are mediated by usefulness. Except reflective thinking, control and active learning as well as cognitive benefits have a direct impact on learning effectiveness.

6.1 Theoretical Implications

There are several theoretical implications provided by this study. First, with two new subdimensions, the VR features that are shown in this model comprehensively characterize the technological features of immersive VR. Researchers are advised to adopt these features as a basis for their immersive VR research in the future.

Second, we have identified how immersive VR enhances learning outcomes through affective and cognitive paths. While the results are consistent with the research by Makransky and Petersen [12], this study further develops their findings. Rather than manipulating the learning outcomes as a second-order construct [18] or ignoring the subjective evaluation of learning effectiveness [12], we also have an advanced understanding of the relationships between specific learning outcomes. We find that motivation is proposed to influence learning performance through learning effectiveness. Prior literature either failed to explain why some psychological factors failed to influence learning performance [12] or identify the precise relationships between them [18]. By unpacking these relationships, we have clearly proposed and confirmed how the affective and cognitive factors influence different learning outcomes and how immersive VR effectively influences learning outcomes.

Third, we also provide a new perspective to explain the contradictory research results on the effects of immersive VR to some extent. We have discovered the indirect impact of immersion and enjoyment on learning effectiveness. For example, immersion is considered to be one of the advantages that contributes to the positive effects of immersive VR on learning outcomes, partially because it allows learners to devote themselves to the virtual learning environment and focus on learning [9, 79–81]. It was predicted that learners should feel that their learning is effective when they are immersed in the learning environment. However, our research indicates the indirect impact of immersion on learning effectiveness, which is aligned with the research by Hamari et al. [82]. The framework shows that immersion needs to increase motivation before positively affecting learning effectiveness. More specifically, if immersive experiences fail to motivate learners, learners will not feel that their learning is effective. In conclusion, our study suggests that it will be difficult for immersive VR to enhance learning outcomes if it cannot influence these affective and cognitive factors.

6.2 Practical Implications

Our work illuminates crucial immersive VR features that play an important role in the learning process. To effectively enhance the positive effects of immersive VR, designers of immersive VR applications should focus on improving the technological features proposed in this research.

Moreover, this research points out the significant psychological factors (i.e., the affective factors and cognitive factors) of learners that immersive VR designers need to pay attention to. The relationships between the psychological factors and learning outcomes are also revealed from the affective and cognitive paths. Based on these findings, effective strategies to increase immersive VR's positive effects on learning will be proposed. For instance, designers are advised to increase learners' motivation through enhancing their immersive and enjoyable feelings. In general, many strategies are concluded from this research to improve learners' immersive VR learning experiences.

6.3 Limitations and Future Directions

This research has limitations, which should be considered in interpreting the findings. The limitations might become future research opportunities. First, given that this study used a single VR application, the generalizability of our findings is not ensured. Researchers are advised to test the model by using diverse immersive VR applications and subject samples. Researchers might benefit from increasing the sample size or recruiting subjects with different educational backgrounds to test the model. Second, although this research included the major psychological factors associated with the learning process in the model, there might be other factors that may mediate the effects of immersive VR on the learning outcomes, such as cognitive load [10], flow [83], and satisfaction [32]. Therefore, researchers will benefit from taking a more comprehensive set of factors into consideration when developing an understanding of immersive VR use in education.

7 Conclusions

This research builds on a two-path model to explain how immersive VR influences learning outcomes through the affective and cognitive paths. Prior studies have shown that VR technologies may or may not enhance learning performances. By investigating how immersive VR features influence learning processes and outcomes, our research provides explanations for the inconsistent findings. We speculate that if immersive VR features do not stimulate the affective and cognitive factors to enhance motivation and learning effectiveness, it will be difficult to eventually improve the learners' performance. We also suggest that the interaction experiences and psychological factors should be considered when designing an immersive VR to enhance learning.

Acknowledgement. This research was supported by the UGC Teaching and Learning Grant titled "Developing Multidisciplinary and Multicultural Competences through Gamification and Challenge-based Collaborative Learning." This research was also partly supported by grant No. CityU 11505118 from the Research Grants Council of the Hong Kong SAR.

Appendix: Measurement Items

Constructs	Items	Sources
Representational fidelity	1. The objects (e.g., the red blood cell) in the VR application seemed real	Self-developed based on [84]
	2. The change of images due to view change and object motion in the VR application seemed smooth	
	3. My actions directly resulted in expected changes in the VR application	
	4. The audio effect of the VR application sounded real	
Immediacy of control	1. The ability to change the viewpoint in the VR application allowed me to learn better	[12]
	2. The ability to change the viewpoint in the VR application made learning more motivating and interesting	
	3. The ability to manipulate the objects (e.g., hit, move, rotate) in the VR application made learning more motivating and interesting	
	4. The ability to manipulate the objects in real time helped to enhance my understanding	
Interactivity	1. I had the impression that I could be active in the virtual environment	[85]
	2. The objects in the VR application gave me the feeling that I could do something with them	
	3. I felt that the objects in the VR application could almost be touched	
	4. There were times during which I felt like I was directly interacting with the objects in the VR application	
	5. I felt the objects in the VR application were aware of my presence	
Aesthetic quality	1. The visual design of the VR application was attractive	[86]
	2. The VR application was aesthetically pleasing	
	3. The VR application displayed a visually pleasant design	
	4. The VR application appealed to my visual senses	
	5. Overall, I found that the VR application was visually appealing	
Usefulness	1. Using the VR application as a tool for learning increased my learning and academic performance	[12]
	2. Using the VR application enhanced the effectiveness of my learning	
	3. The VR application allowed me to progress at my own pace	

(*continued*)

(*continued*)

Constructs	Items	Sources
Immersion	1. I lost track of time while playing the VR application	[12, 82];
	2. I became very involved in the VR application forgetting about other things	
	3. I was involved in the VR application to the extent that I lost track of time	
Enjoyment	1. I found using the VR application enjoyable	[12]
	2. Using the VR application was pleasant	
	3. I had fun using the VR application	
Control and active learning	1. The VR application helped me to have a better overview of the content learned	[12]
	2. The VR application allowed me to be more responsive and active in the learning process	
	3. The VR application allowed me to have more control over my own learning	
	4. The VR application promoted self-paced learning	
	5. The VR application helped to get me engaged in the learning activity	
Cognitive benefits	1. The VR application made the comprehension easier	[18]
	2. The VR application made the memorization easier	
	3. The VR application helped me to better apply what was learned	
	4. The VR application helped me to better analyze the problems	
	5. The VR application helped me to have a better overview of the content learned	
Reflective thinking	1. The VR application enabled me to reflect on how I learned	[12]
	2. The VR application enabled me to link new knowledge with previous knowledge and experiences	
	3. The VR application enabled me to become a better learner	
	4. The VR application enabled me to reflect on my own understanding	
Motivation	1. The VR application could enhance my learning interest	[5, 12, 27]
	2. The VR application could enhance my learning motivation	
	3. The realism of the VR application motivated me to learn	
	4. I was more interested to learn the topics	
	5. I was interested and stimulated to learn more	

(*continued*)

(continued)

Constructs	Items	Sources
Learning effectiveness	1. I learned a lot of biological information in the topics	[18]
	2. I gained a good understanding of the basic concepts of the materials	
	3. I learned to identify the main and important issues of the topics	
	4. The learning activities were meaningful	
	5. What I learned, I could apply in real context	

References

1. Dede, C.: The evolution of constructivist learning environments: immersion in distributed, virtual worlds. Educ. Technol. **35**(5), 46–52 (1995)
2. Biocca, F., Delaney, B.: Immersive virtual reality technology. Commun. Age Virtual Reality **15**, 32 (1995)
3. Parong, J., Mayer, R.E.: Learning science in immersive virtual reality. J. Educ. Psychol. **110**(6), 785 (2018)
4. Chittaro, L., Buttussi, F.: Assessing knowledge retention of an immersive serious game vs. a traditional education method in aviation safety. IEEE Trans. Vis. Comput. Graph. **21**(4), 529–538 (2015)
5. Lee, E.A.-L., Wong, K.W.: A review of using virtual reality for learning. In: Pan, Z., Cheok, A.D., Müller, W., El Rhalibi, A. (eds.) Transactions on Edutainment I. LNCS, vol. 5080, pp. 231–241. Springer, Heidelberg (2008). https://doi.org/10.1007/978-3-540-69744-2_18
6. Makransky, G., Lilleholt, L.: A structural equation modeling investigation of the emotional value of immersive virtual reality in education. Educ. Tech. Res. Dev. **66**(5), 1141–1164 (2018)
7. Ekstrand, C., et al.: Immersive and interactive virtual reality to improve learning and retention of neuroanatomy in medical students: a randomized controlled study. CMAJ Open **6**(1), E103 (2018)
8. Freina, L., Ott, M.: A literature review on immersive virtual reality in education: state of the art and perspectives. In: Proceedings of the International Scientific Conference ELearning and Software for Education, "Carol I" National Defence University (2015)
9. Webster, R.: Declarative knowledge acquisition in immersive virtual learning environments. Interact. Learn. Environ. **24**(6), 1319–1333 (2016)
10. Makransky, G., Terkildsen, T.S., Mayer, R.E.: Adding immersive virtual reality to a science lab simulation causes more presence but less learning. Learn. Instr. **60**, 225–236 (2017)
11. Moreno, R., Mayer, R.E.: Learning science in virtual reality multimedia environments: role of methods and media. J. Educ. Psychol. **94**(3), 598 (2002)
12. Makransky, G., Petersen, G.B.: Investigating the process of learning with desktop virtual reality: a structural equation modeling approach. Comput. Educ. **134**, 15–30 (2019)
13. Van Dam, A., et al.: Immersive VR for scientific visualization: a progress report. IEEE Comput. Graph. Appl. **20**(6), 26–52 (2000)
14. Chavez, B., Bayona, S.: Virtual reality in the learning process. In: Rocha, Á., Adeli, H., Reis, L.P., Costanzo, S. (eds.) WorldCIST'18 2018. AISC, vol. 746, pp. 1345–1356. Springer, Cham (2018). https://doi.org/10.1007/978-3-319-77712-2_129

15. Bailenson, J.N., et al.: The use of immersive virtual reality in the learning sciences: digital transformations of teachers, students, and social context. J. Learn. Sci. **17**(1), 102–141 (2008)

16. Jensen, L., Konradsen, F.: A review of the use of virtual reality head-mounted displays in education and training. Educ. Inf. Technol. **23**(4), 1515–1529 (2018)

17. Papanastasiou, G., et al.: Virtual and augmented reality effects on K-12, higher and tertiary education students' twenty-first century skills. Virtual Reality **23**(4), 425–436 (2019)

18. Lee, E.A.-L., Wong, K.W., Fung, C.C.: How does desktop virtual reality enhance learning outcomes? A structural equation modeling approach. Comput. Educ. **55**(4), 1424–1442 (2010)

19. Whitelock, D., Brna, P., Holland, S.: What is the value of virtual reality for conceptual learning? Towards a theoretical framework. Edições Colibri (1996)

20. Dalgarno, B., Hedberg, J., Harper, B.: The contribution of 3D environments to conceptual understanding (2002)

21. Jennings, M.: Theory and models for creating engaging and immersive ecommerce websites. In: Proceedings of the 2000 ACM SIGCPR Conference on Computer Personnel Research. ACM (2000)

22. Fiorentino, M., et al.: Spacedesign: a mixed reality workspace for aesthetic industrial design. In: Proceedings of the 1st International Symposium on Mixed and Augmented Reality. IEEE Computer Society (2002)

23. Thorsteinsson, G., Page, T., Niculescu, A.: Using virtual reality for developing design communication. Stud. Inform. Control **19**(1), 93–106 (2010)

24. Roussou, M.: A VR playground for learning abstract mathematics concepts. IEEE Comput. Graph. Appl. **29**(1), 82–85 (2008)

25. Ryan, M.-L.: Immersion vs. interactivity: virtual reality and literary theory. SubStance **28**(2), 110–137 (1999)

26. Kaufmann, H., Schmalstieg, D., Wagner, M.: Construct3D: a virtual reality application for mathematics and geometry education. Educ. Inf. Technol. **5**(4), 263–276 (2000)

27. Huang, H.-M., Rauch, U., Liaw, S.-S.: Investigating learners' attitudes toward virtual reality learning environments: based on a constructivist approach. Comput. Educ. **55**(3), 1171–1182 (2010)

28. Patel, K., et al.: The effects of fully immersive virtual reality on the learning of physical tasks. In: Proceedings of the 9th Annual International Workshop on Presence, Ohio, USA (2006)

29. Bliss, J.P., Tidwell, P.D., Guest, M.A.: The effectiveness of virtual reality for administering spatial navigation training to firefighters. Presence Teleoper. Virtual Environ. **6**(1), 73–86 (1997)

30. Pedram, S., Perez, P., Palmisano, S., Farrelly, M.: The factors affecting the quality of learning process and outcome in virtual reality environment for safety training in the context of mining industry. In: Cassenti, Daniel N. (ed.) AHFE 2018. AISC, vol. 780, pp. 404–411. Springer, Cham (2019). https://doi.org/10.1007/978-3-319-94223-0_38

31. Chowdhury, T.I., Costa, R., Quarles, J.: Information recall in a mobile VR disability simulation. In: Proceedings of the 2017 9th International Conference on Virtual Worlds and Games for Serious Applications (VS-Games). IEEE (2017)

32. Stepan, K., et al.: Immersive virtual reality as a teaching tool for neuroanatomy. In: Proceedings of the International Forum of Allergy & Rhinology. Wiley Online Library (2017)

33. Vincent, D.S., et al.: Teaching mass casualty triage skills using immersive three-dimensional virtual reality. Acad. Emerg. Med. **15**(11), 1160–1165 (2008)

34. Han, I., Ryu, J., Kim, M.: Prototyping training program in immersive virtual learning environment with head mounted displays and touchless interfaces for hearing-impaired learners. Educ. Technol. Int. **18**(1), 49–71 (2017)

35. Wong, E.Y.-C., Kong, K.H., Hui, R.T.-Y.: The influence of learners' openness to IT experience on the attitude and perceived learning effectiveness with virtual reality technologies. In: Proceedings of 2017 IEEE 6th International Conference on Teaching, Assessment, and Learning for Engineering (TALE). IEEE (2017)

36. Chen, Y.-L.: The effects of virtual reality learning environment on student cognitive and linguistic development. Asia-Pac. Educ. Res. **25**(4), 637–646 (2016)

37. Bailenson, J., et al.: The effect of interactivity on learning physical actions in virtual reality. Media Psychol. **11**(3), 354–376 (2008)

38. Limniou, M., Roberts, D., Papadopoulos, N.: Full immersive virtual environment CAVETM in chemistry education. Comput. Educ. **51**(2), 584–593 (2008)

39. Cheng, K.-H., Tsai, C.-C.: A case study of immersive virtual field trips in an elementary classroom: students' learning experience and teacher-student interaction behaviors. Comput. Educ. **140**, 103600 (2019)

40. Leder, J., et al.: Comparing immersive virtual reality and powerpoint as methods for delivering safety training: impacts on risk perception, learning, and decision making. Saf. Sci. **111**, 271–286 (2019)

41. Chen, X., et al.: ImmerTai: immersive motion learning in VR environments. J. Vis. Commun. Image Represent. **58**, 416–427 (2019)

42. Kickmeier-Rust, M.D., Hann, P., Leitner, M.: Increasing learning motivation: an empirical study of VR effects on the vocational training of bank clerks. In: van der Spek, E., Göbel, S., Do, E.Y.-L., Clua, E., Baalsrud Hauge, J. (eds.) ICEC-JCSG 2019. LNCS, vol. 11863, pp. 111–118. Springer, Cham (2019). https://doi.org/10.1007/978-3-030-34644-7_9

43. Alavi, M., Leidner, D.E.: Research commentary: technology-mediated learning—a call for greater depth and breadth of research. Inf. Syst. Res. **12**(1), 1–10 (2001)

44. McMahan, A.: Immersion, engagement, and presence: a method for analyzing 3-D video games. In: The Video Game Theory Reader, pp. 89–108. Routledge (2013)

45. Kearney, P.: Affective learning. In: Communication Research Measures: A Sourcebook, pp. 81–85 (1994)

46. Chory-Assad, R.M.: Classroom justice: perceptions of fairness as a predictor of student motivation, learning, and aggression. Commun. Q. **50**(1), 58–77 (2002)

47. Sidelinger, R.J., McCroskey, J.C.: Communication correlates of teacher clarity in the college classroom. Commun. Res. Rep. **14**(1), 1–10 (1997)

48. Witt, P.L., Wheeless, L.R., Allen, M.: A meta-analytical review of the relationship between teacher immediacy and student learning. Commun. Monogr. **71**(2), 184–207 (2004)

49. Buchanan, M.T., Hyde, B.: Learning beyond the surface: engaging the cognitive, affective and spiritual dimensions within the curriculum. Int. J. Child. Spiritual. **13**(4), 309–320 (2008)

50. Bloom, B.S.: Taxonomy of educational objectives. Cogn. Domain **1**, 120–124 (1956)

51. Mandal, S.: Brief introduction of virtual reality & its challenges. Int. J. Sci. Eng. Res. **4**(4), 304–309 (2013)

52. Davis, F.D., Bagozzi, R.P., Warshaw, P.R.: Extrinsic and intrinsic motivation to use computers in the workplace 1. J. Appl. Soc. Psychol. **22**(14), 1111–1132 (1992)

53. Kleinginna, P.R., Kleinginna, A.M.: A categorized list of motivation definitions, with a suggestion for a consensual definition. Motiv. Emot. **5**(3), 263–291 (1981)

54. Hui, W., et al.: Technology-assisted learning: a longitudinal field study of knowledge category, learning effectiveness and satisfaction in language learning. J. Comput. Assist. Learn. **24**(3), 245–259 (2008)

55. Young, M.R., Klemz, B.R., Murphy, J.W.: Enhancing learning outcomes: the effects of instructional technology, learning styles, instructional methods, and student behavior. J. Mark. Educ. **25**(2), 130–142 (2003)

56. Kapralos, B., Moussa, F., Collins, K., Dubrowski, A.: Fidelity and multimodal interactions. In: Wouters, P., van Oostendorp, H. (eds.) Instructional Techniques to Facilitate Learning and Motivation of Serious Games. AGL, pp. 79–101. Springer, Cham (2017). https://doi.org/10.1007/978-3-319-39298-1_5

57. Gerling, K.M., et al.: The effects of graphical fidelity on player experience. In: Proceedings of International Conference on Making Sense of Converging Media. ACM (2013)

58. Lee, J., Kim, M., Kim, J.: A study on immersion and VR sickness in walking interaction for immersive virtual reality applications. Symmetry **9**(5), 78 (2017)

59. Han, S., Kim, J.: A study on immersion of hand interaction for mobile platform virtual reality contents. Symmetry **9**(2), 22 (2017)

60. Chittaro, L., Ranon, R.: Web3D technologies in learning, education and training: motivations, issues, opportunities. Comput. Educ. **49**(1), 3–18 (2007)

61. Zhang, X., et al.: How virtual reality affects perceived learning effectiveness: a task–technology fit perspective. Behav. Inf. Technol. **36**(5), 548–556 (2017)

62. Halabi, O.: Immersive virtual reality to enforce teaching in engineering education. Multimed. Tools Appl. **79**(3), 2987–3004 (2019). https://doi.org/10.1007/s11042-019-08214-8

63. Blackledge, J., Barrett, M.: Development and evaluation of a desktop VR system for electrical services engineers. In: Proceedings of the World Congress on Engineering (2012)

64. Wang, R., et al.: How does web-based virtual reality affect learning: evidences from a quasi-experiment. In: Proceedings of the ACM Turing 50th Celebration Conference, China. ACM (2017)

65. Davis, F.D.: Perceived usefulness, perceived ease of use, and user acceptance of information technology. MIS Q. **13**, 319–340 (1989)

66. Barrett, M., Blackledge, J.: Evaluation of a prototype desktop virtual reality model developed to enhance electrical safety and design in the built environment (2012)

67. Sowndararajan, A., Wang, R., Bowman, D.A.: Quantifying the benefits of immersion for procedural training. In: Proceedings of the 2008 Workshop on Immersive Projection Technologies/Emerging Display Technologies. ACM (2008)

68. Baker, D.S., Underwood III, J., Thakur, R.: Factors contributing to cognitive absorption and grounded learning effectiveness in a competitive business marketing simulation. Mark. Educ. Rev. **27**(3), 127–140 (2017)

69. Cybinski, P., Selvanathan, S.: Learning experience and learning effectiveness in undergraduate statistics: modeling performance in traditional and flexible learning environments. Decis. Sci. J. Innov. Educ. **3**(2), 251–271 (2005)

70. Williams, M.D.: Learner-control and instructional technologies. Handb. Res. Educ. Commun. Technol. **2**, 957–983 (1996)

71. Merrill, M.D.: Learner control: Beyond aptitude-treatment interactions. AV Commun. Rev. **23**(2), 217–226 (1975)

72. Phan, H.P.: An examination of reflective thinking, learning approaches, and self-efficacy beliefs at the university of the south pacific: a path analysis approach. Educ. Psychol. **27**(6), 789–806 (2007)

73. Fitzpatrick, J.: Interactive virtual reality learning systems: are they a better way to ensure proficiency, 15 March 2007 (2007)

74. IJsselsteijn, W., et al.: Virtual cycling: effects of immersion and a virtual coach on motivation and presence in a home fitness application. In: Proceedings of Virtual Reality Design and Evaluation Workshop (2004)

75. Githua, B.N., Mwangi, J.G.: Students' mathematics self-concept and motivation to learn mathematics: relationship and gender differences among Kenya's secondary-school students in nairobi and rift valley provinces. Int. J. Educ. Dev. **23**(5), 487–499 (2003)

76. Waheed, M., et al.: Perceived learning outcomes from Moodle: an empirical study of intrinsic and extrinsic motivating factors. Inf. Dev. **32**(4), 1001–1013 (2016)

77. Novo-Corti, I., Varela-Candamio, L., Ramil-DíAz, M.: E-learning and face to face mixed methodology: evaluating effectiveness of e-learning and perceived satisfaction for a microeconomic course using the Moodle platform. Comput. Hum. Behav. **29**(2), 410–415 (2013)
78. Hair, J.F., et al.: Multivariate Data Analysis, vol. 6. Pearson Prentice Hall, Upper Saddle River (2006)
79. Fassbender, E., et al.: VirSchool: the effect of background music and immersive display systems on memory for facts learned in an educational virtual environment. Comput. Educ. **58**(1), 490–500 (2012)
80. Cheng, M.T., She, H.C., Annetta, L.A.: Game immersion experience: its hierarchical structure and impact on game-based science learning. J. Comput. Assist. Learn. **31**(3), 232–253 (2015)
81. Larsen, R., Reif, L.: Effectiveness of cultural immersion and culture classes for enhancing nursing students' transcultural self-efficacy. J. Nurs. Educ. **50**(6), 350–354 (2011)
82. Hamari, J., et al.: Challenging games help students learn: an empirical study on engagement, flow and immersion in game-based learning. Comput. Hum. Behav. **54**, 170–179 (2016)
83. Riva, G., Castelnuovo, G., Mantovani, F.: Transformation of flow in rehabilitation: the role of advanced communication technologies. Behav. Res. Methods **38**(2), 237–244 (2006)
84. Choi, B., Baek, Y.: Exploring factors of media characteristic influencing flow in learning through virtual worlds. Comput. Educ. **57**(4), 2382–2394 (2011)
85. Bellur, S., Sundar, S.S.: Talking health with a machine: how does message interactivity affect attitudes and cognitions? Hum. Commun. Res. **43**(1), 25–53 (2017)
86. Huang, T.-L., Hsu Liu, F.: Formation of augmented-reality interactive technology's persuasive effects from the perspective of experiential value. Internet Res. **24**(1), 82–109 (2014)

Flip-Flop Quizzes: A Case Study Analysis to Inform the Design of Augmented Cognition Applications

Branden Ogata[✉], Jan Stelovsky, and Michael-Brian C. Ogawa

Department of Information and Computer Sciences, University of Hawaii at Mānoa, 1680 East-West Road, Honolulu, HI 96822, USA
{bsogata,janst,ogawam}@hawaii.edu

Abstract. This paper examines the effectiveness of Flip-Flop quizzes, a pedagogical methodology in which students create quizzes synchronized with lecture videos as part of an inverted classroom. In order to generate quality quizzes, students must understand the material covered in the lecture videos. The results of our study inform the design of augmented cognition applications.

Keywords: Learning by teaching · Inverted classroom · Flipped classroom

1 Introduction

Enrollment in Computer Science programs has increased throughout the United States over the past decade [1–3] even though overall enrollment at universities has remained largely the same [4]. However, this increase in students has not led to a corresponding increase in degrees awarded. The annual Taulbee Survey by the Computing Research Association [3] showed that the rate of enrollment has outpaced the rate of graduations (Fig. 1), and Beaubouef and Mason [5] indicated that 40% of students who started in a Computer Science program ultimately did not receive their bachelor's degrees. Somewhere between 28% [6] and 32.3% [7] of students will drop or fail their introductory programming courses, and [7] suggests that the actual attrition rate may be higher as programs with larger dropout rates would be less inclined to publish that data.

Each student who leaves Computer Science has a reason for doing so. Lewis et al. [8] defined affinity for the program as the most important factor in student retention: a student who is more satisfied with the Computer Science program is more likely to remain in the program. This affinity encompasses both technical skills (the ability of the student to write code and comprehend theory) and soft skills (emotional intelligence and experiences with peers and faculty). The methods that instructors use and the environments in which that education takes place are extremely significant [9–11]; consequently, the rest of this article considers a potential change to those pedagogical methodologies.

© Springer Nature Switzerland AG 2020
D. D. Schmorrow and C. M. Fidopiastis (Eds.): HCII 2020, LNAI 12197, pp. 106–117, 2020.
https://doi.org/10.1007/978-3-030-50439-7_7

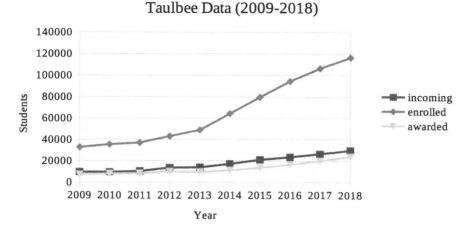

Fig. 1. Increases in enrollment and degrees awarded.

Traditional instructional methods utilize a teacher-centered classroom: the instructor delivers lectures for the students to absorb during class. Those pupils then complete assignments at home before returning to class to take quizzes and tests designed to evaluate their understanding of the material. The inverted or flipped classroom [12] reverses this model: students review materials at home in preparation for exercises during class sessions where they may ask their instructors and peers for assistance. Although this approach is effective in improving student achievement [13–15] and soft skills [16], the inverted classroom depends upon students actually preparing for class in order to effectively participate in the in-class activities. Approximately 20% to 38% of students will complete the assigned readings most or all of the time before class [17], and those who do so may only passively absorb the content without developing a deep understanding of the concepts [14].

Learning by teaching also inverts the traditional model of education in that it places students in the position of actively instructing their peers rather than passively receiving information from a teacher [18]. The Flip-Flop methodology introduced in [19] and expanded upon in [20] utilizes the principles of learning by teaching to add a "flop" component to the flipped classroom: students create multiple-choice questions and answers as part of a Flip-Flop quiz synchronized with a lecture video. This procedure compels students to watch the lecture video at least once to develop the questions and answers. However, the benefits of this process may extend beyond merely ensuring that the students passively view the videos. In order to create a quality quiz, a student must understand the material and identify key concepts to write questions. Writing the correct answer for a question is straightforward; writing the incorrect distractors for a question is non-trivial as those choices must be wrong yet not so absurdly false that those taking the quiz may discount the option entirely.

The pilot study that Ogawa performed for the Flip-Flop methodology found that the use of Flip-Flop improved student performance on a post-test, increased student engagement, and reduced multitasking while watching the lecture video [21]. The study presented below seeks to confirm those results with a larger sample size and an actual course rather than an experimental setting. Specifically, we consider the following questions:

1. Does Flip-Flop improve student achievement?
2. Does Flip-Flop impact the strategies that students utilize when studying?
3. What lessons can be taken from the above questions and applied to a software implementation of the Flip-Flop methodology?
4. How do the results inform augmented cognition design?

2 Methodology

2.1 Student Achievement

We examined the de-identified results for the final examination across two semesters of ICS 101. ICS 101 is a course offered at the University of Hawaii at Mānoa that covers fundamental computer concepts and productivity software usage. The course includes in-person and online lectures that cover theory along with lab sessions for practical guidance; the rest of this paper focuses exclusively on the online video lectures where Flip-Flop was implemented. ICS 101 students in the Spring 2018 semester ($n = 135$) were required to use the Flip-Flop methodology in generating questions and answers for the *Social Informatics* module of the course whereas students in the Spring 2017 semester ($n = 173$) were not exposed to Flip-Flop. If generating Flip-Flop quizzes has any impact on student comprehension of the material, the students who created Flip-Flop quizzes should outperform those who did not write quizzes on the final exam questions covering material pertaining to those quizzes they made. We used an analysis of variance (ANOVA) to determine if such a significant difference existed.

2.2 Student Behaviors

We also considered whether using the Flip-Flop methodology has any impact on how students take notes while studying. 111 students who took ICS 101 in Spring 2018 responded to a survey regarding their experiences while creating questions and answers for the aforementioned *Social Informatics* screencast. Questions 1 and 4 asked students about their note-taking behaviors before (Question 4) and after (Question 1) introduction to the Flip-Flop methodology. Both questions included a multiple-choice component for students to indicate the note-taking approaches they utilized along with an open-ended response to explain their answers in more detail. We used a t-test to determine whether the responses to the multiple-choice component changed significantly. We also performed a content analysis on the open-ended responses to identify patterns in the specific study strategies that students used, then an ANOVA on the results of the content analysis to search for any significant differences in those results.

Question 2 from this survey inquired as to whether students were multitasking while watching the lecture videos. As with the other questions mentioned above, this question included a multiple-choice component (true-false) and an open-ended response for participants to provide more detail. We performed a content analysis on these results as well to identify what exactly students considered as multitasking.

2.3   Future Development

Question 3 in the aforementioned survey asked participants to report the difficulties they encountered when using the Flip-Flop methodology. We performed a content analysis on this data to identify the areas in which students experienced difficulties, then an ANOVA on the results of that content analysis to determine if any of those issues was significantly more problematic for students.

3   Results

3.1   Student Achievement

Only one of the 20 questions on the ICS 101 final exam showed a statistically significant difference between the experimental (Flip-Flop) and control groups ($F(1, 16) = 11.82$, $p = 0.0034$, $\eta^2 = 0.4249$) (Table 1).

Table 1. ANOVA for ICS 101 final exam question 10

	Source	SS	MS	
Between groups	0.6169	1	0.6169	$F = 11.82$
Within groups	0.8350	16	0.0522	
Total	1.452	17		

Question 10 on the ICS 101 final exam was "Dr. Ogawa mentioned from his own research that in a social network, men had a tendency to view other men's profiles and women had a tendency to view other women's profiles because _____." This question tests the ability of the student to recall a fact mentioned during the lecture video and therefore fits into the Knowledge level of Bloom's Taxonomy [22].

3.2   Student Behaviors

Note-Taking Behaviors. Table 2 shows statistically significant changes between the control and experimental groups in the responses to "Create illustrations of concepts" ($t(7) = 2.366$, $p = 0.0499$, $r^2 = 0.4444$) and "Other"

$(t(7) = 2.497,\ p\ =\ 0.0041,\ r^2\ =\ 0.4711)$. The behaviors in both categories decreased: students created fewer illustrations and utilized fewer "other" note-taking techniques.

Table 2. Note-taking behaviors

	Before	After
Create a list of terms and definitions	67	76
Copy text such as bulleted lists from the video (slides)	76	70
Create illustrations of concepts	14	6
Create descriptions of concepts	23	39
Other	16	9

Because the results were provided to us in aggregate as part of the de-identification process, we could not match the multiple-choice and open-ended responses for each student. Consequently, we cannot determine what students who responded with "Other" considered as alternate note-taking strategies. However, a content analysis of the open-ended responses allows us to categorize those answers independent of the quantitative analysis shown in Table 2. The only statistically significant difference in that content analysis (see Table 3) was in the responses to "Create a list of terms and definitions" $(t(7) = -2.966,\ p = 0.0209,\ r^2 = 0.5568)$. This differs from the results in Table 2, especially in the number of responses for the "Other" category. However, the direction of the changes remains the same with the exception of that "Other" category.

Table 3. Note-taking behaviors: content analysis

	Before	After
Create a list of terms and definitions	6	13
Copy text such as bulleted lists from the video (slides)	12	10
Create illustrations of concepts	8	6
Create descriptions of concepts	4	6
Other	37	39

Even though the statistically significant categories targeted creating illustrations of concepts and other note-taking strategies decreasing, the researchers noticed an increase in the creations of definition lists and descriptions of concepts. We believe that this new focus was an adjustment that students made based on the task of creating questions as opposed to studying for a quiz. With little instruction on various types of multiple choice questions that could be created, it appears that students focused on factual recall type questions. This could also serve as a possible explanation for the statistically significant increase in one of the factual questions on the final examination.

3.3 Multitasking

There was no significant difference between the number of students who did and did not multitask while viewing the lecture video (Table 4). However, the content analysis (Table 5) and subsequent ANOVA (Table 6) provide more drastic results $(F(2,21) = 8.586, p = 0.0019, \eta^2 = 0.4499)$.

Table 4. Multitasking - Raw data

Yes, performed other tasks	No, did not perform other tasks
57	54

Table 5. Multitasking - Content analysis

Yes, performed other tasks	No, did not perform other tasks	Not Applicable
39	17	8

Table 6. Multitasking - Content analysis ANOVA

	Source	SS	MS	
Between groups	63.58	2	31.79	F = 8.587
Within groups	77.75	21	3.702	
Total	141.3	23		

A Scheffé post hoc test indicates that the **Yes** condition in Table 5 differs significantly from both the **No** and **Not Applicable** cases. The results from the content analysis thus differ considerably from the multiple-choice responses. The participants who did not multitask may have not thought it worthwhile to provide further details regarding their lack of multitasking. Another possible explanation is that students have a different understanding of multitasking than expected. Some participants were engaged in other activities that clearly distracted them during the lecture video:

- "Social Media and texting"
- "Talking to friends via Discord"

However, other behaviors might not have truly qualified as multitasking. If a participant did not simultaneously engage in some activity while simultaneously watching the lecture video, no additional cognitive load would have burdened the student when they were attending to the video.

- "The video was relatively short so I didn't multi-task throughout but I did take my dog out in the middle."
- "After listening about the 2011 tragedy, I wanted to research it and look at the event."

Some participants also considered the Flip-Flop methodology to be multitasking itself.

- "Creating questions for the assignment while listening to the video"
- "I wrote questions down as i was watching the lecture"

3.4   Future Development

The content analysis of the open-ended responses regarding the difficulties students faced in watching the video to create questions identified eight major themes:

We summarize the challenges that participants reported in Tables 8 and 9 with a **Not Applicable** category added to contain irrelevant responses.

The ANOVA reveals a statistically significant difference between the groups $(F(8, 63) = 4.517, p = 0.0002, \eta^2 = 0.3645)$. The Scheffé post hoc test indicates that the number of responses for the "Generation" group differs significantly from all other categories with "Listening and Attention" differing significantly from "Memory and Cognition" and "Not Applicable". Consequently, we may conclude that generating the questions and answers for the quiz posed the greatest difficulty to students while they took notes; remaining attentive throughout the lecture video was to a lesser extent problematic as well.

Most of the open-ended responses in the "Generation" category emphasized trying to create reasonable questions and answers:

- "Trying to come up with legit, false answers to make the questions challenging for someone that did not put in the effort to take notes."
- "The most difficult aspects were trying to create multiple-choice questions that would seem reasonable and not too off-handish and ridiculous that it's obviously not the correct answer."

However, a few responses focused on the lack of clear expectations for the questions and answers to create:

- "There was no context in terms of what kind of questions we were supposed to create"
- "Honestly the most difficult aspect was wondering if the questions i was making were good enough. I just didn't really know what content to make into a question or not (Table 7)"

Table 7. Common difficulties in taking notes

Category	Description and sample quote
Comprehension	Understanding the content of the lecture video • "Trying to keep up with understanding terms and examples of each topic"
Correctness	Ensuring that the answers are correct • "Making sure that my answer choices weren't contradicting and also true"
Generation	Creating questions and answers • "Coming up with challenging questions"
Identifying Key Concepts	Determining what exactly is important in the screencast • "Knowing what info was most important vs. not as important"
Interface	Technical and usability issues in watching the video lecture • "The video player is a little wonky at times, so when I try to get the timestamps or reverse the video to rehear information, it's a bit of a hassle"
Listening and Attention	Remaining attentive during the lecture video • "The video was too long and I had a hard time staying focused because the material was very bland and seemed like common sense"
Memory and Cognition	Remembering information presented • "multi tasking and remembering the terms"
Process	Issues in following instructions and taking notes • "Personally for me, I tried so hard to listening and jotting down notes at the same time"

Table 8. Difficulties in taking notes

	Total
Comprehension	14
Correctness	6
Generation	34
Identifying Key Concepts	8
Interface	6
Listening and Attention	19
Memory and Cognition	4
Process	15
Not Applicable	5

Table 9. Difficulties in taking notes - ANOVA

	Source	SS	MS	
Between groups	91.78	8	11.47	$F = 4.517$
Within groups	160.0	63	2.540	
Total	251.8	71		

4   Design Implications for Augmented Cognition

The results of this study inform future applications of augmented cognition for the Flip-Flop method of instruction. These applications are described based on the three major areas of results: student achievement and behaviors, note-taking strategies, and the social component of peer learning.

4.1   Student Achievement and Behaviors

The Flip-Flop methodology produced improved student performance on a single question in the social informatics portion of the final exam. Students appeared to adjust their study strategies to focus on a factual/recall type questions. In future studies, we intend to use an experimental approach to better understand student focus and recall information using alternative distractors. In the real-world setting, students indicated that they had a range of distractors that led to multi-tasking. The laboratory setting will allow us to target specific instructions to develop quiz questions using Bloom's Taxonomy as a basis for question development. The experimental distractors will allow us to better understand how students develop questions and identify instructional interventions that can best complement learning for different levels of question development. By isolating instructional strategies that augment question development and subsequent learning, we will be able to design features in the Flip-Flop ecosystem to automate this process. Our long-term goal is to develop the system as an instructional approach that can be used with a large-scale of students to decrease attrition levels in the computer science discipline.

4.2   Note-Taking Strategies

In addition to refining the question development process, the researchers believe that student note-taking strategies influence how they focus their efforts which impacts their performance on examinations. Our initial findings indicated that question generation, listening/paying attention, processing, and comprehending content were the most difficult areas for students to consider while developing notes from video lectures. The researchers posit the possibility that various note-taking strategies will support students in learning concepts at different levels and account for these issues. For example, an active approach to creating illustrations to demonstrate an understanding of a concept could be more useful than copying a list of items from the screen. This strategy could also help students to create a deeper level of understanding and a more profound level of question develop such as the conceptual or application level. Therefore, we would like to research these different strategies to determine how note-taking strategies augment student learning. This will be helpful on a multitude of levels since it can inform practices in various settings including traditional, inverted, on-line, hybrid, and peer instruction instructional practices.

While this study found that students following the Flip-Flop pedagogy tended to use text-based notes, focus on terms and definitions, and multitask during the lecture video, participants in the pilot study [21] drew diagrams, focused on the key concepts in the lecture, and remained on task throughout the screencast. The pilot study used volunteers not enrolled in the ICS 101 course as its participants, whereas this study used current ICS 101 students. The difference in results may thus be due to the difference in participants.

4.3   Group-Based Assessment

Some of the literature illustrates the benefits of peer instruction and the benefits of student discourse in the learning process. Thus, our final recommendation to further the research in this realm is to consider the social aspect of learning through peer discourse. We would like to implement our research strategies listed above in small group settings such as pairs or triads. This will add a layer of complexity to the application of augmented cognition to help us determine how the group dynamics can further student learning or be preventative. There may be specific features in group work that can aid students in creating new understandings of content by learning through multiple perspectives rather than focusing solely on their interpretation of content.

The possibilities for augmented cognition in this area of research are quite vast. We are excited at the possibilities and are hopeful that the benefits will improve computer science education for an increasing number of potential budding professionals in this field.

References

1. Camp, T., Zweben, S., Walker, E., Barker, L.: Booming enrollments: good times? In: Proceedings of the 46th ACM Technical Symposium on Computer Science Education, SIGCSE 2015, pp. 80–81. ACM, New York (2015). https://doi.org/10.1145/2676723.2677333
2. Guzdial, M.: 'Generation CS' drives growth in enrollments. Commun. ACM **60**(7), 10–11 (2017). https://doi.org/10.1145/3088245
3. Zweben, S., Bizot, B.: CRA Taulbee Survey, May 2018. https://cra.org/resources/taulbee-survey/
4. Institute of Educational Sciences: Total fall enrollment in degree-granting postsecondary institutions, by attendance status, sex of student, and control of institution: Selected years, 1947 through 2027, January 2018. https://nces.ed.gov/programs/digest/d17/tables/dt17_303.10.asp

5. Beaubouef, T., Mason, J.: Why the high attrition rate for computer science students: some thoughts and observations. SIGCSE Bull. **37**(2), 103–106 (2005). https://doi.org/10.1145/1083431.1083474
6. Bennedsen, J., Caspersen, M.E.: Failure rates in introductory programming: 12 years later. ACM Inroads **10**(2), 30–36 (2019). https://doi.org/10.1145/3324888
7. Watson, C., Li, F.W.: Failure rates in introductory programming revisited. In: Proceedings of the 2014 Conference on Innovation & Technology in Computer Science Education, ITiCSE 2014, pp. 39–44. ACM, New York (2014). https://doi.org/10.1145/2591708.2591749
8. Lewis, C.M., Titterton, N., Clancy, M.: Using collaboration to overcome disparities in Java experience. In: Proceedings of the Ninth Annual International Conference on International Computing Education Research, ICER 2012, pp. 79–86. ACM, New York (2012). https://doi.org/10.1145/2361276.2361292
9. Lizzio, A., Wilson, K., Simons, R.: University students' perceptions of the learning environment and academic outcomes: implications for theory and practice. Stud. High. Educ. **27**(1), 27–52 (2002). https://doi.org/10.1080/03075070120099359
10. Trigwell, K., Prosser, M., Waterhouse, F.: Relations between teachers' approaches to teaching and students' approaches to learning. High. Educ. (00181560) **37**(1), 57–70 (1999). https://doi.org/10.1023/A:1003548313194
11. Ulriksen, L., Madsen, L.M., Holmegaard, H.T.: What do we know about explanations for drop out/opt out among young people from STM higher education programmes? Stud. Sci. Educ. **46**(2), 209–244 (2010). https://doi.org/10.1080/03057267.2010.504549
12. Bishop, J.L., Verleger, M.A., et al.: The flipped classroom: a survey of the research. In: ASEE National Conference Proceedings, Atlanta, GA, vol. 30, pp. 1–18 (2013)
13. Gilboy, M.B., Heinerichs, S., Pazzaglia, G.: Enhancing student engagement using the flipped classroom. J. Nutr. Educ. Behav. **47**(1), 109–114 (2015). https://doi.org/10.1016/j.jneb.2014.08.008
14. Horton, D., Craig, M.: Drop, fail, pass, continue: persistence in CS1 and beyond in traditional and inverted delivery. In: Proceedings of the 46th ACM Technical Symposium on Computer Science Education, SIGCSE 2015, pp. 235–240. ACM, New York (2015). https://doi.org/10.1145/2676723.2677273
15. Latulipe, C., Long, N.B., Seminario, C.E.: Structuring flipped classes with lightweight teams and gamification. In: Proceedings of the 46th ACM Technical Symposium on Computer Science Education, SIGCSE 2015, pp. 392–397. Association for Computing Machinery, New York (2015). https://doi.org/10.1145/2676723.2677240
16. O'Flaherty, J., Phillips, C.: The use of flipped classrooms in higher education: a scoping review. Internet High. Educ. **25**, 85–95 (2015). https://doi.org/10.1016/j.iheduc.2015.02.002
17. Simon, B., Kohanfars, M., Lee, J., Tamayo, K., Cutts, Q.: Experience report: peer instruction in introductory computing. In: Proceedings of the 41st ACM Technical Symposium on Computer Science Education, SIGCSE 2010, pp. 341–345. ACM, New York (2010). https://doi.org/10.1145/1734263.1734381
18. Duran, D.: Learning-by-teaching. Evidence and implications as a pedagogical mechanism. Innov. Educ. Teach. Int. **54**(5), 476–484 (2017). https://doi.org/10.1080/14703297.2016.1156011
19. Stelovska, U., Stelovsky, J., Wu, J.: Constructive learning using flip-flop methodology: learning by making quizzes synchronized with video recording of lectures. In: Zaphiris, P., Ioannou, A. (eds.) LCT 2016. LNCS, vol. 9753, pp. 70–81. Springer, Cham (2016). https://doi.org/10.1007/978-3-319-39483-1_7

20. Stelovsky, J., Ogata, B., Stelovska, U.: "Flip-Flop" learning by teaching methodologies: "Peer Improvement", "Agile Tooltip", support technology, and next steps. In: Zaphiris, P., Ioannou, A. (eds.) LCT 2018. LNCS, vol. 10925, pp. 391–406. Springer, Cham (2018). https://doi.org/10.1007/978-3-319-91152-6_30

21. Ogawa, M.-B.: Evaluation of flip-flop learning methodology. In: Zaphiris, P., Ioannou, A. (eds.) LCT 2018. LNCS, vol. 10925, pp. 350–360. Springer, Cham (2018). https://doi.org/10.1007/978-3-319-91152-6_27

22. Bloom, B.S., Englehart, M., Furst, E., Hill, W., Krathwohl, D.: Taxonomy of Educational Objectives, Handbook 1: Cognitive Domain, 2nd edn. David McKay Company, New York (1956)

Metastimuli: An Introduction to PIMS Filtering

Rico A. R. Picone[1,2(✉)] ⓘ, Dane Webb[1] ⓘ, and Bryan Powell[2] ⓘ

[1] Saint Martin's University, Lacey, WA 98501, USA
rpicone@stmartin.edu
[2] Dialectica LLC, Olympia, WA 98501, USA
rico@dialectica.io
http://www.stmartin.edu, http://dialectica.io

Abstract. A system design for correlating information stimuli and a user's personal information management system (PIMS) is introduced. This is achieved via a deep learning classifier for textual data, a recently developed PIMS graph information architecture, and a principle component analysis (PCA) reduction thereof. The system is designed to return unique and meaningful signals from incoming textual data in or near realtime. The classifier uses a recurrent neural network to determine the location of a given atom of information in the user's PIMS. PCA reduction of the PIMS graph to $\mathbb{R}^m$, with m the actuator (haptic) dimensionality, is termed a PIMS filter. Demonstrations are given of the classifier and PIMS filter. The haptic stimuli, then, are correlated with the user's PIMS and are therefore termed "metastimuli." Applications of this system include educational environments, where human learning may be enhanced. We hypothesize a metastimulus bond effect on learning that has some support from the analogous haptic bond effect. A study is outlined to test this hypothesis.

Keywords: Design · Human centered design and user centered design · Design · Information design · Technology · Augmented reality and environments · Technology · Haptic user interface · Technology · Intelligent and agent systems · Technology · Natural user interfaces (NUI)

We present a system design to enhance human learning via stimulation generated by a synthesis of textual source material (e.g. text, audio or video with dialog, etc.) with a personal information management system (PIMS) in an information architecture we have previously developed [7,8]. PIMS-augmented stimuli are called *metastimuli*. There are several reasons to expect the proposed system will improve human learning.

1. The new information architecture we have developed in [8] and [7] allows a PIMS to be naturally connected and automatically structured.

© Springer Nature Switzerland AG 2020
D. D. Schmorrow and C. M. Fidopiastis (Eds.): HCII 2020, LNAI 12197, pp. 118–128, 2020.
https://doi.org/10.1007/978-3-030-50439-7_8

2. A PIMS is the result of significant user cognition. Including an effective one as a processing layer for stimuli creates a feedback loop in which the information system reshapes stimuli.
3. Natural language processing technology has now developed enough to allow nearly realtime processing and categorization. However, in many learning environments realtime processing is not required, so established offline natural language processing techniques are sufficient.
4. There is evidence that multimodal stimulation during learning improves human learning rates [12].

In Sect. 1, the system architecture is described at a high level. Some background on haptics and human learning are provided in Sect. 2 along with an hypothesis regarding metastimuli and human learning. This is followed in Sect. 3 by a description of PIMS filtering via PCA of a graph information architecture that includes a demonstration. In Sect. 4, the deep learning textual information stimuli classifier subsystem is described in detail, along with a demonstration thereof. The design of a study to test the hypothesized metastimulus bond effect is outlined in Sect. 5.

The significance of this work is evident in its application: the improvement of human learning through metastimuli. Although there is evidence of a related effect, the metastimulus bond effect requires direct study, such as that proposed in Sect. 5.

1   Overview of the Metastimulus Architecture

A diagram of the metastimulus architecture is shown in Fig. 1.

1.1   Dialectical Architecture for a PIMS

Picone et al. [8] introduced the dialectical information architecture to be a computer medium for human thought through, among other things, automatic structuring of categorized information. This was later extended [7] to include fuzzy categories for the integration of numerical data in the architecture. A PIMS built

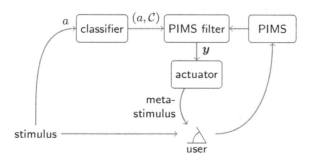

Fig. 1. A diagram of PIMS filtering and metastimulus.

with this architecture has a natural metric to quantify "distance" between atoms of information (any tagged bits of information).[1]

1.2   Filtering Stimuli Through a PIMS

We transform stimuli into metastimuli by "filtering"—in the sense of a dynamic system filter—it through a PIMS. We call this operation a *PIMS filter*. The input to the filter is some categorized atom of information a, with categories in collection $\mathcal{C}$. The output is a real vector $y \in \mathbb{R}^m$ for some $m \in \mathbb{Z}^+$; y can be used as an input to an actuator to generate a metastimulus. A PIMS filter, then, must meaningfully map $a \mapsto y$. A metric on the dialectical architecture enables this meaningfulness. Dimensional compression via principal component analysis (PCA) reduces the dimensionality to the desired real vector space dimensionality m [9]. See Sect. 3 for a detailed description of PIMS filtering.

1.3   Deep Learning for Text Classification

An atom a must be categorized by a *classifier* before the PIMS filter. Of course, for audio and video atoms, natural language processing must precede this classification. Several text classification techniques are available.

The deep learning library Tensorflow is used to accelerate development of the artificial neural networks. Recurrent neural networks (RNN) are able to take advantage of the sequential aspect of natural language. They are capable of discerning meaning in text and once trained, are able to process a large amount of textual data quickly. See Sect. 4 for a detailed description of the classifier subsystem.

2   Metastimuli and Learning

2.1   The Haptic Bond Effect

Haptic bond effect theory [1] suggests that human learning is enhanced by haptic bonding of spatial and temporal sense modalities. The mechanism it suggests is that the sensing of visual stimuli is more spatially oriented and the sensing of auditory stimuli is more temporally oriented; it recognizes the mixing of these in the sensing of haptic stimuli. The mixing in haptic learning "bonds" the visual and auditory, spatial and temporal.

This insight informs our work in two ways: (1) it suggests haptic actuation for metastimuli and (2) by analogy, it suggests that metastimuli may improve human learning. The former is explored in Sect. 2.3 and the latter in Sect. 2.2.

---

[1] The dialectical architecture's structural aspect can be considered to estimate the structure of a *language game*: a communally developed set of language rules of usage [16]. It is, then, important to recognize that *there are many language games* and therefore many structures to be estimated. Furthermore, the rules of these games evolve with use. Therefore, a user's PIMS should not be isolated from others', but neither should there be only one such structure. Additionally, a PIMS should evolve with the language game, a feature that can be detected through collective user estimation.

2.2   The Metastimulus Bond Effect

Since the work of Immanuel Kant, who introduced the notion that the mind itself shapes experience through an *a priori*[2] frame [3], the importance of preconceived categories in human learning has been widely recognized. With the recent developments of classifiers that can categorize stimuli and PIMS that can represent the categorization of human thought, it has become possible (via a PIMS filter and an actuator) to shape stimuli itself through a user's PIMS—that is, to create metastimuli.

By analogy to the haptic bond effect (Sect. 2.1), then, we hypothesize a metastimulus bond effect: "mirroring" the mind's process of categorization and providing corresponding stimuli bonds stimuli and understanding itself—creating a feedback loop that we hypothesize leads to improved human learning.

2.3   Haptic Metastimuli

The output of the PIMS filter is input to an actuator. As we have seen, haptic stimuli can bond spatial and temporal learning, which, especially with the spatial analogies natural to the dialectical information architecture [8], suggests it as a promising method for generating metastimuli.

A haptic device typically stimulates via touch frequency, intensity, position, and pattern. A particularly interesting result for the purposes of metastimuli actuation has been developed in [2]: a sleeve with actuators in a four-by-six array that is worn on the forearm. The sleeve relays to the user 39 English phonemes through frequency and position. In the study, participants learned haptic phoneme representations. All participants were able to learn all 39 distinct haptic phoneme representations within nine days. The findings of Jung et al. [2] demonstrate the promise of human learning through haptic representation of language—essential to the metastimulus bond effect.

We do not conjecture, however, that the metastimulus bond effect will require a direct translation of words to stimuli. For instance, a metastimulus corresponding to an atom of text with categorization X will not need to represent X phonetically. Rather, the metastimulus corresponds to the location of X in the graphical structure of the PIMS itself, which has no relation to the phonetic representation of X.

3   A PIMS Filter via PCA of a Graph Information Architecture

The PIMS filter we present is based on the principle component analysis (PCA) of graphs developed by Saerens et al. [9]. Dimensional reduction of this kind is required for practical application because current methods of haptic

---

[2] Kant claims the mind has *a priori* "intuitions" for space and time, but for our purposes we can take *a priori* to mean pre-existing.

interface (Sect. 2.3) have relatively low "dimensionality": relatively few independent axes of information can be conveyed. For instance, independent two-dimensional arrays of pins on a user's two arms has dimensionality $2 \times 2 = 4$. Such a haptic interface can represent information in $\mathbb{R}^4$. Dimensionality increases when additional haptic modes are included, such as vibration frequency, force, pulsation, etc. However, this dimensionality m is nearly always significantly less than that of the PIMS n. Understanding the meaning of dimensionality in the PIMS requires a brief introduction to the dialectical information architecture.

3.1 The Dialectical Information Architecture

The dialectical information architecture [7,8] includes a graph representation of categories and their relations. Let a category be a set of elements called atoms (e.g. individual paragraphs) associated with it. Then the intersection[3] $\cap \mathcal{C}$ of some collection $\mathcal{C}$ of categories is also a set of atoms. Consider the dictionary $\mathcal{D}$, the set of all categories in a given PIMS. A graph may be constructed with nodes the recursive intersections of the categories of $\mathcal{D}$. Directed edges of the graph represent the relation "has subcategory [subset]." These edges are termed "visible" or "invisible" by Definitions 7 and 8 of [8]. Similarly, by Definition 9 of [8], each atom can be considered "visible" at a given node if and only if it is not contained in a subcategory thereof. This implies that each atom is "visible" at only a single node.

3.2 Adjacency, Distance, and the ECTD Metric

Each atom, then, can be thought to have a *place* in the graph (that is, wherever it is visible). The edges introduce *distance* among atoms. Here we only consider "visible" edges to connect nodes. The metric built from this is called the Euclidean commute time distance (ECTD): it is the square root of the average number of steps ("time") a (Markov-chain) random walk takes from a given node to another, and back [9]. The more-connected one node to another, the shorter the ECTD between the nodes.

This is related to what is called *adjacency*: a node is adjacent to another node if it is connected by an edge. The *adjacency matrix* A encodes adjacency as follows. Let n be the number of nodes in the graph. Then A has dimension $n \times n$ and the element $A_{ij} = 1$ if node i is adjacent to node j and 0 otherwise. For the purposes of PCA, we take edges to be undirected (traversable either direction) and unweighted (constant 1 instead of some variable weight); therefore, $A_{ij} = A_{ji}$ is symmetric and typically sparse.

---

[3] Here we use "crisp" set notation; however, fuzzy set notation can also be used (see [7]).

3.3   ECTD Transformation

Consider, then, that orthonormal standard basis vectors e_i that span the $\mathbb{R}^n$ *node space* each represents a node. Saerens et al. [9] provide a basis transformation of $e_i \mapsto x_i$ such that the transformed vectors x_i are separated precisely by the ECTD. Let $D = \operatorname{diag} A$ and L be the *Laplacian matrix*

$$L = D - A, \tag{1}$$

and let L^+ be the Moore-Penrose pseudoinverse of L. Let U be the orthonormal matrix of eigenvectors of L^+ and Λ be the corresponding diagonal eigenvalue matrix. This transformation is [9]

$$x_i = \Lambda^{1/2} U^\top e_i. \tag{2}$$

Computing the pseudoinverse is generally intractable for large n. Therefore, the following subspace projection is preferred.

3.4   Principal Component Analysis via Subspace Projection

Principal component analysis (PCA) is performed via a projection onto a subspace $\mathbb{R}^m$ of $\mathbb{R}^n$. This preserves only the most significant dimensions of the node space. The technique of Saerens et al. [9] is to compute only the m largest eigenvalues and corresponding eigenvectors of L^+, yielding the transformation of $e_i \mapsto y_i \in \mathbb{R}^m$ via the projection

$$y_i = \widetilde{\Lambda}^{1/2} \widetilde{U}^\top e_i \tag{3}$$

where $\widetilde{\Lambda} : \mathbb{R}^m \to \mathbb{R}^m$ is the truncated diagonal (m largest) eigenvalue matrix and $\widetilde{U}^\top : \mathbb{R}^n \to \mathbb{R}^m$ is the truncated (m largest) eigenvector matrix. This projection approximately conserves ECTD, making this projection optimal with regard to that metric.

3.5   Implementation Considerations

When implementing this projection, a numerical package such as ARPACK [5], which performs an efficient partial eigendecomposition, is key. Convenient wrappers for ARPACK can be found in most programming languages, including Python, SciPy [13] module [10].

An additional key implementation consideration is that the largest eigenvalues and corresponding eigenvectors of L^+ can be computed without computing the pseudoinverse. The $n - 1$ nonzero eigenvalues λ_i^L of L are reciprocally related to those $\lambda_i^{L^+}$ of L^+; that is, [9]

$$\lambda_i^{L^+} = 1/\lambda_i^L \text{ for } \lambda_i^L \neq 0. \tag{4}$$

Therefore, one need only compute the m smallest (nonzero) eigenvalues of L to obtain the m largest of L^+ by Eq. 4. Numerical packages such as ARPACK can perform partial decompositions, seeking only the smallest (or largest) eigenvalues and corresponding eigenvectors.

3.6   The PIMS Filter

The PIMS filter, then, takes a categorized atom that is mapped to its corresponding "place" in a PIMS structure, which is mapped to its node space representation $e_i \in \mathbb{R}^n$, which is projected to $y_i \in \mathbb{R}^m$ via Eq. 3. This process represents the atom's "location" in a user's PIMS in the much smaller space $\mathbb{R}^m$, which can be converted to a metastimulus by an actuator such as a haptic interface.

We have released the software *PIMS Filter* [6], which includes a functional PIMS filter written in Python. Consider contributing to this open-source project.

3.7   PIMS Filter Demonstration

As a demonstration of the use of the *PIMS Filter* software [6], representative atoms of a simple PIMS categorical structure, shown in Fig. 2, are mapped to $\mathbb{R}^2$ and displayed in Fig. 3.

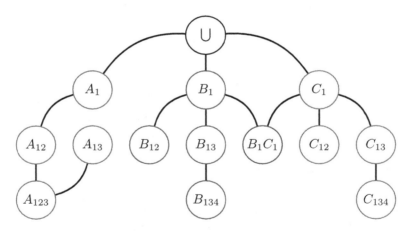

Fig. 2. The dialectical architecture structure for the demonstration PIMS of Sect. 3.7. For clarity, the three primary "branches" of the graph are given colors and the category names A_i, B_j, and C_k. The intersection $A_i \cap A_j$ of categories A_i and A_j is given the shorthand notation A_{ij}. The node $B_1 C_1 = B_1 \cap C_1$ is colored purple to demonstrate that it belongs to both the blue (B) and red (C) branches. The projection of the nodes via the PIMS filter is shown in Fig. 3. (Color figure online)

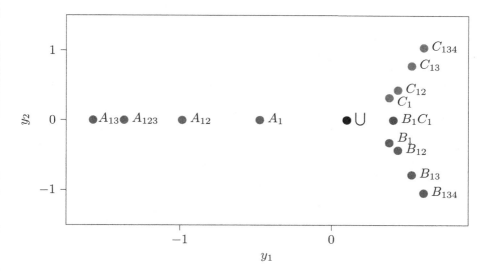

Fig. 3. The PIMS filter output y_i for each node in the demonstration PIMS of Sect. 3.7. Each node of Fig. 2 was projected from node space $\mathbb{R}^{14}$ to the smaller space $\mathbb{R}^2$ using Eq. 3. An atom of information categorized as belonging to a given node i in Fig. 2 is transformed into a value $y_i \in \mathbb{R}^2$ for (haptic) representation to the user.

The graph structure of Fig. 2 has vector space representation in $\mathbb{R}^{14}$; however, even the significant PIMS filter compression down to $\mathbb{R}^2$, shown in Fig. 3, shows significant topological distinction among and in the three primary "branches" of the structure. In fact, it has optimal distinction (in terms of the ECTD), as discussed in Sect. 3.4.

Consider the "mixed" node B_1C_1 in Fig. 2, shared between the B and C branches of the graph. Note how this node appears in $\mathbb{R}^2$ *between* the principal component directions approximating B and C in Fig. 3.

A user familiar with metastimuli from this $\mathbb{R}^2$ representation of their PIMS structure, then, might reasonably be expected to "feel" the semantics of new metastimuli. Furthermore, the metastimulus bond effect hypothesized in Sect. 2 implies this will improve human learning.

4 Textual Information Stimuli Classification

The atoms are input to a neural network where they are classified based on the categories provided by the PIMS. The current atom is tagged with the categories it is placed in, or some null category if it does not have a correlation with any category.

4.1 Classification Method

Atoms are taken to be paragraphs of the textual input. Paragraphs are chosen because a paragraph generally conveys a single concept. Breaking up the textual data this way is similar to most educational course content where everything is taught one concept at a time. For media in which textual data is not formatted at the paragraph level, a method described in [11] may be used to introduce paragraph breaks.

Each atom is processed with a long-short term memory recurrent neural network (LSTM-RNN) that has been previously trained on the same categories provided by the PIMS and similar content. An LSTM-RNN is generally chosen for sentiment analysis in natural language processing. This is because it takes the temporal nature of natural language into account when analyzing text. Further information on LSTM-RNN and sentiment analysis can be found in [17, p. 270ff.].

The output of the network is the atom along with its categories, as identified by the network. When the PIMS's categories change, retraining of the network is required. RNNs are best suited for natural language tasks. They are able to make predictions based on the whole structure of the paragraph. Sentiment analysis of a paragraph is based on the words in the paragraph and where they are located in relation to other words.

A long-short term memory strategy is used to reduce over-fitting. RNNs tend to rely on a small subset of neurons. LSTM employs a strategy of resetting the current weights and biases in a random set of neurons. This strategy forces the network to use all neurons afforded to it to determine outputs.

4.2   An Instantiation in Python

Python includes a powerful deep learning package called Tensorflow, which is used to build the network. The package allows for quicker development because the machine learning architecture is included in the package.

A repository for our text classification tools can be found in Webb et al. [14]. The class within the file `Classifier_RNN.py` is a copy of the class within the file `lstm_rnn.py` found in Karim et al. [4]. This class is used because it is a clean and well-documented way to construct the Tensorflow graph.

4.3   Text Classification Demonstration

For the purposes of this study, a corpus the authors previously created is used. This corpus includes a dataset of mechatronics and system analysis texts, with each paragraph in the dataset manually tagged. The source of this dataset is in the LATEX format. Therefore, we developed a LATEX file classification processor.

The text in the LATEX files needs to be cleaned and organized. A Python class is created to handle this. The class can be found in Webb et al. [15].

The data the network is training on are the processed paragraphs from the corpus. The tags associated with those paragraphs are the label data. The dataset is randomly split 80%/20% between a training set and a testing set, respectively. Once sufficiently trained, the LSTM-RNN is used to classify untagged paragraphs before they are sent to the PIMS filter.

5   Design of a Study for Testing the Metastimulus Bond Effect

In order to determine if the system proposed makes a statistically significant increase in comprehension, we propose a study. Participants will be undergraduate engineering students that meet the prerequisites to take the courses in which the two texts comprising the corpus data from above are taught, but who have not yet begun the courses. The participants will be split between a control group and an experimental group.

Several weeks before the experiment is conducted, participants will be asked to take a test on the field of study. Afterward, they will be asked not to research or seek out knowledge in the field of study until after the experiment is complete. The passage of time between tests mitigates the biasing effect that prior knowledge of test questions may have on test performance.

Both the control and the experimental group will watch an entry level video lecture on the subject. Participants in the same group may watch the video together, but the groups will not be mixed. A common, simple PIMS will be shared among the participants in the experimental group.

Participants in the experimental group will wear haptic sleeves that will be controlled by the program. The sleeve will include a 6-by-4 array of actuators. Haptic feedback will be encoded using location, pattern, frequency, and intensity of vibration.

As the video lecture passes through each atom, the haptic actuators will apply corresponding metastimuli. Upon completion of the video lecture, members of both groups will be tested on their comprehension of the information provided in the video. Additionally, participants will complete a survey in which they describe their subjective experience of the video lecture.

Statistical analysis of the pre- and post-lecture test and survey results of both groups should provide insight into the hypothesis of a metastimulus bond effect.

6   Results

There are several subsystems required to generate metastimuli, as shown in Fig. 1. This work presents the overall architecture (Sect. 1) and develops the PIMS filter and classifier subsystems, presented in Sect. 3 and Sect. 4, respectively. The dialectical information architecture for the PIMS has been previously developed and presented [7,8]. The metastimulus bond effect hypothesis and related literature are given in Sect. 2. Further work will include testing this hypothesis and studying the effect of different actuation modalities for metastimuli. The design of one such study is presented in Sect. 5.

References

1. Fredembach, B., de Boisferon, A.H., Gentaz, E.: Learning of arbitrary association between visual and auditory novel stimuli in adults: the "bond effect" of haptic exploration. PloS one **4**(3), e4844 (2009)
2. Jung, J., et al.: Speech communication through the skin: design of learning protocols and initial findings. In: Marcus, A., Wang, W. (eds.) DUXU 2018. LNCS, vol. 10919, pp. 447–460. Springer, Cham (2018). https://doi.org/10.1007/978-3-319-91803-7_34
3. Critique of Pure Reason. Palgrave Macmillan, London (2007). https://doi.org/10.1007/978-1-137-10016-0_3
4. Karim, M.R.: Deep-learning-with-tensorflow, April 2017. https://github.com/PacktPublishing/Deep-Learning-with-TensorFlow/graphs/contributors
5. Lehoucq, R., Maschhoff, K., Sorensen, D., Yang, C.: Arpack software. https://www.caam.rice.edu/software/ARPACK/
6. Picone, R.A.: ricopicone/pims-filter: Pims filter, January 2020. https://doi.org/10.5281/zenodo.3633355
7. Picone, R.A.R., Lentz, J., Powell, B.: The fuzzification of an information architecture for information integration. In: Yamamoto, S. (ed.) HIMI 2017. LNCS, vol. 10273, pp. 145–157. Springer, Cham (2017). https://doi.org/10.1007/978-3-319-58521-5_11
8. Picone, R.A.R., Powell, B.: A new information architecture: a synthesis of structure, flow, and dialectic. In: Yamamoto, S. (ed.) HIMI 2015. LNCS, vol. 9172, pp. 320–331. Springer, Cham (2015). https://doi.org/10.1007/978-3-319-20612-7_31
9. Saerens, M., Fouss, F., Yen, L., Dupont, P.: The principal components analysis of a graph, and its relationships to spectral clustering. In: Boulicaut, J.-F., Esposito, F., Giannotti, F., Pedreschi, D. (eds.) ECML 2004. LNCS (LNAI), vol. 3201, pp. 371–383. Springer, Heidelberg (2004). https://doi.org/10.1007/978-3-540-30115-8_35
10. Scipy: Sparse eigenvalue problems with arpack. https://docs.scipy.org/doc/scipy/reference/tutorial/arpack.html
11. Sporleder, C., Lapata, M.: Automatic paragraph identification: a study across languages and domains. In: Proceedings of the 2004 Conference on Empirical Methods in Natural Language Processing, pp. 72–79 (2004)
12. Stein, B.E., Meredith, M.A., Wallace, M.T.: Development and neural basis of multisensory integration. In: The Development of Intersensory Perception: Comparative Perspectives, pp. 81–105 (1994)
13. Virtanen, P., et al.: SciPy 1.0-Fundamental Algorithms for Scientific Computing in Python. arXiv e-prints arXiv:1907.10121 (2019)
14. Webb, D.: danewebb/Tag-Classification: Initial release of Tag-Classification, January 2020. https://doi.org/10.5281/zenodo.3633402
15. Webb, D., Picone, R.A.: danewebb/tex-tagging: Initial release of Tex- Tagging, January 2020. https://doi.org/10.5281/zenodo.3633400
16. Wittgenstein, L., Anscombe, G.: Philosophical Investigations: The German Text, with a Revised English Translation. Blackwell, Oxford (2001)
17. Zaccone, G., Karim, M.: Deep Learning with TensorFlow: Explore Neural Networks and Build Intelligent Systems with Python, 2nd edn. Packt Publishing, Birmingham (2018)

How Gamification Increases Learning Performance? Investigating the Role of Task Modularity

Ayoung Suh[(⊠)] [iD] and Mengjun Li

School of Creative Media, City University of Hong Kong, Kowloon Tong, Hong Kong SAR
ahysuh@cityu.edu.hk, mengjunli3@um.cityu.edu.hk

Abstract. Generation Z is technologically aware, achievement focused, and keen on obtaining quick feedback and instant results from learning processes. Members of Generation Z absorb knowledge and information gathered from digital media and actively share their experiences and achievements with others through social media networks. This study explores how learning systems could be designed to enhance student performance considering the characteristics of Generation Z. It contributes to the gamification literature by (1) providing a nuanced understanding of the interplay among gamification affordances, task modularity, and learning performance, (2) developing a framework for a successful gamified learning system, and (3) generating design ideas for gamified learning applications that improve students' learning performance.

Keywords: Gamification · Gamification affordance · Task modularity · Learning performance

1 Introduction

Generation Z expects immediate gratification, quick feedback, and instant results from learning processes; these distinct characteristics of Generation Z "are challenging the traditional classroom teaching structure, and faculty are realizing that traditional classroom teaching is no longer effective with these learners" [1]. To meet the expectations and requirements of Generation Z students, new pedagogical approaches to engaging them in learning activities need to be developed.

One possible solution to this issue is gamification: the use of game design elements and mechanisms in nongame contexts. Research has shown that students try to achieve higher scores (or points) for an activity, reach higher levels, and win badges to demonstrate their performance in gamified contexts, thereby obtaining a sense of accomplishment.

However, despite the benefits of gamification, its effect on learning performance remains unclear. Some studies have found that gamification is effective in enhancing learning performance [2–7], whereas others have failed to prove any significant

© Springer Nature Switzerland AG 2020
D. D. Schmorrow and C. M. Fidopiastis (Eds.): HCII 2020, LNAI 12197, pp. 129–146, 2020.
https://doi.org/10.1007/978-3-030-50439-7_9

effects [8–13]. Consequently, the question of how gamification actually enhances student engagement, through which they perform better in the learning process, has yet to be fully discussed.

According to the gamification literature, task modularization is key to giving students granular and timely feedback in educational gamification. Researchers have suggested that learning activities (including in-class tasks and exercises) should be modularized with carefully developed reward structures. Otherwise, students receive rewards (e.g., points and badges) in an ad hoc manner, often leading to disengagement with gamified learning systems. Nonetheless, few studies have explored how task modularity shapes the effects of gamification on learning performance. To fill the gap in the literature, this study aims to explore how task modularity plays a role in enhancing learning performance in the gamified learning context.

2   Literature Review

2.1   Gamification in Education

The majority of the extant literature finds that gamification has positive effects on the learning performance of students [2, 3, 14, 15], although some researchers have identified insignificant or negative effects [8–10]. Table 1 summarizes the findings from previous studies on the effects of gamification on learning performance.

2.2   Gamification Affordances

Researchers have attempted to explain the mixed findings on the effects of gamification on learning performance based on affordance theory. Affordance theory provides a rationale for why some game elements results in different outcomes. That is, students are motivated to participate in learning activities by what the game elements afford and whether the affordances allow relevant actions to be performed. Accordingly, students' perceived affordances induced by game elements operate in their engagement in learning activities. Researchers have conceptualized major affordances through which students experience game-like playfulness and dynamics via gamification. Table 2 shows the expected gamification affordances in learning environments.

2.3   Task Modularity

Task modularization is one of the most popular and widespread teaching methods [26–28] applied in almost all subjects (e.g., computer science, engineering and medical education) [27–29]. As a pedagogical approach, task modularization organizes curricular materials for learning into relatively short blocks [30]. These short blocks contain various learning activities and emphasize different learning outcomes that serve the curricular objectives [31]. Modularized tasks benefit students who want to reduce the time and effort required to solve complex tasks. Along with game mechanics, modularized tasks enable students to focus on smaller challenges rather than taking on the burden of solving the entire problem.

Table 1. Summary of previous studies

	Game elements	Method	Participants	Focal variables	Key findings	Ref.
Positive effects	- Leaderboard - Badges	Survey	N = 136 university students	- Student engagement in a gamified course - Course type: gamified and non-gamified - Course performance	Performance was higher in the gamified course than in the non-gamified version	[2]
	- Leaderboard - Badges - Levels - Points	Survey	N = 139 university students	- Students' perceptions of gamification - Student performance and achievement	Positive student perceptions of gamification had a positive impact on their performance and achievement	[16]
	- Leaderboard - Badges - Points	Experiment	N = 56 secondary school students	- Course type: gamified and non-gamified - Learning performance	Performance was higher in the gamified course than in the non-gamified version	[4]
	- Leaderboard - Points - Narrative	Experiment	N = 54 elementary school students	- Gamified learning type: competitive, collaborative, and adaptive - Learning performance	- All three learning types contributed to increased student performance in math - The adaptive gamified type prompted the most significant improvement in learning performance	[3]

(continued)

Table 1. (*continued*)

Game elements	Method	Participants	Focal variables	Key findings	Ref.
- Leaderboard - Challenges/Achievement - Levels - Points - Narrative - Time constraints	Experiment	N = 261 university students	- Type of game element integration: Type1: points, levels, challenges/achievement, leaderboard Type2: type1 + narrative Type3: type1 + narrative + time constraint - Learning performance	Gamification had a positive impact on students' learning performance	[15]
- Leaderboard - Badges - Levels - Progress bar	Experiment	N = 96 university students	- Course type: gamified and non-gamified - Learning performance	Learning performance was higher in the gamified course than in the non-gamified version	[7]
- Levels - Points - Avatars - Content unlocking - Visual items "lives" as hearts	Experiment	N = 81 university students	- Learning system type: gamified and non-gamified - Learning outcomes (measured as learners' knowledge comprehension and task performance)	Participants using a gamified e-training system showed improvement in learning outcomes when compared to those using the non-gamified version	[6]

(*continued*)

Table 1. (*continued*)

	Game elements	Method	Participants	Focal variables	Key findings	Ref.
Insignificant effects	- Leaderboard - Levels	Experiment	N = 55 elementary school students	- Learning system type: gamified and non-gamified - Reading comprehension performance	Insignificant difference in reading comprehension performance between the gamified and non-gamified system groups	[10]
	- Leaderboard - Badges - Points	Experiment	N = 63 university students	- Learning platform type: gamified and non-gamified - Academic performance	Insignificant difference in academic performance between the gamified and non-gamified platform groups	[8]
	- Badges - Points - Ranking	Case study	N = 16 secondary school students	- Learning platform type: gamified and non-gamified - Learning performance	Insignificant difference in learning performance between the gamified and non-gamified platform groups	[17]
	- Badges	Experiment	N = 281 university students	- Learning system type: gamified and non-gamified - Academic performance	Insignificant difference in academic performance between gamified and non-gamified system groups	[18]

(*continued*)

Table 1. (*continued*)

	Game elements	Method	Participants	Focal variables	Key findings	Ref.
	- Badges	Experiment	N = 126 university students	- Learning platform type: gamified and non-gamified - Learning performance	Insignificant difference in learning performance between the gamified and non-gamified platform groups	[9]
	- Points	Experiment	N = 1,101 massive open online courses participants	- Learning platform type: gamified and non-gamified - Learning performance	Insignificant difference in learning performance between the gamified and non-gamification platform groups	[19]
Negative Effects	- Leaderboard - Badges	Experiment	N = 80 university students	- Course type: gamified and non-gamified - Academic performance	Students on the gamified course showed less motivation and achieved lower final exam scores than the non-gamified group	[13]
	- Leaderboards	Experiment	N = 80 university students (all female)	- Leaderboard type: type1: men held the majority of the top positions type2: women held the majority of the top positions type3: no leaderboard - Academic performance	A leaderboard on which women held the majority of top positions can negatively influence students' academic performance	[20]

Table 2. Gamification affordances

Affordance	Description	Ref.
Reward	*Reward* refers to an affordance that enables students to obtain points, levels, and badges as a pay-off when they complete pre-designed tasks. As digital form, these game elements give tangible rewards that provide granular and timely feedback as a response to students' activities	[21–24]
Status	*Status* refers to an affordance that encourages students to increase their levels by getting more achievements or reaching targets. Tiered levels provide students with opportunities to track their progress and feel a sense of self-progress when they reach a milestone	[21, 22, 42]
Competition	*Competition* is an affordance that enables students to compare their performance with that of others. Certain game elements provide students with opportunities to compete with others (e.g., leaderboards). Students are expected to try to achieve higher scores in performing their learning activities	[21–23, 25, 42]
Self-expression	*Self-expression* refers to an affordance that encourages students to express who they are. Achievements (badges, trophies, or ranking in a leaderboard) in a gamified learning platform often serve as a means for identity expression that distinguishes them from others	[21, 22, 42]

Many researchers have found that task modularization can bring positive outcomes to students [27, 28, 31, 32]. Specifically, task modularization has positive effects on promoting students' learning, such as improving learning interests [27], motivations [32], participation [27], learning efficiency, [28] and performance [27, 31, 32]. Moreover, task modularization is effective for material science and engineering students' knowledge adaptation, learning experience stimulation, self-study ability cultivation and cognitive ability development [26]. In addition, the modular approach is helpful for students to develop critical [31] and independent thinking [31, 33].

The advantages of task modularization have been identified in a number of education studies. For example, modularized tasks reduce the overall difficulty and improve the completion success ratio [34], and students have been shown to pick up and recognize a module's patterns quickly, allowing them to better understand what they are expected to do in the subsequent modules [35]. Furthermore, task modularity enables a clear and comprehensive view of the content and structure of learning activities and provides flexibility as a result.

Conversely, some studies have reported the disadvantages of task modularization; for example, it can result in knowledge compartmentalization, which makes it difficult for students to build theoretical and methodological links between the modules [29]. In addition, modularity may hinder the creative learning processes because it can inhibit deep comprehension, divergent thinking, risk taking, and reflection [36]. Modular tasks lack the space for slow learning and risk taking because they pressure students into succeeding in the short term by avoiding poor results and moving on [37].

These conflicting findings imply that task modularity may not have a direct influence on students' learning performance but may instead play a moderating role in shaping the effects of gamification on learning processes. Table 3 summarizes the findings of previous studies on the advantages and disadvantages of task modularization.

Table 3. Summary of previous studies

	Subjects	Key findings	Ref.
Advantages	Materials science and engineering	Module learning is effective for students' knowledge adaptation, learning experience stimulation, self-study abilities cultivation and cognitive abilities development	[26]
	Dermatology	Module learning improves students' participation, learning interests, and learning performance	[27]
	Foreign languages	Module learning is effective for students' foreign languages learning, independent and critical thinking development	[31]
	Physics	Module learning improves students' learning outcomes and motivations	[32]
	Entrepreneurship	Module learning improves students' learning outcomes and cultivates self-study abilities	[38]
	Basic electronics	Module learning improves students' learning outcomes	[39]
	Computer Science	Module learning is beneficial in filling students' knowledge gaps	[40]
	Life-long education	Module learning is effective for students' independent thinking development	[33]
Disadvantages	Engineering	Module learning can lead to knowledge compartmentalization	[29]
	General	Module learning is problematic in cultivating students' risk-taking abilities in learning	[37]
	General	Module learning can lead to study overload	[41]
	Creative education	Module learning can lead to the inhibition of deep learning, divergent thinking, risk-taking and reflection	[36]

3   Theory Development

Our literature review reveals that gamification in education should take task modularization into consideration. As students perceive the extent to which tasks are modularized, the concept of task modularity needs to be incorporated into a theoretical model that explores the effects of gamification affordances on learning performance. By combining the gamification affordance model and task modularity, we propose the framework for the successful gamified learning system as shown in Fig. 1.

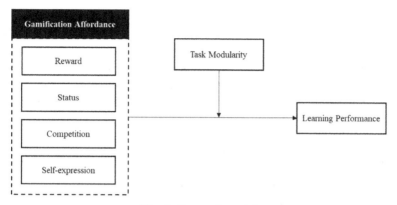

Fig. 1. Research model

3.1 Gamification Affordance and Learning Performance

Building on Suh et al. [42], this study explores the four major gamification affordances (rewards, status, competition, and self-expression) and their effects on learning performance. According to the model, a reward affordance is engendered from game elements, such as points, levels, and badges, providing granular and immediate feedback on the task performed. When students complete a pre-defined task within a gamified learning system, they obtain an immediate reward. Research has found that prompt rewards from a gamified system serve as informational feedback on their short-term performance, and such granular feedback encourages students to set new goals and focus on their current activity [24, 43]. Obtaining rewards enables students to track their performance which helps them experience a sense of achievement. We therefore propose the following hypothesis:

Hypothesis 1: Reward will be positively associated with learning performance.

A status affordance refers to the extent to which a student perceives that he or she can level up his or her standing within a gamified learning system; there is greater status affordance when students are willing to challenge themselves to reach higher levels. Some existing research argues that providing the possibility of status enhancement enables individuals to commit themselves to completing their tasks at hand [10]. When, for example, students feel that they are able to upgrade their game levels by completing given tasks, they are willing to participate in further, more challenging learning activities.

Hypothesis 2: Status will be positively associated with learning performance.

Leaderboards engender competition affordance, enabling students to compare their performance with that of others. Studies have shown that competitive opportunities motivate learners to achieve better personal performance. As such, well-designed competition in gamified contexts can motivate students to engage in learning tasks more actively.

Hypothesis 3: Competition will be positively associated with learning performance.

Last in terms of affordances, self-expression entails the extent to which students feel that they are able to create a unique self-identity. Badges used in a gamified learning platform often serve as a means of identity expression that distinguish individual students from others, and the ability to express a unique self is expected to increase levels of perceived co-presence among other learners, which in turn increase their motivation for learning.

Hypothesis 4: Self-expression will be positively associated with learning performance.

3.2 The Moderating Role of Task Modularity

Successful gamification requires well-structured task modularity components [44]. Task modularization is a process that breaks complex learning tasks down into their modular components. Short modularized task blocks, which contain specific learning activities and emphasize different learning outcomes, reduce the time and effort required by students to solve the overarching complex task. Along with game mechanics, modularized tasks enable students to focus on smaller challenges rather than taking on the burden of solving the entire problem. Gamification in learning requires granular feedback, through which students are supposed to experience a sense of achievement and in turn stimulate their motivation to participate in learning activities. Accordingly, the modularized course curriculum is expected to enable students to have more chances to be involved in getting rewards, upgrade their status, and win the competitions. Thus, we posit that task modularity will strengthen the effects of gamification affordances on learning performance.

Hypothesis 5: Task modularity will positively moderate the effects of gamification affordances on learning performance.

4   Methods

4.1   Data Collection

To test the proposed model, this study collect data using a survey from 56 students enrolled in a gamified course in the subject of data science taught by one of the authors in 2019. Game elements, such as points, badges, and leaderboards, were implemented in an online platform through which students were able to monitor their points, status, and rank through leaderboards. In this course, content was modularized into subtopic chunks, and the students were asked to test their knowledge using an online quiz after completing each module. Once a module was complete, each student received a badge based on their scores received and could monitor their overall performance through the achievement leaderboard. Figure 2 shows the design features of this gamified online learning platform.

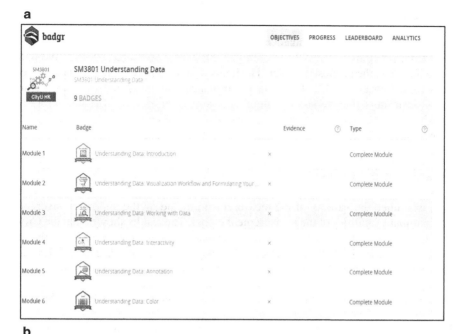

Fig. 2 (a) A screenshot of the modules and badges (b) A screenshot of the leaderboard showing student ranking

4.2 Measures

The measurement items were taken from the existing literature, and the items were adapted from previous work. All research variables were measured using a five-point Likert scale to test the proposed model. The items for the gamification affordances were adapted from Suh et al. [42]. Learning performance was measured by the students' subjective learning effectiveness by adopting from Lee et al. [45]. The items for task modularity were adapted from Saltz et al. [46]. The measurement items are presented in the Appendix.

5 Results and Analysis

We first examined the validity of the measurement, including the internal, convergent, and discriminant validities of the measurement model. The results showed that the Cronbach's alpha values and factor loadings of all the measurement items were above recommended value 0.7 [47]. Then, we examined the average variance extracted (AVE) from each construct and found that all AVEs were higher than the recommended value of 0.5 [48].

Given that our model includes the moderating effects of task modularity between gamification affordances and learning outcomes, we employed hierarchical regression analysis. Before testing the model, we centered the gamification affordance (i.e., reward, status, competition, and self-expression) and task modularity variables before generating the interaction terms [49]. We subtracted the sample mean from each independent variable to cause the variables to have an adjusted mean value of zero with an unchanged sample distribution. Subsequently, we multiplied the centered task modularity score with each of the four centered gamification affordance scores to compute the four interaction terms. We entered each of the four interaction terms in a separate step after examining the direct effects of gamification affordances on the learning performance. The baseline model accounted for 20.7% the variance in learning performance. The results revealed that status and competition affordances were positively associated with learning performance (status: $\beta = 0.212$, $p < 0.01$; competition: $\beta = 0.134$, $p < 0.05$) in the baseline model (see Table 4). The interaction model accounted for 31.2% of the variance in learning performance. The results of the hierarchical regression analysis suggest that the effects of gamification affordances on learning performance were shaped by task modularity.

The moderating effect of task modularity was examined by adding the interaction terms in the regression model (see the interaction model in Table 4). The interaction model would be supported if the addition of the interaction terms resulted in a statistically significant improvement over the regression model containing the main terms. Results show that task modularity positively moderated the relationship between status and learning performance ($\beta = 0.209$, $p < 0.05$) and the relationship between competition and learning performance ($\beta = 0.111$, $p < 0.05$). The results indicate that high task modularity (relative to low task modularity) should be able to exploit the use of game elements more effectively to enhance students' learning motivation and performance.

Table 4. The results of the hierarchical regression analysis

		Baseline model	Interaction model
		Learning performance	Learning performance
Control variable	Age	0.069	0.030
	Gender	0.006	0.014
	Academic year	0.122	0.022
	Usage duration	0.008	0.001
Gamification affordance	Reward	0.048	0.050
	Status	**0.212****	**0.292*****
	Competition	**0.134***	**0.114***
	Self-expression	0.009	0.001
Moderator	Task modularity		**0.202****
Moderating effects	Task modularity × Reward		−0.154
	Task modularity × Status		**0.209***
	Task modularity × Competition		**0.111***
	Task modularity × Self-expression		0.012
R^2		0.207	0.312
R^2 Δ			0.105

*$p < .05$; **$p < .01$; ***$p < .001$

6   Discussion

This study explores the interplay among gamification affordances, task modularity, and learning performance. It aims to highlight the role of task modularity in enhancing learning performance. Our theoretical framework shows that gamified learning systems should be designed in a way that instructors could easily incorporate modularized task components into learning activities through the system. Merely providing gamification affordances without modularizing task activities may not ensure the success of gamified learning systems.

6.1   Theoretical Implications

This study makes several contributions to the gamification literature. First, it responds to the call for a more nuanced understanding of the effects of gamification on learning performance. The findings indicate that students were motivated to engage learning activities when they perceive that they are able to level up their status and compete with others through a gamified learning system. We conjecture that providing proper affordances, such as status and competition, by using game elements (badges and leaderboards) significantly enhances learning performance. Furthermore, our work demonstrates that the reward affordance has an insignificant influence on learning performance, which supports the notion of gamification merely offering rewards does not engage users.

Second, our work elaborates on the concept of task modularity in the gamification context. Extant theories and models that explore the effects of game elements on learning performance have ignored the role of task modularity. Although gamification requires well-structured learning activities to provide immediate informational feedback and sense of achievements, prior literature has not taken a construct that captures the extent to which students perceive their learning tasks are modularized into consideration. Our conceptualization and operationalization of task modularity will help researchers develop, test, and extend the existing gamification models in educational settings.

6.2 Practical Implications

This study has practical implications for instructors seeking new pedagogical ways to encourage students to participate in learning activities. The results imply that instructors should carefully modularize learning content and relevant activities when employing a gamified learning system in their teaching. By breaking complex tasks into small chunk of activities along with game mechanics (e.g., points and badges), instructors are able to provide granular feedback and encourage students to join challenges. System designers may consider including technological functions that help instructors easily modularize learning tasks. The current design features of online learning systems may not be sufficient to help instructors effectively organize learning activities into modularized tasks.

It is noteworthy that status and competition affordances significantly increased learning performance. Our findings suggest that developers of online learning systems should consider certain game elements, such as points, levels, badges, and leaderboards, as effective tools that can stimulate students' motivation to increase their status and join competition. Research has suggested that greater competition affordance can be engendered by proper design features in a gamified system [42]. Rather than merely adding leaderboards that display students' performance into an online learning system, online learning platforms should be designed in a way that students can have diverse opportunities to join challenges display their performance as they want.

Acknowledgement. This research was supported by the UGC Teaching and Learning Grant titled "Devel-oping Multidisciplinary and Multicultural Competences through Gamification and Challenge-based Collaborative Learning." This research was also partly supported by Teaching Development Grant (No. 6000666) from City University of Hong Kong and the grant from the Centre for Applied Computing and Interactive Media (ACIM) of School of Creative Media, City University of Hong Kong.

Appendix

Category	Construct	Items	Reference
All items were measured by using a 5-point Likert scale: 1 = strongly disagree and 5 = strongly agree			
Gamification affordance	Reward	**Prompt:** The online learning system (Canvas) offers me the possibility to: 1. obtain points as a reward for my activities 2. accumulate points I have gained 3. obtain more points if I try harder	Adopted from [42]
	Status	**Prompt:** The online learning system (Canvas) offers me the possibility to: 1. have a higher status than others 2. be regarded highly by others 3. try to increase my status	Adopted from [42]
	Competition	**Prompt:** The online learning system (Canvas) offers me the possibility to: 1. compete with others 2. compare my performance with that of others 3. to threaten the status of others by my active participation	Adopted from [42]
	Self-expression	**Prompt:** The online learning system (Canvas) offers me the possibility to: 1. express my identity through game elements 2. express myself in a way that I want 3. present myself to be distinguished from others	Adopted from [42]
Learning performance		1. I learned a lot of data analysis methods in the topics 2. I gained a good understanding of the basic concepts of data analysis and representation 3. I learned to identify the main and important issues of the topics 4. The learning activities were meaningful	Adapted from [45]
Task modularity		1. The module tasks have a clearly defined goal 2. The module tasks take a reasonable amount of time to complete 3. The module tasks were divided into chunks of appropriate size and scope	Adapted from [46]

References

1. Skiba, D.J., Barton, A.J.: Adapting your teaching to accommodate the net generation of learners. Online J. Issues Nurs. **11**(2), 15 (2006)

2. Tsay, C.H.H., Kofinas, A., Luo, J.: Enhancing student learning experience with technology-mediated gamification: an empirical study. Comput. Educ. **121**, 1–17 (2018)
3. Jagušt, T., Botički, I., So, H.-J.: Examining competitive, collaborative and adaptive gamification in young learners' math learning. Comput. Educ. **125**, 444–457 (2018)
4. Zainuddin, Z.: Students' learning performance and perceived motivation in gamified flipped-class instruction. Comput. Educ. **126**, 75–88 (2018)
5. Ding, L.: Applying gamifications to asynchronous online discussions: a mixed methods study. Comput. Hum. Behav. **91**, 1–11 (2019)
6. Park, J., Liu, D., Mun, Y.Y., Santhanam, R.: GAMESIT: a gamified system for information technology training. Comput. Educ. **142**, 1–19 (2019)
7. Huang, B., Hew, K.F., Lo, C.K.: Investigating the effects of gamification-enhanced flipped learning on undergraduate students' behavioral and cognitive engagement. Interact. Learn. Environ. **27**(8), 1106–1126 (2019)
8. Özdener, N.: Gamification for enhancing Web 2.0 based educational activities: the case of pre-service grade school teachers using educational Wiki pages. Telemat. Inform. **35**(3), 564–578 (2018)
9. Kyewski, E., Krämer, N.C.: To gamify or not to gamify? An experimental field study of the influence of badges on motivation, activity, and performance in an online learning course. Comput. Educ. **118**, 25–37 (2018)
10. Chen, C.-M., Li, M.-C., Chen, T.-C.: A web-based collaborative reading annotation system with gamification mechanisms to improve reading performance. Comput. Educ. **144**, 1–17 (2020)
11. Göksün, D.O., Gürsoy, G.: Comparing success and engagement in gamified learning experiences via Kahoot and Quizizz. Comput. Educ. **135**, 15–29 (2019)
12. DomíNguez, A., Saena-de-Navarrete, J., de-Marcos, L., Fernández-Sanz, L., Pagés, C., Martínez-Herráiz, J.-J.: Gamifying learning experiences: practical implications and outcomes. Comput. Educ. **63**, 380–392 (2013)
13. Hanus, M.D., Fox, J.: Assessing the effects of gamification in the classroom: a longitudinal study on intrinsic motivation, social comparison, satisfaction, effort, and academic performance. Comput. Educ. **80**, 152–161 (2015)
14. Yıldırım, İ., Şen, S.: The effects of gamification on students' academic achievement: a meta-analysis study. Interact. Learn. Environ. 1–18 (2019)
15. Legaki, N.Z., Xi, N., Hamari, J., Assimakopoulos, V.: Gamification of the future: an experiment on gamifying education of forecasting. In: the 52nd Hawaii International Conference on System Sciences (HICSS), USA, pp 1813–1822. IEEE (2019)
16. Davis, K., Sridharan, H., Koepke, L., Singh, S., Boiko, R.: Learning and engagement in a gamified course: investigating the effects of student characteristics. J. Comput. Assist. Learn. **34**(5), 492–503 (2018)
17. Pedro, L.Z., Lopes, A.M.Z., Prates, B.G., Vassileva, J., Isotani, S.: Does gamification work for boys and girls? An exploratory study with a virtual learning environment. In: The 30th Annual ACM Symposium on Applied Computing, Spain, pp. 214–219. ACM (2015)
18. Hakulinen, L.T., Auvinen, T., Korhonen, A.: Empirical study on the effect of achievement badges in TRAKLA2 online learning environment. In: Learning and Teaching in Computing and Engineering, China, pp. 47–54. IEEE (2013)
19. Coetzee, D., Fox, A., Hearst, M.A., Hartmann, B.: Should your MOOC forum use a reputation system? In: The 17th ACM Conference on Computer Supported Cooperative Work & Social Computing, USA, pp. 1176–1187. ACM (2014)
20. Christy, K.R., Fox, J.: Leaderboards in a virtual classroom: a test of stereotype threat and social comparison explanations for women's math performance. Comput. Educ. **78**, 66–77 (2014)

21. Bunchball, I.: Gamification 101: an introduction to the use of game dynamics to influence behavior (2010). https://www.bunchball.com/gamification101
22. Werbach, K., Hunter, D.: For the Win: How Game Thinking Can Revolutionize Your Business. Wharton School Press, Pennsylvania (2012)
23. Suh, A., Wagner, C., Liu, L.: The effects of game dynamics on user engagement in gamified systems. In: The 48th Hawaii International Conference on System Sciences (HICSS), USA, pp. 672–681. IEEE (2015)
24. Hamari, J.: Do badges increase user activity? A field experiment on the effects of gamification. Comput. Hum. Behav. **71**, 469–478 (2017)
25. Liu, D., Li, X., Santhanam, R.: Digital games and beyond: what happens when players compete. MIS Q. **37**(1), 111–124 (2013)
26. Guido, R.M.D.: Evaluation of a modular teaching approach in materials science and engineering. Am. J. Educ. Res. **2**(11), 1126–1130 (2014)
27. Karthikeyan, K., Kumar, A.: Integrated modular teaching in dermatology for undergraduate students: a novel approach. Indian Dermatol. Online J. **5**(3), 266–270 (2014)
28. Sejpal, K.: Modular method of teaching. Int. J. Res. Educ. **2**(2), 169–171 (2013)
29. Palmer, S.: Engineering flexible teaching and learning in engineering education. Eur. J. Eng. Educ. **26**(1), 1–13 (2001)
30. Zahorian, S., Swart, W., Lakdawala, V., Leathrum, J., Gonzalez, O.: A modular approach to using computer technology for education and training. Int. J. Comput. Integr. Manuf. **13**(3), 286–297 (2000)
31. Rybushkina, S.V., Sidorenko, T.V.: Modular approach to teaching ESP in engineering programs in Russia. In: International Conference on Interactive Collaborative Learning (ICL), Italy, pp. 105–108. IEEE (2015)
32. Yuliono, S.N., Sarwanto, S., Cari, C.: Physics-based scientific learning module to improve students motivation and results. J. Educ. Learn. **12**(1), 137–142 (2018)
33. Zhou, W., Liu, B.-B.: The exploration of the skill training modularization in facilitating the life-long education. In: The 5th International Conference on Computer Science & Education, China, pp. 656–657. IEEE (2010)
34. Liu, B.-B., Zhou, W.: Exploration of reverse-based modular skills training system. In: the 7th International Conference on Computer Science & Education (ICCSE), Australia, pp. 1951–1953. IEEE (2012)
35. Chen, Y.-T.: Integrating anchored instructional strategy and modularity concept into interactive multimedia powerpoint presentation. Int. J. Phys. Sci. **7**(1), 107–115 (2012)
36. Dineen, R., Collins, E.: Killing the goose: conflicts between pedagogy and politics in the delivery of a creative education. Int. J. Art Des. Educ. **24**(1), 43–52 (2005)
37. Yorke, M., Knight, P.: Self-theories: some implications for teaching and learning in higher education. Stud. High. Educ. **29**(1), 25–37 (2004)
38. Yulastri, A., Hidayat, H.: Developing an entrepreneurship module by using product-based learning approach in vocational education. Int. J. Environ. Sci. Educ. **12**(5), 1097–1109 (2017)
39. Hamid, M.A., Aribowo, D., Desmira, D.: Development of learning modules of basic electronics-based problem solving in vocational secondary school. J. Pendidik Vokasi **7**(2), 149–157 (2017)
40. Ghemri, L., Yuan, S.: Increasing students awareness of mobile privacy and security using modules. J. Learn. Teach. Digit. Age **4**(2), 1–9 (2019)
41. Dejene, W.: The practice of modularized curriculum in higher education institution: active learning and continuous assessment in focus. Cogent Educ. **6**(1), 1–16 (2019)
42. Suh, A., Cheung, C.M.K., Ahuja, M., Wagner, C.: Gamification in the workplace: the central role of the aesthetic experience. J. Manag. Inf. Syst. **34**(1), 268–305 (2017)
43. Park, J., Kim, S., Kim, A., Mun, Y.Y.: Learning to be better at the game: performance vs. completion contingent reward for game-based learning. Comput. Educ. **139**, 1–15 (2019)

44. Baldwin, C.Y., Clark, K.B.: Design Rules: The Power of Modularity. MIT Press, Cambridge (2000)
45. Lee, E.A.L., Wong, K.W., Fung, C.C.: How does desktop virtual reality enhance learning outcomes? A structural equation modeling approach. Comput. Educ. **55**(4), 1424–1442 (2010)
46. Saltz, J.S., Heckman, R., Crowston, K., You, S., Hedge, Y.: Helping data science students develop task modularity. In: the 52nd Hawaii International Conference on System Sciences (HICSS), University of Hawaii at Manoa, USA, pp 1–10 (2019)
47. Hair, J.F., Black, W.C., Babin, B.J., Anderson, R.E., Tatham, R.L.: Multivariate Data Analysis. Pearson, Upper Saddle River (2006)
48. Fornell, C., Larcker, D.F.: Structural equation models with unobservable variables and measurement error: algebra and statistics. J. Mark. Res. **18**(3), 382–388 (1981)
49. Aiken, L.S., West, S.G., Reno, R.R.: Multiple Regression: Testing and Interpreting Interactions. Sage, New York (1991)

Increasing Engagement in a Cyber-Awareness Training Game

Robert Wray[1(✉)], Lauren Massey[1], Jose Medina[1], and Amy Bolton[2]

[1] Soar Technology, Inc., Ann Arbor, MI 32817, USA
wray@soartech.com
[2] Office of Naval Research, Arlington, VA 33303, USA

Abstract. Today's workforce is generally uneducated in cybersecurity, largely complacent, and fails to embrace the reality that 'a risk to one is a risk to all'. A cyber-aware mindset must be instilled in improved training across the workforce. We are developing a training game designed to improve cyber awareness with the goal of inculcating "cyber mindset" to reduce vulnerabilities to and increase vigilance toward cyber threats. The game stresses critical thinking, intellectual engagement, and countering cognitive biases. We introduce the design and implementation of the training game. Creating effective cyber awareness training is often challenging due to resistance and disinterest from target populations. We outline the current implementation of the training game and introduce additional features or "mechanics" we have developed and also continue to investigate to attempt to improve the game's effectiveness in developing a cyber mindset.

Keywords: Cyber awareness · Cognitive bias · Personalized learning · Adaptive learning

1 Introduction

In today's workforce, most employees operate in the cyber domain, with all systems and all stored data connected to networks at risk. While networked work has contributed to large increases in individual and aggregate productivity [1], networks introduce new risks to individuals and organizations. Every individual now represents a potential attack vector for nefarious actors. A drumbeat of recurring accounts of successful attacks against individuals, companies and organizations, and governments redounds in the media each week [2–12].

While the conduct of cyber warfare increases in scope and sophistication, training for this new reality of risk exposure and vulnerabilities is not keeping pace. The workforce is generally uneducated in cybersecurity, largely complacent, and fails to embrace the reality that 'a risk to one is a risk to all'. To remain vigilant in the face of these threats, a cyber-aware mindset must be instilled in improved training across the workforce [13].

The general workforce typically views cybersecurity as a nuisance that unnecessarily complicates their mission and for which they have little to no direct responsibility. Cybersecurity is "someone else's problem." National and industrial security requires a

© Springer Nature Switzerland AG 2020
D. D. Schmorrow and C. M. Fidopiastis (Eds.): HCII 2020, LNAI 12197, pp. 147–158, 2020.
https://doi.org/10.1007/978-3-030-50439-7_10

more cyber-savvy workforce, where each individual is attuned to the threat, embraces their role in defense, and is able to respond quickly and effectively.

We are researching and developing a training-game prototype designed to improve cyber awareness. We hypothesize that inculcating "cyber mindset" will reduce vulnerabilities to and increase vigilance toward cyber threats without necessitating the development of sophisticated cyber knowledge and expertise. The current training game stresses critical thinking, intellectual engagement, and countering cognitive biases.

This paper introduces the design and implementation of the training game. As we discuss further below, the game allows the player to participate in the game as an attacker, which we hypothesize has a number of benefits for learning. More generally, however, creating effective cyber awareness training is often challenging, for some of the reasons outlined above. Thus, we are investigating design options for subsequent versions of the game that could support greater learner engagement and potential carryover to the everyday activities of learners. Below, we describe the current implementation of the game and then introduce some of the additional features or "mechanics" we are investigating to attempt to improve the game's effectiveness in developing a cyber mindset. We review three types of design options: specialization for player type, customization based on player demography, and active monitoring of engagement.

2   Training for Cyber Mindset

In previous work, we outlined the theoretical and empirical foundations for the design of the cyber mindset game, focusing on general review of the literature and identification of specific training objectives from that review [13]. In this section, we briefly review other attempts to train cyber awareness and how those attempts inform our approach. We summarize some design elements that we have already included in the game to support engagement and effective learning. The following section then describes the implementation of the game itself.

2.1   Creating Effective Cyber-Awareness Training Is Difficult

Organizations have employed various training methods to raise awareness to employees about the dangers of the internet, network intrusions, and social engineering, typically promoting the notion that staff are the first line of defense against such attacks. Commonly, the training that is employed focuses on cyber awareness training that promotes that staff know their roles and responsibilities to protect the organization's information. Most existing training does not ensure that the employees are competent in that role of protection or even than they undertake hands-on training [14].

There are various types of cyber training tools that are currently used for general population consumption, including multimedia static content, on-site presentations, and passive computer-based training (CBT). A company may utilize one or several of these tools in their security training program depending on budget and the amount time allotted to training [15]. Static content and instructor led courses tend to have limited individualized effect and any effect may be highly dependent on the content and delivery [16]. Computer-based training allows for some customization to individual learners, but can typically be completed without significant engagement. These methods typically fail to motivate and engage [15], may not connect learning goals to individual and corporate responsibilities [17], and have been demonstrated to commonly fail to change staff behavior even after "successful" completion of the training [18].

The potential use of games for cyber training is not novel. Multiple games have been developed in research labs, although few commercial serious games have made the transition from research labs [19]. The potential advantage of games to introduce basic cyber principles is that they address the challenges of traditional training methods. Games provide an environment in which trainees can actively participant in the learning process, which may motivate and engage them to pay attention and learn [19, 20]. This motivation and hands-on learning setting could then aid transfer [21] of knowledge to daily adherence without the use of "fear as motivation" that is used in many traditional staff-level cyber training programs.

2.2 Design Elements in Support of Engagement and Learning

The design of our cyber awareness training game draws on two primary training-design elements. First, the game introduces conceptual knowledge about cyber threats, seeking to create and and/or extend learner knowledge structures, which, in turn, modulate attention and metacognition to increase vigilance and awareness of potential of cyber attacks. The primary goal of the game is to facilitate development and transfer (to everyday life) of this new awareness or *cyber-aware mindset*. Examples introduced in the game include the cyber-attack concepts such an attack vector and social engineering [17, 22] and specific examples of higher-level concepts, such as phishing and open-source intelligence gathering (a preliminary activity in social engineering).

A second design element focuses emphasizes the need for attention and practice to support the development of these new knowledge structures and the skill to retrieve and apply them in appropriate context. Most existing cyber awareness tools fail to help users make the leap from awareness to changing behavior, limiting their effectiveness [23]. In contrast, our game adopts turn-based but realistic and familiar depictions of household and office environments, rendered in three-dimensions. A game environment generally supports engagement and immersion; such depictions have been shown to offer increases in attention and retention in comparison to non-immersive environments presenting comparable learning content [24, 25]. Further, one of the advantages of the narrative-based game environment is that it encourages replay and additional exploration, increasing exposure to the core concepts and time on task, both of which are correlated with improved learning outcomes [26].

The use of a simulation game itself, rather than more traditional media, has also been observed to have a marked increase on learner self-efficacy (confidence) [27]. This confidence may help learners transfer knowledge gained in the game to greater cyber awareness in everyday life. Similarly, the game employs tiered badges, to provide explicit performance goals within the game. Performance goals are a self-regulatory strategy known to exert a comparatively large learning impact.

2.3 An Active, Adversarial Role for Learners

In the initial scenario we developed, we employed these practices but noted in piloting that participants' interest quickly flagged. The game allowed users to experience various kinds of cyber intrusions, both directly and indirectly, but the participants role was overly passive; they were "waiting for the attack" and had little agency, even though the game was rendered in an interactive, immersive environment.

To attempt to increase engagement and impact, the latest scenario places the trainee in an adversarial role assigned to attack or compromise various systems (within the game environment). As an aggressor, players view potential cyber events from a vantage point less subject to biases; we hypothesize this vantage will result in greater awareness of system vulnerabilities. This change in perspective supports three distinct goals that are directly (and indirectly) designed to promote engagement:

- **Engagement:** Many players are likely to be curious about the possibility of being a cyber attacker and will be surprised by the opportunity to play this role. These elements help engage the player in a subject that is often viewed as boring.
- **Disposition toward action:** A cyber-aware mindset requires pro-active decision and action. Because the player's role is to undertake many actions, this design approach helps associate a propensity for action with a cyber-aware mindset. In the game, the actions are to attack. However, the player is exposed to the ways in which they may be vulnerable in the real-world. Further, they observe differences in the ways various characters they encounter in the game have and have not limited their exposure to potential cyber attacks.
- **Relatability:** The tools and applications introduced in the game are analogs to applications that are likely to be familiar to many players (e.g., online social media platforms). Having players imagine how to exploit applications without introducing specific vectors or vulnerabilities makes the experience relatable to players' everyday activities without being tied to specific tools and vulnerabilities. The intent behind the relatability of the experience is to amplify the cyber-aware mindset around the players' use of such tools outside of the game.

3 The Cyber-RAMPART Training Game

This section overviews the current implementation of the adversary attack scenario. In this scenario, the user engages within a chat application with "Hacker Mo," a virtual character that cajoles the user into executing an attack and then explains, in a guided dialogue, the various options that can be used for an attack.

Figure 1 shows a screenshot from an early part of the dialogue with Hacker Mo. As outlined above, the game is set in a three-dimensional environment. In addition to the monitor, which is where most of the game takes place, there are also some objects in the environment, such as tablet, USB key, and phone. These become relevant in later stages of the scenario.

The orange, arrowed sidebar on the right side of the screen is used to convey detailed information about the concepts Hacker Mo introduces, as well as historical examples. Information in this sidebar "pops out" when Hacker Mo mentions it in the game and the

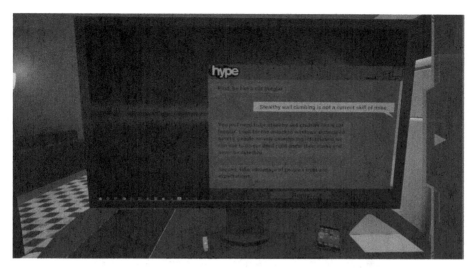

Fig. 1. Hacker Mo describes the attacker mindset.

user can review it at any time as well. This mechanic attempts to balance providing a lot of detailed information about cyber attacks to the player, while also minimizing the dynamic, immersive experience of roleplaying as an attacker.

As the scenario progresses, Hacker Mo describes various kinds of ways attacks can be conducted, what kinds of organizations and people might be more or less vulnerable, etc. The second phase of the game involves the player identifying a specific target to attack (an organization) and a particular person associated with that organization. The player makes these choices by conducting a simulation of open-source intelligence gathering, looking at simulated websites, social media sites, and even a phone book to find information.

Figure 2 illustrates an example from the scenario. The top image shows the "quikpix" social media page of a university student. The user can scroll thru the individual posts and glean information about the character. In this case, this person is a student at a local university, likes classic cars, and recently won an award at the university. The user has also determined her cell phone number.

In order to make the game less memory intense and also not require recording information manually, we added a "tablet" that allows the user to readily capture "notes" from the open-source intelligence gathering. In the game, the user clicks on the content of social media and website information and these are automatically added to the correct entries in the notebook. This approach reduces overall learner cognitive load and is designed to help the learner focus on the concepts of cyber attacks rather than the specific information found for each character.

The third and final phase of the scenario involves actually executing an attack. The scenario supports attacks via a phishing email, a text-based attack (vishing), and malware on a USB (sent by mail). When an attack is successful, Hacker Mo turns on a web camera, allowing the user to visualize the impact of the attack on the target (Fig. 3). The success

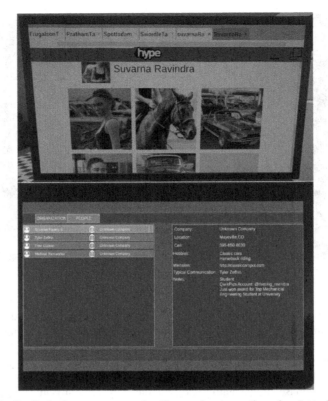

Fig. 2. Gathering open-source intelligence in preparation of an "attack."

of the attack depends on whether the player follows Hacker's Mo guidance in choosing more vulnerable targets as well correctly gathering and using open-source intelligence. For example, for the target illustrated in Fig. 2, a phishing email that focused on "classic cars" or "Congratulations on your recent Mechanical Engineering award" are likely to result in success, while a phishing email about "soccer" (a hobby of another potential target at the university) would fail when sent to this subject.

Two scenarios have been developed for this game to-date and a preliminary evaluation is being undertaken to begin to assess its impact on cyber awareness.

4   Design Alternatives for Increased Engagement

Although initial piloting suggests that the adversarial vantage is helpful and engaging, the training may be perceived as "yet another cyber training exercise," which limits user engagement and thus dampens the potential long-lasting effect on the learner's cyber mindset. In this section, we outline various game-design and adaptation strategies we are currently investigating with goal of enabling still greater engagement and resultant impact. An underlying design philosophy is to align design principles with learner self-regulation functions, so that the game better harnesses the native abilities of players to

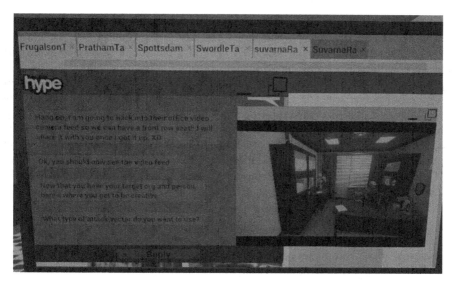

Fig. 3. A successful attack turns on a webcam.

engage "organically," rather than using artificial or coercive methods common in many computer-based training systems. We focus primarily on intrinsic adaptation (adaptation of content within the game itself). However, intrinsic adaptation introduces the need for additional content authoring, which is expensive and can thus inhibit the use of tailored content. We consider three design alternatives: specialization for player type, demographic personalization, and engagement tailoring.

4.1 Specializing for Player Type

Players like to interact with games in different ways, depending on a variety of factors [28]. Some players may want to explore how a game works; others are motivated by earning badges and maximizing their score, etc. Games that can adapt the way they work to a player's preferred playing style may increase engagement and time spent playing the game [29].

There are a number of frameworks that attempt to categorize and rationalize playing styles for video games. The combination of adapting to both playing style and learning style is still relatively novel, but the ADOPTA framework [30] offers one example. ADOPTA (ADaptive technOlogy-enhanced Platform for eduTAinment) introduces four categories of playing style (Competitor, Dreamer, Logician, and Strategist), which are defined in such a way to be correlated with established learning styles. The developers of ADOTA showed that they could learn to recognize certain features of gameplay according to these learning styles and then adapt gameplay according to the observed playing style, improving engagement for that player. For example, for a "Competitor" in a "gold capture" game they developed, the game increases the speed of movement of various gold pieces (and reward when captured). In contrast, for a "Logician," the adjustment of the degree of difficulty results in more hidden gold pieces and more "puzzles" that must be solved to find the gold. Their results show that adjusting difficulty based on player type improves the subjective experience of gameplay.

One of the advantages of the ADOPTA method is that it can be used to determine player style without the use of questionnaires or other explicit methods which makes the player aware of being "categorized." We envision several ways we could use covert observation of player styles in Cyber-RAMPART. For example, we have observed in piloting that some players enjoy exploring the social media sites somewhat exhaustively, while others, seeking to get to the attack, capture the minimum amount of information. We could adjust the narrative to account for these different styles by encouraging different approaches to open-source intelligence gathering, perhaps using a score and badges for the "Competitor" player to encourage that player type to explore more potential attacks and targets before making a commitment.

4.2 Adaptation Based on Demographic Characteristics

As suggested above, a common complaint of cyber awareness training is that it is some-what generic and not specific to the organization or individual [15]. We envision the use of player demography to allow the game to customize player experience to contexts and domains relevant to the player.

Potential demographic elements that could be used for customization include the player's age, their position/rank within their organization, and details about their job roles. For example, Hacker Mo could use several different "dialects, somewhat corre-sponding to age and position. This might result in Mo using text messaging acronyms in conversation much more frequently with younger players than older ones and choosing specific descriptions and word usage mapped to jargon and slang associated with vari-ous generational cohorts. Similarly, the social media content and presentation could be adapted to roughly match the age, positions, and job roles of the player. This matching could increase the identification of the player with the potential targets being researched during the open-source intelligence gathering phase. Familiar contexts are likely to both improve learner confidence and attention.

Adapting the scenario based on demographic information is not technically chal-lenging if there is content available to support personalization. However, the content requirements for adaptive learning has proven to be a significant barrier to widespread use in open-ended domains. Rather than hand-create additional content, as we did for the Hacker Mo scenario, we would instead recommend the use of various automated content generation [31] and content repurposing methods to support demographic-based adaptation.

4.3 Engagement Tailoring

Learner engagement has direct corollaries with self-regulatory mechanisms such as attention and persistence [27]. In past work, we have explored how various kinds of passive observation of learner activity can be used as proxies or "markers" to estimate on-going engagement as an experience unfolds [32–34]. Examples of passive observa-tion include assessing the learner's responsiveness to prompts, eye tracking via webcam, and observations of the patterns of mouse movements and hand gestures. Passive observation is less intrusive (and evident) to learners and generally requires little/no additional hardware, making it reasonably inexpensive and easy to deploy.

Via combination of such instrumentation and dynamic content selection, we envision a system capable of producing prompts within the game itself that will encourage sustained engagement and intervene when engagement begins to flag. Consider the conceptual illustration in Fig. 4. As the player participates in the game, the system tracks the estimated level of engagement. When engagement begins to trend toward some minimum arousal threshold, the system introduces an in-game or *intrinsic* prompt such as an unusual comment or unexpected question from Hacker Mo via chat message. For example, Hacker Mo might interrupt a player response by saying, "Hey, you are taking a while here… can we move on?" Intrinsic tailoring (also known as pedagogical experience manipulation) offers the benefit of providing scaffolding and support for the learner without interrupting the immersive experience [35].

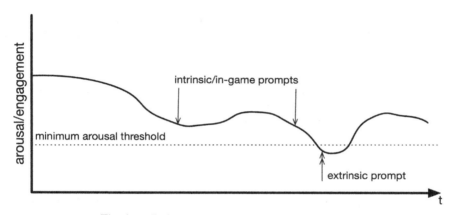

Fig. 4. Tailoring via prompts to sustain engagement.

The intent of these intrinsic prompts is to arrest the falling engagement level, which, in the figure is successful following the initial prompt. As the game progresses, these in-game prompts may lose some power (e.g., the first time Mo surprises a player with a question, it will be a surprise, but less so on the third or fourth use of the same technique). When the intrinsic prompts fail, the game can switch to an extrinsic prompt, such as interrupting the dialogue with Mo to ask the player if they want to continue or take a break.

There are two technical challenges that would need to be addressed to realize this capability in Cyber-RAMPART. First, the level of engagement would need to be estimated, which requires not only the choice of some sensor(s), but also mechanisms for calibrating thresholds for individual players, which may differ substantial from person to person. Second, similar to the previous technique, engagement tailoring also requires additional content authoring. In this case, however, it may be possible to generate prompts and questions somewhat independent of the context at the moment in which they are introduced. This approach would make the authoring requirements straightforward but may be difficult to achieve truly intrinsic prompting with more generic prompting.

5 Conclusions

Effective cyber awareness training is an acute need for today's workforce. Cyber threats are increasing and increasingly calamitous but existing training options commonly are ineffective. In this paper, we introduced an innovative training game that focuses on the development of a cyber mindset by allowing a player to plan and to execute notional examples of various cyber attacks. The attacker perspective offers several advantages that encourage attention and engagement essential for learning, including an engaging and relatable narrative that encourages player action. We also described a number of potential elaborations of the game mechanics to further encourage player engagement and learning.

Acknowledgements. This work was supported by the Office of Naval Research project N68335-18-C-0724 Cyber-RAMPART. The views and conclusions contained in this document are those of the authors and should not be interpreted as representing the official policies, either expressed or implied, of the Department of Defense or Office of Naval Research. The U.S. Government is authorized to reproduce and distribute reprints for Government purposes notwithstanding any copyright notation hereon. We thank colleagues and collaborators who have provided insights and operational perspectives in the development of the Cyber-RAMPART including Kelly Neville, Behrooz Mostafavi, Alex Nickels, Larry Flint, Julie Marble, Alice Jackson and Justin Bonny.

References

1. Cortada, J.W.: The Digital Hand: How Computers Changed the Work of American Manufacturing, Transportation, and Retail Industries. Oxford University Press, Oxford (2003)
2. Varol, O., Ferrara, E., Davis, C.A., Menczer, F., Flammini, A.: Online human-bot interactions: detection, estimation, and characterization. In: Proceedings of the Eleventh International AAAI Conference on Web and Social Media (ICWSM 2017). AAAI Press (2017)
3. Schreckinger, B.: How Russia Targets the U.S. Military. Politico (2017)
4. Kettani, H., Wainwright, P.: On the top threats to cyber systems. In: 2019 IEEE 2nd International Conference on Information and Computer Technologies (ICICT), pp. 175–179 (2019)
5. Howard, P.N., Ganesh, B., Liotsiou, D., Kelly, J., François, C.: The IRA, Social Media and Political Polarization in the United States, 2012–2018. Project on Computational Propaganda, p. 46. Oxford University, Oxford (2018)
6. Adebiaye, R., Alryalat, H., Owusu, T.: Perspectives for cyber-deterrence: a quantitative analysis of cyber threats and attacks on consumers. Int. J. Innov. Res. Sci. Eng. Technol. **5**, 12946–12962 (2016)
7. Suraj, M., Singh, N.K., Tomar, D.S.: Big data analytics of cyber attacks: a review. In: 2018 IEEE International Conference on System, Computation, Automation and Networking (ICSCA), pp. 1–7. IEEE (2018)
8. Sumi, F.H., Dutta, L., Sarker, F.: A review on cyberattacks and their preventive measures. Int. J. Cyber Res. Educ. (IJCRE) **1**, 12–29 (2019)

9. Chowdhury, A.: Recent cyber security attacks and their mitigation approaches – an overview. In: Batten, L., Li, G. (eds.) ATIS 2016. CCIS, vol. 651, pp. 54–65. Springer, Singapore (2016). https://doi.org/10.1007/978-981-10-2741-3_5

10. Mezzour, G., Carley, K.M., Carley, L.R.: An empirical study of global malware encounters. In: Proceedings of the 2015 Symposium and Bootcamp on the Science of Security, pp. 1–11 (2015)

11. Thornton-Trump, I.: Malicious attacks and actors: an examination of the modern cyber criminal. EDPACS **57**, 17–23 (2018)

12. Mezzour, G., Carley, L., Carley, K.M.: Global mapping of cyber attacks. SSRN 2729302 (2014)

13. Neville, K., Flint, L., Massey, L., Nickels, A., Medina, J., Bolton, A.: Training to instill a cyber-aware mindset. In: Schmorrow, Dylan D., Fidopiastis, Cali M. (eds.) HCII 2019. LNCS (LNAI), vol. 11580, pp. 299–311. Springer, Cham (2019). https://doi.org/10.1007/978-3-030-22419-6_21

14. Amankwa, E., Loock, M., Kritzinger, E.: A conceptual analysis of information security education, information security training and information security awareness definitions. In: The 9th International Conference for Internet Technology and Secured Transactions, London, pp. 248–252. IEEE Press (2014)

15. Aldawood, H., Skinner, G.: Reviewing cyber security social engineering training and awareness programs - pitfalls and ongoing issues. Future Internet **11**, 73 (2019)

16. Ghafir, I., et al.: Security threats to critical infrastructure: the human factor. Supercomputing **74**, 4956–5002 (2018)

17. Hadnagy, C.: Social Engineering: The Art of Human Hacking. Wiley, Hoboken (2010)

18. Caldwell, T.: Making security awareness training work. Comput. Fraud Secur. **2016**, 8–14 (2016)

19. Hendrix, M., Al-Sherbaz, A., Victoria, B.: Game based cyber security training: are serious games suitable for cyber security training? Int. J. Serious Games **3**, 53–61 (2016)

20. McGonigal, J.: Reality Is Broken: Why Games Make Us Better and How They Can Change the World. Penguin Press, New York (2011)

21. De Corte, E.: Transfer as the productive use of acquired knowledge, skills, and motivations. Curr. Dir. Psychol. Sci. **12**, 143–146 (2003)

22. Krombholz, K., Hobel, H., Huber, M., Weippl, E.: Advanced social engineering attacks. J. Inf. Secur. Appl. **22**, 113–122 (2015)

23. Bada, M., Sasse, A., Nurse, J.R.: Cyber security awareness campaigns: why do they fail to change behaviour? In: International Conference on Cyber Security for Sustainable Society, Coventry, UK, pp. 118–131 (2015)

24. Hwang, G.-J., Wu, P.-H., Chen, C.-C., Tu, N.-T.: Effects of an augmented reality-based educational game on students' learning achievements and attitudes in real-world observations. Interact. Learn. Environ. **24**, 1895–1906 (2016)

25. Hamari, J., Shernoff, D.J., Rowe, E., Coller, B., Asbell-Clarke, J., Edwards, T.: Challenging games help students learn: an empirical study on engagement, flow and immersion in game-based learning. Comput. Hum. Behav. **54**, 170–179 (2016)

26. Koedinger, K.R., Booth, J.L., Klahr, D.: Instructional complexity and the science to constrain it. Science **342**, 935–937 (2013)

27. Sitzmann, T., Ely, K.: A meta-analysis of self-regulated learning in work-related training and educational attainment: what we know and where we need to go. Psychol. Bull. **137**, 421–442 (2011)

28. Heeter, C., Winn, B.: Implications of gender, player type and learning strategies for the design of games for learning. In: Kafai, Y., Heeter, C., Denner, J., Sun, J. (eds.) Beyond Barbie and Mortal Kombat: New Perspectives on Gender and Gaming. MIT Press, Cambridge (2008)

29. Heeter, C., Magerko, B., Medler, B., Fitzgerald, J.: Matching game mechanics to player motivation. In: Meaningful Play Conference (2008)
30. Bontchev, B., Vassileva, D., Aleksieva-Petrova, A., Petrov, M.: Playing styles based on experiential learning theory. Comput. Hum. Behav. **85**, 319–328 (2018)
31. Summerville, A., et al.: Procedural content generation via machine learning (PCGML). IEEE Trans. Games **10**, 257–270 (2018)
32. Wearne, A., Wray, R.E.: Exploration of behavioral markers to support adaptive learning. In: Kurosu, M. (ed.) HCI 2018. LNCS, vol. 10903, pp. 355–365. Springer, Cham (2018). https://doi.org/10.1007/978-3-319-91250-9_28
33. Workshop on Brain, Body and Bytes: Psychophysiological User Interaction at CHI 2010, Atlanta, GA (2010)
34. Wray, R.E., Woods, A.: A cognitive systems approach to tailoring learner practice. In: Klenk, M., Laird, J. (eds.) Proceedings of the 2013 Advances in Cognitive Systems Conference, Baltimore, MD (2013)
35. Lane, H.C., Johnson, W.L.: Intelligent tutoring and pedagogical experience manipulation in virtual learning environments. In: Cohn, J., Nicholson, D., Schmorrow, D. (eds.) The PSI Handbook of Virtual Environments for Training and Education, vol. 3. Praeger Security International, Westport (2008)

Augmented Cognition for Well-Being, Health and Rehabilitation

Acceptability and Normative Considerations in Research on Autism Spectrum Disorders and Virtual Reality

Anders Dechsling(✉) ⓘ, Stefan Sütterlin ⓘ, and Anders Nordahl-Hansen ⓘ

Østfold University College, Halden, Norway
{anders.dechsling,Stefan.sutterlin,
anders.nordahl-hansen}@hiof.no

Abstract. The number of publications on Autism Spectrum Disorders (ASD) and Virtual Reality (VR) has increased considerably the past few years. Interventions using VR show promising effect in improving social skills and other daily living activities for persons with ASD. Researchers have expressed concerns regarding the acceptability among people with ASD towards using Head-Mounted Displays (HMD) due to sensory oversensitivity. In this study we provide data from 155 publications on the reported acceptability by participants with ASD. Over 80% of the studies reported using VR as positive. The negative sentiment towards using VR across studies was below 1%. These findings indicate that VR and other computer-based tools are broadly accepted amongst participants with ASD. We suggest normative considerations that researchers and clinicians should take into account when planning and conducting research and clinical interventions. Focusing on the social validity, user involvement, and individual considerations.

Keywords: Autism spectrum disorder · Virtual Reality · Acceptability · Normative considerations

1 Introduction

Autism spectrum disorder (ASD) is a neurodevelopmental diagnosis defined by persistent deficits in social interaction and communication and by restricted, and repetitive patterns of behavior, interests, or activities [1]. An overall prevalence of ASD estimate it to be around 1% [2]. Psychosocial deficits and challenges are common in ASD. Various aspects such as less friendships with peers [3], and increased loneliness and isolation in their teens [4] have been reported by adolescents with ASD. Young people with ASD display more school refusal behavior compared to others [5], and children report having experienced anxiety in various settings [6] in addition to bullying [7].

Enhancing social skills is a key target for interventions for this group [8], and there has been a durable trend in empirical research on social skills interventions in ASD [9]. Virtual Reality (VR) as a tool in research on ASD has gained more attention in recent years. In fact, the number of published research articles on using Virtual Reality

© Springer Nature Switzerland AG 2020
D. D. Schmorrow and C. M. Fidopiastis (Eds.): HCII 2020, LNAI 12197, pp. 161–170, 2020.
https://doi.org/10.1007/978-3-030-50439-7_11

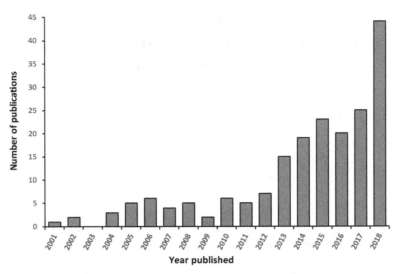

Fig. 1. Number of peer-reviewed publications on ASD and the use of VR or other computer-based tools, each year since 2001–2018.

or other computer-based assessments and interventions on ASD in 2018 alone exceeded all publications produced in the consecutive years of 2001–2011 (see Fig. 1).

The majority of interventions using VR aims at improving social or emotional skills [10, 11], and using VR show promising treatment effects [12]. Several researchers describe VR as having advantages over other forms of tools, such as combining ecological validity provided by real-world simulation [11], with a less stressful environment [13].

There are various forms of VR equipment that enable different levels of immersions [12]. For instance, Ip et al. [14] used a four-side CAVE VR system, in which big screens and projectors halfway surround the participant. Despite some promising treatment effects reported [14, 15], implementation of CAVE VR systems in ASD interventions suffer from practical requirements/challenges related to space, equipment, and its immobility [16].

The highest level of immersion provided by mobile equipment is realized in VR Head-Mounted Displays (HMD), where the goggles provide a sense of a virtual environment completely surrounding the user. An alternative to VR HMD is glasses with Augmented Reality (AR) technology. AR is an enhanced version of reality created by the use of technology to overlay digital information on an image of something being viewed through a camera device [17], and AR-glasses differ from VR HMD as visual information is superimposed on the actual environment [18]. Henceforth, we include both VR and AR-glasses when referring to the abbreviation HMD. Other computer-based interventions use computers, laptops, tablets, mobile phones, Kinect®, or other technology such as for example social robots [19].

In regard to the usability and acceptability of HMDs, previous research suggests that sensory oversensitivity among participants with ASD may cause irritation and thus low acceptability [20–22]. This criticism might lead to some reluctance towards the use of HMDs. However, there is little empirical research that support these claims and the references used in the aforementioned studies were published before 2004 e.g., [23, 24] and do not account for recent developments in usability, comfort and improved motion sickness prevention. Findings from a more recent study indicate that in tests for acceptability and sensitivity of participants using HMDs, participants express both acceptance and enjoyment of HMDs [25]. These results where further corroborated in a literature review by Bozgeyikli et al. [26], which also suggested guidelines in design of VR as a training tool for participants with ASD. (We argue that) the question of acceptability is of central importance for the future development of AR/VR-supported intervention research in ASD. Lacking acceptability of an intervention method can affect common factors central for intervention effects [27]. Participants' outcome expectations, collaboration, affirmation and motivational factors contributing to a sustainable treatment alliance are known to be central to increase effects of psychological interventions [27]. Although the usability and acceptability of HMD is reported in a number of studies [15, 25], a systematization of data across studies reporting acceptability (i.e., participant reports on usability, enjoyment, likeability, tolerability and so on) amongst participants with ASD is lacking. This chapter addresses the gap in knowledge to provide a more contemporary and comprehensive conclusion regarding the acceptability of HMD-supported AR/VR interventions in ASD.

All interventions represent an active and targeted manipulation of the individual's cognitive and emotional status with the goal to reach sustainable change in how one perceives, interpret, experiences and reacts to the environment. These intended psychosocial changes are to be reached via empirically validated methods. As all intervention can have unintended side effects, which also can apply to VR-interventions, it is important for researchers, clinicians, and practitioners to be aware and account for these. The use of VR and computer-based tools has undergone numerous iterations during its relatively short history and there is a need to maintain a focus on social validity when developing and using new technology.

In this chapter, we provide a number of normative considerations of how the VR interventions can relate to the non-epistemic values of the relevant party (i.e., considerations of how the participant experiences the purpose of and the VR intervention itself). Using these considerations, and report these judgements and decisions in research should contribute to the focus on social validity in research, and ensures that individual consideration is always accounted for.

2   Method

2.1   Literature Search and Screening

We searched PsycInfo, Pubmed, ERIC, Education Research Complete, Web of Science, and IEEE Xplore, between August 23rd 2019 and August 27th 2019, using the thoroughly selected Boolean operators for the search strings: (Pervasive development disorder OR pdd OR pdd-nos OR pervasive developmental disorder not otherwise specified OR autism

OR autistic OR Autism Spectrum Disorder OR autism spectrum disorders OR Asperger OR asd OR autism spectrum condition* OR asc) AND (Virtual Reality OR vr OR hmd OR Head-mounted display OR Immersive Virtual Environment OR Augmented Reality OR Artificial Reality OR Oculus OR Immersive Technolog* OR Mixed Reality OR Hybrid Reality OR Immersive Virtual Reality System OR 3D Environment* OR htc vive OR cave OR Virtual Reality Exposure OR vre).

After removing duplicates, we manually screened the results according to pre-defined inclusion criteria: The articles are peer-reviewed and written in English, include participants with ASD, and include the use of virtual reality/environment equipment. We excluded meta-analyses, reviews, conceptual articles, and design articles that were not original research articles providing new empirical data.

2.2 Coding Acceptability Data

In order to discover acceptability data in all the included publications, we searched for the following terms in each publication: accept*, question*, usability*, evaluat*, ask*, enjoy*, valid*, and tolerab*. As the publications does not use a general standard of assessing and reporting acceptability data, we extracted qualitative statements from participants, data provided by questionnaires or other forms for reporting acceptability. If the articles included acceptability data, we coded the data as either; 0 = negative sentiment towards the VR, 1 = inconclusive, and 2 = positive sentiment towards VR. If different participants within a study reported acceptability data that contradicts each other, the data would be coded as inconclusive.

3   Results

The initial screening and removing of duplicates resulted in 226 publications. Figure 2 (flow chart) shows that the number of publications we included according to our inclusion and exclusion criteria where 155. Sixty-three of these publications included acceptability data.

A descriptive analysis of the data shows that 63 (40.6%) publications reported data on the acceptability from participants with ASD. See Table 1 for the summary of the evaluation from the participants. Out of the total number of publications, 22 (14.2%) used HMD or AR-glasses, and the remaining used other types of computer-based apparatus (e.g., CAVE, laptop, tablet etc.). Sixteen (72.7%) of the 22 publications using HMD reported acceptability data, three of which was inconclusive and the rest was positive. An independent coder that was unfamiliar with the project coded a randomly selected sample of 16 publications (10.3%) to check for inter-rater reliability (IRR) with the first author. The IRR agreement in this sample was 100%.

4   Discussion

The aim of this research was to provide an indication of whether VR and HMDs are accepted amongst participants with ASD across studies. Our findings indicates that VR

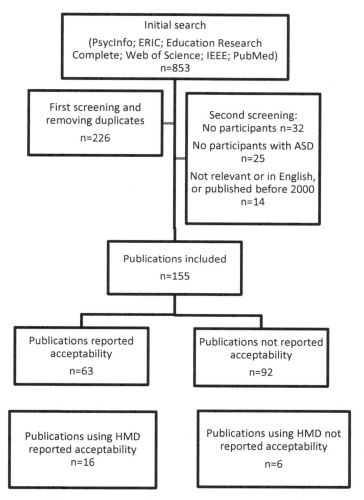

Fig. 2. Flow chart of the search and screening process.

interventions are welcomed by participants with ASD as over 80% of the publications reporting evaluation data report positive evaluations. This is in line with suggestions from Newbutt et al. [25]. The majority of the overall studies including both HMD and other computer-based apparatus does not report evaluation data on acceptability. Most of the publications using HMD, however, have reported acceptability data. We encourage researchers to include these kinds of evaluations when doing research and clinical work. As the positive results are shown both across studies and within studies such as Newbutt et al. [25] it seems that there is no evidence suggesting to avoid HMDs in ASD research and clinical practice. It is a vast amount of studies not reporting the evaluation so there might be some unrecorded negative evaluations. However, individual variation

Table 1. The table shows descriptive statistics on the number of publications (n) reporting acceptability data. The lower section in the table shows the similar data in the publications using HMDs. The numbers inside the parenthesis is the percent of n its respective unit.

Reported n(%)	Positive	Negative	Inconclusive	Using HMD
Yes				Yes
63 (40.6)	56 (88.9)	1 (.6)	6 (9.5)	22 (14.2)
No				No
92 (59.4)				133 (85.8)
HMD and reported	Positive	Negative	Inconclusive	
Yes				
16 (72.2)	13 (81.3)	0 (0)	3 (18.7)	
No				
6 (27.3)				

will always apply to specific situations and necessary adaptions is needed whenever using VR apparatus in clinical research.

4.1 Future and Normative Considerations

Where the clinical assessment concludes the need for an intervention with the goal of sustainable behavior change, normative considerations are important to determine the appropriateness of the additional burden for the client. Decisions based on evidence-based conclusions regarding the effectiveness and efficacy of an intended intervention should therefore always differentiate between descriptive and normative premises and undergo an additional normative consideration to ensure the participant's interests.

As shown amongst others by Gillespie-Lynch et al. [28], people with ASD may often be considered experts on autism, and there is an increased and pronounced expectation amongst persons within the autism spectrum to be heard in respect to societal issues, and whether one should change people's personality, or the construct of society. The increased awareness towards this, emphasize the importance of cooperating with people with ASD in both research and clinical practice.

Normative Considerations According to Løkke and Salthe's [29] Checklist.[1] The first step in any clinical interventions is to collect the facts about the participant's behavior, situation, and somatic and psychological health. Before continuing, one should verify the participant's values, interests and opinion regarding the topic [29]. Then, after this assessment and before anything else, one should ask the question: is there any reason to

---

[1] The checklist is originally published in Norwegian, in the OpenAccess Norwegian Journal of Behavior Analysis. An English version of the complete checklist of Jon A. Løkke and Gunnar Salthe can be provided on request to the corresponding author of this chapter.

do anything? If no, then do not. However, if the answer is yes, one can go on checking evidence, effects and possible side effects of various interventions before collecting data.

It is also important to do normative considerations of the specific intervention, and our paper provides some evidence that VR and HMDs are a tolerable tool in interventions. Even though our results suggest that HMDs and other computer-based tools are accepted amongst the majority of participants with ASD, it is not a justification towards using it against someone's will. One should principally make sure that the intervention does not contain force/compulsion, reduction of choices, or other unwanted elements for the participant.

In addition to the acceptability towards the intervention method, the normative issues of the possible outcomes of the interventions for the individual is important. According to the philosopher James Griffin [30], one can categorize values that make life worth living into five categories of wishes for our lives: *Good experiences, freedom, knowledge, relations*, and to *have accomplished something*. When setting up an interventions, the researcher/clinician should think carefully about if the interventions itself provides a *good experience*, and if the possible outcome expands the individual's opportunities later in life for more good experiences. These two considerations may contradict one another; however, the latter may oust the first in many cases. Actually, in regards to important measurements and outcomes, happiness is ranked, by parents of children with ASD, as the most important outcome that researchers should measure [31].

Freedom, to both make one's own decisions and be able to withstand unwanted decision made unto by others, may also be important. This wish should be considered both prior to and during the interventions, as well as when evaluating outcomes. User involvement is an example which is important prior and during, but one should try as an outcome to give participants enhanced opportunities to increase their own sense of freedom in life. *Knowledge* in itself is of value, and also the notion of *having accomplished something*. That might be on a smaller or bigger scale, e.g., from learning how to read or raising happy children. The fifth wish, *relations*, can often seem to be underrated by the surrounding of people with ASD. Maybe because children with ASD tends to orient themselves more towards non-social stimuli rather than social [32], this might be mistaken to be interpreted as people with ASD not wanting any social relations. A lot of research has refuted this [7]. Again, it is important to emphasize that research does not show that all persons with ASD want social relationships either, just as with people "outside" the autism spectrum.

4.2 Limitations

The analysis of the existing study laid out some systematic knowledge gaps to be addressed by future research, as they are not uncommon in new research fields such as VR interventions and research. There is no coherent way of how to report data on acceptability even though attempts have been made to develop questionnaires for researchers and teachers [33]. The various ways of assessing and reporting evaluation data, may have led to variations and/or errors in our appraisal of the evaluations provided in the different studies. However, the results from IRR suggests agreement and thus reduce the margin of errors. The results would have been strengthened by coding more or all the data with two independent coders.

Many of the studies do not report acceptability data. There are also instances where parents, caregivers, or teachers are asked to evaluate the participants' acceptance, which may increase the risk of socially desired answers. The fact that participation in studies is based on ethically approved informed consent, may lead to a self-selection effect limiting the generalization of conclusions beyond settings where participants with ASD receive detailed information prior to HMD exposure.

5   Conclusion

This study shows that almost half of the studies on VR and ASD report acceptability data. Of the publications that reported such data, the majority overall and studies using HMD report positive sentiments on acceptability. These data suggest that persons with ASD can readily use computer-based and virtual reality technology, HMDs included. However, over half of the studies we have checked has not reported evaluation data. This indicates that there might be a number of unrecorded cases of negative evaluations. Either way, we emphasize that individual considerations must be made in all cases. Further, we have suggested normative considerations that is eligible to use prior and during interventions in regards to aims and outcome.

References

1. American Psychiatric Association: Diagnostic and Statistical Manual of Mental Disorders: DSM-5, 5th edn. American Psychiatric Association, Washington, D.C. (2013)
2. Lord, C., et al.: Autism spectrum disorder. Nat. Rev. Dis. Primers 6(1), 1–23 (2020). https://doi.org/10.1038/s41572-019-0138-4
3. Chamberlain, B., Kasari, C., Rotheram-Fuller, E.: Involvement or isolation? The social networks of children with autism in regular classrooms. J. Autism Dev. Disord. 37(2), 230–242 (2007). https://doi.org/10.1007/s10803-006-0164-4
4. Lasgaard, M., Nielsen, A., Eriksen, M.E., Goossens, L.: Loneliness and social support in adolescent boys with autism spectrum disorders. J. Autism Dev. Disord. 40, 218–226 (2010). https://doi.org/10.1007/s10803-009-0851-z
5. Munkhaugen, E.K., Gjevik, E., Pripp, A.H., Sponheim, E., Diseth, T.H.: School refusal behaviour: are children and adolescents with autism spectrum disorder at a higher risk? Res. Autism Spectr. Disord. 41, 31–38 (2017). https://doi.org/10.1016/j.rasd.2017.07.001
6. Adams, D., Simpson, K., Keen, D.: Exploring anxiety at home, school, and in the community through self-report from children on the autism spectrum. Autism Res. 13(5), 1–12 (2019). https://doi.org/10.1002/aur.2246
7. Skafle, I., Nordahl-Hansen, A., Øien, R.A.: Short report: social perception of high school students with ASD in Norway. J. Autism Dev. Disord. 50(2), 670–675 (2019). https://doi.org/10.1007/s10803-019-04281-w
8. Wolstencroft, J., Robinson, L., Srinivasan, R., Kerry, E., Mandy, W., Skuse, D.: A systematic review of group social skills interventions, and meta-analysis of outcomes, for children with high functioning ASD. J. Autism Dev. Disord. 48(7), 2293–2307 (2018). https://doi.org/10.1007/s10803-018-3485-1
9. Ke, F., Whalon, K., Yun, J.: Social skill interventions for youth and adults with autism spectrum disorder: a systematic review. Rev. Educ. Res. 88, 3–42 (2018). https://doi.org/10.3102/0034654317740334

10. Mesa-Gresa, P., Gil-Gómez, H., Lozano-Quilis, J., Gil-Gómez, J.-A.: Effectiveness of virtual reality for children and adolescents with autism spectrum disorder: an evidence-based systematic review. Sensors **18**, 2486 (2018). https://doi.org/10.3390/s18082486
11. Lorenzo, G., Lledó, A., Arráez-Vera, G., Lorenzo-Lledó, A.: The application of immersive virtual reality for students with ASD: a review between 1990–2017. Educ. Inf. Technol. **24**(1), 127–151 (2018). https://doi.org/10.1007/s10639-018-9766-7
12. Miller, H.L., Bugnariu, N.L.: Level of immersion in virtual environments impacts the ability to assess and teach social skills in autims spectrum disorder. Cyberpsychol. Behav. Soc. Netw. **19**, 246–256 (2016). https://doi.org/10.1089/cyber.2014.0682
13. Didehbani, N., Allen, T., Kandalaft, M., Krawczyk, D., Chapman, S.: Virtual reality social cognition training for children with high functioning autism. Comput. Hum. Behav. **62**, 703–711 (2016). https://doi.org/10.1016/j.chb.2016.04.033
14. Ip, H.H.S., et al.: Enhance emotional and social adaptation skills for children with autism spectrum disorder: a virtual reality enabled approach. Comput. Educ. **117**, 1–15 (2018). https://doi.org/10.1016/j.compedu.2017.09.010
15. Halabi, O., El-Seoud, S.A., Alja'am, J.H., Alpona, H., Al-Hemadi, M., Al-Hassan, D.: Design of immersive virtual reality system to improve communication skills in individuals with autism. Int. J. Emerg. Technol. Learn. **12**(5), 50–64 (2017). https://doi.org/10.3991/ijet.v12 i05.6766
16. Yuan, S.N.V., Ip, H.H.S.: Using virtual reality to train emotional and social skills in children with autism spectrum disoder. London J. Prim. Care **10**(3), 110–112 (2018). https://doi.org/ 10.1080/17571472.2018.1483000
17. The Merriam-Webster.com Dictionary. https://www.merriam-webster.com/dictionary/aug mented%20reality. Accessed 29 Jan 2020
18. Rodriguez, S., Munshey, F., Caruso, T.J.: Augmented reality for intravenous access in an autistic child with difficult access. Pediatr. Anesth. **28**(6), 569–570 (2018). https://doi.org/10. 1111/pan.13395
19. Kim, E.S., et al.: Social robots as embedded reinforcers of social behavior in children with autism. J. Autism Dev. Disord. **43**(5), 1038–1049 (2013). https://doi.org/10.1007/s10803-012-1645-2
20. Saadatzi, M.N., Pennington, R.C., Welch, K.C., Graham, J.H., Scott, R.E.: The use of an autonomous pedagogical agent and automatic speech recognition for teaching sight words to students with autism spectrum disorder. J. Spec. Educ. Technol. **32**(3), 173–183 (2017). https://doi.org/10.1177/0162643417715751
21. Saiano, M., et al.: Natural interfaces and virtual environments for the acquisition of street crossing and path following skills in adults with autism spectrum disorders: a feasibility study. J. Neuroeng. Rehabil. **12**(1), 17 (2015). https://doi.org/10.1186/s12984-015-0010-z
22. Wallace, S., Parsons, S., Westbury, A., White, K., White, K., Bailey, A.: Sense of presence and atypical social judgments in immersive virtual environments: responses of adolescents with autism spectrum disorders. Autism **14**(3), 199–213 (2010). https://journals.sagepub.com/doi/ pdf/10.1177/1362361310363283
23. Tanaka, N., Takagi, H.: Virtual reality environment design of managing both presence and virtual reality sickness. J. Physiol. Anthropol. Appl. Hum. Sci. **23**(6), 313–317 (2004). https:// doi.org/10.2114/jpa.23.313
24. Regan, E.C., Price, K.R.: The frequency of occurrence and severity of side-effects of immersion virtual reality. Aviat. Space Environ. Med. **65**(6), 527–530 (1994)
25. Newbutt, N., Sung, C., Kuo, H., Leahy, M.J.: The potential of virtual reality technologies to support people with an autism condition: a case study of acceptance, presence and negative effects. Annu. Rev. Cyberther. Telemed. **14**, 149–154 (2016)

26. Bozgeyikli, L., Raij, A., Katkoori, S., Alqasemi, R.: A survey on virtual reality for individuals with autism spectrum disorder: design considerations. IEEE Trans. Learn. Technol. **11**(2), 133–151 (2018). https://doi.org/10.1109/TLT.2017.2739747

27. Wampold, B.E.: How important are the common factors in psychotherapy? An update. World Psychiatry Off. J. World Psychiatr. Assoc. (WPA) **14**(3), 270–277 (2015). https://doi.org/10.1002/wps.20238

28. Gillespie-Lynch, K., Kapp, S.K., Brooks, P.J., Pickens, J., Schwartzman, B.: Whose expertise is it? Evidence for autistic adults as critical autism experts. Front. Psychol. **8**, 438 (2017). https://doi.org/10.3389/fpsyg.2017.00438

29. Løkke, J.A., Salthe, G.: Guidelines for goal-directed interventions: from normative and descriptive assumptions to intervention and evaluation. Norw. J. Behav. Anal. **39**, 17–32 (2012)

30. Griffin, J.: Well-Being. Its Meaning, Measurement and Moral Importance. Oxford University Press, Oxford (1986)

31. McConachie, H., et al.: Parents suggest which indicators of progress and outcomes should be measured in young children with autism spectrum disorder. J. Autism Dev. Disord. **48**(4), 1041–1051 (2017). https://doi.org/10.1007/s10803-017-3282-2

32. Gale, C.M., Eikeseth, S., Klintwall, L.: Children with autism show a typical preference for non-social stimuli. Sci. R. **9**, 10355 (2019). https://doi.org/10.1038/s41598-019-46705-8

33. Chia, N.K.H., Li, J.: Design of a generic questionnaire for reflective evaluation of a virtual reality-based intervention using virtual dolphins for children with autism. Int. J. Spec. Educ. **27**(3), 45–53 (2012)

Feedback Control for Optimizing Human Wellness

Bob Hanlon[1]($\boxtimes$), Monte Hancock[2], Chloe Lo[1], John Grable[3], Kristy Archuleta[3], Alexander Cohen[4], Chris Mazdzer[4], Sandra Babey[1], Eric Miller[5], and Alex Nunez[5]

[1] Living Centerline Institute, Morristown, USA
bob.hanlon@livingcenterlineinstitute.com
[2] 4Digital Inc., Los Angeles, USA
[3] University of Georgia, Athens, USA
[4] U.S. Olympic Committee, Colorado Springs, USA
[5] Innovarius, Orlando, USA

Abstract. For most people, decisions about "wellness" are made by default. Without specific evidence-based guidance, most find it difficult to make specific long-term commitment to achieving and maintaining wellness. Proposed here is an approach to filling the information gap through "Informed spectatoring" using appropriate instrumentation and advice: a type of "feedback control".

Human beings are complex systems having many interoperable subsystems. Neither these subsystems nor their interactions are completely understood. Further, while there are broad principles describing these subsystems and their interactions, the specifics are personal. Taken together, these factors make optimization of the human being by achieving and maintaining Wellness a challenging problem.

We describe a cloud-based system intended to provide actionable recommendations to users for improving their wellness. The recommendations are personalized for the user demographic and history. This is accomplished in three steps:

1) Assessment: a knowledge-based expert system ingests data from the user, and positions them on an objective, numeric multi-dimensional wellness scale in the general areas of Psychology, Physiology, and Finanaces. This assessment of wellness is informed by the user demographic, history, and a wearable monitoring device.

2) Projection: A bank of feedback controllers (pid controllers) is used to estimate user future wellness states by projecting the user wellness state forward in time. This is a *prospective*, rather than merely *retrospective* analysis. It is focused on where the user *will be* rather than merely where the user *has been*.

3) Recommendation: Based upon the assessment of user wellness, and the forward projection of the user wellness, a knowledge-based expert system selects a few areas of wellness that can be addressed in a unified way by simple user actions. An innovation is that the expert system makes these recommendations by automatically crafting a report in colloquial prose that reads like it was written by a human.

© Springer Nature Switzerland AG 2020
D. D. Schmorrow and C. M. Fidopiastis (Eds.): HCII 2020, LNAI 12197, pp. 171–190, 2020.
https://doi.org/10.1007/978-3-030-50439-7_12

An operational, cloud-based prototype has been built and tested on simulated data. We describe an upcoming human trial using members of the U.S. Olympic Team as subjects.

Keywords: Wellness · Psychophysiological economics · Intelligent recommender systems

1   Assumptions Underlying the Model

As a process, Wellness is a path through a sequence of wellness states. Mathematically, this path is a trajectory through the wellness state space.

Our treatment of Wellness addresses three fundamental components: Mental Wellness, Physical Wellness, and Financial Wellness.

The components of Wellness are interdependent: what affects one affects the others, often in complex ways that vary from person to person, and for a particular person over time. Therefore, Wellness must be achieved as a balance among all three components.

Wellness is a dynamic homeostasis. People are not static, but constantly changing as they respond to the changing circumstances of their mental, physical, and financial environment, environment. Maintaining Wellness requires making appropriate adjustments so that balance is maintained. (Allostasis is a state other than a state of Wellness.)

The sequence of wellness states occupied by a person through time constitutes their Wellness Trajectory. A path that is a sequence of optimum wellness states for a person is their centerline. Achieving wellness means following a stable trajectory that is near one's own centerline.

Mathematically, maintaining wellness can be modeled as interoperable systems having interactions that involve non-observability, discontinuity, non-linearity, time-latency, feedback, habituation, and non-stationarity.

2   Background: Psychophysiological Economics

The notion that behavioral, cognitive, and physiological mechanisms interact in a way that shapes wellness is closely associated with an emerging field of study called psychophysiological economics (PE). PE theory presupposes that behavior and cognitive processing are indivisible and that behavioral, cognitive, and physiological tools and techniques can be combined to create interventions that improve the physical, mental, and health outcomes of individuals.

As noted by Grable (2013), PE differs from behavioral economics/finance (BE/F) and functional magnetic resonance imaging (FMRI) applications. Grable noted that BE/F theory and usages are based are observational data (e.g., measuring performance, assessing both objective and subjective responses, etc.), whereas FMRI is typically applied exclusively to the study of brain activity (e.g., mapping neural activity with changes in blood flow). PE, on the other hand, emphasizes the role of the peripheral nervous system as it relates to economic behavior. The peripheral nervous system includes the spinal and cranial nerves. Of specific relevance to those interested in the connection between PE and wellness is the Automatic Nervous System (ANS). ANS regulates visceral structures (i.e., glands and other internal organs). Together, these structures control involuntary physiological activities and behavior. When a PE approach is used, researchers can directly measure the sympathetic nervous system as a person responds to stressors within the environment. It is the manner in which someone reacts physiologically to specific stressors, rather than how they plan to react, that influences the person's behavioral tendencies.

An assumption imbedded in the PE hypothesis is that stress shapes how individuals react, both physically and mentally, to stressors. A secondary assumption is that stress cannot easily be evaluated and assessed using traditional observational methods. A third, and critical assumption, associated with PE models is that distress can alter behavior by causing changes in a person's visceral structures. These changes occur involuntarily, and as suggested by Grable (2013), unless someone understands how they cognitively process stress information, their behavior may appear to be erratic and not rational.

Stress is idiosyncratic. As such, the types of stressors that lead to stress (both the level and type of stress) differ based on the person asked and the person's perceptions of the likely threat associated with an event or circumstance. Once stressors have been identified, it is important to take this knowledge and apply it clinically. This means, simply, designing psychophysiological tools and techniques that can be used to help individuals and professional help providers identify stressors and how to deal with such stressors.

PE researcher tend to focus on evaluating and assessing physiological aspects of household level economic and financial behavior. Specific tools used by PE researchers include the following:

- Surface Electromyography: Muscle Tension Patterns
- Electrodermal Activity: Skin Conductance
- Peripheral Skin Temperature: Sympathetic Activation
- Respiratory Feedback: Breathing Patterns
- Cardiovascular Activity: Heart Rate Variability

These tools allow PE researchers to evaluate directly, in real time, the relationship between cognition and behavior, and ultimately degrees of wellness. In general, someone experiencing stress (either acute or chronic) should exhibit the following characteristics as they relate to any given economic stressor: (a) muscle tension, (b) increased sweat production, (c) a decrease in peripheral skin temperature, (d) increased breathing, and (e) increased heart rate variability (Grable 2013). Because these factors are sometimes difficult to recognize, and because these factors do not occur at the same rate or intensity for any person, it is essential to measure such factors objectively. For those who are experiencing stress, specific interventions can be used to reduce such stress, with the likely outcome being an improved willingness to engage in longer-term positive behaviors resulting in physical, mental, and financial wellness.

3   Technical Formalisms: Automating a Wellness Assessment-Recommender System

The authors have implemented a prototype web-based Wellness Recommender System. It ingests data from a user; assesses their wellness in the categories of:

- Psychology (modes of thinking that affect overall wellness)
- Physiology (direct monitoring using a wearable monitors of heart-rate variability (HRV))
- Finances (financial state and behavior); and produces a prose report recommending immediate short-term behaviors likely to improve overall wellness.

Considered together, these three elements of wellness constitute the interrelated elements of the emerging discipline of Psychophysiological Economics.

The mathematics supporting the approach is an application of many formalisms. Some are conventional, while others are implementations of advanced or emerging technology. The technology draws from:

- Data Science:
  - Data ingestion and formatting
  - Formulation of data architecture (analysis, flows, archiving)
- Data Mining for-
  - Data Conformation
  - Data Visualization
  - Data Repair (outlier correction, imputation of missing values)
  - Parametric modeling of populations
  - Data Segmentation and Clustering
  - Data Quantization/Coding
  - Data Normalization/Registration
- Data Modeling for-
  - Feature Extraction and Enhancement
  - Assignment of Ground Truth
  - Definition of Sampling Strategy
- Closed-Loop Control Theory for-
  - PID Control Strategy for Behavior Feedback (Wellness Corrections)
- Spectral Analysis for-
  - Vectorization of Wellness Concepts
  - Mapping Wellness Corrections to recommended behaviors
- Machine Learning for-
  - Interview of Domain Experts
  - Ontology Development
  - Establishment of distributions characterizing wellness in its various aspects
  - Encapsulation of Domain Knowledge
  - ML Paradigm Selection

- Decision Subsystem Training/Testing
- Auto-Report Generation to recommend user behaviors.

4 Processing Philosophy

The output produced by the Recommender for each user is a combination of numeric and nominal data presented in an annotated prose report (an example is provided at the end of this paper).

During ingest and processing, data are represented in numeric form. Some data elements are stored as single numbers (e.g., chronological age), while others require multiple values (Average Blood Pressure as a pair: a Diastolic value, and a Systolic value). Text (e.g., country of residence) is numericized.

Automating the assessment of human wellness using descriptive statistics requires embedded knowledge. The authors have prototyped such an assessment ontology consisting of a trainable, knowledge-based expert system which performs user wellness assessment as a side-effect (in the software sense). The assessment is personalized, objective, quantitative, and includes an explanation of the findings.

Projecting the user wellness state forward in time is accomplished using a bank of 25 P.I.D. controllers (one for each of the key components of wellness; see the list below). Each controller contributes its projection estimate to a simple fusion algorithm, which rolls them up into the aggregate assessment of the user wellness state.

The most challenging aspect of system development has been creating a the knowledge-based expert system which ingests numeric wellness metrics and state projections into the system output: the users' personalized recommendations in a prose report that reads like it was written by a human author. Initially, recommender data have been drawn from clinical and financial literature, and simulations informed and reviewed by domain experts. The results of the Centerline Study (described below) will be will be used to refine the decision processes within the Centerline System.

All of the formalisms called out above are be supported by a vector space approach to data modeling. Each entity (person, behavior, state, outcome) is represented as an ordered list of numeric values in a finite-dimensional vector space. This is a standard approach used in many (most) intelligent systems.

4.1 The 25 Components of Wellness

This will begins with the collection of feature vectors containing indicators in each of the three wellnesses. At present, these data are held in a repository of 10,000 simulated users (simulants). Each has data for the 25 areas of wellness:

- Psychological:
  - Positivity=wellness 1
  - Engagement=wellness 2
  - Relationships=wellness 3
  - Meaning=wellness 4
  - Accomp=wellness 5

- Emo_Stability=wellness 6
- Optimism=wellness 7
- Resilience=wellness 8
- Self_Esteem=wellness 9
- Vitality=wellness 10
- Physical:
  - weight=wellness 11
  - BP=wellness 12
  - Sugar=wellness 13
  - Heart_Rate=wellness 14
  - Age=wellness 15
  - Sleep=wellness 16
  - Diet=wellness 17
- Financial:
  - Spend<Earn=Spending is less than income=wellness 18
  - Timely_Bills=Pay all bills on time=wellness 19
  - Short_Savings=Sufficient liquid savings=wellness 20
  - Long_Savings=Sufficient long term savings=wellness 21
  - Managed_Debt=Manageable debt load = wellness 22
  - Solid_Credit=Prime credit score=wellness 23
  - Insured=Have appropriate insurance=wellness 24
  - Expense_Plan=Plan ahead for expenses=wellness 25.

4.2   The Assessment and Recommendation Process

A person's Centerline Wellness is the collection of their numeric wellness values in the wellness areas listed above. Initially, target values (Centerline Wellness values) will be established by the judgement of domain experts in each wellness area; by reference to the published literature and population norms; and by one or more empirical case studies. Servicing Centerline Users is a linear process conducted in two phases. Phase 1 consists of abstract data processing, while Phase 2 consists of knowledge-intensive analytic processing.

Phase 1: Mathematical Assessment and Optimization

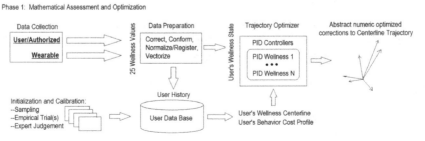

Raw data for a person's instantaneous wellness state (a vector of their wellness values) will be collected/inferred from information they provide, information collected from sources authorized by the user, and from one or more wearable devices on the user's person.

This raw data will be conditioned and converted into a set of deviations from the person's Centerline Wellness. Because the various wellness factors impact the user's life in different ways, and by different amounts, and with different latencies, the aggregate deviation from Centerline Wellness will be a weighted sum of the wellness deviations. Both instantaneous and long-term average wellness values will be created.

Given the deviation from Centerline Wellness, a collection of PID controllers will determine the correcting signals needed to restore the person to their Centerline Wellness. These correcting signals are abstract numeric representations of the adjustments needed to return the user's to a Centerline Trajectory in the Wellness state space.

Phase 2: Translation and Presentation

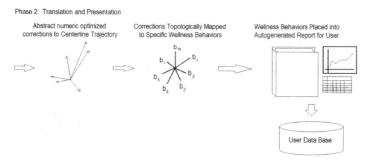

To make the abstract adjustments actionable, they must be translated into a coherent, comprehensible set of recommended behaviors. The intention is that, by undertaking the recommended behaviors, the user's deviation from Centerline Wellness will be reduced over time to nominal levels. The software we have implemented we call the Recommender.

The prototype Recommender currently uses two knowledge-based expert systems and a bank of 25 PID controllers. This prototype is being used to determine whether and how wellness can be restored to a homeostatic set point using a closed-loop controller. A case study using volunteer(s) from the U.S. Winter Olympic Team, and moderated by experts from the University of Georgia, is currently underway.

The principal use-case for this Recommender employs two expert systems, and a PID controller bank. The first expert system assesses the user wellness state; the second expert system formulates recommendations (in colloquial natural language) that address the user's demographic, and areas of greatest deviation from wellness; and the PID controllers project the user state forward in time to characterize the benefits of following the recommendations. Two compliance factors specific to the individual user are applied to select recommendations most likely to be adopted by the user.

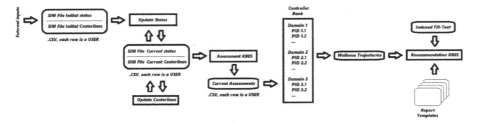

While this approach allows the determination the kinds of corrections that are required in a vector space sense, it is still necessary to determine what specific real-world actions must be undertaken to produce the indicated changes. This will be accomplished through the use of an aspect of AI based upon the use of trainable expert systems. These are state machines.

Different users will experience levels of difficulty carrying out specific recommendations. This means that the "terrain of action" will be specific to the individual: what is difficult ("uphill") for one person might be relatively easy ("level or downhill terrain") for another.

Therefore, the recommendations ("control signals") offered to move a person back to their centerline are individual in nature. These can be individualized over time. Initially, population norms can be used to establish these factors.

5   The Prototype System

The figure below depicts a functional architecture of the existing prototype. This prototype has been hosted on the web, and has a simple web-accessible user interface that facilitates data input and the delivery of reports created by the Recommender.

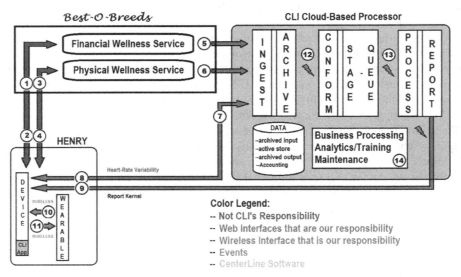

Notional CenterLine Institute Principal Use Case   MFH, 05/18/19
Rev.,MFH, 06/13/19

The legend below describes the interfaces depicted in the figure. (HENRY is term referring to the user.)

"Best-O-Breeds" are independent service applications that currently contract with user's to provide information services using as input data uploaded synchronously/asynchronously by users through an HMI web interface. CLI will partner with some set of these ("Best of Breeds", which will change from time to time) to provide to CLI the information they currently serve to their population of users.

The user side of the web interface resides on the user's computer. The CLI App data and software reside on the user computer.

Embedded data collection functionality resides on a User-Wearable device.

Data are exchanged synchronously/asynchronously wirelessly between the CLI App on the user Computer and the User-Wearable device.

- Circle 1: Financial Wellness Input from the user
  - Service Request Record (e.g., INSTALL App, generate report)
  - Account Management Request (CRUD Record having Customer ID and password, other request record(s))
- Circle 2: Financial Wellness Output to User CLI App
  - INSTALL/UPDATE/REMOVE account data and software for CLI App
  - Service Request Record (e.g., upload data request, app status request(s), billing/account information)
  - Information Services Report (i.e., product content)
- Circle 3: Physical Wellness Input from the user
  - Service Request Record (e.g., INSTALL App, generate report)
  - Account Management Request (CRUD Record having Customer ID and password, other request record(s))
- Circle 4: Physical Wellness Output to the user App
  - INSTALL/UPDATE/REMOVE account data and software for CLI App
  - Service Request Record (e.g., upload data request, app status request(s), billing/account information)
  - Information Services Report (i.e., product content)
- Circle 5: Financial Wellness Output to CLI Ingest Web Interface
  - Account Management Record (e.g., upload data request, app status request(s), billing/account information)
  - Annual Income, Credit Score, Short Save, Long Save, Health Insure, Life Insure, Budgeting, Mortgage Debt, College Debt, Non-Educational Unsecured Debt
- Circle 6: Physical Wellness Output to CLI Web Interface.

5.1 The Prototype Implementation

The implemented software starts with data input after login. The information required for the Knowledge Based Expert System to work is split into two parts: 1) some demographic information about the person to be evaluated, and 2) the measurement, score of importance, and score of compliance for each component in 3 wellness domains.

The demographic information is used to determine the ideal value of each component and the set of suitable recommendations and action plan tailor-made for the user. A few important information that influence heavily the outcome by the algorithms are, for example, age, marital status, and gender.

User Demographics

Name	JACQUELINE_LONG
Age	30
Gender	○ Male ● Female
Marital Status	● Single ○ Married
Education	2
Region	2

The measurement for each component of all wellness domains are required. In the prototype, the software requires direct input from the users. The final system should take the measurements directly from wearable health devices or obtain the score by presenting the user a survey. The score of importance reflects how important the user subjectively thinks the component contributes to his or her overall wellbeing. The score of compliance is the computed probability that a user may follow the recommendation given by the system to improve the overall wellness. Again, the prototype currently accepts direct input, but it should be computed automatically by the system basing on the frequency of the user following the recommendation in the final product.

Health Data

balance sheet	Value	Importance	Compliance	body	Value	Importance	Compliance	mind	Value	Importance	Compliance
annual income	16.096	0.1933	0.7646	bmi	15.489	0.5825	0.7541	positivity	0.2419	0.5660	0.7712
short save	0.0243	0.6648	0.2100	bp systolic	149.61	0.8249	0.18121	engagement	0.6233	0.5370	0.7821
long save	0.6751	0.7647	0.8740	bp diastolic	75.281	0.3337	0.9737	relationships	0.4856	0.3687	0.8396
mortgage debt	269.80	0.3048	0.0654	wb sugar	102.39	0.3628	0.7888	meaning	0.7385	0.4619	0.7507
college debt	46.038	0.4077	0.9282	resting hr	82.509	0.2159	0.1729	accomplishment	0.2981	0.5002	0.8603
unsec debt	10.178	0.6399	0.9440	act hr	116.65	0.6433	0.2446	emo stability	0.4531	0.2072	0.2364
credit score	613.62	0.2297	0.2478	age	29.646	0.5132	0.2441	optimism	0.2287	0.3818	0.7808
health insure	0.8913	0.4412	0.8666	sleeps	4.1947	0.4937	0.7902	resilience	0.5466	0.4033	0.7748
life insure	0.7941	0.5457	0.7796	tot cals	2299.6	0.3163	0.2485	self esteem	0.7807	0.5466	0.8220
budgeting	1.3296	0.3211	0.7675	carb cals	305.08	0.5466	0.8481	vitality	0.5560	0.6983	0.8116

The information is sent to the algorithm after collection to go through an evaluation engine, which gives a diagnosis on each component of all wellness domain, and a recommendation engine, which gives a recommendation to improve each wellness domain by targeting the most urgent, i.e. influential, component to be improved within. The recommendation engine also provides a forecast that shows the percentage of improvement the user will have over a period of time if he or she follows the given recommendation.

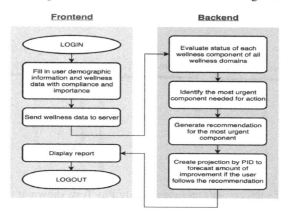

6   The Centerline Study

The authors are preparing a study to assess and optimize the Assessment-Recommender Prototype involving human volunteers.

Purpose:

- Collect information to assess and refine the performance of the Centerline Wellness system
- Collect empirical data to support the completion of a scientific paper to be presented at an international technical conference.

Duration of Study: 3 months

- On-boarding can occur in Week 1 of any month.
- In-Brief includes a survey instrument (next slide)
- Subjects will interact with Financial Counselors at the beginning, middle and end of each month.
- It is anticipated that all Financial Counselors will be Ph.D. Students/Candidates acting under the direction of one of the study Principal authors.

Out-Brief includes a summary interview

At the Financial Counselor Check-ins, subjects will be advised using recommendations created by the Centerline Wellness system. The figure below characterizes the study architecture.

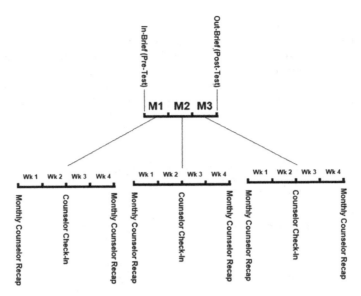

On-Boarding of subjects will include a Study Pre-Test, shown at right. Information will be collected during the In-Brief.

Subjects will complete the psychological assessment(s) online and report the results within the week of their In-Brief. (No specific assessment has yet been selected.)

The information being collected is only that which is required for the operation of the Centerline Wellness System.

Study Pre-Test

Psychological: STAI: State-Trait Anxiety Inventory
               Keirsey's Four Temperaments
               MBTI: Myers-Briggs Type Indicator
               Yerkes-Dodson Law

Physiological:

_____ Gender

_____ Marital Status

_____ Age (years, months)

_____ Weight (pounds)

_____ Height (feet, inches)

_____ Resting Blood Pressure (Systolic)

_____ Resting Blood Pressure (Diastolic)

Chronic Health Conditions (list) _____

Financial:

_____ Monthly Income (USD)

_____ Credit Score

_____ Long-term Savings (USD)

_____ Following a Budget? (yes/no)

_____ Have Health Insurance? (yes/no)

_____ Have Life Insurance? (yes/no)

_____ Total Mortgage Debt (USD)

_____ Total College Debt (USD)

_____ Total Unsecured Debt (e.g., credit cards, personal loans, USD)

Some information will be collected using a wearable device that subjects will wear at all times during the study. Some information will be collected during Check-ins with Financial Counselors, and in a post-study survey completed by subjects.

Monitored/Inferred Items

Heart Rate Variability (Mean, Variance, Sigma)

Resting Heart Rate (-0.5 Z)

Active Heart Rate (1.0 Z)

Minutes of Inactivity in last 24 hours (minutes Z<-0.5)

Minutes of elevated activity in last 24 hours (minutes Z>1)

Menstrual Cycle

Interview Feedback from Financial Counselors

Participant Surveys

Subjects will complete a checklist once a day (upon arising). These will be collected at the Financial Counselor Check-ins and/or monthly recaps. The information being collected is only that which is required for the operation of the Centerline Wellness System.

Name: _____   Date: _____   ☐ Morning   Evening ☐

Psychological Twice-Daily Checklist Items

(Check a box in each row, or leave blank...)
(- means more negative feeling, + means more positive feeling, leave blank if "just OK")

−   +

OK	Positivity:	I think today is a GOOD day!
OK	Engagement:	I feel like I am part what is going on around me today.
OK	Relationships:	I am satisfied with my relationships with others today.
OK	Meaning:	I believe that what I am doing today matters.
OK	Accomplishment:	I feel GOOD about this week's plans and accomplishments.
OK	Emo_Stability:	My emotions are steady and under control today.
OK	Optimism:	I am looking forward to the rest of this week.
OK	Resilience:	I am overcoming today's challenges.
OK	Self_Esteem:	I feel GOOD about myself today.
OK	Vitality:	I have plenty of energy today.

In at most three words: Best thing about today ⇨ [_____]

In at most three words: Worst thing about today ⇨ [_____]

Physiological Twice-Daily Checklist Items

[____] ⇦ Hours of Sleep in the last 12 hours (write in box)

[____] ⇦ Blood Sugar (if measured) (write in box)   ☐YES   NO☐   **Menstruating?**

Total Calories consumed since waking up today: (check one box)
☐ 0 - 1000   ☐ 1000 - 2000   ☐ 2000 - 3000   ☐ Over 3000

Total Colories of Carbohydrates consumed since waking up today: (check one box)
☐ 0 - 1000   ☐ 1000 - 2000   ☐ 2000 - 3000   ☐ Over 3000

Financial Twice-Daily Checklist Items

(check one box in each row)

☐YES   NO☐   **Short-term Savings: I have more than enough money to meet this month's expenses.**

☐YES   NO☐   **Budgeting: I am sticking to a budget today.**

At the conclusion of the study, subjects will complete a Summary Interview. The Summary Interview includes a survey about the study experience. The Out-brief Survey has not yet been developed

6.1 Handling "missing data": A Degapping Algorithm for Numeric Data

Any decision-making system that operates in a dynamic, real-world environment must have a mechanism in place for coping with missing data: data that, for whatever reason, is not available at the time a decision must be made. This is a common challenge when data must be volunteered by a human user.

One possible approach is to construct the decision system so that it can operate with partial (missing) input fields. Expert Systems are particularly good at this. Another possible approach is to use the data available to infer ("impute") values for missing fields. This section describes the latter process as we have implemented it.

One simple inter-vector imputation method is to replace missing values with their population means, a $O(n)$ process. This naïve approach is simple, but ignores context within the record. For numeric data, a more sophisticated method is the nearest neighbor normalization technique. This can be applied efficiently even to large data sets having many dimensions (in a brute force approach this is a $O(n^2)$ process).

The following is an explanation of how the nearest neighbor normalization method works. This technique proceeds in the following manner for each missing feature in a given vector, V_1:

1. From a degapping set of feature vectors, find the one, V_2, which:
   a. Shares a sufficient number of populated fields with the vector to be degapped (this is to increase the likelihood that the nearest vector is representative of the vector being processed).
   b. Has a value for the missing feature, F_m.
   c. Is nearest the vector to be degapped (possibly weighted).
2. Compute the weighted norms of the vector being degapped, V_1, and the matching vector found in step 1, V_2, in just those features present in both.
3. Form the normalization ratio $Rn = |V_1|/|V_2|$.
4. Create a preliminary fill value $P = R_n * F_m$.
5. Apply a clipping (or other) consistency test to P to obtain F'_m, the final, sanity checked fill value.
6. Fill the gap in V_1 with the value F'_m.

This method is based upon the idea that someone who is expensive/cheap in several areas is likely to be expensive/cheap in others, and by about the same ratio. The nearest neighbor normalization technique can be applied to nominal data, but in that application the available symbol in the matching vector is usually copied directly over without further processing.

6.2 Data Conformation for Human Instrumentation

The authors have developed a software application which monitors a human user's wellbeing from the interconnected aspects of psychological, physiological, and financial "wellness". The application uses one knowledge-based expert system to assess the user's wellness state, and a second knowledge-based expert system to make specific actionable suggestions for optimizing the user's wellness by addressing problem areas. This recommender system creates these suggestions periodically as part of a monitored feedback loop. The recommendations are dynamic, informed by the user's demographic, and personalized for the user's goals.

Collecting data for assessing the physiological and financial aspects of wellness is fairly straightforward. More difficult is the characterization and collection of data for objectively assessing a user's psychological state.

Heart-Rate Variability (HRV) is regarded by many experts as an informative, non-invasive, passively accessible physiological indicator of psychological stress (Thayer et al. 2012). It also carries information about physical activity in terms of tempo and intensity, which can be used to infer sleep and activity cycles. For this reason, the authors are using HRV as a salient indicator of psychological wellness, and drawing upon the significant research in this area (Hancock 2020).

Continuous HRV monitoring can be done in a variety of ways. In recent years, a variety of devices for this purpose have been produced for direct sale to the general public. These wearable devices include chest strap monitors, wrist-mounted monitors, and monitors worn as rings. While chest straps are the most accurate and are relatively inexpensive, they have not enjoyed wide-spread popularity with the public; most people would rather use a wrist-wearable device.

The Conformation Problem

This variety of collection devices poses a possible obstacle to adoption of systems that use HRV data. Existing devices themselves are very different in quality, operation, and cost. Any add-on system that relies upon HRV data must accommodate this variation, since users are unlikely to adopt a system that requires them to purchase a new monitoring device.

A Conformation Solution

A solution to this problem is the construction of a device-agnostic front-end, which ingests data from a wide range of devices, and conforms it to a device agnostic standard for subsequent processing. The authors have built such a conformation transform. This transform has been calibrated using data collected in a live trial for a variety of input devices. The conformation transform uses a multi-variate regression-(parameterized radial basis function, RBF).

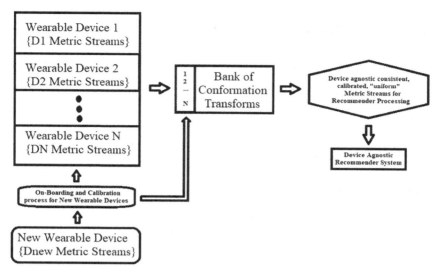

Ground-truth data are collected by instrumenting a "standardizing user" with a chest strap heart rate monitor. These devices achieve high accuracy by monitoring cardio-electric signals by direct contract through the chest wall.

Simultaneously, the standardizing user wears various wrist-mounted heart monitors [list here]. These concomitant data serve as the independent and dependent variables for the regression, respectively. The RBF is numerically optimized to transform the wrist data (which is of variable sampling rate, sensitivity, precision, bio-coupling effectiveness, etc.) to best-estimates of the corresponding chest-strap data.

7 A Sample Report Generated Automatically Without Human Involvement by the Recommender Expert System

ANTONE MCMAHON is a 31-year-old single male.

Evaluation Report
Financial State
ANTONE MCMAHON's overall financial wellness level is 0.004, which is at the Centerline for his demographic.
He has reached or exceeded Centerline Financial Wellness in 8 out of 10 areas.
The Financial Wellness factors that meet or exceed the centerline for his demographic are:

Assessment	Value	Z-Score	Set Point	Importance	Compliance
A medium level of annual income is fine:	31.1	-0.485	35.002	0.848	0.881
A medium level of short save is fine:	0.057	0.337	0.05	0.681	0.796
A medium level of long save is fine:	0.461	-0.173	0.496	0.257	0.235
A medium level of college debt is fine:	28.875	-0.383	31.786	0.876	0.812
A medium level of unsec debt is fine:	5.785	0.086	5.218	0.544	0.248
A medium level of credit score is fine:	603.143	-0.282	610.03	0.671	0.151
A medium level of life insure is fine:	0.995	-0.022	1.003	0.501	0.758
A low level of mortgage debt is good:	11.517	-1.293	100.995	0.539	0.222

His Financial Wellness factors that need improvement are:

Assessment	Value	Z-Score	Set Point	Importance	Compliance
A low level of health insure is not healthy:	0.751	-0.612	1.0	0.325	0.793
A low level of budgeting is not healthy:	0.411	-1.444	1.011	0.682	0.756

Physical State
ANTONE MCMAHON's overall physical wellness level is -0.112, which is under the Centerline for his demographic.
He has reached or exceeded Centerline Physical Wellness in 9 out of 10 areas.
The Physical Wellness factors that meet or exceed the centerline for his demographic are:

Assessment	Value	Z-Score	Set Point	Importance	Compliance
A medium level of bmi is fine:	22.07	0.087	21.554	0.178	0.783
A medium level of wb sugar is fine:	92.474	0.914	84.864	0.605	0.963
A medium level of act hr is fine:	128.178	-0.112	129.594	0.199	0.766
A medium level of age is fine:	31.588	-0.203	31.588	0.171	0.216
A medium level of tot cals is fine:	2711.334	0.971	2395.075	0.484	0.177
A medium level of carb cals is fine:	497.583	0.593	402.509	0.563	0.202
A very high level of bp diastolic is quite unhealthy:	97.501	2.131	79.816	0.171	0.765
A very low level of resting hr is quite unhealthy:	45.935	-3.184	71.723	0.207	0.248
A very low level of sleeps is quite unhealthy:	3.775	-2.574	8.009	0.698	0.753

His Physical Wellness factors that need improvement are:

Assessment	Value	Z-Score	Set Point	Importance	Compliance
A high level of bp systolic is not healthy:	137.864	1.095	119.667	0.507	0.22

Psychological State
ANTONE MCMAHON's overall psychological wellness level is -0.067, which is under the Centerline for his demographic.
He has reached or exceeded Centerline Psychological Wellness in 6 out of 10 areas.
The Psychological Wellness factors that meet or exceed the centerline for his demographic are:

Assessment	Value	Z-Score	Set Point	Importance	Compliance
A medium level of positivity is fine:	0.325	0.159	0.308	0.436	0.196
A medium level of engagement is fine:	0.585	-0.105	0.61	0.397	0.784

A medium level of meaning is fine:	0.679	0.328	0.618	0.739	0.951
A medium level of emo stability is fine:	0.38	-0.148	0.412	0.547	0.889
A medium level of optimism is fine:	0.368	0.258	0.323	0.411	0.17
A medium level of resilience is fine:	0.594	-0.042	0.617	0.663	0.804

His Psychological Wellness factors that need improvement are:

Assessment	Value	Z-Score	Set Point	Importance	Compliance
A low level of relationships is not healthy:	0.291	-0.89	0.513	0.408	0.842
A low level of accomplishment is not healthy:	0.011	-1.873	0.309	0.463	0.155
A low level of self esteem is not healthy:	0.572	-0.783	0.708	0.563	0.755
A low level of vitality is not healthy:	0.503	-1.21	0.725	0.899	0.237

Recommendation Report
For domain financial, the significant issue area is budgeting:

Significance score	0.409
Importance to user	0.682
User compliance	75.6%
Centerline deviation	−0.6

Recommendations:
Increase: Plan! Develop weekly monthly and annual budgets you can live with. Be realistic but hold your resolve.

If your absolute essential expenses overshoot 50 percent of your income you may need to dip into the wants portion of your budget for a while. Its not the end of the world but youll have to adjust your spending. Even if your necessities fall under the 50 percent cap revisiting these fixed expenses occasionally is smart. You may find a better cell phone plan an opportunity to refinance your mortgage or less expensive car insurance. That leaves you more to work with elsewhere.

Forecast:
If he follows the recommendation above, he may expect an 82.343% improvement in 5 weeks.

[placeholder for projection visualisation]

For domain physical, the significant issue area is bp_systolic:

Significance score	9.228
Importance to user	0.507
User compliance	22.0%
Centerline deviation	18.197

Recommendations:
Decrease: Enroll in a gym for weekly exercise. Take an afternoon and evening walk. Spend less time on sedentary activities.

When it comes to resting heart rate lower is better. It usually means your heart muscle is in better condition and doesnt have to work as hard to maintain a steady beat. Studies have found that a higher resting heart rate is linked with lower physical fitness and higher blood pressure and body weight.

Forecast:

If he follows the recommendation above, he may expect an 92.892% improvement in 5 weeks.

[placeholder for projection visualisation]

For domain psychological, the significant issue area is relationships:

Significance score	0.09
Importance to user	0.408
User compliance	84.2%
Centerline deviation	−0.221

Recommendations:

Increase: Discuss relationships with someone older who cares about you. Read books about improving relationships.

Cuddling kissing hugging and sex can all help relieve stress. Positive physical contact can help release oxytocin and lower cortisol. This can help lower blood pressure and heart rate both of which are physical symptoms of stress. Interestingly humans arent the only animals who cuddle for stress relief. Chimpanzees also cuddle friends who are stressed.

Forecast:

If he follows the recommendation above, he may expect an 73.517% improvement in 5 weeks.

References

Grable, J.E.: Psychophysiological economics: introducing an emerging field of study. J. Finan. Serv. Prof. **67**(5), 16–18 (2013)

Thayer, J., Abs, F., Fredrikson, M., Sollers, J., Wager, D.: A meta-analysis of heart rate variability and neuroimaging studies: implications for heart rate variability as a marker of stress and health. Neurosci. Biobehav. Rev. **36**(2), 747–756 (2012)

Hancock, M.: Non-monotonic bias-based reasoning under uncertainty. In: Proceedings of the 14th International Conference on Augmented Cognition, Copenhagen, Denmark, July 2020

The Case for Cognitive-Affective Architectures as Affective User Models in Behavioral Health Technologies

Eva Hudlicka[✉]

Psychometrix Associates & Therapy 21st, Amherst, MA, USA
hudlicka@ieee.org

Abstract. As technology proliferates into all aspects our lives, affect-adaptive interaction becomes increasingly important. A growing area where this is particularly critical is behavioral health technology: mental health (MH) apps, serious therapeutic games, virtual agents and social robots, and virtual reality environments. These technologies aim to support mental health and wellness via psychoeducation, behavior coaching, supportive and motivational interventions, and opportunities to practice coping skills. Affect-adaptive interaction requires detailed affective user models, capturing the range of the user's affective states, their triggers, expressive manifestations, and consequences on behavior, within the specific interaction context. Cognitive-affective architectures have been proposed as a possible approach to affective user modeling, due to their ability to represent detailed information about internal processing and support complex what-if and abductive reasoning. However, their construction is highly labor-intensive, requiring significant knowledge engineering and parameter tuning. In this paper I argue that the unique interaction context offered by MH apps helps address these issues. By facilitating self-reporting and non-intrusive sensing, mobile apps support user-driven construction of individual- and context-specific cognitive-affective architectures. The focus on emotions makes the explicit collection of affective data central, thereby encouraging direct user involvement, which is also enhanced by engaging the user in the on-going architecture refinement. I describe an approach to constructing a cognitive-affective architecture-based affective user model, providing several illustrative examples which demonstrate how the model would support more personalized assessment, intervention and progress tracking in mental health apps. The paper concludes with a discussion of challenges and opportunities.

Keywords: Affective HCI · Cognitive-affective architectures · Affective user models · Behavioral health technology · Mental health apps

1 Introduction

As technology proliferates into all aspects our lives, affect-adaptive interaction becomes increasingly important. A growing area where this is particularly critical is behavioral

© Springer Nature Switzerland AG 2020
D. D. Schmorrow and C. M. Fidopiastis (Eds.): HCII 2020, LNAI 12197, pp. 191–206, 2020.
https://doi.org/10.1007/978-3-030-50439-7_13

health technology: mental health (MH) apps, serious therapeutic games, virtual agents and social robots, and virtual reality environments. These technologies aim to support mental health and wellness by providing psychoeducation (e.g., causes of depression, coping strategies for anxiety), behavior coaching (on-going support to establish healthy behavior, e.g., exercise), supportive and motivational interventions (directly providing emotional support; targeting specific components of motivation), and opportunity to practice specific skills and coping strategies (e.g., practice assertiveness skills in gaming environments).

Affect-adaptive interaction requires detailed affective user models, capturing the range of the users' affective states, their triggers, expressive manifestations, and consequences on behavior, within the specific interaction context. In therapeutic contexts, additional information is required, including preferences for the type of support the user prefers to manage distressing emotions and maintain motivation for change, and information about the relative effectiveness of different emotion regulation strategies; e.g., User A may prefer a display of empathy when feeling discouraged, whereas User B may prefer a "tough love" approach to be pushed to a higher performance level.

Cognitive-affective architectures have the necessary representational and inferencing capabilities to capture the affective user information outlined above, and to serve as dynamic, simulation-capable affective user models. However, their construction is highly labor-intensive, and infeasible for most current contexts where affect-adaptive interaction is explored (e.g., intelligent tutoring, affective gaming).

However, the emergence of mobile apps, and their capability to rapidly and easily collect large quantities of individual user data, have created the possibility for dynamic, user-driven construction of individual- and context-specific cognitive-affective architectures. In this paper I make a case for exploring the effectiveness of cognitive-affective architectures as affective user models, and describe how a cognitive-affective architecture could be constructed from user data collected via mobile devices. I describe examples of the resulting affect-adaptive interactions in the context of behavioral health technologies; specifically, in the context of an anxiety management MH app. Although the focus here is on behavioral health technologies, the approach is applicable more broadly to any context where affect-adaptive interaction is beneficial and where the required user data can be readily collected.

The paper is organized as follows. Section 2 provides the relevant background. Section 3 makes the case for using cognitive-affective architectures as dynamic affective user models. Section 4 describes how such models would be created from user data, collected via a mobile MH app, and provides several examples of improved affect-adaptive interaction. Section 5 briefly addresses some challenges and limitations of the proposed approach, including ethical issues.

2   Background Information

This section provides a brief overview of the areas relevant for the proposed approach: affect-adaptive interaction and affective user modeling, behavioral health technologies, computational affective modeling, and cognitive-affective architectures.

2.1 Affect-Adaptive Interaction and Affective User Modeling

Affect-Adaptive Interaction. Affect-adaptive interaction refers to the ability of a system to detect the user's affective state (*'emotion recognition'*), understand the consequences of this state for the interaction and user task performance (*'emotion understanding'*), and adapt the system functionality to ensure the desired user behavior (e.g., safe driving, successful learning or gameplay) and subjective state (e.g., minimal frustration, satisfaction or flow). The ultimate objective is to establish an 'affective loop' (Sundstrom 2005) between the user and the system: an engaging interaction, where the system responds to the users' emotions to optimize both the user performance and his/her mental (cognitive+affective) and physical state. Examples of such interactions include players being highly engaged in a gameplay and experiencing flow states (Csikszentmihalyi 1990), or users interacting with a virtual agent acting as a coach, and experiencing the agent as caring and empathic over the course of long-term interaction.

Affect-adaptive interaction is a growing area of research involving Affective Computing, AI and HCI, across several application areas: affect-adaptive gaming (Karpouzis and Yannakakis 2016; Yannakakis and Togelius 2018); intelligent tutoring and training (e.g., AffectiveAutoTutor (D'Mello and Graesser 2013)); artificial social agents (e.g., Tielman et al. 2019) and decision-support systems (e.g., adapting to driver stress levels (Healy and Picard 2005; Nasoz et al. 2010)). In gaming, tutoring and decision-support systems the typical affective states of interest are boredom and interest, frustration or anger, satisfaction and disappointment, and fear and anxiety. Interactions with artificial social agents represent the most challenging contexts for affect-based adaptations. These virtual agents and robots are increasingly called up to display emotional and social intelligence, whose core feature is the recognition of, and adaptation to, their interaction partners' affective states. Social agents or robots should be able to detect a variety of human emotions, relevant to the specific task context, and display appropriate emotions to the human user, at the appropriate level of intensity, to establish an affective loop. The specific adaptations differ across the contexts (e.g., complexity in tutoring systems, content and difficulty in games, goal difficulty and type of emotional support in behavior coaching).

Affective User Modeling. To achieve affect-adaptive interaction the system needs information about the user's affective behavior within the specific interaction context: an *affective user model* (Hudlicka and McNeese 2002). Ideally, these models would represent detailed affective profiles of the users, including: typical emotions experienced, within the interaction context (e.g., game, tutoring session); their triggers (e.g., losing or gaining game points); their manifestations (e.g., slower rate of interaction); user's affective goals (e.g., reduce anxiety, increase satisfaction); possible adaptations; and strategies for emotion regulation.

Typically, affective user models (also referred to as learner models, or player models, depending on the context) are more limited in scope, focusing on information facilitating user emotion recognition, and selection of a particular adaptation. In symbolic user models, this information is often represented by Bayesian belief nets (Conati and MacLaren 2009; Grawemeyer et al. 2017) or rules (D'Mello and Graesser 2013). Increasingly, machine learning methods (supervised, unsupervised and reinforcement learning) are being used to develop user models (e.g. Holmgard et al. 2014), based on user data collected automatically during the interaction, from a variety of sensors (e.g., level of arousal from sensors detecting autonomic nervous system activation, or emotional facial expressions). Although some of these models are generative (i.e., capable of simulating aspects of user behavior), they do not represent the hypothesized structures of the underlying cognitive and affective processes, and thus do not lend themselves to the psychotherapeutic applications described below.

2.2 Behavioral Health Technologies and Mobile Mental Health Apps

The term 'behavioral health technologies' (BH technologies) refers to a variety of systems and applications whose aim is to promote and support mental health and wellness, by providing psychoeducation (e.g., causes of depression, coping strategies for anxiety), behavior coaching (on-going support to establish healthy behavior, e.g., exercise), supportive and motivational interventions (directly providing emotional support; targeting specific components of motivation), and opportunity to practice specific skills and coping strategies (e.g., practice assertiveness skills in gaming environments). BH technologies range from the relatively simple display of information about a particular topic (e.g., "chronic stress can cause depression"; "exercise reduces anxiety") (Wahle et al. 2017), through interactions with chatbots and embodied conversational agents (Provoost et al. 2017; Gaffney et al. 2019), to therapeutic serious games and interactions with virtual characters (Hudlicka 2016; Hudlicka et al. 2008); and virtual reality settings that provide immersive, realistic environments within which to develop new skills (Garrett et al. 2018; Oing and Prescott 2018).

Increasingly, these technologies are becoming available on mobile devices and often enable the users to monitor their affective states, to better understand the triggers and causes of both negative and positive emotions. These data then become the basis for improved understanding of the individual's affective response patterns, the factors that cause and maintain emotional distress, as well as factors that help induce positive emotions and promote mental health. This information can be used by the user alone, or by the user in conjunction with a psychotherapist.

Given the rapid growth in these technologies over the past decade, including the development of thousands of mobile mental health apps and web sites, it is beyond the scope of this paper to provide an in-depth review. A recent overview of these technologies can be found in Luxton (2016).

2.3 Computational Affective Modeling

The past two decades have witnessed a rapid growth in symbolic computational models of emotion and affective architectures (Hudlicka 2014). Although some emotion models are being developed for basic research purposes, aiming to elucidate the mechanisms that mediate emotions in biological agents (e.g., Hudlicka 2008) the majority of existing emotion models have been developed to enhance agent and robot affective realism and social competence (e.g., Paiva et al. 2005; Becker et al. 2005).

These emotion models typically implement emotion generation via cognitive appraisal, most often using the OCC model of appraisal (Ortony et al. 1988), and model the effects of the dynamically generated emotions on the robot or agent's expressions and behavior selection, via direct mapping of a particular emotion on the configurations of the available expressive channels in the agent or robot (e.g., face, gestures), as well as behavioral choice (e.g., approach vs withdraw).

I have previously suggested that in order to advance the state-of-the-art in emotion modeling, it would be helpful to deconstruct the high-level term 'emotion modeling' by (1) suggesting that we view emotion models in terms of two fundamental categories of processes: *emotion generation* and *emotion effects;* and (2) identifying some of the fundamental computational tasks necessary to implement these processes (Hudlicka 2011). These 'model building blocks' can then provide a basis for the development of more systematic guidelines for emotion modeling, theoretical and data requirements, and representational and reasoning requirements and alternatives. Figure 1 provides a summary of the processes (generic computational tasks) involved in emotion generation and emotion effects modeling.

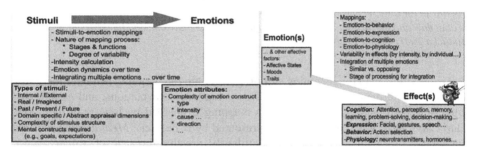

Fig. 1. The generic computational tasks required for modeling emotion generation via cognitive appraisal (left) and emotion effects (right). (Hudlicka 2011)

2.4   Cognitive-Affective Architectures

Emotion models are typically embedded within a broader agent architecture which then controls the behavior of an associated virtual agent or robot. A number of symbolic cognitive-affective architecture have been developed, primarily in the context of agent research (e.g., FaTIMa (Paiva et al. WASABI (Becker-Asano et al. 2008), but also for basic research (Sloman et al. 2005). Below I briefly describe the MAMID architecture, to provide a concrete example of a symbolic cognitive-affective architecture (Hudlicka 2007).

MAMID was developed to model the effects of emotions on cognitive processing, via a series of parameters controlling the processing within the individual modules (refer to Fig. 2). MAMID implements a sequential see-think/feel-do processing sequence, consisting of the following modules: *Attention (*filters incoming cues and selects a subset for processing); *Situation Assessment* (integrates individual cues into an overall situation assessment); *Expectation Generation* (projects current situation onto possible future states); *Affect Appraiser* (derives a valence and four of the basic emotions. from external and internal elicitors); *Goal Management* (identifies high-priority goals); and *Behavior Selection* (selects the best actions for goal achievement).

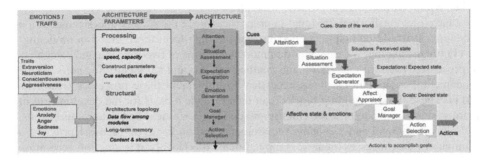

Fig. 2. MAMID Cognitive-Affective Architecture. *Left side of the figure illustrates the relationship of the emotions and traits, the parameters controlling the architecture processing, and the structure of the architecture. Right side shows a more detailed view of the architecture, illustrating the data flow and the distinct constructs manipulated by the individual modules (in yellow stickies) (Hudlicka 2007)* (Color figure online)

The modules map the incoming stimuli (cues) onto the outgoing behavior (actions), via a set of intermediate internal structures (situations, expectations, and goals), collectively termed *mental constructs*. This mapping is enabled by long-term memories (LTM) associated with each module, represented by belief nets. *Mental constructs* are characterized by their attributes (e.g., familiarity, novelty, threat level, valence, etc.), which influence their processing; that is, their rank and the consequent likelihood of being processed by the associated module within a given execution cycle; (e.g., cue will be attended, situation derived, goal or action selected). MAMID models both emotion generation and emotion effects, but focuses on affective biases on cognition.

3   The Case for Cognitive-Affective Architectures as Affective User Models in Behavioral Health Technologies

In the context of behavioral health technologies, which often directly address the recognition, regulation and management of affective states, affect-adaptive interaction plays a central role, and can 'make or break' the usefulness of a particular application. For example, consider a mobile app whose aim is to provide support for individuals with depression and a situation when the user indicates that they are feeling "sad and hopeless". There is great variability in the specific response that different individuals would find useful and supportive. For some people, a statement showing empathy and understanding helps; e.g., "Yes, that sounds very distressing. I'm sorry things are so difficult for you right now." For others, a statement combining empathy with psychoeducation and hope can be helpful; e.g., "Sorry things are so tough for you right now, but you know that these feelings won't last and you just have to get through this brief period." For yet others, a 'downward social comparison' can be helpful, where they are reminded that others are worse off and, by comparisons, their situation is not so dire; e.g., "Sorry things are so difficult for you right now but at least your work is going ok and you don't have to worry about getting food on the table."

What enables psychotherapists to 'say the right thing' in these situations is their knowledge of the specific individuals: their personality, their history (what has or has not worked in the past), and their general preferences regarding emotional support. Returning to the example above: for some individuals, a reminder that they have some things for which to be grateful is all they need to feel some relief, whereas for others this could exacerbate their depression, make them feel even more disconnected and hopeless. Clearly, providing a mismatched response could seriously exacerbate the situation and severely increase the individual's distress.

Thus the key to successful emotional support interventions in behavioral technologies, and more broadly, effective affect-adaptive interaction in general, is a detailed, accurate affective model of the user. The model should capture the range of the user's affective states, their associated triggers and idiosyncratic expressive manifestations, and the consequences of each state on the user's behavior, within the specific interaction context. For technologies whose aim is to provide emotional support and help the users with emotion regulation, additional information would need to be included regarding preferences for specific emotion regulation strategies.

Cognitive-affective architectures have the ability to represent highly detailed information about the user's internal processing, and thus provide the basis for more accurate and useful adaptations to the user's changing affective and cognitive states. The fact that they are executable representations, capable of supporting simulations and predictions, further augments their utility.

Although the use of cognitive-affective architectures as affective user models has been suggested (Martinho et al. 2000), the labor-intensive knowledge engineering effort required for their construction have made this approach difficult or infeasible. However, the increasing, pervasive use of mobile apps that have the ability to implement momentary ecological assessments (i.e., quickly obtain user data regarding a specific attribute, such as current emotional state), significantly increases the feasibility of dynamically constructing user-specific cognitive-affective architectures (Hudlicka 2019).

The availability of specific user data (affective, cognitive, behavioral) that can be collected via mobile devices (both via self-reports and automatic sensing) provides a basis for constructing an executable, cognitive-affective architecture based, affective user model that represents the task-relevant aspects of the user's cognitive-affective processing (e.g., anxiety management). The architecture structure and processing can dynamically evolve and become increasingly accurate as the user remains engaged with the app, and continues to provide data to populate and refine the model. The user also has the ability to inspect the model and correct any inaccuracies. In the case of behavioral health technologies, the fact that an architecture-based user model is executable provides another distinct advantage, by supporting more personalized assessment, treatment planning and outcome monitoring, as illustrated by the examples below.

4 User-Driven Dynamic Construction of a Cognitive-Affective Architecture User Model

Below I describe how an affective user model, based on a user-specific cognitive-affective architecture, would be constructed from user data. I then provide several examples illustrating how the architecture would support enhanced affect-adaptive interaction with the user, and improve the app's effectiveness. Anxiety management is used as the specific context, and the MAMID cognitive-affective architecture as the architecture framework. However, the proposed approach applies more broadly to other mental health conditions and other symbolic cognitive-affective architectures. Figure 3 provides a high-level overview of the overall app system architecture. (Note that the app represents a conceptual design and has not yet been implemented.)

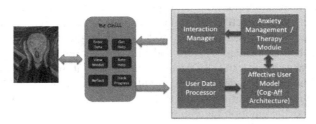

Fig. 3. High-level overview of the app system architecture

4.1 Functionality of the Anxiety Management MH App

Since the specific objective of the app would be anxiety management, the focus would be on helping the user identify triggers and manifestations of anxiety, track and become familiar with the idiosyncratic dynamics of the anxious states, and track the impact of anxiety management strategies and progress over time. The MH app would thus provide the following functionalities:

Architecture-Based Affective Model Construction and Refinement

- Help identify anxiety triggers (specific events, individuals, thoughts, sensations)
- Help track the dynamics of an evolving anxiety episode; anxiety onset and decay
- Track the effects of anxiety on thoughts, behavior and subsequent affective states

Delivery and Customization of Interventions

- Provide reminders of anxiety reducing strategies; emotion regulation strategies
- Suggest new emotion regulation strategies, based on user-specific information

Track Progress

- Help track the impact of the emotion regulation strategies

- Track progress over time by showing changes in the model reflecting new coping strategies, their ability to reduce anxiety, and more positive expectations and self-evaluations

Make Predictions and Suggest Causes

- Predict the effects of new stimuli (e.g., events, individuals) and provide this information to the user; both positive ("X could make you feel less anxious") and negative ("Better be careful in situation Y, as it could make you feel anxious")
- Use abductive reasoning to help identify causes of problematic behavior or anxiety

4.2 Construction and Refinement of an Architecture-Based Affective User Model

During the initial phase of using the app, the interaction would focus on the construction and refinement of the affective user model, as represented by a particular architecture framework (in this case the MAMID architecture). The model building process would begin by asking the user to input anxiety triggering cues. These could be: specific thoughts, events or situations, individuals, locations, physical sensations (e.g., pain), or emotions (e.g., awareness of anger could trigger anxiety, and the awareness of feeling anxious could further increase anxiety, as is the case in panic attacks). The user would also indicate the intensity of the experienced anxious state, as well as any specific effects on behavior. For example, a college student might indicate that going to see a professor during office hours to ask for homework help produces a high degree of anxiety. The effects on the student's behavior could include inability to discuss the homework issues, procrastination and missing the office hours, avoiding seeing the professor altogether, or avoiding working on the homework (refer to Fig. 4).

This user-provided information would be encoded in the form of directed graphs, which would be available to the user to visualize the emerging model structure, as well as to support further data entry during the model refinement phase. The data in these graphs would then be processed by the "User Data Processor" module (refer to Fig. 3), and mapped onto the cognitive-affective architecture modules and the knowledge-bases associated with each module (represented by the dashed vertical arrows in Fig. 4). For example, the node in the graph "See Prof. during office hours" would be mapped onto a *situation construct* represented in the belief nets associated with the Situation Assessment module of the architecture, and the nodes "Expect Criticism" and "Expect Put Downs" would be mapped onto *expectation constructs* associated with the Expectation Generation module.

The app would also support subsequent, and on-going, refinement of the model, by allowing the user to enter additional data about the felt anxiety, triggers and effects on behavior, and by modifying the existing model structures. This could be done either during an active anxiety episode, or during a reflective period, while analyzing past anxious episodes. For example, upon noticing a feeling of increasing anxiety, the user would select the 'Enter Data' button. The app would then display the existing trigger-emotion-behavior diagram, and allow the user to either select an existing node in the graph, to indicate another instance of the same trigger or effect, or enter new data (new node in the graph). The user would also indicate the level of felt anxiety. For

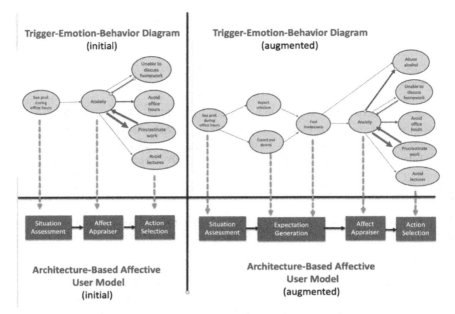

Fig. 4. Two versions (initial on the left, augmented on the right) of the "Trigger-Emotion-Behavior" diagram representing user data (top) and corresponding architecture-based affective user model (bottom)

example, while getting ready to see the professor during office house, the user might notice that he expects to be criticized or demeaned by the professor, and these thoughts further increase his anxiety. Upon further reflection, while not anxious, the user might realize that receiving criticism makes him feel inadequate, and that the belief of being inadequate further contributes to his feeling of anxiety. In addition, the user might then notice that it is during the periods of low-confidence that s/he turns to alcohol. These additional data would then result in an augmented model (Fig. 4, right), which would more accurately and extensively represent the elements of the user's anxious episodes, either by augmenting existing elements of the model structure, or by augmenting the model with additional triggers and behaviors.

Note that this process of affective model construction and refinement is very similar to the process that takes place in psychotherapy, when the therapist works with a client to identify and articulate the problematic affective states, their causes and dynamics, and their consequences for the client's functioning.

4.3   Delivery and Customization of Interventions

The app would be initialized with a generic set of interventions to suggest to the user, based on existing evidence regarding strategies for reducing anxiety; e.g., take a few deep breaths, go for a walk, watch a funny video clip, call a friend, remind yourself that the feeling is transient, remind yourself of your strengths, challenge catastrophizing thinking, etc. Although many of these strategies may appear self-evident, the challenge for many individuals who are experiencing an acute anxiety episode is to actually remember these strategies. A simple reminder of a coping strategy, possibly including specific instructions (e.g., for deep breathing), can thus be very helpful.

The app could be configured to display these suggestions in one of several modes: via an explicit request when the user clicks the 'Get Help' button; at any time, when the user indicates that s/he is experiencing anxiety, possibly anxiety above a specific threshold; or linked to some other sensor monitoring the user's physiological state or behavior (e.g., when the heart rate reaches a specific threshold). The third alternative would need to be carefully coordinated across multiple sensors, to avoid situations such as the app encouraging the user to take a walk while the user was on his/her daily 5K run, which caused the increased heart rate.

The user would have the option to provide feedback regarding the effectiveness of the suggestion, by clicking the 'Rate Help' button. The affective user model would thus be refined over time, as the emotion regulation strategies would be ranked by their effectiveness. The app would then suggest strategies that have proved helpful in the past, thereby providing more individualized suggestions to the user. The strategies would also be organized by the intensity of anxiety for which they were effective. For example, anxiety management strategies that emphasize cognition, such as cognitive re-appraisal of the situation or challenging a particular distorted thinking style (e.g., challenging catastrophizing thought patterns) may be more effective at lower anxiety levels. Higher anxiety levels may require more 'active' strategies, which directly involve the autonomic nervous system (e.g., deep breathing, brisk walk).

The model would also support customizing the offered strategies, as well as entering additional user-specific coping strategies, to augment the default list. An example of the former would be implemented by asking the user for additional details. For example, generic suggestions such as "Call a friend" could be customized for a specific person, e.g., "Call Bob", and generic suggestions such as "Try taking a walk" could be customized to a specific place. In many cases, more concrete suggestions are more likely to be implemented. An example of the latter would be querying the user for specific anxiety management strategies that have been helpful in the past, and adding these to the default list; e.g., "Take a bath", "Ride your bike", "Play the guitar".

In addition to these relatively simple user-adapted interactions, the architecture-based affective model would support the creation of more sophisticated, user-specific emotion regulation strategies, based on the information provided by the user during the model-refining interaction with the app. To generate these customized coping strategies, the 'Therapy Module' would use templates based on cognitive-behavioral therapy interventions to analyze the affective user model for specific anxiety-inducing situation interpretations, negative self-appraisals, and counterproductive or harmful coping strategies, and possibly query the user for additional information; e.g., previous successes to counter the negative self-assessments. For example, the augmented user model (refer to Fig. 4) indicates that one of the factors contributing to anxious feelings is a negative self-assessment ("Feeling Inadequate") and that one of the undesirable coping strategies is substance use ("Abuse Alcohol"). The 'Therapy Module' would construct from this information new, customized suggestions for emotion regulation, such as: "Criticism does not mean you are a bad student", "Remember that you did well on the first two tests", and, to try and re-direct the user from harmful coping strategies: "Although it may seem easier to get a drink, you will feel better in the long run if you take a walk when you feel anxious."

4.4 Tracking of Progress

With continued app use, the user would develop more awareness of his/her emotional states, and begin to explore different anxiety reduction coping strategies. The app would thus support the process that takes place in psychotherapy: increased ability to make explicit, symbolic representations of the distressing emotional states, and development of more adaptive coping behavior (e.g., 'go for a walk') instead of the maladaptive avoidant behavior ('procrastinate') or destructive behavior ('abuse alcohol'). During the process the users would continue to input data, and be able to visualize their progress. Figure 5 shows a segment of the initial model shown in Fig. 4, but augmented with a symbolic representation of the distressing feeling state (anxiety) as an explicit situation ('Feeling anxious'), which would then serve as a direct trigger for the newly acquired coping strategy ('Go for a walk'). Note that although the explicit awareness of anxiety also contributes to the feeling of anxiety (hence the arrow to the 'Anxiety' node in the diagram), the link to the coping strategy is much stronger, indicating that this is now the more frequent behavior and that the link from the coping behavior to anxiety is negative, indicating that the coping strategy reduces anxiety.

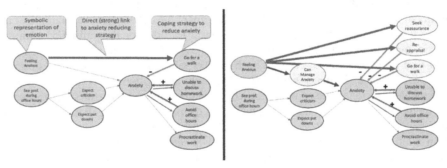

Fig. 5. Augmented user diagram showing the development of anxiety-reduction coping strategies (left) and further refinements (right)

The user would continue refining the model with additional coping strategies, represented by additional nodes representing alternative coping strategies (e.g., 'Thought Stopping', 'Re-appraisal', 'Seek reassurance') (Fig. 5, right). On-going user input would help define the strength of the links between different anxiety intensities and specific coping strategies. This would in turn support more customized affect-adapted interaction, by enabling the app to suggest specific strategy based on the current level of anxiety.

As the user would develop increased confidence in his ability to manage his anxiety, additional expectation nodes would be introduced; e.g., an expectation 'Can manage anxiety' would be introduced between the 'Feeling Anxious' and 'Anxiety' nodes, with a negative value on the link between the expectation and the anxiety intensity. As the user would gain increased confidence, the strength of these links would increase; in other words, a feeling of anxiety would immediately lead to an expectation that this anxiety can be managed, which would then reduce the felt sense of anxiety.

4.5 Make Predictions and Suggest Causes

The ability to support simulation is one of the key benefits of using a cognitive-affective architecture as an affective user model. The architecture can perform what-if reasoning, predicting the effects of particular situations or expectations on emotions, and the effects of emotions on behavior, based on the data previously entered by the user. Additional capabilities, such as the ability to generalize across situations, would further enhance the architecture's predictive abilities, enabling it to predict the effects of new situations and warn the user of possible increased anxiety (e.g., a critical colleague or partner), or point out that the new situation may reduce anxiety.

The architecture would also support abductive reasoning, helping the user identify possible causes of anxiety or problematic behavior. In addition to suggesting emotion regulation strategies, the 'Get Help' button would also access predictive and diagnostic (abductive) functionalities. For example, the user would indicate that they are struggling with procrastination. Using the existing model (refer to Fig. 4), the app would use abductive reasoning to identify possible causes of procrastination. Suggesting first that the user may be avoiding anxiety and that the anxiety could be due to an expectation of criticism. The user could then reflect on these suggestions, and thereby gain additional insight into the potential causes of the problematic behavior or distressing emotion. Again, this process resembles the type of systematic analysis of the causes of problematic behavior and emotions that takes place in psychotherapy.

5 Challenges and Limitations

The feasibility and utility of the proposed user-driven construction of an architecture-based affective user model would need to be demonstrated via empirical studies with implemented MH apps. A number of challenges exist, including: maintaining user engagement required to construct an accurate model; limitations of self-reports in regards to affective data; and appropriateness of the specific theories implicit in the cognitive-affective architecture framework.

Maintaining engagement is a recognized challenge in MH apps, and various methods are being explored to address this issue, including gamification, intelligent virtual agents, and use of augmented and virtual reality. On-going user engagement is essential for the development of the architecture-based user model. However, it is possible that direct user engagement in the construction of the customized user model, and the resulting enhancements of the affect-adaptive interactions, would enhance user motivation to continue active engagement with the app, and further motivate the user to continue refining of the model.

Self-reports cannot access automatic processing, which plays an important role in the maintenance and treatment of mental health disorders. This limitation could be partially addressed by augmenting self-reports with on-going, non-intrusive tracking of physiological signals reflecting arousal level, facial expression recognition, and behavior, to detect specific emotions, and identify triggers and effective coping strategies.

The architecture framework necessarily embeds a particular theory about cognitive and affective processing. These theories might be wrong, or simply inapplicable to the specific user context. Here again, however, the proposed approach resembles psychotherapy, for better or worse: each therapeutic approach is (hopefully) guided a specific theory, which may or may not be applicable to the client's problem. The on-going, user-driven model construction offers the means of identifying inaccuracies and refining the model structure, and thereby also has the potential to refine the associated theories regarding the processes involved in particular mental health disorders.

It should be pointed out that this approach would not be applicable for all types of users addressing all types of mental health conditions. For example, individuals suffering from major depressive disorders would likely lack the motivation required to construct the model, and individuals with attention deficit disorders might find it difficult to maintain the necessary to engagement.

Last but not least affect-adaptive interaction in general, and behavioral health technologies in particular, raise a number of ethical considerations. In addition to the serious issue of affective privacy, risks exist that are unique to the MH app context, including: possible induction of distressing emotions (perhaps via inappropriate affect-adaptive strategies); uncovering some disturbing MH issue or vulnerability during the model construction process; the possibility that the model makes a distressing prediction. In addition, there is the possibility that the user may rely on technology to address a serious mental health issue that should be addressed by a professional. It is beyond the scope of this paper to address the ethical issues in behavioral health technologies. A discussion of some of these issues can be found in Hudlicka (2016).

6   Conclusions

In this paper I argued that cognitive-affective architectures can serve as affective user models, and support enhanced affect-adaptive interaction with behavioral health technologies. The argument is based on the premise these interaction contexts are characterized by a unique set of features that help address the challenges associated with the construction of cognitive-affective architectures. Specifically: user- and task-specific context, explicit focus on emotion, ability to collect large amounts of user data (cognitive, affective, behavioral), and ability to support the refinement and corrections of the evolving model. The ability to represent detailed symbolic models of cognitive and affective processes, and support what-if and abductive reasoning, makes cognitive-affective architectures particularly well suited for affect-adaptive interactions with mental health apps. Examples describing the construction and use of an architecture-based user model illustrated this approach within the context of a notional mobile MH app for anxiety management, and highlighted its benefits.

As is always the case, the devil is in the details, and the ultimate feasibility and utility of the proposed approach can only be determined empirically. However, I believe that the approach outlined in this paper represents a promising area of research, and one with the potential to advance the state-of-the-art in affective user modeling, cognitive-affective architecture development, and behavioral health technologies.

References

Becker-Asano, C., Kopp, S., Pfeiffer-Leßmann, N., Wachsmuth, I.: Virtual humans growing up: from primary toward secondary emotions. KI Zeitschift (Ger. J. Artif. Intell.) **1**, 23–27 (2008)

Becker, C., Nakasone, A., Prendinger, H., Ishizuka, M., Wachsmuth, I.: Physiologically inter-active gaming with the 3D agent Max. In: Proceedings of the International Workshop on Conversational Informatics (JSAI-05), Kitakyushu, Japan (2005)

Conati, C., MacLaren, H.: Empirically building and evaluating a probabilistic model of user affect. User Model. User-Adap. Interact. **19**, 267–303 (2009). https://doi.org/10.1007/s11257-009-9062-8

Csikszentmihalyi, M.: Flow: Psychology of Optimal Experience. Harper & Row, New York (1990)

D'Mello, S., Graesser, A.: AutoTutor and affective AutoTutor: learning by talking with cognitively and emotionally intelligent computers that talk back. ACM Trans. Interact. Intell. Syst. **2**(4), 1–38 (2013)

Gaffney, H., Mensell, W., Tai, S.: Conversational agents in the treatment of mental health problems: mixed-method systematic review. JMIR Ment. Health **6**(10), e14166 (2019)

Garrett, B., Taverner, T., Gromala, D., Tao, G., Cordingley, E., Sun, C.: Virtual reality clinical research: promises and challenges. JMIR Serious Games **6**(4), e10839 (2018)

Grawemeyer, B., Mavrikis, M., Holmes, W., Gutierrez-Santos, S., Wiedmann, M., Rummel, N.: Affective learning: improving engagement and enhancing learning with affect-aware feedback. User Model. User-Adap. Interact. **17**(1), 119–158 (2017). https://doi.org/10.1007/s11257-017-9188-z

Healey, J., Picard, R.: Detecting stress during real-world driving tasks using physiological sensors. IEEE Trans. Intell. Transp. Syst. **6**(2), 156–166 (2005)

Holmgard, C., Liapis, A., Togelius, J., Yannakakis, G.: Generative agents for player decision modeling in games. In: Proceedings of 9th Foundations of Digital Games (2014)

Hudlicka, E.: Cognitive-Affective Architectures as Clinical Case Formulations. ISRE (2019)

Hudlicka, E.: Virtual companions, coaches, and therapeutic games in psychotherapy. In: Luxton, D.D. (ed.) AI in Mental Healthcare. Academic Press/Elsevier, Waltham (2016)

Hudlicka, E.: Affective BICA: Challenges and Open Questions. Biol. Inspired Cogn. Archit. **7**, 98–125 (2014)

Hudlicka, E.: virtual training and coaching of health behavior: example from mindfulness meditation training. Patient Educ. Couns. **92**(2), 160–166 (2013)

Hudlicka, E.: Guidelines for developing computational models of emotions. Int. J. Synth. Emot. **2**(1), 26–79 (2011)

Hudlicka, E.: Modeling the mechanisms of emotion effects on cognition. In: Proceedings of the AAAI Fall Symposium on Biologically Inspired Cognitive Architectures, TR FS-08-04, pp. 82–86. AAAI Press, Menlo Park (2008)

Hudlicka, E.: Reasons for emotions. In: Gray, W. (ed.) Advances in Cognitive Models and Cognitive Architectures. Oxford University Press, New York (2007)

Hudlicka, E., Lisetti, C., Hodge, D., Paiva, A., Rizzo, A., Wagner, E.: Artificial agents for psychotherapy. In: Proceedings of the AAAI Spring Symposium on Emotion, Personality and Social Behavior, TR SS-08-04, pp. 60–64 AAAI, Menlo Park (2008)

Hudlicka, E., McNeese, M.: User's affective & belief state: assessment and GUI adaptation. User Model. User Adapt. Interact. **12**(1), 1–47 (2002)

Karpouzis, K., Yannakakis, G.N. (eds.): Emotion in Games. SC, vol. 4. Springer, Cham (2016). https://doi.org/10.1007/978-3-319-41316-7

Luxton, D.D.: Artificial Intelligence in Mental Healthcare. Academic Press/Elsevier, Waltham (2016)

Martinho, C., Machado, I., Paiva, A.: A cognitive approach to affective user modeling. In: Paiva, A. (ed.) IWAI 1999. LNCS (LNAI), vol. 1814, pp. 64–75. Springer, Heidelberg (2000). https://doi.org/10.1007/10720296_6

Nasoz, F., Lisetti, C., Vasilakos, A.V.: Affectively intelligent and adaptive car interfaces. Inf. Sci. **20**(15), 3817–3836 (2010)

Oing, T., Prescott, J.: Implementations of virtual reality for anxiety-related disorders: systematic review. JMIR Serious Games **6**(4), e10965 (2018)

Ortony, A., Clore, G., Collins, A.: The Cognitive Structure of Emotions. Cambridge, New York (1988)

Paiva, A., Dias, J., Sobral, D., Aylett, R., Woods, S., Hall, L., et al.: Learning by feeling: evoking empathy with synthetic characters. Appl. AI **19**(34), 235–266 (2005)

Provoost, S., Ming Lau, H., Ruwaard, J., Riper, H.: Embodied conversational agents: a scoping review. JMIR **19**(5), e151 (2017)

Sloman, A., Chrisley, R., Scheutz, M.: The architectural basis of affective states and processes. In: Fellous, J.-M., Arbib, M.A. (eds.) Who Needs Emotions?. Oxford, New York (2005)

Tielman, M.L., Neerincx, M.A., Brinkman, W.-P.: Design and evaluation of personalized motivational messages by a virtual agent that assists in post-traumatic stress disorder therapy. JMIR, **21**(3), e9240 (2019). https://doi.org/10.2196/jmir.9240

Wahle, F., Bollhalder, L., Kowatsch, T., Fleisch, E.: Toward the design of evidence-based mental health information systems for people with depression: a systematic literature review and meta-analysis. JMIR **19**(5), e191 (2017)

Yannakakis, G., Togelius, J.: Artificial Intelligence and Games. Springer, Heidelberg (2018). https://doi.org/10.1007/978-3-319-63519-4

Sundstrom, P.: Exploring the affective loop. Ph.D. dissertation. Stockholm University, Stockholm (2005)

Gathering People's Happy Moments from Collective Human Eyes and Ears for a Wellbeing and Mindful Society

Risa Kimura[(✉)] and Tatsuo Nakajima

Department of Computer Science and Engineering, Waseda University, Tokyo, Japan
{r.kimura,tatsuo}@dcl.cs.waseda.ac.jp

Abstract. Information technologies are dramatically and rapidly changing our world. Human well-being is one of the most important factors in our everyday lives. Although the importance of well-being is widely discussed, there are very few actual digital services to facilitate human well-being. Remembering happy moments significantly increases human well-being, so memorizing happy moments from the past and presenting those moments in the present is a promising future direction for promoting well-being in our everyday lives. In this paper, we present *CollectiveEyes*, an invention that allows us to share collective people's eyes and ears, and use *CollectiveEyes* to memorize and present visuals and sound recordings of various people's happy moments to explore human well-being. We first analyze happy moments and introduce an analysis framework for designing happy moments. We also conduct experiments using a scenario-based analysis to investigate how to gather happy moments with *CollectiveEyes*.

Keywords: Collective human eyesight · Virtual reality · Happy moments · Positive psychology · Social media · Human wellbeing · Scenario-based analysis · Smart city and society

1 Introduction

Human well-being is one of the most important factors in our everyday lives [20]. Although the importance of well-being is widely discussed, there are a very few actual digital services to facilitate human well-being. In particular, recording people's happy moments and watching those moments later among friends or family significantly increase their well-being. As shown in [3], remembering happy moments increases human well-being significantly, so recording happy moments from the past and presenting those moments at an opportune time are part of a promising future direction for increasing well-being in our everyday lives, and recent information technologies offer new opportunities to enhance human well-being.

In our research, we have developed various case studies, technologies and design methods to design alternate reality experiences [8, 15–17, 23]. Alternate reality experiences are typically achieved by modifying our eyesight or replacing our five senses for others, and these experiences make our world interactive by implicitly influencing

© Springer Nature Switzerland AG 2020
D. D. Schmorrow and C. M. Fidopiastis (Eds.): HCII 2020, LNAI 12197, pp. 207–222, 2020.
https://doi.org/10.1007/978-3-030-50439-7_14

human attitudes and behaviors. In this paper, we present a new case study named *Collec-tiveEyes* that allows us to share collective people's eyes and ears like a sharing economy service [6]. The tool can be applied to various use cases, one of which makes it possible to enhance our everyday life with recorded visuals containing various people's happy moments.

We define happy moments as visual and auditory images or videos that are recorded at the moment when we feel happy. In our current everyday life, people take many photos when they feel happy and upload them to social media sites such as Facebook, as shown in Fig. 1. The photos that record individuals' happy moments are effective representations of their well-being. While we have fragmented spare time every day and occasionally have negative feelings, we may look back on our own happy moments from the past to recover from those negative emotions and to make us feel more positive and happier. Additionally, if social media automatically and frequently presents us with our past happy moments, we are reminded that we have experienced a large number of happy moments in the past. This approach is essential because people can easily decrease their positive feelings by experiencing negative feelings [12]. So, frequent presentation of happy moments and viewing those happy moments allows us to experience higher levels of well-being.

Fig. 1. Happy moments

In this study, we adopt *CollectiveEyes* to semiautomatically record and present our happy moments by sharing collective people's eyes and ears. Our current prototype makes it difficult to conduct a practical experiment with which we can extract useful insights by demonstrating the results to the experiment's participants. Therefore, in the current study, we adopt a scenario-based analysis to use *CollectiveEyes* to gather and present happy moments in order to determine the potential opportunities and pitfalls in our approach. We analyze the three scenarios developed by participants to discuss happy moments in our everyday lives, then we discuss how to use *CollectiveEyes* to capture happy moments based on the analysis.

The rest of the paper is structured as follows. In Sect. 2, we present an overview of *CollectiveEyes*. In Sect. 3, we introduce an analysis framework to investigate happy moments in our everyday lives. The framework organizes these moments into five categories. The framework can be used to analyze the scenarios developed in the next section. Section 4 presents three scenarios: in the first scenario, we use *CollectiveEyes* to memorize a user's personal happy moments and occasionally remind that user to increase his/her positive emotions. In the second scenario, we use *CollectiveEyes* to memorize the happy moments of a user's family members and occasionally remind that user to increase his/her positive emotions. In the final scenario, we use *CollectiveEyes* to memorize the happy moments of a user's friends and occasionally remind that user to increase his/her positive emotions. The purpose of the analysis is to reveal a good strategy with which our tools can gather and present appropriate happy moments at the opportune time. In Sect. 5, we present some related work, and, finally, Sect. 6 concludes the paper.

2 CollectiveEyes

CollectiveEyes is a digital platform used to capture collective human visual perspectives [11]. We assume that each person is equipped with a wearable device containing a camera and microphone, typically wearable glasses. The current version of *CollectiveEyes* uses a head-mounted display (HMD) and puts a camera and microphone in front of the HMD. The HMD projects the view captured by a camera. However, the future version of *CollectiveEyes* will use lightweight smart glass to improve the ease of daily use. One popular existing smart glass is Google Glass[1], and eSense [10] is a research oriented smart earphone platform.

2.1 Finding an Appropriate Eye View

While each person's eye view is captured by his/her HMD through the camera attached to the device, the person periodically submits appropriate keywords that can be seen in his/her view as hashtags for *CollectiveEyes*. The future version of the platform will extract the keywords automatically from the captured visuals by using a method described in [14] or annotate the videos and find similarities in them by using a crowdsourcing platform similar to [24]. However, for the purpose of discussing the feasibility of the proposed approach, in the current prototype platform, the person whose visuals are captured registers the keywords into the platform manually.

The extracted keywords are clustered by using Related Words[2] and presented through Word Cloud[3]. A user can watch multiple clustered word clouds extracted from the visuals of crowds of people who wear HMDs in a virtual space, as shown in Fig. 2. A user chooses one of the word clouds, and he or she sees the selected word cloud through his/her HMD. Finally, he/she chooses a word in the word cloud to present a related view in front of him or her.

[1] https://www.google.com/glass/.

[2] http://relatedwords.org/.

[3] https://www.wordclouds.com/.

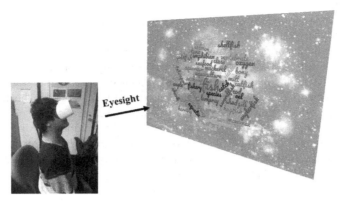

Fig. 2. Finding an appropriate view

2.2 Gaze-Based Gesture

One of the most essential characteristics of *CollectiveEyes* is the use of gaze-based gestures for all controls. We defined the following five basic commands. The first command is the *select command*. This command is employed by the user to select a target person by moving the user's eyesight from top to bottom. The second command is the *deselect command*. This command is used to return to the previous view by the user moving his/her eyesight from bottom to top. The third command is the *hijack command*. This command is employed by the user to choose another person whose perspective the user wants to hijack by watching a selected person and moving the user's eyesight from top to bottom around the person. The fourth command is the *change command*. This command is used to change the current view to the view of another randomly selected person near the user by the user moving his/her eyesight from top to bottom in the current view. The fifth and final command is the *replace command*. This command is used to remove a view that the user wants to replace by moving the user's eyesight from bottom to top on the view.

2.3 Watching Multiple Eye Views

When showing multiple views, the views are shown in a virtual space. *CollectiveEyes* offers two modes with which to present the multiple eye views. The first mode is the *spatial view mode*. The second mode is the *temporal view mode*, as shown in Fig. 3. When using the spatial view mode, the four views are automatically selected and shown in the virtual space. Another view is shown instead of the removed view if one of views does not interest a user. Additionally, the change command can replace all views at the same time. When using the temporal mode, one view is selected and displayed. The view can be successively changed to another one until the most desirable view can be found.

2.4 Layered Multiple Eye Views Representation

The proposed approach requires diving into a virtual reality space to access multiple views. To use this approach in our everyday lives, a more lightweight method for

Spatial View

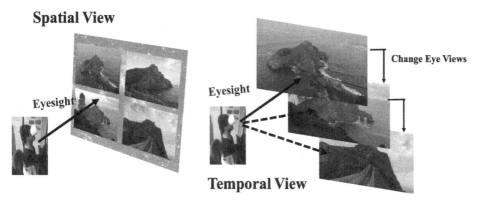

Temporal View

Fig. 3. Watching multiple eye views

accessing a virtual reality space is desirable. Our approach can use layered representation, such as [19], which provides a new way to present multiple people's perspectives simultaneously, as shown in Fig. 4.

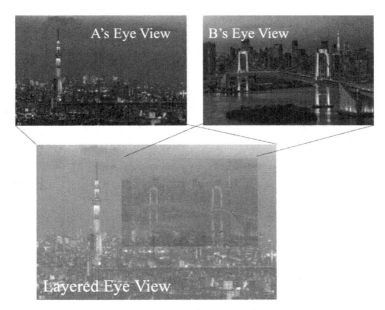

Fig. 4. Layered representation

3   An Analysis Framework for Examining Happy Moments

In this section, we present an analysis framework for examining happy moments in our everyday lives. First, we want to show how our framework is introduced. A person who offers data in the experiment has tried to publish various photos taken at moments when he felt happy on his Facebook timeline. His happy moments are recorded from January 1, 2019 to December 31, 2019 for one year. The total number of published happy moments within the one year is 296.

After collecting his happy moments, we tried to categorize them by using affinity diagrams [13]. We finally classified them into the following five categories: *personal belongings, foods, curiosity, families/communities* and *landscape*. Thus, the proposed analysis framework contains the five categories, as shown in Fig. 5. The first category is *personal belongings*, which typically shows people's newly acquired goods or their favorite goods that are used every day. Some examples recorded in the one year experiment are shown on the left side in Fig. 6. The second category is *foods*, which typically shows some fantastic foods eaten by people. Some examples are shown on the right side of Fig. 6. The third category is *curiosity*, which shows some things that aroused curiosity in people regarding the circumstances surrounding those things. Some examples are shown on the left side of Fig. 7. The fourth category is *families/communities*, which typically shows some events with people's families or close communities. Some examples are shown on the right side of Fig. 7. The last category is *landscape*, which typically includes photos taken while visiting pleasant places. Some examples are shown in Fig. 8.

Fig. 5. Happy moment analysis framework

Figure 9 shows the number of happy moments in each category from every month from January to December 2019. This person mainly submitted happy moments related to food and landscapes, but he also published various photos on his Facebook timeline. Therefore, the data are sufficient to extract materials for the necessary categories.

The extracted analysis framework is used to examine happy moments in our everyday life. As shown in Sect. 4, we can use the framework to analyze scenarios in our everyday life. In particular, we can discuss which scene in the scenario encourages happy

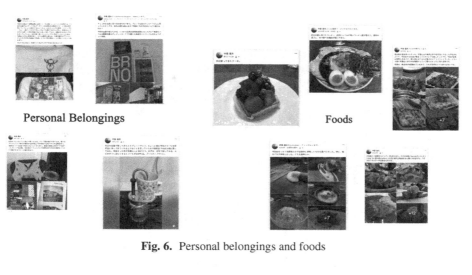

Fig. 6. Personal belongings and foods

Fig. 7. Curiosity and communities/families

moments and which category is essential in each scenario. The analysis is also useful for understanding appropriate strategies indicating how to record and remind us of each happy moment in *CollectiveEyes*.

4   Analyzing Scenarios with the Happy Moment Analysis Framework

In the experiment to investigate insights to capture and record the appropriate happy moments, we asked participants to develop three scenarios. In the first scenario, they used our service to memorize personal happy moments and remind them to increase their positive emotions. This scenario describes a person's favorite idol group. In the

Landscape

Fig. 8. Landscapes

second scenario, they used our service to memorize their happy family moments and occasionally remind them to increase their positive emotions. In the final scenario, they used our service to memorize happy moments involving their friends on social media and occasionally remind them to increase their positive emotions. With this scenario-based analysis based on the research through design method [4], *CollectiveEyes* has been designed to access people's recorded happy moments. The aim of our investigation is to determine how these happy moments can be gathered in order to increase our well-being. Additionally, the analysis discusses which strategies are appropriate for capturing and remembering happy moments in order to increase a user's positive emotions.

Based on the scenario analysis, we hope to answer the following two research questions in this paper.

- What is the best way to record happy moments?
- When should happy moments be remembered?

The following subsections show several scenarios and analyze them with the happy moment framework. Finally, we will summarize the answers to the above research questions.

4.1 Personal Scenario and Its Analysis

Personal Scenario: *Akari is a young woman who likes an idol group named "Chain." Chain is a trio of male idols who perform singing and dancing. In May of this year, Chain was performing their concert for two days at Tokyo Dome for the first time. Akari went to Tokyo Dome to see the live concert of Chain with her friend. Tokyo Dome is the most famous baseball field in Tokyo, where only the most popular musicians can perform their*

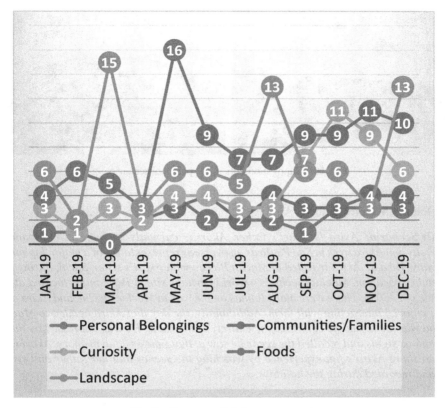

Fig. 9. Every month's happy moments

live concerts. She was the first to go to see Chain's live concert. Her favorite song "Lily" was performed in front of her, and she enjoyed the live performance. After watching the live show, she bought some of the group's merchandise to remember the concert. Then, she recorded her thoughts on the scenes that she saw during the show and held on to that fun memory. A week later, Akari replayed the scenes recorded during school holidays and recalled her feelings of fun when "Lily" was performed.

Scenario Analysis: In this scenario, Akari visited her favorite idol group's concert and enjoyed her good memories from the concert. In this case, for her, various scenes before, during, and after the concert became her happy moments because the scenes allowed her to recall her positive feelings during the concert. The landscapes shown in Fig. 10 that are related to the concert, personal belongings bought at the concert, foods that she ate before/after the concert with her friends, and friends who went to the concert with her all became part of her happy moments. Additionally, some specific announcements (ex. future event schedules) during the concert became, for her, strong happy moments.

Fig. 10. Happy moments in the personal scenario

4.2 Family Scenario and Its Analysis

Family Scenario: *Akira is Akari's father. Akari is currently studying at a Japanese university, but Akira lives in Sao Paulo as an overseas representative of a major Japanese company, Takada. Akira traveled to Machu Picchu with Akari's sister, Sachi, during his December vacation. Machu Picchu is a world heritage site in Peru and is located at an altitude of 2430 m. There were many llamas on the way to Machu Picchu, and Akira and Sachi enjoyed interacting with them. Additionally, the day that Akira headed to Machu Picchu was clear, and he could see the entirety of the ruins. He was impressed by these miraculous sights and recalled the scene he saw at that time as a fun memory. Akari was able to share Akira's fun experience by watching the scenes that her father and sister saw and recorded during the holidays.*

Scenario Analysis: In this scenario, Akari empathized with her father's happy moments during his holiday travel. The most important aspect of the scenario is that she also wanted to visit to Machu Picchu with her father. Therefore, the scenes from her father's travels became happy moments for her as well; the landscapes that her father saw became her happy moments because she is strongly interested in Machu Picchu. For other categories in the happy moment framework, we found that the following two aspects are essential. The first aspect is that the scenes contain her father's favorite things, such as his personal belongings or his favorite foods. Additionally, some specific moments related to her father's hobbies may become her happy moments if she knows her father's favorite things well. In particular, if the scene contains her father, she may particularly empathize with that scene, as shown in Fig. 11. The second aspect is that the scenes containing her father's favorite things may become her happy moments. If the scene contains his favorite foods and goods, the scenes become her happy moments. Additionally, the scenes containing things related to her hobbies may become her happy moments as well.

4.3 Friend Scenario and Its Analysis

Friend Scenario: *Makoto is an Akari's friend on Facebook. Makoto took a holiday in November and went to the Christmas market in Nuremberg, Germany. Nuremberg has an annual Christmas market during the advent of Hauptmarkt in the central square of*

Fig. 11. Happy moments in the family scenario

the old town and in the adjacent squares and streets. The Christmas market is full of ornaments, wood crafts, food stalls, and other accessories for decorating a Christmas tree. Makoto went around the stalls selling miniature ornaments and recorded the scenes he liked. Akari watched Makoto's recorded scene from the "Christmas" tag and shared in Makoto's fun experience by imagining that she would someday visit various stalls at the Nuremberg Christmas Market.

Scenario Analysis: The happy moments in this case seem to be different depending on whether the person in the scenario is Akari's close friend or not. If Makoto is her very close friend, and she knows his favorite things well, the various scenes witnessed by Makoto have a high possibility of becoming Akari's happy moments, similar to her father's travel scenario. However, if Makoto is not her close friend, whether a scene becomes a happy moment or not is determined by whether or not she empathizes with that scene. For example, the landscape of a scene she has wanted to visit for a long time or one she sees when she comes to visit may become her happy moment. Similarly, the scenes containing her favorite/curious goods, foods or events may become her happy moment. For example, Akari is strongly interested in the food stalls shown in Fig. 12, so that scene might become Akari's happy moment.

4.4 Investigating Research Questions and Future Issues

In this section, we would like to discuss how to detect and gather happy moments captured from people's perspectives through *CollectiveEyes*. From the scenario analyses presented in the above subsections, the most important aspect when gathering happy moments depends on how people empathize with certain scenes.

What Is the Best Way to Record Happy Moments? For each individual, there are many factors influencing which visuals become that user's happy moments. In particular, when they feel happy—for example, when they attend some of their favorite events or stay with

Fig. 12. Happy moments in the friend scenario

people close to them such as good friends or families—the landscape, foods, goods, or curious things all become happy moments. Additionally, the view containing the people close to them during their happy time may become their happy moments.

On the other hand, when observing other people's views, whether the views become happy moments depends on whether that person is close to the user. If the person is close to the user, such as a good friend or a family member, the views containing that person's favorite things or landscapes may become the user's happy moment. On the other hand, when observing the views of people who are not close, whether those views become happy moments depends on whether the view contains the user's favorite things or landscapes.

Based on the above discussions, in order to develop a practical happy moment service, the service needs to know whether a person offering views feels happy, whether the person is close to the user, and whether views are interesting to the user.

To extract the user's happy moments from others' views, the views are classified as their respective events, i.e., travelling or attending ceremonies. Then, the service categorizes the views based on whether they contain the user's favorite things or landscapes.

When Should Happy Moments Be Remembered? The user wants to recall happy moments while they have a short period of spare time, such as when they are waiting for a train. One of the important issues is that recalling happy moments frequently may decrease any present negative feelings. Thus, the service should use the user's fragmented spare time to watch happy moments. Personal happy moments are the most effective for helping users maintain their positive feelings, but the happy moments of those close to users are useful because these happy moments also remind users of their social relationships with close friends and family.

On the other hand, others' happy moments can be useful for stimulating the user's happiness and curiosity because these happy moments allow the user to see landscapes, goods and foods that they do not know well.

In the next study, we need to investigate an actual algorithm to detect the user's happy moments and to evaluate the accuracy of the algorithm.

4.5   What Is the Best Way to Use *CollectiveEyes* to Collect Happy Moments?

CollectiveEyes can capture and gather users', their families' and their friends' views anytime and anywhere. Extracting happy moments from the gathered visuals is necessary, but it is hard to automate the gathering process with only image processing techniques. Of course, we may use various sensors to retrieve real-world information to capture happy moments, but the collected visuals become useless if they contain many moments that do not make people happy. In particular, anything that might recall negative emotions should be avoided. With the current image processing technologies, it is hard to avoid the incorrect collection of happy moments.

When pursuing a happy moment to put on Facebook, a user usually takes many photos and chooses desirable photos from those taken. Users are very conscious of this process, and they very carefully select photos to capture moments that are important to them. However, this process is not easy for typical, more casual users, so it is necessary to consider a simpler method for capturing happy moments.

One promising approach is to use a hand gesture to select happy moments from all captured eye views, but in many cases, users may forget to perform gestures when they feel a deep emotion in the moment, as shown in the left of Fig. 13. An alternative approach is to capture a user's eye views when he/she feels deep emotion by using biosensors. Then, *CollectiveEyes* presents multiple views by using the spatial view mode. The user chooses appropriate happy moments from the captured and simultaneously shown eye views through the gaze-based gesture, as shown in the right of Fig. 13.

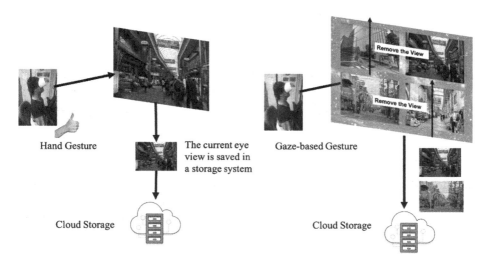

Fig. 13. Capturing happy moments with *CollectiveEyes*

In the next step, we will compare the above two approaches through user studies and capture user experiences collecting happy moments with *CollectiveEyes*.

5   Related Work

5.1   Positive Psychology

Positive psychology is the scientific study of human flourishing and an applied approach to optimal functioning. It has also been defined as the study of the strengths and virtues that enable individuals, communities, and organizations to thrive. The critical positivity ratio is a largely discredited concept in positive psychology positing an exact ratio of positive to negative emotions that distinguishes flourishing people from languishing people [1]. Fredrickson et al. hypothesize that positive emotions undo the cardiovascular effects of negative emotions [2]. If people do not regulate heart rate, blood sugar, and changes to immunosuppression once stress has passed, they may experience illness, such as coronary disease, and heightened mortality. Research indicates that positive emotions help people who were previously under stress relax back to their physiological baseline. Seligman proposes a well-being theory that consists of five elements of "well-being," which fall under the mnemonic PERMA: Positive emotion, Engagement, Relationships, Meaning, Achievement [20].

5.2   Jack-In Head

Jack-In is a concept for augmenting human capability and human existence [9]. The concept enables immersive connection between humans and other artifacts or between humans. The technique of hijacking another person's eyesight that is used in our approach is similar to the Jack-In concept.

5.3   KinecDrone

A KinecDrone enhances our somatic sensation of flying in the sky [7]. A video captured by drone is transmitted to a user's smart glasses. While a user behaves as though they are flying in the sky, they can watch the scene captured by the flying drone while still in a room. Thus, the user feels as though they are truly flying in the sky.

5.4   HoloMoL

HoloMoL (HoloLens's Method of Loci) supports a human memorization technique that combines the method of loci (MoL) and MR [22]. The main objective of the HoloMoL prototype is 1) to support MoL with MR technologies based on the Microsoft HoloLens and 2) to enable utilizing the MoL with minimal training.

5.5   Micro-crowdfunding

The approach proposes a new mobile social media infrastructure for motivating collective people to participate in the flourishing of our society [5, 18]. The proposal is to use a layered approach in which an entire community consists of many microcommunities. If each independent community is encouraged to contribute to its society, the entire community will be motivated to achieve a flourishing society. The most important idea is that microtasks that allow community members to achieve higher levels of well-being are automatically added, then the members increase their sense of well-being by performing the microtasks.

6   Conclusion and Future Direction

In this paper, we presented *CollectiveEyes*, a system that allows us to share various people's eyes and ears and reports how to use those observations to memorize and present eye views containing various people's happy moments. We analyzed happy moments and used five categories to organize them. We conducted a user study to investigate the potential possibilities of our approach based on the scenario-based analysis.

In [21], the authors found that the negativity bias in reframing attenuated as age increased. The result shows that presenting happy moments is more effective for elderly people. In the future, research on an aging society is essential in accordance with the increase in the number of elderly people. In particular, elderly people's well-being is an essential issue because people will live for a longer period after retirement in the future. We will investigate the effects of our approach on elderly people in the next study.

References

1. Fredrickson, B.L., Losada, M.F.: Positive affect and the complex dynamics of human flourishing. Am. Psychol. **60**, 678–686 (2005)
2. Fredrickson, B.L.: Updated thinking on positivity ratios. Am. Psychol. **68**, 814–822 (2013)
3. Fredrickson, B.: Love 2.0: Finding Happiness and Health in Moments of Connection. Plume, New York (2013)
4. Giaccardi, E., Stappers, P.J.: Research through design (chap. 43). In: The Encyclopedia of Human-Computer Interaction, 2nd edn. Interaction Design Foundation (2017)
5. Gushima, K., Sakamoto, M., Nakajima, T.: Community-based crowdsourcing to increase a community's well-being. In: Proceedings of the 18th International Conference on Information Integration and Web-based Applications and Services (iiWAS 2016), pp. 1–6 (2016). https://doi.org/10.1145/3011141.3011173
6. Hamari, J., Sjöklint, M.: Ukkonen a: the sharing economy: why people participate in collaborative consumption. J. Assoc. Inf. Sci. Technol. **67**(9), 2047–2059 (2015). https://doi.org/10.1002/asi.23552
7. Ikeuchi, K., Otsuka, T., Yoshii, A., Sakamoto, M., Nakajima, T.: KinecDrone: enhancing somatic sensation to fly in the sky with Kinect and AR.Drone. In: Proceedings of the 5th Augmented Human International Conference (AH 2014), 2 p. Article 53 (2014). http://dx.doi.org/10.1145/2582051.2582104
8. Ishizawa, F., Sakamoto, M., Nakajima, T.: Extracting intermediate-level design knowledge for speculating digital-physical hybrid alternate reality experiences. Multimedia Tools Appl. **77**(16), 21329–21370 (2018). https://doi.org/10.1007/s11042-017-5595-8
9. Kasahara, S., Rekimoto, J.: JackIn: integrating first-person view with out-of-body vision generation for human-human augmentation. In: Proceedings – 3 – of the 5th Augmented Human International Conference. Article 46 (2014). http://dx.doi.org/10.1145/2582051.2582097
10. Kawsar, F., Min, C., Mathur, A., Montanari, A.: Earables for Personal-Scale Behavior Analytics. IEEE Pervasive Comput. **17**(3), 83–89 (2018). https://doi.org/10.1109/mprv.2018.03367740

11. Kimura, R., Nakajima, T.: A ubiquitous computing platform for virtualizing collective human eyesight and hearing capabilities. In: Novais, P., Lloret, J., Chamoso, P., Carneiro, D., Navarro, E., Omatu, S. (eds.) ISAmI 2019. AISC, vol. 1006, pp. 27–35. Springer, Cham (2020). https://doi.org/10.1007/978-3-030-24097-4_4

12. Ledgerwood, A., Boydstun, A.E.: Sticky prospects: loss frames are cognitively stickier than gain frames. J. Exp. Psychol. Gen. **143**(1), 376–385 (2014). https://doi.org/10.1037/a0032310

13. Lucero, A.: Using affinity diagrams to evaluate interactive prototypes. In: Abascal, J., Barbosa, S., Fetter, M., Gross, T., Palanque, P., Winckler, M. (eds.) INTERACT 2015. LNCS, vol. 9297, pp. 231–248. Springer, Cham (2015). https://doi.org/10.1007/978-3-319-22668-2_19

14. Mahajan, D., Girshick, R., Ramanathan, V., Paluri, M., von Der Maaten, L.: Advancing state-of-the-art image recognition with deep learning on hashtags (2018). https://code.fb.com/applied-machine-learning/advancing-state-of-the-art-image-recognition-with-deep-learning-on-hashtags/. Accessed 28 Sept 2018

15. Nakajima, T., Lehdonvirta, V.: Designing motivation using persuasive ambient mirrors. Pers. Ubiquit. Comput. **17**(1), 107–126 (2013). https://doi.org/10.1007/s00779-011-0469-y

16. Sakamoto, M., Nakajima, T., Alexandrova, T.: Enhancing values through virtuality for intelligent artifacts that influence human attitude and behavior. Multimedia Tools Appl. **74**(24), 11537–11568 (2015). https://doi.org/10.1007/s11042-014-2250-5

17. Sakamoto, M., Nakajima, T., Akioka, S.: Gamifying collective human behavior with gameful digital rhetoric. Multimedia Tools Appl. **76**(10), 12539–12581 (2017). https://doi.org/10.1007/s11042-016-3665-y

18. Sakamoto, M., Gushima, K., Alexandrova, T., Nakajima, T.: Designing human behavior through social influence in mobile crowdsourcing with micro-communities. In: Kő, A., Francesconi, E. (eds.) EGOVIS 2017. LNCS, vol. 10441, pp. 189–205. Springer, Cham (2017). https://doi.org/10.1007/978-3-319-64248-2_14

19. Saraiji, M.H.D.Y., Sugimoto, S., Fernando, C.L., Minamizawa, K., Tachi S.: Layered telepresence: simultaneous multi presence experience using eye gaze based perceptual awareness blending. In: ACM SIGGRAPH 2016 Emerging Technologies (SIGGRAPH 2016), 2 p. Article 14 (2016). https://doi.org/10.1145/2929464.2929467

20. Seligman, M.: Flourish: A New Understanding of Happiness and Wellbeing: The Practical Guide to Using Positive Psychology to Make You Happier and Healthier. Nicholas Brealey Publishing, London (2011)

21. Sparks, J., Ledgerwood, A.: Age attenuates the negativity bias in reframing effects. Pers. Soc. Psychol. Bull. **45**(7), 1042–1056 (2019). https://doi.org/10.1177/0146167218804526

22. Yamada, Y., Irie, K., Gushima, K., Ishizawa, F., Al Sada, H., Nakajima T.: HoloMoL: human memory augmentation with mixed-reality technologies. In: Proceedings of the 21st International Academic Mindtrek Conference (2017)

23. Yamabe, T., Nakajima, T.: Playful training with augmented reality games: case studies towards reality-oriented system design. Multimedia Tools Appl. **62**(1), 259–286 (2013). https://doi.org/10.1007/s11042-011-0979-7

24. Von Ahn, L., Dabbish, L.: Designing games with a purpose. Commun. ACM **51**(8), 58–67 (2008). https://doi.org/10.1145/1378704.1378719

Examination of Stammering Symptomatic Improvement Training Using Heartbeat-Linked Vibration Stimulation

Shogo Matsuno[1,2(✉)], Yuya Yamada[3], Naoaki Itakura[3], and Tota Mizuno[3]

[1] Hottolink, Inc., Tokyo, Japan
18kz011@ms.dendai.ac.jp
[2] Tokyo Denki University, Tokyo, Japan
[3] The University of Electro-Communications, Tokyo, Japan

Abstract. In this paper, we propose a training method to improve the stammering symptom, which automatically adjusts the rhythm of speech using vibrational stimulation linked to heart rate through a smart watch. We focus on the rhythm control effect by vibration, confirmed by the tactile stimulation training, and propose a training method to improve the symptoms of stammering while automatically adjusting the rhythm of the utterance based on the heartbeat-linked vibration stimuli. In addition, a system using the proposed method is constructed, and its effects on the heart rate by providing vibration stimulation to stutterers and on stammering symptoms are investigated. We present the effectiveness of stammering improvement training through vibration stimuli by experimenting with eight subjects.

Keywords: Stammering · Computer-aid · Delayed auditory feedback · Smartwatch

1 Introduction

In this paper, we propose a training method to improve the stammering symptom, which automatically adjusts the rhythm of speech using vibrational stimulation linked to the heart rate through a smart watch. A disorder in which a person is not able to speak smoothly is called stammering. There are training methods that use delayed auditory feedback (DAF), a hooked metronome, and other mechanisms to improve the symptoms of stammering [1]. In addition, tactile stimulation training is proposed as a method to improve stammering using something other than hearing [2]. We focus on the rhythm control effect through vibration confirmed by tactile stimulation training, and propose a training method to improve the symptoms of stammering while automatically adjusting the rhythm of the utterance based on heartbeat-linked vibration stimuli. In addition, a system using the proposed method is constructed, and the effects of providing vibration stimulation to stutterers on the heart rate and stammering symptoms are investigated. It is examined whether stammering improvement training through vibration stimuli is effective.

© Springer Nature Switzerland AG 2020
D. D. Schmorrow and C. M. Fidopiastis (Eds.): HCII 2020, LNAI 12197, pp. 223–232, 2020.
https://doi.org/10.1007/978-3-030-50439-7_15

Stammering symptoms include repeating a particular sound involuntarily during speech, or uttering with a stammer or timid hesitancy. In these symptoms, speech fluency is compromised. This condition is said to affect approximately 1% of adults, regardless of nationality and language. Symptoms can be broadly divided into three types: problems with words and phrases that are difficult to start with, prolongation of repetitive sounds, and repeated occurrences of the same sound. Stuttering occurs most often occurs in childhood at elementary school and is often repeated at first. However, as a child grows, the block reaction to the utterance gradually becomes stronger due to anxiety after being pointed out by others, and an uncomfortable feeling about their utterance. The rate of departure increases; therefore, stuttering in adults is difficult. Because stutterers are more likely to have social anxiety disorder (SAD) or depression due to insecurity about interpersonal relationships, which may hinder social life, effective improvement measures are desired.

Training methods to improve or reduce stammering include DAF that enables speaking while listening to the uttered voice with a momentary delay, shadowing (repetition), and sound emitted at a constant tempo from a hearing aid-type device. There are ear-hung metronomes that speak together. However, DAF is difficult to use in situations where first voice is difficult to emit in the first place, shadowing is not used in situations where speech is actually required, and ear-hung metronomes inhibit the feedback of a speaker's own voice by voice. In addition, similar to DAF, there is a problem that metronomes have to be placed on the ears; hence, young people are particularly resistant to its daily use.

In this study, we focus on the symptoms that are common in adults, and propose a training method that solves the above problems. In addition to the degree of stuttering symptoms, using the heart rate as biological information related to the state of speech and considering large individual differences, we show the efficacy of the oscillator in improving stuttering, and at the same time, we conduct an oscillator training in a more practical way. Furthermore, we conduct an experiment in which not only vibration is generated at a fixed tempo but also the tempo of vibration is changed by the fluctuation of heart rate, and confirmed whether the heart rate can be stabilized through this.

2 Proposed Method

Several methods have been proposed to improve the symptoms of stammering. In [1], stammering training using an ear-hung metronome was conducted for one subject for 100 days. In this study, stuttering symptoms began to decrease about one month after use. However, a comparison between the cases of not wearing and wearing an ear-hung metronome suggests that the effect of wearing a metronome is obtained from the beginning of use; therefore, in this study (which was a short-term experiment), it is observed even when using vibration stimulation. Similarly, it is important to compare the non-mounted state with the mounted state to obtain the effect. Other experimental conditions differ when a plurality of subjects is considered. The utterance is considered a speech instead of a telephone response, and heart rate information is used in addition to stuttering symptoms.

In [3], improvements of stuttering symptoms by conducting a short-term shadowing training are discussed. This study analyzed the effects on the stuttering symptoms and psychological aspects of 16 adult stutterers who listened to model sounds that were continuously heard and played back orally in parallel. The short-term stuttering symptoms can be expected to improve and alleviate, and the frequency of stuttering in shadow-reading tasks after shadowing is lower than that before shadowing. This differs from this study in that it uses speech tasks that are considered more practical than reading-aloud tasks.

The relationship between heart rate variability (HRV) and mental stress has also been studied. Stuttering symptoms are thought to be greatly affected by psychological pressure. Therefore, the result of measuring heart rate in a speech situation is important. The researchers in [4] investigated the correlation between the intensity of mental stress and the magnitude of HRV at rest, at the time of presentation practice, research presentation, and question-and-answer sessions in subjects involved in research. In any of the factors representing the magnitude of the fluctuation of the heart rate, a negative correlation is shown with the intensity of mental stress. In other words, the fluctuation of the heartbeat reduces as the mental stress increases, and the fluctuation increases as the mental stress reduces.

In this study, we focus on the symptoms that are common in adults, and propose a training method that solves the above problems. Figure 1 shows the overview of the proposed method. The system proposed in this paper consists of a sensor that acquires heart rate information, and a vibrator that generates a vibration stimulus. The acquired heart rate information is linked to the vibration stimulation interval, and the users are provided with a vibration at a constant rhythm to improve stuttering. The heart rate information is obtained using the Android app "Heart Rate Monitor Ware". In addition, the smartwatch "HUAWEI WATCH 2" is used as a device. The proposed system generated a vibration stimulus at a constant tempo or automatically changed the tempo of a vibration stimulus according to the change in heart rate from a preset reference value. The vibration tempo of the latter type was designed to decrease by 1 bpm as the heart rate increased by 1 bpm. The vibration tempo range is the vibration reference value ± 10 bpm. The heart rate I at this time is represented by the following equation:

$$I = k - (HB - i)$$

where i is the reference heart rate, k is the reference vibration tempo, and HB is the current heart rate.

3 Experiment

An experiment was conducted to evaluate whether the proposed system is effective in improving stuttering, which included eight subjects who stutter (five men and three women).

3.1 Experimental Method

The experiment was conducted in a closed conference room. There were occasions when external sounds were heard, but there was no problem with recording. The experimenter and the subject were seated equally in a rectangular conference room consisted of multiple tables.

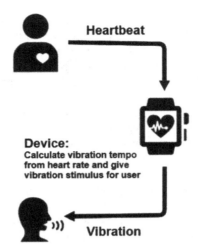

Fig. 1. Overview of the method of a system in which vibration tempo changes automatically.

The subject reads aloud a manuscript prepared by the experimenter for about 1 min. At this time, we examined the effect of the heart rate-linked vibration stimulus generated by the proposed system on the generation of stuttering. First, each subject introduced himself/herself to other subjects to get accustomed to the speech. Subsequently, the following three experimental tasks were performed sequentially. For speech to be read aloud, the subjects selected different speech for each task from multiple speech manuscripts prepared by the experimenter.

- Task 1: Read the selected speech aloud.
- Task 2: Generate a vibration stimulus at a constant tempo from a smartwatch, and read aloud according to the tempo.
- Task 3: Generate a vibration stimulus that changes the tempo according to the change in heart rate, and read aloud according to the tempo.

During this, the heart rate of the subject was measured using a smartwatch, and at the same time as the beginning of the measurement, a timer was used to record the heart rate data and the start time of the utterance. Simultaneously, the heart rate was measured, and the time was measured using a stopwatch. Next, the time of the stopwatch was noted at the same time as the beginning of the utterance to know the point of the heart rate data at which the utterance started. In Tasks 1 and 2, the utterance started 20 s after the beginning of the heart rate measurement, but in Task 3, the utterance started after about 1 min for preparation, as described later. The utterance content was selected by the subject from each of the 15 speech themes prepared by the experimenter. The reason why we chose speech as a task was that we thought that it was an important scene in social life, and it was easy to prepare an experimental environment, especially for stutterers who struggled with this scenario. Although the speech time was set to be 1 min as a guide, it varied in the experimental results because the subject gave priority to giving speech from beginning to end rather than time. Table 1 shows the prepared speech themes. All utterances in these experimental tasks were recorded using an IC recorder.

Table 1. List of speech themes.

Themes
Current goals and efforts
Episode that was moved
What I am good at
The interesting features of the school I came from
Hobbies I have brought
How to spend the year-end and New Year holidays
Memories of club activities (circles)
If you use 10 million in a day
City where you want to live
Life failure stories
Scary experience
My favorite (dislike) place
Life success experience
Commitment that cannot be yielded
Favorite (dislike) food

Furthermore, after performing three types of tasks, all subjects were asked to respond to an experiment evaluation questionnaire distributed in advance. Table 2 shows the contents of the questionnaire. We asked them to respond to the stuttering situation when not wearing an ear-hung metronome, evaluate this device using the 7-step semantic differential (SD) method [5], and freely give their specific impressions and hopes for the experiment in the free description column.

3.2 Experiment Results

The stuttering frequency was calculated by dividing the number of stutters in the recorded speech content by the total number of phrases. Symptoms were classified into the following four groups based on the stuttering method [6]:

1) Rarely-onset symptoms (prevention, discontinuity, preparation, duration, withdrawal)
2) Repeated symptoms (sound repetition, partial word repetition)
3) Insertion symptoms (insertion, repeated words)
4) Other non-fluid symptoms (excluding the other three groups of symptoms)

In addition, repeated words were not added as phrases, and when two or more symptoms appeared at the same time, the one with the strongest intensity was selected. For

Table 2. Contents of the questionnaire of SD method.

Left answer	Question items	Correct answer
Became painful	1) Did the utterance change?	It became easier
Deteriorated	2) Did you notice any improvement while using the system?	Improved
Could not control speed	3) Did you feel that you could control the speed of the utterance yourself?	Did it
Could not calm down	4) Did you speak calmly because the tempo of the vibration was adjusted automatically?	Calm down
Felt uncomfortable	5) Did you feel uncomfortable while feeling the vibration on your wrist when speaking?	There was no discomfort
Dissatisfied	6) Are you satisfied with this training method?	Is pleased

the heart answers of each subject to the questions described in the questionnaire of the SD method rate data, the average and variance of the heart rate during the utterance of each subject in each task were obtained, and then compared for the three tasks.

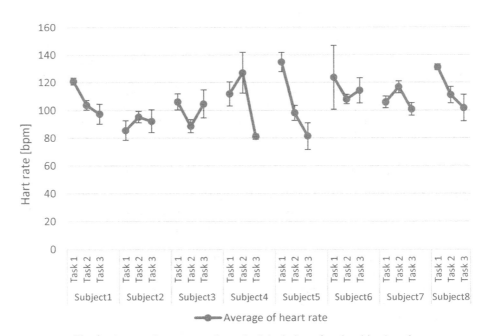

Fig. 2. Average heart rate and standard deviation of each subject's task.

For the heart rate data, the average heart rate and standard deviation during utterance in each subject for all subjects were obtained, and then compared for the three tasks. Figure 2 shows the results.

Table 3. Answers of each subject to the questions described in the questionnaire of the SD method.

Subjects	Q1	Q2	Q3	Q4	Q5	Q6
Sub1	4	4	4	4	4	4
Sub2	4	4	4	4	4	4
Sub3	6	7	3	6	7	7
Sub4	4	4	4	4	3	3
Sub5	4	4	3	5	5	6
Sub6	4	4	4	4	6	5
Sub7	4	4	2	3	5	3
Sub8	4	4	5	5	4	4
Avg.	4.25	4.38	3.63	4.38	4.75	4.50

*values: 1: disagree ~ 7: agree

Table 3 shows the answers of each subject to the questions described in the questionnaire of the SD method.

4 Discussion

As a result of the experiment, the frequency of stuttering in Task 2 and Task 3 was lower than in Task 1 among five of the eight subjects (Subjects 1, 2, 5, 6, and 8). Four of them (subjects 1, 5, 6, and 8) showed a decrease in stuttering frequency as the task progressed. In Task 3, the frequency of stuttering decreased in all seven subjects, except Subject 7, who failed to record. In the stuttering system with stuttering as the core symptom, and in the continuous system, at least seven out of eight subjects recorded the lowest symptom incidence in Task 2 or Task 3 with vibration. In particular, in the refractory system in which adult stutterers tended to have the highest symptom, six out of seven subjects, excluding subject 7, showed the lowest symptom incidence in Task 3.

As shown in Fig. 2, the average heart rate of six subjects out of eight is lower in Task 3 than in Task 2, and three subjects show a decrease in the heart rate as the task progresses. It was suggested that the height of the number may be related to the frequency of stuttering, and in the standard deviation of heart rate, Task 3 showed the highest value (the heart rate fluctuated greatly). Many subjects showed less mental stress in Task 3 because the variation was large. As mentioned above, by changing the vibration tempo during speech, it is suggested that there are two effects of a greater decrease in the mental stress relaxation of the stuttering frequency.

As shown in Table 3, many subjects feel it difficult to control the speed of utterance. Many subjects answered that there was not much discomfort at the time of wearing. In the open-ended description, the points that were popular in this experiment were those when they used a wristwatch, unlike the ear-hung metronome, and that they could relax by attaching the device. On the other hand, there are many criticisms and points of improvement. It can be observed from the results in Table 3 that it is difficult to adjust to the tempo because subjects were not familiar with the experiment method. Hence, more training is required, and it is better to shorten the experiment.

Effectiveness of Vibration Stimulation: Five subjects showed that the frequency of stuttering in Tasks 2 and 3 with vibration was lower than that in Task 1 without vibration, suggesting that vibration stimulation has the effect of reducing stuttering. It is possible that the effect of the habit of having performed a number of speeches may be involved, but it is considered that the influence of habituation was suppressed to some extent during the self-introduction by each subject before starting the experiment. The two subjects whose stuttering frequency in Task 2 exceeded the stuttering frequency in Task 1 had the highest stuttering frequency among the subjects, and the effect of vibration stimulation may differ depending on the severity of stuttering. Was suggested. However, in both cases, the frequency of stuttering in Task 3 was clearly lower than that in Task 1, and it is conceivable that either the stuttering effect was slightly slower than that of mild stutterers or the effect was specific to the condition of Task 3. In any case, the utterance at the tempo of the vibration stimulus is expected to have an effect of improving the stuttering symptoms of many stutterers.

Effect of Changes in Vibration Tempo: Compared to Task 2 in which the vibration tempo was constant, Task 3 in which the vibration tempo changed was more effective for difficult-to-treat symptoms. The frequency was lower, and six out of eight subjects showed a lower average heart rate in Task 3 than in Task 2, so that changing the vibration tempo was more effective in reducing the effect. This possibility was suggested. In addition, many subjects showed a high standard deviation (variation) of heart rate in Task 3. A previous study [5] showed that the lesser the mental stress, the greater the fluctuation of the heart rate. It can be said that there were many subjects who had less mental stress in Task 3. In addition, although the standard deviation of the heart rate of subject 4 in Task 3 was small, the average value of the heart rate showed a considerably low value, indicating that the heart rate in Task 3 was always low and stable. It is determined that Task 3 was also more effective mentally in Task 3 than in Task 2. These results suggest that changing the vibration tempo during speech has two effects: a greater reduction in stuttering frequency and a reduction in mental stress.

Relationship Between Stammering Frequency and Heart Rate:
Although no direct comparison between stuttering frequency and heart rate has been found, a previous study [7] that examined the relationship between stuttering and anxiety found that stuttering frequency increased during higher anxiety situations, that is, greater mental stress. A previous study [8] that examined the relationship between high heart rate and mental stress showed that higher the mental stress, the higher the heart rate. Together, these studies predicted that higher the heart rate, the higher the frequency of stuttering. Based on these facts, in this experiment, five out of eight subjects showed similar trends in the transition of stuttering frequency between tasks and the transition of the average value of heart rate. It was suggested that this might be related to the frequency

of stuttering. The correlation between the heart rate standard deviation and stuttering frequency was not as high as the average heart rate (only two people showed similarity). Considering that the target stress is less, it is difficult to say that there exists a relationship between the magnitude of heart rate variability and the likelihood of stuttering.

Practicality of Vibration Stimulation Training Based on Subject Impressions:
Finally, we will evaluate the practicality of vibration stimulation training for stuttering symptoms based on the results of the experimental evaluation questionnaire of subjects. According to the results of the experimental evaluation questionnaire, many subjects felt it difficult to speak according to the vibration tempo. However, it was found that stuttering frequency reduced by speaking according to the rhythmic vibration. Therefore, as one becomes accustomed to the utterance timing through long-term training, further improvement can be expected, as shown in a previous study [2]. In addition, because many subjects did not feel discomfort from the vibration stimulus, their mental discomfort was less and they could be used for long-term training without any problems. Furthermore, some subjects were interested in using a wristwatch as a device. Therefore, it was suggested that it could be a tempo stimulator instead of the ear-hung metronome. All three subjects who responded favorably to this vibration stimulus in the free description column (Subjects 3, 5, and 8) had a large standard deviation of the heart rate in Task 3; therefore, they were subjective and objective. It can be observed that the effect was also obtained in from the above. It is considered that vocal training in which the vibration stimulus tempo is changed by a change in the heart rate is an effective and practical method for improving stuttering.

5   Conclusion

In this study, we aimed to improve stuttering and conducted an experiment to utter at the tempo of vibration stimulation. Consequently, it was confirmed that stuttering decreased by changing the vibration tempo according to the change in heart rate during an utterance. This means that providing a vibration stimulus to the stuttering person based on the heart rate information is more effective in reducing stuttering symptoms than simply providing vibration at a constant tempo.

In future studies, it will be necessary to consider its use in long-term training. In this study, although it was shown that changing the vibration tempo according to the change in heart rate was more effective in reducing stuttering than a constant tempo, it was not possible to examine the appropriate tempos for individuals. A tendency was observed in the subjective evaluation of subjects and the stuttering reduction effect of the vibration stimulus with the changing tempo. Therefore, if a subject adjusts to an appropriate tempo by subjective judgment during long-term training, a more appropriate experimental design can be realized. Consequently, it can be used as effective stuttering improvement training that replaces the ear-hung metronome.

References

1. Sakai, N., Mori, K., Ozawa, E., Mochida, A.: Stuttering treatment with a behind-the-ear type metronome for an adult stutterer. Jpn. J. Logop. Phoniatr. **47**(1), 16–24 (2006)
2. Doseki, K.: Global and dynamic structuring methodology in aphasic therapy. Jpn. J. Commun. Disord. **23**(1), 17–22 (2006)
3. Rong-Na, A., Sakai, N., Mori, K.: Short-term effects of speech shadowing training on stuttering. Jpn. J. Logop. Phoniatr. **56**(4), 326–334 (2015)
4. Takatsu, H., Munakata, M., Ozeki, O., Tokoyama, K., Watanabe, Y., Takata, K.: An evaluation of the quantitative relationship between the subjective stress value and heart rate variability. IEEJ Trans. Electron. Inf. Syst. **120**(1), 104–110 (2000)
5. Ichihara, S.: The perspective of the research for the semantic differential and the problems to be solved. Jpn. J. Ergon. **45**(5), 263–269 (2009)
6. Ito, S., Umehara, M., Matumoto, S., Murakami, H.: Trials toward estimate stuttering-by examining the inspection method of stuttering. Jpn. J. Logop. Phoniatr. **25**(3), 243–251 (1984)
7. Homma, T., Aoki-Sasaki, S., Yamada, J., Imaizumi, S.: Anxiety response, vowel space and stuttering rate for persons who stutter and do not stutter under high and low anxiety conditions. Jpn. J. Logop. Phoniatr. **52**(1), 19–25 (2011)
8. Omori, S., Kobayashi, T.: Observers' effects on subjective emotion and physiological response during mental stress task. J. Res. Inst. Bunkyo Gakuin Univ. **14**, 159–161 (2012)

Guided Mindfulness: Using Expert Schemas to Evaluate Complex Skill Acquisition

Mina Milosevic[✉], Katherine Rau, Lida P. Ponce, Nicholas A. Moon, Nisha Quraishi,
Alexandria Webber, and Richard L. Griffith

Florida Institute of Technology, Melbourne, FL 32901, USA
jmilosevic2015@fit.edu

Abstract. This paper presents qualitative data regarding the experiential learning process, including how this process can be enhanced through an individualized learning approach we refer to as Guided Mindfulness (GM). GM works as a facilitating mechanism where specific self-regulated learning opportunities are a) identified as learning events, b) reflected on, and probing questions on the event are c) asked to heighten experiential learning. Specifically, frequent experiential learners (FELs) were interviewed regarding their learning process, comparing that process between expert and novices and providing preliminary identification of the differences between expert and novice schemas. Finally, the five GM intervening processes (i.e., situational awareness, self-awareness, social awareness, sensemaking, simulation) were mapped into each of these stages. Future directions are proposed regarding understanding these expert schemas more thoroughly and applying this understanding across cultures.

Keywords: Guided mindfulness · Experiential learning · Cognitive schema · Complex skills

1 Introduction

Today's global work environment is characterized by an unprecedented pace of technological change and increasingly complex jobs. To keep up, workers must acquire and develop a variety of highly complex skills, ranging from analytical and critical-thinking skills to interpersonal and leadership skills. There is a growing consensus that the development of such complex skills requires learning beyond the classroom environment, specifically in settings where there are opportunities to experience the context in which skills are applied, refined, and practiced [1]. In fact, research supports the conventional wisdom that complex skill acquisition is greatly enhanced with the addition of experiential learning [2].

Experiential learning, the practice of learning through reflecting on experiences [3], has several notable advantages compared to traditional classroom instruction: a) learning is inherently more relevant as it based on personal experiences, b) more transferable as it occurs in the settings where skills are immediately applied, and c) more generalizable as it exposes the learner to variety of real-world situations where acquired skills are

© Springer Nature Switzerland AG 2020
D. D. Schmorrow and C. M. Fidopiastis (Eds.): HCII 2020, LNAI 12197, pp. 233–256, 2020.
https://doi.org/10.1007/978-3-030-50439-7_16

repeatedly tested, refined, and adapted to different circumstances [4, 5]. In practice, however, these benefits of experiential learning are not easy to achieve. First, exposure to potential learning events is unplanned and is not uniform, resulting in learning that is opportunistic, unstructured, and difficult to evaluate. Second, the idiosyncratic nature of experience-based learning means that self-directed learners are likely to selectively attend to personally relevant or emotionally charged moments and not necessarily to those experiences directly tied to job performance. Finally, "learning while doing" puts additional demands on cognitive and motivational resources, and, when resources are stretched, the need to complete the task often outweighs the need to learn. In fact, much of the effectiveness of experiential learning depends on the learner's ability to self-regulate available resources [6, 7].

Self-regulation, the inhibition or activation of affective, behavioral, and cognitive processes, allows the learner to focus attention, reflect, and achieve goals [8, 9]. In essence, effective self-regulation is a cornerstone of learning in general and when of acquiring complex skills in particular. Thus, it is reasonable to conclude that promoting and strengthening self-regulatory processes can greatly enhance experiential learning. We have previously proposed a new individualized learning approach that improves the effectiveness of experiential learning through enhanced self-regulation [10]. Our proposed artificial intelligence (AI)-assisted platform, Guided Mindfulness (GM), is envisioned to direct an individual through a learning experience with prompting questions and activities before, during, and after a specific experiential learning event [11]. By prompting event-based preparation and reflection, GM is expected to trigger self-regulatory processes necessary for self-paced directed learning of any type of competency or targeted complex skill.

The aim of the current study is to better understand event-based preparation and reflections processes, identify strategies that enhance self-directed experiential learning, and map these processes and strategies onto five intervening GM processes: 1) self-awareness, 2) situational awareness, 3) social awareness, 4) sensemaking, and 5) simulation. To accomplish this, we interviewed and analyzed how experts in self-directed learning approach complex skill acquisition and then modeled GM's event-based preparation and reflection questions and activities onto experts' cognitive schemas related to learning. Next, we aligned the five proposed GM intervening processes with related self-regulation processes at different stages of complex skill acquisition. Finally, we highlighted differences in instructional pathways effective in improving the competencies of novice versus more experienced learners.

2 Complex Skills

Complex skills or tasks are sparsely defined in the literature, and these terms are often used as hypernyms to a wide variety of skills and tasks with commonalities among them. These commonalities of complex skills and tasks are described by Wulf and Shea [12] as those that "generally cannot be mastered in a single session, have several degrees of freedom, and perhaps tend to be ecologically valid," (p. 186). Distinctions or certain types of complex skills have been specified in the literature, such as complex motor skills and technical versus non-technical [12, 13]. Due to the nature of today's work (i.e., the shift towards automation and an enhanced focus on STEM work), our focus is on enhancing the acquisition of both complex technical and non-technical skills through GM. Technical skills refer to those that require specialized knowledge in a specific area of expertise [13, 14] and include skills such as software coding, medical billing, or statistical analysis. Non-technical skills, conversely, do not require specialized knowledge and are often unrelated to science or technology [13]. While the definition of non-technical complex skills may signal a lack of importance or value to some, non-technical skills also include leadership, communication, interpersonal, decision-making, situational awareness, professionalism, and teamwork skills, making up a large portion of what are often referred to as "soft" skills that are highly important in today's workforce [13]. Despite the distinctions between technical and non-technical skills that have been made in the literature, evidence suggests that these distinctions may be arbitrary when evaluating the effects of interventions on complex skill acquisition [15]. As such, we instead focus our current research efforts on complex skills (i.e., whether technical or non-technical) that are best suited to be learned experientially. Such skills are those where preliminary knowledge might be taught in formal settings but are highly reliant on actual hands-on practice, simulation, or observation of others in order for a learner to become proficient in that particular skill. Thus, acquiring such skills typically occurs over a period of time, rather than at one learning event, specifically through multiple phases discussed in the following section.

2.1 Complex Skill Acquisition

In order to facilitate learning of complex skills, it is important to understand the process of complex skill acquisition. A prominent framework of skill acquisition is that of Van-lehn [16], which was originally framed for cognitive skill acquisition but fits the broader context of complex skills. To clearly convey the skill acquisition stages, we will discuss them in the context of cognitive skills (i.e., the ability to attain and solve problems when it comes to intellectual tasks, specifically) [16]. This framework explains that skill acquisition occurs in three phases: a) the early stage, b) the intermediate stage, and c) the late stage. These phases blend into each other (i.e., as an individual strengthens in one, they enter the following stage). The first stage, or early stage, is marked by comprehending the knowledge domain without actually applying what has been learned yet, where individuals tend to experience cognitive overload by the amount of information being exposed to them. As proficiency in the cognitive skill increases, a learner enters the intermediate phase, whereby problem solving occurs because the learner begins to call forth examples of problems that have been solved. By now, the learner has acquired some relevant knowledge on how to solve problems but also has some flaws in the knowledge domain (i.e., missing knowledge or incorrect knowledge). The final stage, or the late stage, is where learners continue to improve in accuracy and speed of cognitive skills as they practice. During this stage, the same style of gathering knowledge and problem solving is still employed; however, transfer and practice effects are the main occurrence [16]. As expected, the stages of complex skill acquisition are only neatly defined and organized when written about theoretically, and for many working individuals, learning new skills rarely follows any structured patterns. In fact, practitioners suggest that as much as 70% of learning should happen through on-job experiences in comparison to the 10% suggested to occur in formal settings [17]. Since experiential learning is highly recommended and often utilized, it is imperative that any tool that wishes to aid an individual's experiential learning considers the stages of complex skill acquisition to better assist the learner.

3 Experiential Learning

Experiential learning refers to the practice of learning through reflection on previous experiences [3]. Particularly, experiential learning can take place while on-the-job. It is often one of the most effective ways that adults capitalize on the outside-of-the-classroom learning opportunities that comprise a significant portion of adult learning [17]. While experiential learning is an effective way in which adults can learn, there are also drawbacks to learning through experience. Specifically, being exposed to experiences does not inherently translate into learning. The learning from an experience greatly depends on the learner's self-regulatory processes [6, 7].

Learning from experience is not a new concept; however, experiential learning theory was not conceptualized until recent years [3]. Despite advances with the development of the modern classroom, learning from experience is still a very effective method of learning. Experiential learning comprises two ways of extracting experiences, concrete (i.e., observations) and abstract (i.e., implications of observations) conceptualization, and two ways of transforming experiences, reflective observation (i.e., reflecting on observations) and active experimentation (i.e., testing these experiences in new situations) [18]. These processes encompass the four stages of experiential learning (i.e., concrete experience, reflective observation, abstract conceptualization, and active experimentation) [19].

Taken together, experiential learning can play an instrumental role in developing complex skills. Specifically, at each stage of complex skill acquisition (early stage, intermediate stage, and late stage), different resources need to be allocated for learning to occur. For example, experts, in the late stage, might need fewer concrete experiences and more abstract conceptualization and active experimentation, whereas novices, in the early stage, need more concrete experience and reflection. Effective learners are expected to self-regulate throughout this experiential learning cycle by allocating resources to optimize learning and subsequent performance. Thus, in order to successfully learn from experiences, learners must employ self-regulatory processes.

When considered with other types of learning utilized to prepare the workforce (e.g., classroom settings), experiential learning has a number of benefits during complex skill acquisition. First, experiential learning allows learners to understand and practice complex skills in greater depth than what is possible in traditional classroom settings. Second, experiential learning enables the learner to be more directly involved in his or her learning experience [3]. This approach is inherently more relevant because each learner is directed by their unique personal experiences allowing them to engage with information and tasks in different ways; thus, this learning approach is more representative of the real world. Third, experiential learning offers opportunities for learners to apply their newly acquired knowledge and skills into real-life situations or challenges that they may encounter. Experiential learning encourages deep reflection and provides learners with the opportunity to seek and receive instant feedback. Lastly, this learning approach allows learners to identify their mistakes and encourages them to focus their attention on action and reflection [4, 5].

Despite these benefits of experiential learning, learning from experience is not guaranteed and therefore has limitations. First, learners can experience the learning event in several ways, eliciting unstructured learning. Second, due to the lack of certainty and structure in learning opportunities, experiential learning can sometimes be difficult to assess. Lastly, *learning* and *doing* act as competitors when using cognitive and self-regulatory resources. Learners must use self-regulatory processes to effectively learn by doing [6, 7]. Thus, a successful approach to experiential learning of complex skills must consider and attempt to overcome the barriers associated with experiential learning.

4   A Self-Regulatory Approach to Experiential Learning

Self-regulatory processes are critical for the effectiveness of learning and developmental intervention [6, 7]. Self-regulation is rooted in control theory whereby an individual acts within a negative feedback loop where he or she compares their current state to their desired state. If a discrepancy occurs between the two states, the individual will adjust his or her behavior to reach the desired one [20]. Complex learning environments require individuals to have an awareness of their current state via both internal and external sources. The internal source determines their skill and competency state, while the external source refers to the social and situational factors. For the individual to be successful in reaching their desired state, they must gather all relevant feedback from the assessment of their state, reflect on it, and effectively integrate it into their repertoire of learning experiences to further improve and attain their goals [10]. However, in the context of fast-paced work environments, many individuals lack the resources (e.g., time, energy, self-awareness, knowledge) to engage in these self-regulatory processes, which is ultimately to the detriment of their own learning. As such, taking a targeted, self-regulatory approach to experiential learning can enable an individual to capitalize on opportunities to learn beyond the traditional instructional setting. Next, we discuss how our proposed GM approach does just that: enhances learning from events through insightful preparation and reflection that ultimately develops and strengthens self-regulatory processes.

5   Guided Mindfulness

GM is an individualized approach to experiential learning designed to enhance complex skill acquisition by facilitating and strengthening the self-regulatory processes of the learner. More specifically, GM is envisioned as a technology-assisted platform that guides an individual through a learning experience with prompting questions and activities before, during, and after a specific experiential learning event [11]. Thus, the central activity of GM approach is event-based learning involving preparation and reflection around these events [21].

Event-based preparation and reflection is expected to trigger and, over time, strengthen self-regulatory processes by prompting a learner to engage in 5 intervening processes: (1) self-awareness, (2) situational awareness, (3) social awareness, (4) sense-making, and (5) simulation [10, 21, 22]. *Self-awareness* refers to the understanding of one's beliefs, assumptions, bias, emotions, skills, values, strengths, and weaknesses [23]. In general, self-awareness enables an individual to accurately assess their current state in terms of their abilities, skills, or emotions, and to identify where resources should be allocated to obtain desired outcomes. *Situational awareness* represents a dynamic understanding of a situation in a way that recognizes and incorporates environmental and situational changes quickly and effectively thus continually updating one's understanding of the situation and of a larger context in which a particular situation occurs [24, 25]. Similarly, *social awareness* refers to the ability to recognize tacit social cues in order to understand social dynamics of an interpersonal environment and select behaviors appropriate for the given social situation [26]. Learning complex skills often involves navigating socially dynamic interpersonal environments regardless of whether the skills per se are inherently social (e.g., leadership) or simply learned or practiced in a social environment. For example, on-the-job learning may occur in an environment comprised of individuals with diverse backgrounds, competing motives, and different behavioral styles, all of which can impact the process of skill acquisition. Consider an example of a novice surgeon. Although the complex skill being learned is technical in nature (i.e., performing surgery), the operating room in which the skill is learned, performed, and perfected represents a dynamic social environment that still needs to be navigated by the surgeon.

Improved self-awareness, situational awareness, and social awareness leads to more effective *sensemaking*, the key process in experiential learning [27]. Sensemaking is a process by which people infer meaning from an event and then use this newly derived connotation to guide their future decisions. Effective sensemaking requires extrapolation of meaning from a complex combination of situational, social, and internal cues [28]. Cues are organized, clarified, and interpreted using existing mental models; in turn, the existing mental models are continually tested against new interpretations. Thus, sensemaking is an active process that, over time, evolves our understanding of the stimuli around us and enables us to adapt and prosper in complex and changing environments.

Finally, *simulation* refers to enactment or re-enactment of perceptual, motor, and introspective states that are either anticipated or have already been experienced during a learning episode. In experiential learning context, simulation can take the form of mental rehearsal prior to the event and it typically includes an "if, then" internal dialogue. Simulation can also occur after the event in the form of re-enactment. In this case, the learner may think through, or simulate, alternative courses of action that lead to different outcomes. In many ways, simulation helps the learner connect one learning episode to the next by testing out mental models which have been updated based on prior experiences before subsequently deciding on the best course of action in the next episode.

5.1 How GM Operates

We propose that our GM approach can trigger these five intervening processes with prompting questions and activities before and after specific experiential learning events [11]. The core mechanisms underpinning five intervening processes is self-regulation.

Practicing self, situational, and social awareness essentially requires that learners deliberately and conscientiously process information from multiple sources to determine their current state. Post-event reflection allows for a comparison to pre-event expectations of the event itself and anticipated versus actual state in the event. During sensemaking, learners synthetize feedback from themselves and the environment, identify discrepancies between current and desired state and outcomes, and determine where next to direct resources and how much effort to expend. Simulation allows learners to test out alternative solutions thus optimizing the use of available resources and maximizing desired outcomes in the future episodes.

There are several ways in which our proposed GM approach is expected to address known barriers to experiential learning. First, GM can assist with allocation of cognitive and self-regulatory resources. During the preparation phase, GM will direct the learner's attention to relevant self, situational, and social cues, which will, in turn, free cognitive and self-regulatory resources during the actual experiential learning episode. Research has shown that one of the greatest barriers to experiential learning is cognitive overload that occurs when a learner is tasked with performing and learning at the same time [29]. Attempts to simultaneously process significant amounts of information and extract what is relevant for both performance and learning puts additional strain on self-regulatory resources and often result in a failure to learn. With GM, learners can identify before the event what is relevant for learning and what to pay attention to, thus reducing information processing and a need for increased self-regulation during the event.

Next, GM can provide structure and organization for the information received during the learning episode. Specifically, through event-based preparation, GM activates mental models which helps with encoding of the information during the event and decoding after the event. In other words, GM assists learners to store and later retrieve information effectively enabling learners to memorize information obtained during the experience.

Finally, GM facilitates self-feedback during the event-based reflection. Feedback is a core self-regulatory process that drivers learning and performance improvements [30]. Feedback provides individuals with standing on their competencies, helps identify gaps, and prompts the learner to engage in behaviors that reduce discrepancies and ultimately close the gaps [31]. In the context of experiential learning, feedback may be difficult to obtain as learning opportunities are often unpredictable and the lack of clearly established metrics makes it difficult to evaluate performance. GM addresses these challenges by facilitating and guiding self-feedback. Guided self-feedback is similar to the team-level intervention known as guided team self-correction [32], which has been shown to greatly improve team processes and performance. With GM, an individual learner is encouraged to continually self-assess performance by integrating various cues from self and the environment. Additionally, GM prompts the learner to evaluate the match between pre-event expectations and post-even insights thus setting the stage for further action. Finally, when mistakes are identified, GM enables learners to self-correct by testing out alternatives through simulation.

Through our current research efforts, we aim to explore the cognitive schemas of self-directed learning experts to validate the theoretical underpinnings of the GM platform. As such, we employed qualitative methods to describe the learning process of experts, specifically how they prepare for an event, how they engage in learning during an event, how they reflect on their learning following an event throughout the lifespan of learning (i.e., early, intermediate, and late stage) [16], and barriers or challenges they face when learning in or from an event. By identifying the successful and unsuccessful learning processes and strategies of self-directed learning experts, it allows us to more effectively guide learners using purposeful and targeted preparation and reflection tactics that are rooted within the five underlying processes of GM.

6 Method

6.1 Development of Protocol and Procedure

Twelve Ph.D. graduate students in Industrial-Organizational Psychology developed interview questions based on experiential learning and GM literature. A total of 31 interview questions were developed. Through discussion of the relevance, research aims, and practical considerations (e.g., time allotted for interviews, ability of participants to effectively answer each question), the final interview protocol was reduced to include 11 unique questions with additional prompting questions as necessary. The interview questions were framed into 3 categories regarding experiential learning and complex skill acquisition: learning from experience, experiential learning across skill levels, and barriers to experiential learning. The frequent experiential learners (FELs) who were interviewed were instructed to think of a skill specific to their occupation as they answered each question. The goal of the interview questions was to identify an expert schema of how experts learn from experience, similar to a cognitive task analysis (CTA) or a structured recall activity [33]. The categories of interview questions, the description and purpose of each category, and example questions for each can be found in the Appendix (Table 2).

Following the development of the interview protocol, the researchers compiled a list of occupations that met the criteria of high-stakes, highly-complex positions. Researchers then utilized the existing social networks of the team to identify individuals who could be considered FELs within their respective fields. The list of contacts was prioritized by the occupations that were of most interest to the current research effort (i.e., with highest priority given to the occupations that were considered to be the most complex and have the highest stakes). Researchers then personally contacted each individual FEL to introduce the study and schedule a 1-hour interview session. Each interview was conducted by a researcher, recorded and subsequently content analyzed using thematic analysis methods [34]. FELs were continuously contacted and interviewed until a point of saturation was reached [35], where further interviews were beyond the scope of this manuscript and would not contribute additional subject matter.

6.2 Coding Process

Once each of the interviews were transcribed, data was prepared for coding. Specifically, seven Ph.D. graduate students in Industrial-Organizational Psychology coded and transcribed the interview data. First, the coders reviewed the data and met to discuss the initial themes through an initial pilot test. Although this approach is not ideal, it is considered an acceptable alternative to coding [36]. Second, the coders met together to discuss definitions and example themes and subthemes. Third, the coders were paired together for each of the interviews where each coder independently coded the interview. Finally, these themes were combined and are presented in the Results section below.

Specifically, these themes were centered around the mental models of these experts. Mental models refer to a network of representations of knowledge about an individual's environment and experience [37–39]. In this case, these themes were designed to outline the process of experiential learning and developing expertise. Specifically, these themes were examined through the differences between experts and novices, such that experts have specific mental models and processes that are unique [40]. The expert mental model is meant to contain diagnostic clues and be qualitatively superior to the novice mental model [41].

6.3 Sample

Twelve FELs were invited to participate in the semi-structured interview process. Of these twelve, ten FELs representing each occupation invited were subsequently interviewed using the questions referenced above (i.e., 83.33% response rate). FELs occupations included pilots (e.g., airline pilots and flight school instructors), health-care providers (e.g., X-ray technicians, emergency medicine physicians, pediatricians), entrepreneurs, and consultants. The interviews ranged in experience level in their respective fields, but each interviewee was considered an expert in experiential learning based on their experiences within their occupation.

7 Results

After each interview was coded by two raters, a total of 25 broad themes were found that consisted of multiple sub-themes grouped under the broad themes. A total of 140 themes were found including both broad and sub-themes. See the Appendix for a full list of themes (Table 3).

7.1 Learning from Experience Process

Four general themes were found for the process of experiential learning: *attending to cues, mindset, process-based learning, and task-based learning. Attending to cues* represents responses that involve paying attention to one's environment, situation, and/or the individuals in the situation or environment in order to learn from experience. Sub-themes included examining context, adapting behavior, focusing on details, learning from peers, etc. *Mindset* represent responses that entail getting into the right headspace to learn from experience. Sub-themes include staying focused, dealing with ambiguity, staying open-minded, remaining calm, etc. Responses entailed a "need to be alert, have the right mindset, and have good attention to detail." *Process-based learning* represents steps taken to facilitate the experiential learning process. *Task-based learning* represents preparing for specific tasks to facilitate the experiential learning process.

7.2 Learning Across Skill Levels

Experiential learning across skill levels was broken down into two categories: novice and expert. This helped identify and highlight the differences in the experiential learning process between these two stages. Ten broad themes were found for experiential learning from a novice's perspective: *learning and practicing steps, goals, mindset, feedback, time spent preparing and reflecting, checking understanding against others, focusing on fun, self-motivated, and commitment. Learning and practicing steps* represents steps taken at the novice level to facilitate the experiential learning process. *Goals* represent responses that include goals that novices are trying to accomplish to better facilitate the experiential learning process. For example, one FEL stated, "If you look at these micro episodes, each step has a goal, it has a method, and has common mistakes. If you can break down things like that and attack one little thing at the time, I think it helps the learning." *Mindset* involves the types of mindset novices have during the experiential learning process. *Feedback* consists of the feedback process a novice takes into account when trying to facilitate experiential learning. Results regarding how much time novices spent preparing and reflecting were inconclusive; some interviewees indicated that, as novices, they spent more time whereas others suggested they spent less time on preparation and reflection compared to experts. Additionally, some FELs felt that the difference between novices and experts was in how much time they spent on preparing versus reflecting. For example, one FEL stated, "As a novice, you may take more time preparing, but less time reflecting because you're still overwhelmed by the learning experience." *Checking understanding against others* represents how novices attempt to validate their knowledge. *Focus on fun* represents responses that discuss seeking elements of fun in the learning experience. *Self-motivated* represents responses that entail motivating oneself to facilitate the experiential learning process for novices. *Commitment* entails responses that require commitment to a task or situation to better facilitate the experiential learning process for novices.

Five broad themes were found for experiential learning from an expe"rt's perspective: goal change, process improvement, spending less time preparing and reflecting, seeking feedback on the process, and having more resources to draw on when preparing and reflecting. *Goal change* represents responses that entail goal shifts in the experiential learning process for expert learners. Sub-themes include goal or focus becoming efficient once the task is extremely familiar, goal becomes self-improvement (more analytical, forward thinking), and honing what you know. *Process improvement* represents responses that allude to a process shift in experiential learning once a novice becomes an expert. Sub-themes include applying skills to new situations, questioning the system, adapting quicker, etc. *Spending less time preparing and reflecting* represent responses that suggest experts spending less time preparing and reflecting versus novices. *Seeking feedback on the process* represents responses that depict how experts seek feedback in their experiential learning process. *Having more resources to draw on prepare and reflect* represent responses where experts described a cognitive shift to more resources once one gets better at preparing and reflecting.

7.3 Barriers to Experiential Learning

Six general themes were found for barriers to experiential learning: *learning from others, speed of the situation, lacking feedback, forced performance, lack of guidance, and within-person barriers. Learning from others* represents barriers to the experiential learning process when it comes to learning from others. Sub-themes include being dependent on those around you and your awareness. *Speed of the situation* includes responses that represent time constraints as a barrier to the experiential learning process. *Lacking feedback* represents responses that discuss lack of feedback as a barrier to the experiential learning process. Sub-themes include not knowing when a mistake was made and if feedback was not framed as mistakes or was too broad (i.e., not in step-by-step fashion as preferred). For example, one FEL said, "In a situation where a patient dies, it's hard to decipher whether you made a mistake or what else may have been at fault." *Forced performance* represents practicing when one is not ready as a barrier to the experiential learning process. Sub-themes include being forced to execute without having prepared (either through simulation or practice) and performing an action you do not want to do. For example, another FEL described the problem with having to perform a task that is "beyond your skill set." *Lack of guidance* represents responses that discuss a need for more guidance to better capitalize on the experiential learning process. *Within-person barriers* represent responses that vary within individuals that may act as barriers to the experiential learning process. Sub-themes include physical exhaustion, mental exhaustion, anxiety, ego, distractions, and pressure.

8 Discussion

The aim of the current study was to better understand event-based preparation and reflections processes, identify strategies that enhance self-directed experiential learning, assess if and how experts' processes and strategies map onto our five proposed GM processes, and identify barriers to experiential learning. Our interviews with experts in experiential learning revealed a number of relevant themes resulting in several conclusions we will discuss next.

8.1 Process of Experiential Learning

First, we were able to confirm that our five proposed GM processes (i.e., self-awareness, situational awareness, social awareness, sensemaking, and simulation) are indeed key drivers of effective experiential learning. As part of the event preparation, our FELs frequently discussed paying attention to internal states and processes, to the anticipated elements of the upcoming situation, or to the specific aspects of the task, activity, or behavior that they were about to perform. Thus, awareness seems to play a critical part in the preparation stage of experiential learning. When reflecting on experiences, the experts seemed to focus their attention on making sense of the past events. In particular, FELs discussed identifying patterns, reinforcing the existing knowledge, incorporating new information, and making necessary adjustments. The reflection phase also had a clear forward-thinking component as FELs frequently described developing an action plan for future events as well as simulating step-by-step how a new event could or should unfold.

Second, another important finding was that effective learners spend time thinking about both the process of learning and the behavior needed to be performed. This was particularly evident in the intermediate stage of learning when learners had compiled enough knowledge about the skill that they seemed to benefit from focusing on specific details and engaging in repeated practices. Interestingly, our FELs also indicated that they spend considerable time during this phase thinking about the learning process and developing and implementing strategies to aid in learning from the events. For example, they described taking notes, at times even mental notes, creating diagrams, practicing visualization, and breaking learning into smaller episodes.

Third, the temporal distinction between pre- and post-event does not translate neatly into a process distinction. In other words, processes taking place during preparation and reflection seem to form a continuous loop, where reflection feeds into preparation for the next event and preparation for the future events builds on the reflection of the past event(s). Similarly, we found that, while perhaps more pronounced at certain stages, all five intervening processes played an important role throughout the experiential learning process. When designing a GM system, there may be trade-offs between enhancing all intervening processes simultaneously and exposing the learner to potential cognitive overload, which is precisely what GM is designed to avoid. Thus, a successful approach to enhancing experiential learning would likely need to promote the right processes at the right time to maximize benefits to the learner.

However, one finding that disconfirmed our initial assumption was the lack of significant differences between learners of technical versus non-technical skills. In general, our FELs described similar processes when learning from experiences, regardless of the skill that was being learned. There were also quite a few similarities in strategies that they used to optimize their learning. However, we did find some occupational differences in terms of the systems put in place to facilitate experiential learning. Pilots, more than any other professionals we interviewed, seemed to have a well-designed system that enhances experiential learning. For example, during training, pilots have pre-brief and debrief after every flight, regardless of whether it's in a simulator or in the actual plane. Most of the pre-brief and debrief activities are centered around the same themes we propose in GM: self, situational, and social awareness, sensemaking, immediate feedback, and plan of action. Additionally, pilots engage in simulator training designed to allow them to practice their skills, identify errors, and self-correct in the next event.

Overall, the process of experiential learning employed by FELs can be described as an iterative process, entailing all five intervening processes in a continuous feedback loop and is largely driven by self-regulation.

8.2 Experiential Learning Across Skill Levels

Our results suggest that there are significant differences in what makes experiential learning effective for novices compared to experts. Although both novices and experts engaged in the five processes over the course of event preparation and reflection, there were differences in what their overall goals were, what they focused on, which strategies they used and the level of detail in their preparation/reflection activities. Additionally, we concluded that the three-phased skill acquisition framework proposed by Vanlehn [16; i.e., early, intermediate, and late stage) was well suited for the experiential learning process and that it captured progression of learning better than just the two categories (i.e., novice and expert).

In the early stages of learning, our interviewees predominantly discussed focusing on one's internal states. In the preparation phase, they seemed to be primarily concerned with reducing anxiety, tolerating ambiguity, bolstering an open mindset, invoking calmness, and gearing themselves toward learning. Possibly since learners at this stage have limited information, if any, about the upcoming event, their thoughts are turned mostly inward. These findings suggest that self-awareness is the key process in the early stages of the event-based preparation and that GM questions should be centered around this type of insight. In the reflection phase, novice learners relied heavily on sensemaking to extract relevant information, identify patterns, and derive meaning. FELs frequently discussed identifying what's important, identifying relevant tasks, and analyzing available and necessary resources. Thus, it seems that in the event reflection stage, novices might benefit from GM questions that assist them in identifying critical elements of the past experience. The questions will then guide novice learners to use these critical elements as foundations for preparing for the future learning experience.

The next, intermediate, phase was described as the time when learners had some prior learning experiences to draw from but have not yet reached the proficiency level necessary to be considered experts. In this phase, the focus has clearly shifted from the self to the situation and task at hand. FELs described focusing their attention on the critical elements identified in the prior reflect stage. This stage was characterized by increased specificity and greater level of detail. Our learners discussed breaking down experiences into micro-episodes of learning and, for each micro-episode, identifying what the goal is, the steps needed to be performed, and the potential mistakes. Experiential learners in this phase seemed to integrate internal, situational, and social cues to make sense of the experienced events and prepare for the next event. Several interviews mentioned visualization and mental practice, thus supporting our fifth process, simulation. They made a clear connection between identifying critical elements, practicing those elements (e.g., acting out specific behaviors, mentally testing alternatives and related consequences, imagining "if, then" scenarios) and then paying attention to these elements during the next event. When reflecting on the event, learners in the intermediate stage focused on evaluating how they performed during the event and making sure they auto-corrected. Specifically, the learners described thinking about the event or re-enacting the event in their mind (i.e., simulation) with the purpose of evaluating their performance (i.e., self-feedback), identifying discrepancies, and determining what they needed to pay attention to in the next event (i.e., sensemaking). One particularly interesting finding was that learners in this phase seemed preoccupied with mistakes made and focused their attention on correcting those mistakes in future events. Overall, the intermediate phase of experiential learning seemed to be the most active phase when most learning seems to occur. In this stage, learners are relying on self, situational, and social awareness to make sense of the events and learn from the experiences. This phase was characterized by the great amount of detail regarding specific strategies learners use to facilitate and expedite experiential learning. Thus, learners in this stage might benefit from GM prompts and questions centered around evaluating one's performance, developing specific strategies to enhance performance, and then practicing through simulation.

Later stages of experiential learning are characterized by refinement of skills and expansion of skill application. At this point, most learners have mastered the skill enough to be considered experts. For them, preparing for the event entails focusing on specific elements, or minor details that can help them perfect the skill. Several FELs described thinking about and anticipating infrequent, atypical situations and then preparing for them mentally. When reflecting on the event, experts are still performing self-evaluation and assessing if their competencies are a match for the situation they encountered. Some are looking for information that may challenge the assumption of expertise. In many ways, experts seem to be focused on refining their skills by testing them, or on expanding their skills by trying to apply them beyond the original context. Thus, elements of GM that facilitate self-evaluation and assist learners in integrating and synthesizing experiences would still be quite beneficial in these later stages of learning. It is worth noting that some experts turn their focus inward again, similar to the early stages of learning, but this time from a deeper meta-cognitive perspective. However, there were several experts in our sample that described turning attention to others and becoming mentors, informal coaches, or advisors. Therefore, we conclude that GM can also be beneficial in promoting broader critical thinking, self-regulation, and mindfulness beyond the specific event to helping others (Table 1).

8.3 Barriers to Experiential Learning

The final objective of the current research effort was to identify the biggest challenges to experiential learning for this particular demographic. Our purpose in identifying challenges to experiential learning was twofold: not only would these perspectives help us better understand the cognitive schemas of experts, but these answers would also assist us in addressing the needs of GM's target population. FELs responses regarding barriers to experiential learning revolved primarily around 6 themes: the ability and willingness to learn from others, the speed of the situation, lack of feedback, performance without adequate preparation, guiding one's own improvement, and within-person barriers.

The challenge of having the ability and willingness to learn from others can be explained by the inherent nature of learning on the job; a person who is learning-by-doing by watching or working with others is highly dependent on those around them. For example, a surgeon who is observing a colleague perform a new procedure is at the mercy of the colleague and the colleague's ability to effectively discuss the actions of the procedure, the underlying thought processes, nuances of the task, and so forth. Further, the learner has to be willing to learn from another person or the circumstance itself, which requires paying attention to cues, effectively organizing the information that is being gathered, and the inclination to learn, especially in situations that may cause feelings of nervousness, stress, or anxiety. GM has the opportunity to alleviate this broad category of challenges by addressing specific needs. For example, while GM may not be able to choose who a person shadows or works alongside, the use of guided preparation and reflection provides insight to a learner into ways to better guide their own learning (e.g., guiding a user to identify his or her own needs to effectively learn and how to enact necessary learning strategies). For those who are less willing to try to learn experientially, GM may also aid in improving motivation by enhancing self-regulatory processes and boosting self-efficacy in specific competencies, either through recorded progress and successes in past events or through gamified assessments where users will receive badges in coins for making progress or completing competencies.

Table 1. Mapping main themes on stages of experiential learning and intervening processes

	Prepare	Reflect	Guided mindfulness processes
Early	Reducing anxiety Tolerating ambiguity Bolstering open mindset Invoking calmness Gearing up for learning	Sensemaking to extract relevant information, identify patterns, and derive meaning from information given Identifying what is important Identifying relevant tasks Analyzing available and necessary resources	Self-Awareness Sensemaking
Intermediate	Breaking down experiences into micro episodes of learning Identifying goals, necessary steps to be performed, and identifying potential mistakes within each learning episode	Evaluating performance during the event and making sure performance is auto-corrected Thinking about the event or re-enacting the event in their mind with the purpose of evaluating their performance Identifying discrepancies and determining what they needed to pay attention to in the next experience	Situational Awareness Simulation Social Awareness Self-Awareness
Late	Focusing on specific elements, or minor details that can help them perfect the skill Thinking about and anticipating infrequent, atypical situations and then preparing for them mentally	Self-evaluation and assessing if competencies are a match for the situation encountered Looking for information that may challenge the assumption of expertise Refining skills by testing them Expanding skills by trying to apply them beyond the original context	Situational Awareness Simulation Social Awareness Self-Awareness Sensemaking

In a similar vein, the speed of the situation was consistently identified as a barrier to effectively learning from experience. Learning that occurs during experience, especially in the context of highly complex and high-stakes environments, happens quickly and often without time to adequately process all elements or information. Guided reflection within the GM platform allows designated time for a user to contemplate their experience in an event, identifying successes, areas for improvement, ultimately reinforcing the accurate memory of the learning event by asking a user to recall and once again encode the event through the lens of the particular reflection questions asked of them.

An additional challenge mentioned by FELs was the lack of feedback one typically receives during experiential learning episodes. Interestingly, this theme contradicts popular experiential learning literature, where extracting feedback in real time is noted as an advantage to taking an experiential approach to learning [3]. FELs indicated that the lack of feedback is particularly harmful in contexts where feedback is needed for process improvement and a lack of feedback fails to inform oneself of their own performance. GM's underlying processes are specifically designed to aid a user in soliciting feedback, whether it is by identifying unique outlets to ascertain feedback from or by reflecting inward to use internal cues as a source of performance or process feedback.

Another challenge faced by our interviewees was having to perform during a learning event despite not being ready to do so, either because of knowledge or experience level or a lack of preparation for that particular event. GM's underlying processes are intended to aid synthesis of existing knowledge across domains and situations to result in better performance in novel situations. Through the use of GM, users should be better prepared to access necessary knowledge and skills to apply in new learning situations or performance episodes where they do not have prior experience.

Guiding one's own improvement, an additional challenge noted by FELs, is characterized by having to know how to improve, potentially without the help from others. Experiential learning is a very active form of learning that requires proactivity by the learner. However, learners may find themselves to be complacent, passively learning by not practicing outside of learning episodes, or lacking specific goals without effective guidance or knowledge of how to effectively improve. The GM platform provides a roadmap of guidance and structure that prompts users to proactively learn but also gives them effective tools to do so.

Within-person barriers such as physical exhaustion, mental exhaustion, anxiety, ego, distractions, and pressure comprised the final theme of barriers to experiential learning. These elements impede on learning through challenging effective functioning of self-regulatory processes or by diverting one's focus away from learning. As discussed, GM has the opportunity to quell mental unrest through reinforcing a calm and collected mindset for the user and also providing effective tools to boost self-efficacy. The platform is also designed to require minimal time from participants outside of learning episodes in an effort to not further drain energetic resources of the user.

9 Future Directions

Our current research effort, while it produced invaluable insight, also spurred directions for future research. First, validation of the themes we found in our qualitative interviews against a larger, more diverse sample is needed. While we purposely chose to interview individuals who are employed in high-stakes and highly complex occupations, we ultimately believe these findings should generalize to other highly complex jobs. Further testing this concept by interviewing and surveying experts in experiential learning will provide insight into the usefulness and utility of the GM concept. Additionally, further research using a larger, more diverse sample may help answer the question as to whether or not the GM platform is better suited to aid in the learning of particular types of skills (e.g., technical versus non-technical) and why. Usefulness of GM should also be assessed for the type of skills that have long been considered best acquired through practice, such as cross-cultural competence, global leadership, and interpersonal skills.

Second, further research utilizing qualitative methods and cognitive task analysis is needed to capture the cognitive schemas of individuals across the continuum of skill acquisition stages to better inform the GM platform. Specifically, in an effort to make any future use of AI viable, we want to continue mapping GM's underlying processes (i.e., situational awareness, self-awareness, social awareness, sensemaking, and simulation) against the processes of learning complex skills. Doing this will allow us to validate the assessment metrics of GM as it pertains to evaluating a user's learning progress within a particular competency. As such, expert schemas of specific competencies become a standard in which novices or intermediate learners are quantifiably compared against. Creating such a standard informs how GM will effectively interact with an individual and ultimately guide a user's learning, which is paramount to the success of the GM platform in aiding skill acquisition.

Finally, one of the significant areas for future research would be exploring how GM can be supported with technology, and ultimately with Artificial Intelligence, so that the platform can be optimized to prompt both the processes that are most effective in each stage of the learning and those that are most useful to the individual learner. If so, AI technology can further increase the effectiveness of GM in enhancing self-regulatory processes required to become a more effective learner.

10   Conclusion

We observed and analyzed how experts in self-directed learning approach complex skills acquisition and subsequently modeled GM's event-based preparation and reflection questions and activities on experts' cognitive schemas related to learning. By engaging in guided preparation and reflection, one can create stronger cognitive schemas to better facilitate the experiential learning process to better learn from on the job experiences. Next, we aligned the five proposed GM intervening processes with related self-regulation processes in different stages of complex skill acquisition. Finally, we highlighted differences in instructional pathways effective in improving the competencies of novice learners versus more experienced learners. Future research directions include validating the findings of the current study in a larger, more diverse sample and further use of cognitive schemas as a way to measure and more effectively guide the learning process.

Appendix A

Table 2. Interview Questions by Categories

Interview topic	Description and purpose	Example questions
Process of Experiential Learning	Identify process of experiential learning Identify elements of preparation before and reflection after learning events Identify success metrics for learning	Think of a time when you learned something on-the-job; what does that look like? How do you prepare for the learning experience? What do you do during the learning episode? What do you do after the learning episode? How do you know you're actually successfully learning from your experiences?
Experiential Learning Across Skill Levels	Identify differences in learning between novices and experts	Think of a time when you were not experienced in a particular skill: was your learning different at this stage? How much time do you spend preparing and reflecting at your skill level? Is it different at a novice-level in comparison to an expert-level?
Barriers to Experiential Learning	Identify challenges of complex skill acquisition that are unique to experiential learning Identify needs and opportunities for an experiential learning tool	What are some barriers or challenges to learning from experience? Think of a time you didn't successfully learn from experience or you missed an opportunity to learn: what made it difficult to learn from experience? In hindsight, what would have been helpful to you before to avoid missing an opportunity to learn?

Table 3. Interview topic and resulting themes

Interview topic	Resulting themes
Process of experiential learning	1. Attending to cues • Environmental • Examine context • Identify the purpose • Adapt behavior • Examining others' behavior • Observing behavioral mannerisms • Learning from peers • Take mental notes • Drawing attention to cues that will provide feedback • Looking at experience at higher level • Focusing on the bigger picture • Focusing on the details • Situational awareness 2. Mindset • Staying focused • Developing cognitive path • Identifying critical steps • Dealing with ambiguity • Open-minded • Remaining calm • More metacognitive • Self-evaluation 3. Process-based learning • Seek feedback • Reflect/debrief on past experiences • Identifying patterns • Develop action plan • Identifying mistakes • Reinforce existing knowledge • Prioritizing • Breaking down steps into smaller episodes • Process is automatic • Take your time • Put effort • Practice process • Stay current 4. Task-based learning • Simulating tasks (more for expert) • Preparing for specific experience (task) • Anticipating tasks

(*continued*)

Table 3. (*continued*)

Interview topic	Resulting themes	
Experiential learning across skill levels	Novice	5. Learning and practicing in steps • Break task down into procedural steps as way to organize information • As a way to really learn the task; • As way to prioritize energy • Prioritize attention/energy at perfecting doable steps (can't be immediately perfect); • Smaller chunks/components • Simulating step-by-step how a new event would unfold • Identify what's important • Identifying relevant tasks 6. Goals • Accomplish the task • Track goals • Watch procedures • Improving skills • Longer • Self-aware • Focus on building skills 7. Mindset • Nervousness • Anxiety as splitting cog faculties • Stress • Being open-minded 8. Feedback (wherever you can get it) • Asking questions is really important • More practice and then ask questions 9. Spend more time preparing/reflecting • Focus on building skills • Analyze available resources • Taking notes during/after learning • Slower 10. Checking understanding against others • Examine mistakes 11. Might spend less time reflecting than preparing (overwhelmed, need break as novice) • Apply what you learn • Less risk taking • Want to win game • Trial and error 12. Focused on fun • Forget your problems 13. Self-motivated 14. Commitment

(*continued*)

Table 3. (*continued*)

Interview topic	Resulting themes	
	Expert	15. Goal change • Goal/focus becomes efficiency once the task is extremely familiar • Goal becomes self-improvement (more analytical, forward thinking) • Honing what you know 16. Process improvement • Apply skills to new situation • Build on previous experiences • Question the system • Clearly identify things • Adapt quicker • Less stressed out • Multitask • Applying critical thinking • More about application • More clarity • More confidence 17. Spend less time preparing/reflecting • Strategy might be the same • Self-reflection automatic process • Improving errors • Self-awareness • Situational awareness • Maintaining new info • Incorporating new info 18. Seek feedback on process • Know how/where to seek out feedback • Know which feedback matters 19. Have more resources to draw on to reflect/prepare • Connect present knowledge to past • Focusing on bigger picture and • Key takeaways

(*continued*)

Table 3. (*continued*)

Interview topic	Resulting themes
Barriers to experiential learning	20. Learning from others • Dependent on those around you and your awareness • Level of motivation to learn • Too much material 21. Speed of situation • Time constraints 22. Feedback is lacking • May not know when a mistake was made • Feedback – not framed as mistakes or too broad/need step-by-step 23. Forced performance • Being forced to execute without having prepared (either through simulation or practice) • Perform action you don't want to do 24. Lack of guidance • Requires thinking of how to improve (potentially without feedback or help from others) • Complacent • Not practicing • Lacking goals • Close-mindedness • Failing to prepare 25. Within-Person barriers • Physical exhaustion • Mental exhaustion • Anxiety • Ego • Distractions • Pressure

References

1. London, M., Mone, E.M.: Continuous learning. In: Ilgen, T.D.R., Pulakos, E.D. (eds.) The Changing Nature of Performance: Implications for Staffing, Motivation and Development. Jossey-Bass, San Francisco (1999)
2. McCall, M.W., Lombardo, M.W., Lombardo, M.M., Morrison, A.M.: Lessons of Experience: How Successful Executives Develop on the Job. Simon and Schuster, New York (1988)
3. Kolb, D.A.: The process of experiential learning. In: Experiential Learning: Experience as the Source of Learning and Development, pp. 20–38. Prentice-Hall, Inc. (1984)
4. Gupta, A.K., Govindarajan, V.: Cultivating a global mindset. Acad. Manag. Perspect. **16**(1), 116–126 (2002)
5. Ng, K.Y., Van Dyne, L., Ang, S.: From experience to experiential learning: cultural intelligence as a learning capability for global leader development. Acad. Manag. Learn. Educ. **8**(4), 511–526 (2009)
6. Kanfer, R.: Self-regulatory and other non-ability determinants of skill acquisition (1996)

7. Sitzmann, T., Ely, K.: A meta-analysis of self-regulated learning in work-related training and educational attainment: what we know and where we need to go. Psychol. Bull. **137**(3), 421 (2011)

8. McCall, M.W.: Leadership development through experience. Acad. Manag. Exec. **18**, 127–130 (2004)

9. McCall, M.W.: Recasting leadership development. Ind. Organ. Psychol. **3**(1), 3–19 (2010)

10. Griffith, R.L., Steelman, L.A., Wildman, J.L., LeNoble, C.A., Zhou, Z.E.: Guided mindfulness: a self-regulatory approach to experiential learning of complex skills. Theoret. Issues Ergon. Sci. **18**(2), 147–166 (2017)

11. Griffith, R.L., Sudduth, M.M., Flett, A., Skiba, T.S.: Looking forward: meeting the global need for leaders through guided mindfulness. In: Wildman, J.L., Griffith, R.L. (eds.) Leading Global Teams, pp. 325–342. Springer, New York (2015). https://doi.org/10.1007/978-1-4939-2050-1_14

12. Wulf, G., Shea, C.H.: Principles derived from the study of simple skills do not generalize to complex skill learning. Psychon. Bull. Rev. **9**(2), 185–211 (2002). https://doi.org/10.3758/BF03196276

13. Nestel, D., Walker, K., Simon, R., Aggarwal, R., Andreatta, P.: Nontechnical skills: an inaccurate and unhelpful descriptor? Simul. Healthc. **6**(1), 2–3 (2011)

14. Medina, R.: Upgrading yourself-technical and nontechnical competencies. IEEE Potentials **29**(1), 10–13 (2010)

15. Yule, S., Flin, R., Paterson-Brown, S., Maran, N.: Non-technical skills for surgeons in the operating room: a review of the literature. Surgery **139**(2), 140–149 (2006)

16. VanLehn, K.: Cognitive skill acquisition. Annu. Rev. Psychol. **47**(1), 513–539 (1996)

17. Lindsey, E.H., Holmes, V., McCall Jr., M.W.: Key Events in Executives' Lives. Center for Creative Leadership, Greensboro (1987)

18. Kolb, D.A., Boyatzis, R.E., Mainemelis, C.: Experiential learning theory: previous research and new directions. Perspect. Think. Learn. Cogn. Styles **1**(8), 227–247 (2001)

19. Kolb, A.Y., Kolb, D.A.: Learning styles and learning spaces: enhancing experiential learning in higher education. Acad. Manag. Learn. Educ. **4**(2), 193–212 (2005)

20. Vancouver, J.B.: Self-regulation in organizational settings: a tale of two paradigms. In: Handbook of Self-Regulation, pp. 303–341. Academic Press (2000)

21. Griffith, R.L., Steelman, L.A., Moon, N., al-Qallawi, S., Quraishi, N.: Guided mindfulness: optimizing experiential learning of complex interpersonal competencies. In: Schmorrow, D.D., Fidopiastis, C.M. (eds.) AC 2018. LNCS (LNAI), vol. 10916, pp. 205–213. Springer, Cham (2018). https://doi.org/10.1007/978-3-319-91467-1_17

22. Quraishi, N., et al.: Guided mindfulness: new frontier to augmented learning. In: Schmorrow, D.D., Fidopiastis, C.M. (eds.) HCII 2019. LNCS (LNAI), vol. 11580, pp. 586–596. Springer, Cham (2019). https://doi.org/10.1007/978-3-030-22419-6_42

23. Goleman, D.: Working with Emotional Intelligence. Bantam, New York (1998)

24. Cooke, N.J., Kiekel, P.A., Helm, E.E.: Measuring team knowledge during skill acquisition of a complex task. Int. J. Cogn. Ergon. **5**(3), 297–315 (2001)

25. Rico, R., Sánchez-Manzanares, M., Gil, F., Gibson, C.: Team implicit coordination processes: a team knowledge–based approach. Acad. Manag. Rev. **33**(1), 163–184 (2008)

26. Hilton, R.M., Shuffler, M., Zaccaro, S.J., Salas, E., Chiara, J., Ruark, G.: Critical Social Thinking and Response Training: A Conceptual Framework for a Critical Social Thinking Training Program (ARI Research Report). Army Research Institute for the Behavioral and Social Sciences, Arlington (2009)

27. Sandberg, J., Tsoukas, H.: Making sense of the sensemaking perspective: its constituents, limitations, and opportunities for further development. J. Organ. Behav. **36**(S1), S6–S32 (2015)

28. Weick, K.E., Sutcliffe, K.M., Obstfeld, D.: Organizing and the process of sensemaking. Organ. Sci. **16**(4), 409–421 (2005)

29. Kanfer, R., Ackerman, P.L., Sternberg, R.J.: Dynamics of skill acquisition: building a bridge between intelligence and motivation. In: Advances in the Psychology of Human Intelligence, vol. 5, pp. 83–134 (1989)

30. London, M., Smither, J.W.: Can multi-source feedback change perceptions of goal accomplishment, self-evaluations, and performance-related outcomes? Theory-based applications and directions for research. Pers. Psychol. **48**(4), 803–839 (1995)

31. London, M.: Job Feedback: Giving, Seeking, and Using Feedback for Performance Improvement. Psychology Press, New York (2003)

32. Smith-Jentsch, K.A., Cannon-Bowers, J.A., Tannenbaum, S.I., Salas, E.: Guided team self-correction: Impacts on team mental models, processes, and effectiveness. Small Group Res. **39**(3), 303–327 (2008)

33. Crandall, B., Klein, G., Klein, G.A., Hoffman, R.R.: Working Minds: A Practitioner's Guide to Cognitive Task Analysis. MIT Press, Cambridge (2006)

34. Braun, V., Clarke, V.: Using thematic analysis in psychology. Qual. Res. Psychol. **3**(2), 77–101 (2006)

35. Mason, M.: Sample size and saturation in PhD studies using qualitative interviews. In: Forum Qualitative Sozialforschung/Forum: Qualitative Social Research, vol. 11, no. 3, August 2010

36. Lombard, M., Snyder-Duch, J., Bracken, C.C.: Practical resources for assessing and reporting intercoder reliability in content analysis research projects (2010)

37. Ford, M.: Mental Models: Towards a Cognitive Science of Language, Inference, and Consciousness (1985)

38. Johnson-Laird, P.N.: Mental Models: Towards a Cognitive Science of Language, Inference, and Consciousness, no. 6. Harvard University Press, Cambridge (1983)

39. Langan-Fox, J., Code, S., Langfield-Smith, K.: Team mental models: Techniques, methods, and analytic approaches. Hum. Factors **42**(2), 242–271 (2000)

40. Ford, J.K., Kraiger, K.: The application of cognitive constructs and principles to the instructional systems model of training: implications for needs assessment, design, and transfer. In: International Review of Industrial and Organizational Psychology, vol. 10, pp. 1–48 (1995)

41. Lesgold, A., Rubinson, H., Feltovich, P., Glaser, R., Klopfer, D., Wang, Y.: Expertise in a complex skill: diagnosing x-ray pictures (1988)

An Overview of Virtual Reality Interventions for Two Neurodevelopmental Disorders: Intellectual Disabilities and Autism

Anders Nordahl-Hansen[1](✉) (iD), Anders Dechsling[1] (iD), Stefan Sütterlin[1,2] (iD),
Line Børtveit[1], Dajie Zhang[3,4,5], Roald A. Øien[6,7] (iD), and Peter B. Marschik[3,4,5,8] (iD)

[1] Østfold University College, Halden, Norway
{anders.nordahl-hansen,anders.dechsling,
stefan.sutterlin}@hiof.no, linebortveit@gmail.com
[2] Division of Clinical Neuroscience, Oslo University Hospital, Oslo, Norway
[3] Child and Adolescent Psychiatry and Psychotherapy,
University Medical Center, Göttingen, Germany
{Dajie.marschik,peter.marschik}@med.uni-goettingen.de
[4] Leibniz Science Primate Cognition, Göttingen, Germany
[5] Division of Phoniatrics, Medical University of Graz, Graz, Austria
[6] School of Medicine, Yale University, Child Study Center, New Haven, USA
roald.a.oien@uit.no
[7] The Arctic University of—University of Tromsø, Tromsø, Norway
[8] Center of Neurodevelopmental Disorders, Department of Women's and Children's Health,
Karolinska Institutet, Stockholm, Sweden

Abstract. In this overview of the two neurodevelopmental disorders, intellectual disabilities and autism spectrum disorders we systematically searched the literature for scientific publications of group-based designs that tested various interventions through the use of Virtual Reality technology. After screening of a total of n = 366 publications, n = 13 studies (intellectual disabilities n = 7, autism spectrum disorders n = 6) were included in the final analyses. We present descriptive data in terms of type of intervention content for the various studies as well as information regarding research design, number of participants enrolled in the studies, age cohorts, and outcome measures. We discuss the findings as a whole but also by comparing the studies that are published within each of the two neurodevelopmental disorder groups. Finally we discuss some challenges and opportunities for future research.

Keywords: Autism Spectrum Disorder · Intellectual disabilities · Virtual Reality · Measurement · Intervention

1 Introduction

Virtual Reality (VR) is defined as 'artificial environment which is experienced through sensory stimuli (such as sights and sounds) provided by a computer and in which one's actions partially determine what happens in the environment' [1].

© Springer Nature Switzerland AG 2020
D. D. Schmorrow and C. M. Fidopiastis (Eds.): HCII 2020, LNAI 12197, pp. 257–267, 2020.
https://doi.org/10.1007/978-3-030-50439-7_17

VR has not only become popular for the gaming industry but has been continuously developed to implement educational and interventional approaches, with a strong focus on certain psychiatric disorders. VR interventions for individuals with psychiatric diagnoses and its effectiveness have repeatedly been reported for diagnostic groups such as anxiety and depression [2]. VR interventions allow for controllable and safe exposure that are difficult to construct in real world settings.

Typical intervention settings for children and adolescents with neurodevelopmental disorders such as autism spectrum disorder (ASD) and intellectual disability (ID), are one-to-one training sessions. Target of these interventions for neurodevelopmental groups vary but usually include specified training within areas of language, non-verbal communication and social skills, learning of specific academic skills as well as daily living skills. All these areas of training can be supported by using VR technology [3]. Common outcome measures assessed in VR-supported therapeutic interventions include behavioral and emotional functioning levels, typically obtained via observation, questionnaires, or parental reports. Although there is no shortage of available measures [4], no consensus exists regarding which measures should be used to appraise therapeutic success [5]. For a first discussion and overview of intervention research within neurodevelopmental disorders we chose to focus on autism spectrum disorders (ASD) and intellectual disability (ID) as two major conditions of interest to be studied in VR research (Table 1).

Table 1. Diagnostic criteria for ASD and ID from DSM-5 [6]. *IQ may not always be an exact feature, has to be seen in relation to the qualitative indications of the other functions.

Autism Spectrum Disorders (ASD)	Intellectual Disability (ID)
Deficits in communication and social interaction	Deficits in intellectual functions
Repetitive patterns in behaviour and activities	Deficits in adaptive functions compared to developmental standards
No criteria regarding IQ, however, the IQ may indicate level of severity	Onset of the above mentioned deficits during the developmental period
	IQ>70*

VR systems are not yet widely used in clinical interventions and when VR is being used in clinical studies it is rather an implementation of an already existing therapeutic concept. Further, significant outcomes are not always expected [7] as most interventions are short-term pilot-studies. Also, most behavioral intervention studies for individuals with neurodevelopmental disorders indicate that although there might be an effect of the intervention at short-term follow-ups, this effect tends to fade out when longer term follow-ups are carried out [8].

In their review of the 31 published articles on VR in children and adolescents with ASD, Mesa-Gresa and colleagues [7] mentioned over 25 different standardized evaluation forms, and an additional 46 specifically constructed/newly designed evaluation forms of effectiveness and outcome measures. The findings extracted from Mesa-Gresa et al. illustrate the lack of consensus in regard of the use of common outcome measures. As noted earlier, this is however not only a problem within VR-research and ASD but the intervention research in general. In this study we conduct a systematic search for empirical intervention studies that used VR technology in children and adolescents with neurodevelopmental disorders. We investigated what type of designs, intervention types, sample size, age cohorts and outcome measures that have been used in intervention trials using group-based research designs for individuals with ASD and ID. Our intentions for conducting this review is to get an overview of intervention studies within two of the most common neurodevelopmental disorder groups. We address implications for future research using VR in interventions studies with individuals with neurodevelopmental disorders.

2   Method

2.1   Literature Search

We systematically searched for empirical VR-based intervention studies. The literature search was conducted in the second week of January 2020. We searched the databases PubMed, PsycINFO, and ERIC that are broad-based but with a focus on articles related to medicine, health, psychology, and education. The search string below with Boolean operators was used to search titles and abstracts: autism OR autistic OR asd OR asperger* OR pervasive development* disorder* OR pdd OR pdd-nos AND virtual reality OR vr OR virtual world OR cyberspace OR hmd OR head-mounted display* OR virtual learning environment OR immersive virtual environment OR augmented reality OR artificial reality OR oculus OR display technolog* OR immersive technolog* OR mixed reality OR hybrid reality OR virtual environment OR immersive virtual reality system OR 3d environment OR htc vive OR cave OR virtual reality exposure AND intervent* OR treat* OR therap* OR train* AND experiment OR randomized controlled trial OR randomised controlled trial OR controlled clinical trial OR group* OR quasi experiment. We used the same search string only substituting the first part related to autism diagnosis with: intellectual impairment OR intellectual disability* OR intellectual dysfunction OR developmental disabilit* OR intellectual developmental disorder OR mental deficiency OR mental* retard* OR mental* handicap* OR mental* disab* OR mental* insufficiency OR mental impair* OR mental* subnormality OR learning disabilit*. In addition to the structured search, we conducted an ancestry search of all authors of relevant papers. A search string like this enabled us to compare empirical evidence across the autism spectrum diagnosis and intellectual disability.

2.2   Inclusion Criteria

Following the literature search relevant articles on VR and therapeutical interventions were selected based on the following inclusion criteria: publications had to include therapeutic interventions with human participants with a diagnosis of ASD or intellectual disability (if participants had syndromic variations such as Down Syndrome but were categorized and met general criteria for an ID diagnosis, these studies were also included); studies had to be in English language and published in peer-reviewed journals; we included group comparison studies, with participants assigned to at least two groups (case-control designs).

2.3 Screening and Study Selection

The total number of studies from the initial search was N = 366. We removed all duplicates and one of the authors (LB) screened the remaining titles and abstracts. Another author (AN-H) double-screened 45 titles and abstracts (approx. 15%) of the titles and abstracts for reliability purposes. Agreement was met on 43 of the 45 abstracts and the two publications where there was disagreement were included for closer full-text inspection. Full-texts were assessed for eligibility by two authors (AN-H and AD). After full-text screening 13 (ASD n = 6, ID n = 7) publications were included in the final review and included in Tables 2 and 3. The screening process is depicted in Fig. 1.

2.4 Criteria for Included Participant Numbers

The number of participants that were included in the final analyses are reported separately for the respective studies. We report the number of participants in the intervention after dropouts were excluded. This leads to reporting the true size of the study sample on which results are based.

2.5 Analysis

In this article we report descriptive results of the included studies. For each of the studies reported we list study design, number of participants, type of intervention, age range of participants, as well as outcome measures used. We compare the studies conducted for the two different disorders with each other to indicate trends within the intervention research.

3 Results

Details of the included studies are reported in Tables 2 and 3 below. A total of six intervention studies for persons with ASD and seven studies for persons with ID were included for final analyses (Fig. 1). The prevailing study design was of pretest-posttest structure with non-randomized groups, with small sample sizes (range 8–105 meaning smallest to largest N in ASD and ID and overall). The total number of participants in all ID-studies were n = 502, whereas for the ASD-studies the total number of participants were n = 182. The age range in ID-studies was 7–60 years, and 2–60 years in ASD-studies. Only one out of seven ID-studies, and half of the ASD-studies (3/3) were randomized controlled trials. We found a broad spectrum in terms of scope in the included research work. In most of the applications, the intervention target aimed to alter a specific well-defined behaviors, such as physical fitness, motor-skills or particular skills such as job interview training or space navigation. The typical measures of outcome for the majority of these studies were measures that can be considered proximal to the intervention target, meaning that there is a great overlap of the intervention content and the operational definition of outcome. Three of the ASD-studies [24, 25, 27] were considered broader than the remaining ones, as the intervention type and content addressed social communication in general and as such per definition targeted a core symptom and diagnostic criterion of the disorder. The outcome measures used in the three studies were also more general as the measures used are designed to capture a more global, as opposed to specifically defined skill set within social communicative behaviors.

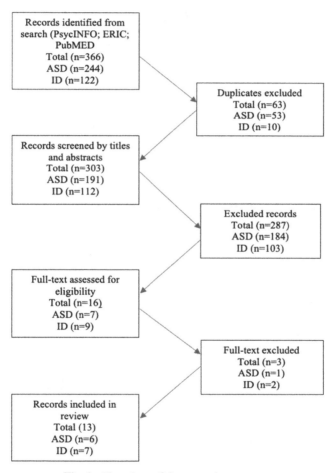

Fig. 1. Flowchart of the screening process.

Table 2. Included studies with participants with intellectual disabilities

Author (Year)	Design	Intervention type	Diagnostic label(s)	Intervention/control	Age cohort (years)	Outcome measure(s)
Lin et al. [15]	RCT	Physical exercise training	Down syndrome, intellectual disability	46/46	10.9 (mean)	Muscle strength, agility performance
Lotan et al. [16]	Pre-post-test w. two groups	Physical exercise training	Intellectual developmental disorder	30/30	35–60	Heart rate, energy expenditure index, modified Cooper test
Mengue-Topio et al. [17]	Learning phase -Post-test w. two groups	Space navigation virtual environment	intellectual disability Non-ID controls	18/18	29.39 (mean)	Mean walked distance between two points and number of attempts need to navigate correctly
Rose et al. [18]	Learning phase-post-test w. three groups	Fine motor skills (steadiness tester)	Learning disabilities	n=45 in three groups (no info on group size)	16–46	Errors during steadiness tester device
Tam et al. [20]	Pre-post-test quasi experiment w. two groups	Psychoeducational supermarket-shopping training	Intellectual disability	8/8	17–23	Checklist for supermarket shopping skills
Wuang et al. [21]	Pre-post-test quasi experiment w. two groups	Sensori-motor skills	Down syndrome	52/53	07-Dec	Motor proficiency, visual motor integration, sensory integration

Table 3. Included studies with participants with intellectual disabilities with Autism Spectrum Disorder

Author (Year)	Design	Intervention type	Diagnostic label(s)	Intervention/control	Age cohort (years)	Outcome measure(s)
Humm et al. [22]	RCT	Job interview training	ASD, Schizophrenia, PTSD	32/64	20–60	Role play interview scores
Josman et al. [23]	SSD & group comparison	Pedestrian training	ASD, typical development	6/6	8–16	Controlling an avatar in traffic
Lorenzo et al. [24]	Pre-post quasi experimental	Social interaction	ASD	5/6	2–6	Autism spectrum inventory, augmented reality social interaction score
Maskey et al. [25]	RCT	Cognitive behaviour and phobia treatment	ASD	17/16	8–14	Behavioural rating, social communication questionnaire, anxiety disorders interview schedule
Self et al. [26]	Pre-post-test w. two groups	Fire and tornado safety skills training	ASD	4/4	6–12	Que response (Yes/no)
Strickland et al. [27]	RCT	Job interview training	ASD	11/11	16–19	Interview skills rating, response delivery scale, social responsiveness scale

4   Discussion

Even though there has been a substantial increase in the use of VR in studies with neurodevelopmental disorders (e.g. Autism; [9]). Empirical evidence to support the efficacy of its implementation is however still scarce and unsystematic. Only one out of seven ID-studies and half of the ASD-studies were randomized controlled trials, which also results in the need to interpret all findings on a cautionary note. Future research with greater sample sizes will shed light on details regarding this new approach, to date the total number of participants in all ID-studies were n = 502, whereas for the ASD-studies the total number of participants were n = 182. In addition, to the age range of VR application studies ranges from 2–60 years (ID-studies: 7–60 years; ASD-studies: 2–60 years).

There was a large difference of participants enrolled in the ID-studies (n = 502) compared to ASD-studies (n = 182). A closer inspection of the age ranges in the studies also shows that, to a large extent, the studies were conducted with older children, adolescents and young adults. It is perhaps natural that the youngest children are under-represented, especially since head-mounted wearables can be difficult to use for a long time for this group. Still, Lorenzo et al. [24] did include participants as young as 2 years of age. Thus, augmented and virtual reality studies are also possible to conduct with very young children.

Using VR-devices for persons with neurodevelopmental disorders such as ASD and ID is not necessarily straight forward or easily applicable. For instance, head mounted displays can be difficult to use if the participant is reluctant or feel anxious. There can be many reasons for this, such as feelings of unease of wearing such equipment. For instance, different age-groups might react differently. In particular, as noted, with young children it can be challenging to use VR-equipment. Individuals with neurodevelopmental disorders can also experience unease such as sensory issues. However, there are individual differences in terms of the acceptability for use among this group. In fact, as is reported elsewhere in this volume [9] many studies report that acceptability of use of VR-technology is high for groups with neurodevelopmental disorders such as ASD.

Few of the included studies had older adults as participants. This age cohort are generally under-researched within the neurodevelopmental field [10], and hence in need of more focus from researchers whether conducting VR-studies or not. It is also worth noting that, in particular within the ID-studies, the studies are quite old and mainly conducted in the 2000s. For the ASD-studies, there were also some older studies but newer studies such as Lorenzo et al. and Maskey et al. indicate that interventions that in one way or another use augmented reality and virtual reality studies and different studies are in the pipeline.

4.1   Intervention Type and Outcome Measures

Assessing the studies reported here some trends emerge regarding intervention type and outcome measures deployed in the studies. All publications with ID-participants included interventions that targeted a well-defined specific skill or behavior. The measures used to assess the outcome from these studies were proximal to the intervention target, meaning that the outcome measure used highly resembles what was being trained in the intervention (e.g. fine-motor skill task as measure of outcome in an intervention targeting fine-motor skills). Specifically defined skills as intervention targets and overlapping measures of outcome were also used in the studies conducted by Humm et al., Josman et al., and Self et al. in the included ASD-studies. Such studies have high value as they go on to show that training particular skills can be taught and many of these skills can be lifesaving on its own, as well as many of the studies give clear indices that using VR for training skills can be used effectively. The studies of Lorenzo et al., Maskey et al., and Strickland et al. are somewhat different as the outcome measure are more distally, as opposed to proximal, to the intervention target [11]. Further, a dimension is added in that these studies aim to increase social communicative abilities more broadly and as such alter core diagnostic symptoms more globally [12].

5 Future Considerations

There is a need for more intervention studies using VR-technology as the body of research to date is too small. Most of the studies reported in this review have small sample sizes. Also, the studies go over a short time span and there is a particular need for studies that investigate longer terms effects following intervention. Further, the development of measures that capitalize on VR technology can potentially be used for objective outcome assessments and may compensate observational measures or parent reports. For instance, in head mounted displays used in VR-interventions it is possible to develop and capitalize on measures such as eye-tracking through the built-in technology which records for instance eye-gaze "live" during intervention [13]. However, little attention has been given to investigate the strengths and weaknesses of such assessments, nor to their potential in intervention research for children and adolescents with neurodevelopmental disorders.

6 Conclusion

Although there are some intervention studies published using VR-technology for individuals with ASD and ID, and that there are some promising results it is still difficult to conclude on the efficacy and effectiveness of such interventions. This is not surprising given the relatively short time span that VR-technology has been available for use in interventions.

Technological advancements are moving fast and opportunities for non-traditional measures are being developed. These advancements can lead to more precise outcome measures with potential clinical relevance available. However, creating a need for more systematic research on their respective validity, reliability and practical-clinical appropriateness. This may lead to better precision in terms of lower measurement error as for instance eye-tracking devices [14] and similar measures do not rely on arbitrariness in terms of subjectivity such as can be the case when observers score behavioral acts in observational scoring paradigms. This may in turn pave the way for more solid conclusions regarding the relevance and sustainability of effects following VR interventions in individuals with neurodevelopmental disorders.

References

1. The Merriam-Webster.com Dictionary. https://www.merriam-webster.com/dictionary/aug mented%20reality. Accessed 29 Jan 2020
2. Fodor, L.A., Coteţ, C.D., Cuijpers, P., Szamoskozi, Ş., David, D., Cristea, I.A.: The effectiveness of virtual reality based interventions for symptoms of anxiety and depression: a meta-analysis. Sci. Rep. **8**(1), 1–13 (2018)
3. Parsons, S.: Authenticity in virtual reality for assessment and intervention in autism: a conceptual review. Educ. Res. Rev. **19**, 138–157 (2016)
4. Bolte, E.E., Diehl, J.J.: Measurement tools and target symptoms/skills used to assess treatment response for individuals with autism spectrum disorder. J. Autism Dev. Disord. **43**(11), 2491–2501 (2013)
5. Nordahl-Hansen, A., Hart, L., Øien, R.A.: The scientific study of parents and caregivers of children with ASD: a flourishing field but still work to be done. https://doi.org/10.1007/s10 803-018-3526-9
6. American Psychiatric Association: Diagnostic and Statistical Manual of Mental Disorders: DSM-5, 5th edn. American Psychiatric Association, Washington (2013)
7. Mesa-Gresa, P., Gil-Gómez, H., Lozano-Quilis, J., Gil-Gómez, J.A.: Effectiveness of virtual reality for children and adolescents with autism spectrum disorder: an evidence-based systematic review. Sensors **18**, 2486 (2018). https://doi.org/10.3390/s18082486
8. Bailey, D., Duncan, G.J., Odgers, C.L., Yu, W.: Persistence and fadeout in the impacts of child and adolescent interventions. J. Res. Educ. Effectiveness **10**(1), 7–39 (2017)
9. Dechsling, A., Sütterlin, S., Nordahl-Hansen, A.: Acceptability and normative considerations in research on autism spectrum disorders and virtual reality. In: Lecture Notes in Computer Science (LNCS) (2020). ISSN 0302-9743
10. Howlin, P., et al.: Research on adults with autism spectrum disorder: roundtable report. J. Intellec. Dev. Disabil. **40**(4), 388–393 (2015)
11. Yoder, P.J., Bottema-Beutel, K., Woynaroski, T., Chandrasekhar, R., Sandbank, M.: Social communication intervention effects vary by dependent variable type in preschoolers with autism spectrum disorders. Evid.-Based Commun. Assess. Interv. **7**(4), 150–174 (2013)
12. Nordahl-Hansen, A., Fletcher-Watson, S., McConachie, H., Kaale, A.: Relations between specific and global outcome measures in a social-communication intervention for children with autism spectrum disorder. Res. Autism Spectrum Disord. **29**, 19–29 (2016). https://doi.org/10.1016/j.rasd.2016.05.005
13. Lutz, O.H.M., Burmeister, C., dos Santos, L.F., Morkisch, N., Dohle, C., Krüger, J.: Application of head-mounted devices with eye-tracking in virtual reality therapy. Curr. Direc. Biomed. Eng. **3**(1), 53–56 (2017)
14. Townend, G.S., Marschik, P.B., Smeets, E., van de Berg, R., van den Berg, M., Curfs, L.M.: Eye gaze technology as a form of augmentative and alternative communication for individuals with Rett syndrome: experiences of families in The Netherlands. J. Dev. Phys. Disabil. **28**(1), 101–112 (2016)

*Studies included in systematic review alphabetically and marked with an asterisk**

15. *Lin, H.-C., Wuang, Y.-P.: Strength and agility training in adolescents with down syndrome: a randomized controlled trial. Res. Dev. Disabil. Multidisc. J. **33**(6), 2236–2244 (2012). http://dx.doi.org/10.1016/j.ridd.2012.06.017

16. *Lotan, M., Yalon-Chamovitz, S., Weiss, P.L.: Improving physical fitness of individuals with intellectual and developmental disability through a virtual reality intervention program. Res. Dev. Disabil. **30**(2), 229–239 (2009). https://doi.org/10.1016/j.ridd.2008.03.005

17. *Mengue-Topio, H., Courbois, Y., Farran, E.K., Sockeel, P.: Route learning and shortcut performance in adults with intellectual disability: a study with virtual environments. Res. Dev. Disabil. Multidisc. J. **32**(1), 345–352 (2011). http://dx.doi.org/10.1016/j.ridd.2010.10.014

18. *Rose, F., Brooks, B., Attree, E.: An exploratory investigation into the usability and usefulness of training people with learning disabilities in a virtual environment. Disabil. Rehabil. Int. Multidisc. J. **24**(11–12), 627–633 (2002). https://doi.org/10.1080/0963828011011405

19. *Passig, D.: Improving the sequential time perception of teenagers with mild to moderate mental retardation with 3D immersive virtual reality (IVR). J. Educ. Comput. Res. **40**(3), 263–280 (2009). https://doi.org/10.2190/EC.40.3.a

20. *Tam, S.-F., Man, D.W.-K., Chan, Y.-P., Sze, P.-C., Wong, C.-M.: Evaluation of a computer-assisted, 2-D virtual reality system for training people with intellectual disabilities on how to shop. Rehabil. Psychol. **50**(3), 285–291 (2005). https://doi.org/10.1037/0090-5550.50.3.285

21. *Wuang, Y.-P., Chiang, C.-S., Su, C.-Y., Wang, C.-C.: Effectiveness of virtual reality using Wii gaming technology in children with down syndrome. Res. Dev. Disabil. Multidisc. J. **32**(1), 312–321 (2011). http://dx.doi.org/10.1016/j.ridd.2010.10.002

22. *Humm, L.B., Olsen, D., Be, M., Fleming, M., Smith, M.: Simulated job interview improves skills for adults with serious mental illnesses. Annu. Rev. CyberTherapy Telemed. **12**, 50–54 (2014). http://ovidsp.ovid.com/ovidweb.cgi?T=JS&CSC=Y&NEWS=N&PAGE=fulltext&D=psyc11&AN=2015-00374-009

23. *Josman, N., Ben-Chaim, H.M., Friedrich, S., Weiss, P.L.: Effectiveness of virtual reality for teaching street-crossing skills to children and adolescents with autism. Int. J. Disabil. Hum. Dev. **7**(1), 49–56 (2008). http://ovidsp.ovid.com/ovidweb.cgi?T=JS&CSC=Y&NEWS=N&PAGE=fulltext&D=psyc6&AN=2008-18791-009

24. *Lorenzo, G., Gómez-Puerta, M., Arráez-Vera, G., Lorenzo-Lledó, A.: Preliminary study of augmented reality as an instrument for improvement of social skills in children with autism spectrum disorder. Educ. Inf. Technol. **24**(1), 181–204 (2019). http://dx.doi.org/10.1007/s10639-018-9768-5

25. *Maskey, M., et al.: An intervention for fears and phobias in young people with autism spectrum disorders using flat screen computer-delivered virtual reality and cognitive behaviour therapy. Res. Autism Spectr. Disord. **59**, 58–67 (2019). https://doi.org/10.1016/j.rasd.2018.11.005

26. *Self, T., Scudder, R.R., Weheba, G., Crumrine, D.: A virtual approach to teaching safety skills to children with autism spectrum disorder. Top. Lang. Disord. **27**(3), 242–253 (2007). https://doi.org/10.1097/01.tld.0000285358.33545.79

27. *Strickland, D., Coles, C., Southern, L.: JobTIPS: a transition to employment program for individuals with autism spectrum disorders. J. Autism. Dev. Disord. **43**(10), 2472–2483 (2013).https://doi.org/10.1007/s10803-013-1800-4

Human Cognition and Behavior in Complex Tasks and Environments

Effect of Robotic Surgery Simulators in Training Assessed by Functional Near-Infrared Spectroscopy (fNIRs)

Mehmet Emin Aksoy[1,2], Kurtulus Izzetoglu[3], Atahan Agrali[1(✉)], Dilek Kitapcioglu[2], Mete Gungor[4], and Aysun Simsek[5]

[1] Biomedical Device Technology Department, Acibadem Mehmet Ali Aydinlar University, Istanbul, Turkey
atahanagrali@gmail.com
[2] Center of Advanced Simulation and Education (CASE), Acibadem Mehmet Ali Aydinlar University, Istanbul, Turkey
[3] School of Biomedical Engineering Science and Health Systems, Drexel University, Philadelphia, USA
[4] Obstetrics and Gynecology Department, Medical School, Acibadem Mehmet Ali Aydinlar University, Istanbul, Turkey
[5] Department of General Surgery, Haydarpasa Numune Training and Research Hospital, Istanbul, Turkey

Abstract. In the last decade, robotic surgery has enhanced doctors' capabilities and enabled surgeons to operate complex procedures even through smaller incisions. Due to the increasing trend of using of robotic surgery there is a need for training the surgeons of different disciplines including urology, general surgery, gynecology, cardiovascular surgery, endocrine surgery and thoracic surgery. Training surgeons with the robotic surgery simulator (RSS) is the common and initial procedure. This study focuses on monitoring training effect of robotic surgery simulators with neurophysiological assessment via functional near-infrared spectroscopy (fNIRS) and with the embedded scoring systems of the RSS. fNIRS allow researchers to record hemodynamic responses within the prefrontal cortex (PFC) in response to stimuli. Cortical oxygenation changes of the PFC from participants were monitored while they were performing suturing tasks with various difficulty levels. Twenty four resident surgeons from two different disciplines without prior robotic surgery experience (mean age $\pm$ SD = 28.25 $\pm$ 1.98 years19 OB&GYN residents, 5 general surgery residents) completed the experimental protocol consisting of two standard training blocks. On both blocks, participants completed suture sponge tasks on three different difficulty levels. Simulator scoring provided task performance assessment of each trainee. Participants' oxygen consumption levels were higher on the first block, where they familiarized themselves with the suturing task using RSS. On the second block, oxygen consumption levels decreased while performance scores significantly increased compared to the first block.

Keywords: Robotic surgery · fNIRS · Training · Medical education

© Springer Nature Switzerland AG 2020
D. D. Schmorrow and C. M. Fidopiastis (Eds.): HCII 2020, LNAI 12197, pp. 271–278, 2020.
https://doi.org/10.1007/978-3-030-50439-7_18

1 Introduction

In the last decade, there is an increasing demand for use of robotic surgery that allows surgeons to operate several complex procedures even through smaller incisions, and has been successfully implemented in many hospitals around the globe. 650,000 surgical procedures were performed worldwide in 2015 [1]. This technique provides advantages for the patients such as shorter hospitalization, reduced pain and discomfort, faster recovery time and reduced risk of infections [2]. The major advantages of robotic surgery for the surgeons are higher image quality, better depth perception, comfort, better surgical ergonomics, dexterity, precision of motion, speed of motion, and range of motion [3–5]. These advantages are acquired by three-dimensional vision with up to ten times magnification, with instruments with seven degrees of freedom, a software allowing motion scaling avoiding tremor and intuitive hand–eye alignment [6].

Due to the increasing trend of using of robotic surgery different disciplines including but not limited to, urology, general surgery, gynecology, cardiovascular surgery, endocrine surgery and thoracic surgery there is a demand for training surgeons [1]. A structured curriculum including basic robotic skills and complex maneuvers, allows is required for the development of these skills in a safe environment to increase patient safety [1].

Appropriate robotic surgery training ensuring surgical competency is essential to perform safe surgery [7]. Before performing operations with robotic surgery the surgeons have to complete course modules like Robotic Training Network (RTN), the Fundamentals of Robotic Surgery (FRS) Program, the Fundamental Skills of Robotic Surgery (FSRS) Program and the Morristown Protocol [8, 9].

FRS program is the most widely used robotic surgical skills education, training, and assessment program. FRS is a multi-specialty, proficiency-based curriculum of basic technical skills to train and assess surgeons to perform robotic-assisted surgery [7, 8].

Training surgeons with robotic surgery simulators (RSS) is the first important step of all these training protocols. The available simulators for the (RSS) have embedded scoring systems for each training scenario. This study focuses on measuring differences of hemodynamic responses of trainees during the trainings accompanied by the embedded scoring systems of RSSs. Functional Near Infrared Spectroscopy (fNIRS) has been chosen as the appropriate modality for this purpose among similar commonly used neuroimaging techniques like Electroencephalography (EEG), Functional Magnetic Resonance Imaging (fMRI), Magnetoencephalography (MEG), Positron Emission Tomography (PET). fNIRS is a non-invasive, affordable, and practical cerebral hemodynamic monitoring system technique and measures neuronal activities in human brain by using near-infrared light by recording differentiation in blood volume, and the average hemoglobin-oxyhemoglobin equilibrium [10]. fNIRS monitoring system uses light emitting diodes (LED) as light sources and photodiodes for collection of radiated light from the brain. The amount of absorbed and radiated light from the brain tissue can be calculated by using modified Beer-Lambert Law [11, 12].

The fNIRS is a noninvasive, safe and affordable optical brain imaging modality which allow researchers to record hemodynamic responses within the prefrontal cortex (PFC) in response to stimuli [13, 19]. Optical brain imaging techniques have been used in healthcare and surgical training simulations to assess trainees' cortical oxygenation changes [14, 15]. The objective of this study was to assess the effectiveness of the robotic surgery training simulator and explore differences of hemodynamic responses associated with the different task difficulty levels.

2 Methods

2.1 Participants

Twenty-four surgeons were recruited to participate our study (mean age $\pm$ SD $=$ 28.25 $\pm$ 1.98 years, 18 female, 6 male). The participants provided voluntary written informed consent for participation in the study, which has been reviewed and approved by the Ethical Committee of Acıbadem Mehmet Ali Aydınlar University. All participants were resident surgeons from two different disciplines, 19 OB&GYN residents and 5 general surgery residents. Participants had no prior robotic surgery training or experience.

2.2 Experimental Protocol

Prior to the experimental protocol begins, all participants had a familiarization session, in which an expert robotic surgery instructor showed participants basic operations of the surgeon console. Robotic arms control, surgeon console adjustment, camera targeting, zoom in and zoom out were the operations that were covered during the device familiarization session, which lasts between 10–12 min. Participants then performed a ring walk exercise on easiest level – level 1 – in order to acclimate themselves to the simulator. Finally fNIRS headband is placed on the forehead of the participants before the protocol begins.

Experimental protocol consists of two blocks. Both blocks have the same sponge suturing tasks, however each block presented tasks on different order. First block presented the tasks on an ascending difficulty order, participants started with easy, continued with medium and completed the block one with hard level suture sponge tasks. Second block presented same easy, medium and hard level tasks, however on second block tasks were presented in a randomized order (see Fig. 1).

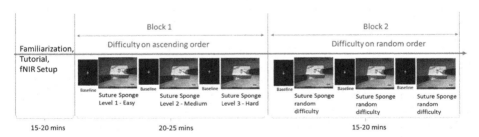

Fig. 1. Experimental protocol timeline

Robotic Surgery Simulator (RSS). Simulator, which was utilized in this study is an add-on component to the surgeon console of the Da Vinci Surgical System (Intuitive Surgical Inc., Sunnyvale, CA, USA). RSS has its own scoring system, which provided objective assessment for each task performance. Simulator scoring system collects and evaluates data on 7 different aspects of the procedure, these are time to complete exercise, economy of motion, instrument collisions, excessive instrument force, instruments out of view, master workplace range, instrument drops and missed targets. Upon finishing a task, RSS combines the scores from above aspects and generates a composite score, which shows the overall performance of the participant. A maximum of 1440 points could be obtained from the RSS composite scoring system.

Functional Near-Infrared Spectroscopy. fNIRS is a non-invasive optical brain imaging modality. It is a safe and a portable modality, which allows researchers to acquire data from the natural environment of the participant, rather than performing on the experimental protocol in a lab environment. The hemodynamic response from the prefrontal cortex of the participants were recorded with fNIRS Imager 1200 continuous wave system (fNIRS Devices LLC, Potomac, MD, USA). A headband with 4 LEDs and 10 detectors, a control box for electronic circuitry and a computer for acquiring data were the three components of the fNIRS system (Fig. 2).

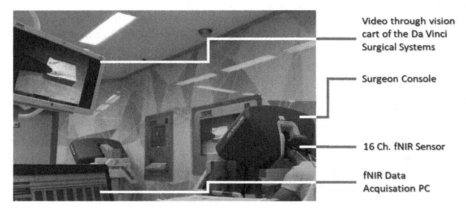

Video through vision cart of the Da Vinci Surgical Systems

Surgeon Console

16 Ch. fNIR Sensor

fNIR Data Acquisation PC

Fig. 2. A participant is performing on a task while fNIRS headband is collecting data

2.3 Data Analysis

This study combined three different metrics (simulator scores, tasks completion times and cortical oxygenation changes from the prefrontal cortex) to evaluate the performances and monitor improvements on the performances of the participants while they were using robotic surgical system.

Statistical Analysis. Descriptive statistics were presented using mean, standard deviation, median, minimum, maximum scale variables. Wilcoxon test were utilized for the comparison of two non-normally distributed dependent groups. While comparing two normally distributed dependent groups Paired Samples t-test was used. Statistical significance was accepted, if the two-sided p value was lower than 0.05. MedCalc Statistical Software version 12.7.7 (MedCalc Software, Ostend, Belgium, www.medcalc.org) were utilized to run statistical analysis tests.

fNIRS Data. A finite impulse response (FIR) filter was applied to the light intensity data (raw fNIRS data) to eliminate high frequency changes. With the aim of clearing noise caused by the motion artifacts, sliding window motion artifact rejection filter were also utilized to the fNIRS data [13, 14]. The filtered light intensity data from each of the 16 channels were then processed with the Modified Beer Lambert Law [19]. For the data analysis of fNIRS, spatial average of the 16 channels of the OXY values (OXY = oxyHb − deoxyHb) were used.

3   Results

In this study, the differences in participants' performance (simulation given scores and completion times) and the differences in participants' hemodynamic responses between two blocks were investigated (Table 1).

Table 1. Comparisons of block measurements.

	Block 1 Mean + Std.Dev. *Med.(Min − Max)*	Block 2 Mean + Std.Dev. *Med.(Min − Max)*	P value
OXY (HbO − HbR)	1.28 ± 1.1 *0.92 (−0.18 − 4.24)*	−0.07 ± 0.59 *−0.04 (−1.12 − 1.16)*	**<0.002***
Simulation score	480.17 ± 179.1 *468.23(171.23 − 921.6)*	637.29 ± 227.76 *577.04(328.3 − 1195.2)*	**< 0.001**
Completion time (ms)	493.14 ± 127.45 *504.96(233.67 − 829.77)*	303.75±69.03 *302.72(202.33 − 442.77)*	**<0.001**

Paired Samples t test, *Wilcoxon test

There is statistically significant difference between Block 1 and Block 2 measurements in terms of Oxy, Score and Completion time (p < 0.05). The average of Oxy and Completion Time is higher for Block 1 measurement. The average of Score is higher for Block 2 measurement (see Fig. 3).

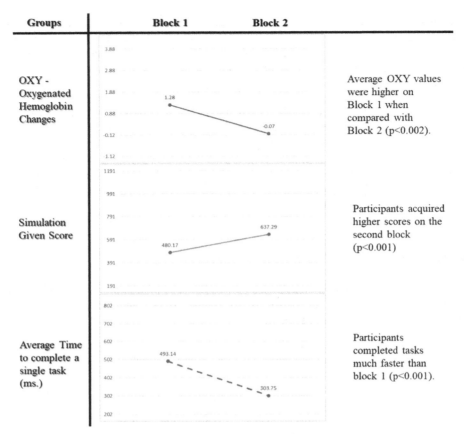

Fig. 3. Graphical comparison for the averages of oxygenated hemoglobin changes, simulation given scores and task completion times between blocks.

4 Discussion and Conclusions

Optical brain imaging is nowadays accepted as an additional tool to measure changes of oxygenation from PFC in order to measure mental and cognitive workload of surgical novices [16–18]. Although existing studies reported use of the fNIRS measurements during laparoscopic surgery trainings [16, 17], our study focuses on revealing the training effect of robotic surgery simulation techniques by neurophysiological assessment via fNIRS and embedded RSS scores. As reported in the previous studies for the laparoscopy trainings, our results also indicated training effect of the RSS. Both the performance scores obtained from the RSS and fNIRS data from the prefrontal cortex revealed the increase of performance scores and while decrease in the oxygen consumption when the trainees become proficient in use of the RSS. We hypothesized that the RSS should be effective and beneficial while the participants increase their performance score yet utilizes less oxygenation in the PFC region following RSS's standard tasks. These preliminary results were in agreement and indicated that the participants required and consumed less oxygen on the second block, while they completed tasks faster and more accurately.

Based on these promising results, we aim to increase the sample size in our upcoming study. We plan to classify the data into the groups based on task difficulty and compare the tasks across training blocks. Finally, the future study will include modification of the experimental protocol in order to study the effects of workload induced by the RSSs.

Disclosure. fNIR Devices, LLC manufactures the optical brain imaging instrument and licensed IP and know-how from Drexel University. Dr. K. Izzetoglu was involved in the technology development and thus offered a minor share in the startup firm, fNIR Devices, LLC.

References

1. Sridhar, A.N., Briggs, T.P., Kelly, J.D., Nathan, S.: Training in robotic surgery—an overview. Curr. Urol. Rep. **18**(8), 58 (2017)
2. Leddy, L., Lendvay, T., Satava, R.: Robotic surgery: applications and cost effectiveness. Open Access Surg. **3**, 99–107 (2010)
3. Van Koughnett, J.A., Jayaraman, S., Eagleson, R., Quan, D., van Wynsberghe, A., Schlachta, C.M.: Are there advantages to robotic-assisted surgery over laparoscopy from the surgeon's perspective? J. Robot. Surg. **3**(2), 79–82 (2009)
4. Hanly, E.J., Talamini, M.A.: Robotic abdominal surgery. Am. J. Surg. **188**(4), 19–26 (2004)
5. Nunes, F.F., Kappaz, G.T., Franciss, M.Y., Barchi, L.C., Zilberstein, B.: Applications and economics aspects of robotic surgery. Int. J. Adv. Robot. Autom. **1**(1), 1–3 (2016)
6. Jung, M., Morel, P., Buehler, L., Buchs, N.C., Hagen, M.E.: Robotic general surgery: current practice, evidence, and perspective. Langenbeck's Arch. Surg. **400**(3), 283–292 (2015)
7. Mizota, T., Dodge, V.G., Stefanidis, D.: Fundamentals of robotic surgery. In: Palazzo, F. (ed.) Fundamentals of General Surgery, pp. 215–225. Springer, Cham (2018). https://doi.org/10.1007/978-3-319-75656-1_16
8. Schreuder, H.W., Persson, J.E., Wolswijk, R.G., Ihse, I., Schijven, M.P., Verheijen, R.H.: Validation of a novel virtual reality simulator for robotic surgery. Sci. World J. **2014** (2014)
9. Andolfi, C., Umanskiy, K.: Mastering robotic surgery: where does the learning curve lead us? J. Laparoendosc. Adv. Surg. Tech. **27**(5), 470–474 (2017)
10. Jöbsis, F.: Noninvasive, infrared monitoring of cerebral and myocardial oxygen sufficiency and circulatory parameters. Science **198**, 1264–1267 (1977)
11. Rolfe, P.: In vivo near-infrared spectroscopy. Annu. Rev. Biomed. Eng. **2**, 715–754 (2000)
12. Cope, M., Delpy, D.T.: System for long-term measurement of cerebral blood and tissue oxygenation on newborn infants by near infra-red transillumination. Med. Biol. Eng. Comput. **26**(3), 289–294 (1988)
13. Izzetoglu, M., Bunce, S.C., Izzetoglu, K., Onaral, B., Pourrezaei, K.: Functional brain imaging using near-infrared technology. IEEE Eng. Med. Biol. Mag. **26**, 38–46 (2007)
14. Ayaz, H., Shewokis, P.A., Bunce, S., Izzetoglu, K., Willems, B., Onaral, B.: Optical brain monitoring for operator training and mental workload assessment. Neuroimage **59**(1), 36–47 (2012)
15. Aksoy, E., Izzetoglu, K., Baysoy, E., Agrali, A., Kitapcioglu, D., Onaral, B.: Performance monitoring via functional near infrared spectroscopy for virtual reality based basic life support training. Front. Neurosci. **13**, 1336 (2019)

16. Singh, H., et al.: Robotic surgery improves technical performance and enhances prefrontal activation during high temporal demand. Ann. Biomed. Eng. **46**(10), 1621–1636 (2018)
17. Nemani, A., et al.: Assessing bimanual motor skills with optical neuroimaging. Sci. Adv. **4**(10), eaat3807 (2018)
18. Khoe, H.C., et al.: Use of prefrontal cortex activity as a measure of learning curve in surgical novices: results of a single blind randomised controlled trial. Surg. Endosc. 1–12 (2020)
19. Ohuchida, K., et al.: The frontal cortex is activated during learning of endoscopic procedures. Surg. Endosc. **23**(10), 2296–2301 (2009)

Preparing for Cyber Crisis Management Exercises

Grethe Østby[✉] ⓘ and Stewart James Kowalski ⓘ

Norwegian University of Science and Technology, Gjøvik, Norway
{grethe.ostby,stewart.kowalski}@ntnu.no

Abstract. In this paper the authors discuss how to create a preparation schedule for exercises (PSE) to support EXCON-teams and instructors for full-scaled combined crisis management and cyber-exercises. The process to create the preparation schedule starts by performing vulnerability analysis to identify the most relevant and likely threats to the organization, before processing historical threats and attacks to further focus our simulation scenario development by planning and designing a socio-technical scenario. Moreover, a plan for simulation that are realistic and based on the organization's maturity will be considered, and finally, in terms of a societal crisis impact exercise necessary lectures will be prepared.

After this framework has been reviewed by the HCI International 2020, we plan to test the model when planning for exercises at the Norwegian Cyber Range (NCR) environment. NCR will be an arena where testing, training, and exercise will be used to expose individuals, public and private organizations, government agencies to simulate socio-technical cyber security events and situations in a realistic but safe environment.

Keywords: Exercices · Cyber exercices · Cyber management exercices · Cyber crises · Cyber crises management exercices

1 Introduction

The Norwegian Directorate for Civil Protection (DSB) recommends that full scale crisis management exercises consist of two major components: an exercise directive, and a scenario (DSB 2016). An exercise directive sets the framework for the exercise, whilst the scenario sets the content and timeline for the exercise.

The *ENISA Good Practice Guide on National Exercises* outlines in the initiation and planning phases of exercises that an organization should prepare an exercise directed to the needs of the organization but does not give guidelines how these needs should be identified (ENISA 2009).

The authors have from years of experience of planning for crisis management exercises, done vulnerability analysis on the organizations, and prepared and executed lectures beforehand the exercises based on such analyses. Recent research has also suggested the need to use maturity modelling to prepare for exercises, to plan the exercise for

© Springer Nature Switzerland AG 2020
D. D. Schmorrow and C. M. Fidopiastis (Eds.): HCII 2020, LNAI 12197, pp. 279–290, 2020.
https://doi.org/10.1007/978-3-030-50439-7_19

an appropriate level for the participating organizations (Østby and Katt 2019). Additionally, other recent research suggests preparing scenarios in a socio-technical root cause analytical context to prepare for different types of exercises (Østby et al. 2019).

By providing a clear step-by-step guide to follow, the authors suggest that such planning framework can provide a more effective and efficient learning environment for exercises.

After the introduction we present background and relevant literature in Sect. 2, before presenting the research approach in Sect. 3. In Sect. 4 we present the suggested preparation schedule for exercises, and in Sect. 5 we conclude and present our future plans on the topic.

2 Background and Relevant Literature

The scope of the authors' research is to investigate information security awareness and cyber security preparedness in society and public organizations like municipalities and counties and to investigate cyber-management in the public emergency organizations in both the organizations themselves and cyber-operations centers. To meet the scope, we will arrange cyber-incidents exercises at the Norwegian Cyber Range (NCR) (NTNU 2019).

At the NCR, we want to develop and offer near to real life exercises, i.e. full-scaled exercises in a secure environment, to train organizations on strategic, tactical and operational levels together. We plan to copy – paste the organizations socio-technical control structure into a safe environment at the cyber range, and train the teams on system incident handling, incident information escalation and crisis management.

To prepare for such exercises, we plan to test our suggested Preparation Schedule for Exercises (PSE) - framework as presented in this paper. Preparation for cyber exercises often centers around the cyber test bed and a fictive scenario (Micco et al. 2002; Vykopal et al. 2017), and the cyber exercises are often executed as competitions (Bei et al. 2011; Patriciu and Furtuna 2009). The author's approach is however, to focus on status in the organization that will be trained, to make the exercise as realistic as possible, but also aligned with the organization's level of awareness and knowledge.

In Jason Kick's Cyber Exercise Playbook (2014), the training audience is divided into 5 different challenge levels, and suggest impact and resolution on how to address the audience based on these challenges. In this paper the author suggests using vulnerability analysis and maturity modelling to find the training audiences/organizations level of expertise or lack of expertise.

System vulnerability analysis involves discovering a subset of the input space with which a malicious user can exploit logic errors in an application to drive it into an insecure state (Sparks et al. 2007). Vulnerability analysis in this paper also includes physical/material, social/organization and motivational/attitudinal analysis similar to those presented by Twigg (2001), and will be comparable with Norwegian guidelines for risk – and vulnerability analysis made by The Norwegian Directorate for Civil Protection (DSB) (DSB 2014). The author also consider Shah and Mehtre's Vulnerability Assessment and Penetration Testing (VAPT) relevant for some organizations of which is competent to consider and relate to such approach (Shah and Mehtre 2013).

In preparation for full-scale exercises it is also relevant to investigate the organization's experience with crisis in general and cyber-crisis in particular. Additionally, societal trends of which may impact the organization should be considered. To run such investigations, the use of threats - and opportunities analysis can be justified. In a study by Jackson and Dutton (1988) designed to investigate the use of threats and opportunities analysis among decision makers, the authors suggests that managers are being more sensitive to issue characteristics associated with threats than to those associated with opportunities. We suggest however, that by combining threat and opportunity analysis with vulnerability analysis like those previously presented, or together with a vulnerability functional assessment analysis as presented by Depoy et al. (2005), opportunities and vulnerabilities can be presented together.

According to Liao et al., Gartner define Cyber Threat Intelligence (CTI) as "evidence-based knowledge, including context, mechanisms, indicators, implications and actionable advice, about an existing or emerging menace or hazard to assets that can be used to inform decisions regarding the subject's response to that menace or hazard" (Liao et al. 2016). Research on managing CTI by CTI-sharing (Brown et al. 2015; Burger et al. 2014), and use of SIEM's systems to present and evaluate threats (Al Sabbagh and Kowalski 2015), are relevant practice to consider when preparing for cyber exercises. The authors consider in this paper how to bring this information from ICT-management to the organizations top management to prepare for cyber incidents and exercises which affects the organizations ability to still run their daily business or activity and thereby require top management involvement.

Since the beginning of 2011 DSB has published annual description of possible crisis scenario that could have major impact on Norway (DSB 2019). There are three major developments in the society, which are presented in the 2019 analysis, and one of these is the security consequences from rapidly increasing digitalization. Such development has led to a number of security analysis and techniques (Mahmood and Afzal 2013), and it is difficult for organizations which do not have ICT-security as their main tasks, to keep track with these trends. As a preparation for cyber incidents and exercises, the author suggests conducting trend analysis targeting the organizations to be trained as a part of the overall CTI-analysis.

In previous research Østby et al. suggests that socio-technical scenario building can be useful in understanding and defining training scenarios as it gives a good indication on both social and technical challenges from real life cases (Østby et al. 2019). A socio-technical system considers both social and technical aspects of change as presented in Fig. 1.

In preparation for exercises the socio-technical approach is useful for making scenarios to highlight possible imbalance between the component and may give the organizations directions of how to bring their system back into balance. On the social side of the approach ISO standards like ISO 27005 (ISO 27005 2018) and ISO 27035 (ISO 2016) can be used to investigate culture and structure, and on the technical side standards like presented by the Telecommunication Standardization sector of ITU (ITU-T) (ITU-T 2019) about protection assurance (Chapter 5) can be used to investigate the methods and machines in the organization.

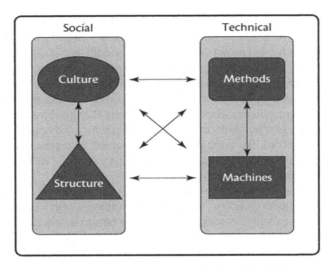

Fig. 1. Socio-technical approach (Kowalski 1994)

Recent research by Walhgren and Kowalski in Sweden indicates that there is a lack of cyber security incident readiness and that most organizations are at such a low level of maturity to deal with information security incidents that it may not even be beneficial for the organization to start off by running full scale exercise (Wahlgren and Kowalski 2016). Østby and Katt (2019) recently tested Wahlgren & Kowalski's model (Wahlgren and Kowalski 2016) in the Norwegian Inland Hospital trust, and the results indicated a diversity in needs and knowledge on strategic, tactical and operational layers in the organization.

Van Laere and Lindblom (2018) suggest theoretical education sessions via table-top discussions to role-playing, to give the trainees a fair chance of building skills and confidence before the exercise starts. This is also supported by the authors experience of running crisis management exercises both with and without theoretical lectures beforehand the exercises, and we suggest better learning from the exercise when preparing with lectures.

The Poorvu Center for Teaching and Learning at Yale (Yale 2019) presents how to write intended learning outcomes from lectures, and suggests that by writing specific, measurable takeaways, learning outcomes improves (Richmond et al. 2016). This is from a cyber security perspective also supported by ENISA's guidelines on assessing key objectives for operators of essential services (OES) and for the digital service providers (DSP) (ENISA 2018). The assessment is presented in the order of 1) security measures, 2) questions and 3) evidence.

In this paper we present an adaption of the Backward Design framework, presented by The Porvu Center for Teaching and Learning at Yale (Yale 2019).

3   Research Approach

In this paper, we approach the cyber security exercise design and execution challenge by using the design science research in information systems (DSRIS) (Kuechler and Vaishnavi 2012). Design science research (DSR) is a methodology which can be conducted when creating innovations and ideas that define technical capabilities and product through which the development process of artifacts can be effectively and efficiently accomplished (Kuechler and Vaishnavi 2012).

How to work on DSR was presented in a thesis written by G. R. Karokola (Karokola et al. 2011). He visualized this approach as outlined in Fig. 2.

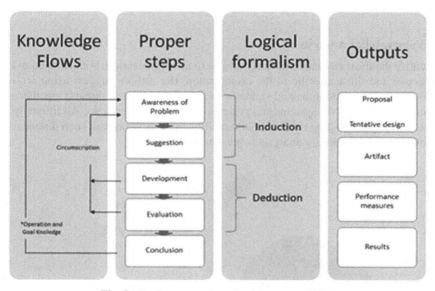

Fig. 2. Design research methodology - modified

The main goal of the research is to develop a step by step preparation framework for planning for cyber full-scale exercises.

The authors approach the goal by what can be referred to as a naive inductivist approach. The naive inductivist approach starts by first observing a phenomenon and then generalizing about the phenomenon which leads to theories that can be falsified or validated (Kowalski 1994).

Our proposed artifact in this paper is a framework to prepare for exercises which involves

1) vulnerability analysis to identify the most relevant and likely threats to the organization, both from an overall perspective and those specific to the organization,

2) work with the cybersecurity teams to understand historical threats and attacks to further focus our relevant simulation scenario development,

3) plan and design a socio-technical scenario for the exercise,

4) plan for simulation that are realistic and based on the organization's maturity,

5) and finally, in terms of a societal crisis impact exercise; look into the organization's responsibility (laws and regulations), crisis management roles and responsibilities, and suggested escalation continuity plans (involving information continuity plans), to prepare for exercise lectures.

4   Preparation Schedule for Exercises (PSE)

The proposed artifact is based on relevant literature presented and practical experience in planning for exercises. In this section we present the five-step preparation schedule for exercises (PSE) for full-scale cyber-incident exercises that will be executed for both strategic, tactical and operational participants from the organizations that are being trained.

4.1   Vulnerability Analysis

To identify the most relevant and likely threats to the organization, both from an overall perspective and those specific to the organization, the authors suggest using an overall SWOT-analysis (strengths and weaknesses, opportunities and threats) together with information from SIEMS systems and systems architecture overviews. Additionally, we want to investigate existing risk- and vulnerability analysis or prepare such if none exist. The suggested vulnerability analysis is presented in Fig. 3.

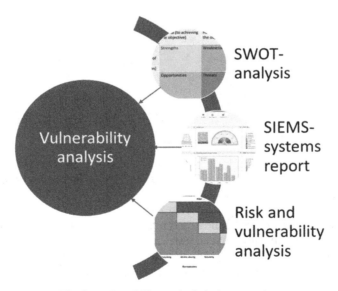

Fig. 3. Vulnerability analysis before exercises

We also suggest executing penetration testing as a part of the vulnerability analysis before exercises. Such reports will outline all the system and cyber vulnerabilities in a more detailed context and will give the organization more detailed results to work with after the exercise.

4.2 Historical Threats and Attacks

To approach analysis of historical threats and attacks both from within and from the outside of the organization, the authors suggests implementing reporting escalation processes from Security Incident and Event Management Systems (SIEMS) - to improve information flow and indicate what relevant information should be provided from a cyber incident to the crisis information management systems used.

In future research, a socio-technical escalation framework (STEF) to support synchronizing Security Incident and Event Management Systems (SIEMS) and Crisis Information Management Systems (CIMS) as suggested in Østby et al. (2019) could be implemented.

It is also important to evaluate trends in threats both within the organization and in the society in a national and international context similar to the DSB's incidents analysis (DSB 2019).

4.3 Socio-Technical Scenario Building

In Østby et al. (2019), different socio-technical models are suggested for different types of exercises. In our planning for exercises at the NCR, we will test how this approach works compare them to other scenario-building models. The socio-technical models, however, needs measurement standards when setting up the scenarios. We intend to use learning outcomes as described in Sect. 4.5 as measurement. We want to use ISO-standards to measure the status of the social part of the organization, and the ITU-T-standards to measure the technical part of the organization. Our approach can be visualized like presented in Fig. 4.

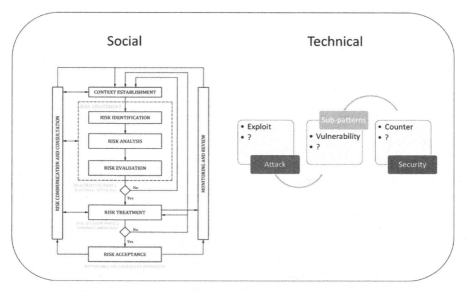

Fig. 4. Socio-technical standardization scenario measurement

The scenario will combine the results from the social and technical measurements to provide a total-concept scenario for the full-scale exercises. By this approach it will be possible to also measure the scenario incident handling process during the exercise.

4.4 Escalation Maturity Analysis

When analyzing the mentioned maturity escalation study executed at the Inland Hospital trust in Norway (Østby and Katt 2019), the authors focused on the weakest scores. However, the authors suggest that it is also important to focus on the high scores, to find the organization's strengths, and find the prioritization to the management to find an action strategy within the regulations of crisis management in the organization.

This is especially important when preparing for the exercise, to give the participants the possibility to perform successfully on their strengths. In planned research we want to suggest improvement-work during and after the exercise (Østby and Katt 2019).

It is however important to give the participants the possibility to train on their maturity weaknesses, and when preparing for the exercises, there will be a need to provide suggestions on how to handle the organizations biggest challenges on both strategic, tactical and operational layers. This process can be presented as in Fig. 5.

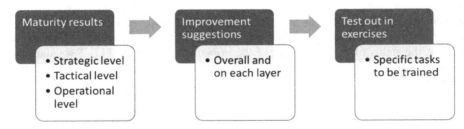

Fig. 5. Maturity results to be trained

For the exercise itself it can be organized with time-outs when these specific tasks are put in focus. That is, time to reflect on specific tasks and to document lessons learned experiences.

4.5 Lectures

In an evaluation after a recent table-top exercise for the tactical/emergency ICT-management team at our university, most of the participants answered relevant, very relevant and huge relevance when asked about relevance in lectures beforehand the exercise. The results are presented in Table 1.

When planning the lectures for the organization participating in exercises at the Norwegian Cyber Range, it will be necessary to plan the lectures to support both the strategic, tactical and operational teams. However, the focus in the lectures will still be responsibilities, roles and escalation procedures. In this research we suggest preparing lectures as executed in this mentioned exercise, of which is a modified version of Backward Design, in a socio-technical context, as presented in Fig. 6.

Table 1. Relevance in lectures beforehand exercises[a]

	No relevance	Some relevance	Relevant	Very relevant	Huge relevance
Regulations in security and emergency at Universities and Colleges (laws, regulations and guidance's), and other tasks crises management should be prepared for	0%	16,7%	50%	16,7%	16,7%
Crisis management and work in crisis staff: Situational analysis, need of recourses (personnel and material), roles in crises and operative management	0%	0%	33,3%	50%	16,7%
Information in emergencies, crisis management brief, crises communication and CIM	0%	0%	16,7%	66,7%	16,7%
Emergency plans, task lists for roles in crisis (critical analysis of the team's emergency plan)	0%	0%	33,3%	50%	16,7%

6 out of 10 participants answered the evaluation form.

By using this approach, the learning experience may flow into the exercise. That is that we don't need to "stop" the lectures when starting the exercise, as much of the learning experience will take place in the actual exercise in so called teachable moments.

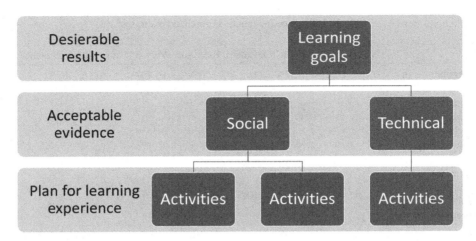

Fig. 6. Backward Design, modified in a socio-technical context (Wiggins et al. 2005)

5   Conclusion and Future Research

In this paper the author discusses a work in progress, to create a preparation schedule for exercises (PSE) to support exercise control (EXCON) teams and instructors for full-scaled combined crisis management and cyber-exercises.

After this framework has been reviewed and presented at the HCI International 2020 we plan to implement, test and evaluate the framework when setting up cyber crises exercises in the Norwegian Cyber Range (NCR) environment. We shall test the relevance of the framework for different types of organizations in Norwegian public sector and help develop interoperability standards so that scenario and exercises can be exchange both with in Norway and around the world.

References

Al Sabbagh, B., Kowalski, S.: Multidisciplinary Security (2015). www.computer.org/security

Bei, Y., Kesterson, R., Gwinnup, K., Taylor, C.: Cyber defense competition: a tale of two teams* (2011)

Brown, S., Gommers, J., Serrano, O.: From cyber security information sharing to threat management. In: WISCS 2015 - Proceedings of the 2nd ACM Workshop on Information Sharing and Collaborative Security, co-located with: CCS 2015, pp. 43–49. Association for Computing Machinery, Inc. (2015). https://doi.org/10.1145/2808128.2808133

Burger, E.W., Goodman, M.D., Kampanakis, P., Zhu, K.A.: Taxonomy model for cyber threat intelligence information exchange technologies. In Proceedings of the ACM Conference on Computer and Communications Security, vol. 2014, pp. 51–60. Association for Computing Machinery (2014). https://doi.org/10.1145/2663876.2663883

Depoy, J., Phelan, J., Sholander, P., Smith, B., Varnado, G.B., Wyss, G.: Risk assessment for physical and cyber attacks on critical infrastructures (2005)

DSB: Veileder til helhetlig risiko og sårbarhetsanalyse i kommunen (2014). https://www.dsb.no/globalassets/dokumenter/veiledere-handboker-og-informasjonsmateriell/veiledere/veileder-til-helhetlig-risiko-og-sarbarhetsanalyse-i-kommunen.pdf

DSB: VEILEDER I PLANLEGGING, GJENNOMFØRING OG EVALUERING AV ØVELSER Metodehefte: Fullskalaøvelse (2016)

DSB: Analyser av krisescenarioer (2019). https://www.dsb.no/globalassets/dokumenter/rapporter/p1808779_aks_2018.cleaned.pdf

ENISA: Good Practice Guide on National Exercises Enhancing the Resilience of Public Communications Networks Good Practice Guide on Exercises 2 Good Practice Guide on National Exercises (2009). http://www.enisa.europa.eu/act/res

ENISA: Guidelines on assessing DSP and OES compliance to the NISD security requirements (2018). https://doi.org/10.2824/265743

ISO: ISO 27035 - 1 (2016). https://www.standard.no/nettbutikk/sokeresultater/?search=ISO+27035&subscr=1

ISO 27005: ISO 27005 (2018). www.iso.org

ITU-T: ITU-T FG-DFC telecommunication standardization sector of ITU protection assurance use case for a payment transaction Security Working Group Deliverable Focus Group. Technical report (2019). https://www.itu.int/en/ITU-T/focusgroups/dfc/Documents/DFC-O-009_Securitydeliverable_Report_ProtectionAssuranceUseCaseforaPaymenttransaction.pdf

Jackson, S.E., Dutton, J.E.: Discerning threats and opportunities. Sour.: Adm. Sci. Q. **33,** 370–387 (1988)

Karokola, G., Kowalski, S., Yngström, L.: Secure e-government services: towards a framework for integrating IT security services into e-government maturity models. In: 2011 Information Security for South Africa - Proceedings of the ISSA 2011 Conference (2011). https://doi.org/10.1109/ISSA.2011.6027525

Kick, J.: Cyber Exercise Playbook (2014)

Kowalski, S.: IT Insecurity: A Multi-disiplinary Inquiry. Stockholm University (1994)

Kuechler, W., Vaishnavi, V.: A framework for theory development in design science research: multiple perspectives. J. Ass. Inf. Syst. **13**, 3 (2012)

Liao, X., Yuan, K., Wang, X., Li, Z., Xing, L., Beyah, R.: Acing the IOC game: toward automatic discovery and analysis of open-source cyber threat intelligence. In: Proceedings of the ACM Conference on Computer and Communications Security, 24–28 October-2016, pp. 755–766. Association for Computing Machinery (2016). https://doi.org/10.1145/2976749.2978315

Mahmood, T., Afzal, U.: Security analytics: big data analytics for cybersecurity a review of trends, techniques and tools. In: 2nd National Conference on Information Assurance (NCIA). https://ieeexplore.ieee.org/stamp/stamp.jsp?tp=&arnumber=6725337

Micco, M., Ed, D., Rossman, H.: Building a Cyberwar Lab: Lessons Learned. Teaching cybersecurity principles to undergraduates (2002). http://penguin.nsm.iup.edu/security

NTNU: The Norwegian Cyber Range (2019). https://www.ntnu.no/ncr

Østby, G., Berg, L., Kianpour, M., Katt, B., Kowalski, S.: A socio-technical framework to improve cyber security training: a work in progress (2019). https://ntnuopen.ntnu.no/ntnu-xmlui/handle/11250/2624957

Østby, G., Katt, B.: Maturity modelling to prepare for cyber crisis escalation and management (2019). https://orcid.org/0000-0002-7541-6233

Østby, G., Yamin, M.M., Asabbagh, B.: SIEMS in Crisis Management: Detection, Escalation and Presentation – A Work in Progress (2019). https://www.researchgate.net/profile/Stefan_Suetterlin/publication/334139727_Team_learning_in_cybersecurity_exercises/links/5d1a241e299bf1547c8eec06/Team-learning-in-cybersecurity-exercises.pdf#page=40

Patriciu, V.-V., Furtuna, A.C.: Guide for designing cyber security exercises. In: WSEAS International Conference on Information Security and Privacy. WSEAS Press (2009). http://www.wseas.us/e-library/conferences/2009/tenerife/EACT-ISP/EACT-ISP-28.pdf

Richmond, A.S., Boysen, G.A., Gurung, R.A.R.: AN Evidence-Based Guide to College and University Teaching (2016)

Shah, S., Mehtre, B.M.: A modern approach to cyber security analysis using vulnerability assessment and penetration testing. Int. J. Electron. Commun. Comput. Eng. **4**. www.ijecce .org

Sparks, S., Embleton, S., Cunningham, R., Zou, C.: Automated vulnerability analysis: leveraging control flow for evolutionary input crafting. In: Proceedings - Annual Computer Security Applications Conference, ACSAC, pp. 477–486 (2007). https://doi.org/10.1109/ACSAC.200 7.27

Twigg, J.: Sustainable livelihoods and vulnerability to disasters. Disaster Management Working Paper, vol. 2 (2001)

van Laere, J., Lindblom, J.: Cultivating a longitudinal learning process through recurring crisis management training exercises in twelve Swedish municipalities. J. Contingencies Crisis Manag. (2018). https://doi.org/10.1111/1468-5973.12230

Vykopal, J., Vizvary, M., Oslejsek, R., Celeda, P., Tovarnak, D.: Lessons learned from complex hands-on defence exercises in a cyber range. In: FIE Frontiers in Education (2017)

Wahlgren, G., Kowalski, S.: A maturity model for measuring organizations escalation capability of IT-related security incidents in Sweden. Assosiation for Information Systems (2016)

Wiggins, G., Wiggins, G.P., McTighe, J.: Understanding by Design. ASCD (2005). https:// books.google.no/books?hl=no&lr=&id=N2EfKlyUN4QC&oi=fnd&pg=PR6&dq=Wiggins+ GP,+McTighe+J.+(2005).++Understanding+by+Design.&ots=gpcyn4UH5x&sig=HmWITi tQ3nVTu1XKcvtKGTibJfA&redir_esc=y#v=onepage&q&f=false

Yale: Intendent learning outcomes (2019). https://poorvucenter.yale.edu/IntendedLearningOu tcomes

Tracking Technostress: A Task Interruption of Data Entry Study

Bruce W. Barnes III[✉] and Randall K. Minas

Shidler College of Business, University of Hawaii at Manoa, 2404 Maile Way,
Honolulu, HI 96822, USA
{brucewb,rminas}@hawaii.edu

Abstract. The prevalence of information systems and the resulting increase in continuous notifications have blurred the lines of work and leisure, resulting in increased stress. These changes in the work environment have had detrimental effects on workers ability to sustain attention and remain productive. Despite academic interest in both IT-mediated interruptions and technostress, there has been little research on the juncture of both of these while also utilizing eye tracking. We propose an experimental design on a sampling of undergraduate students in order to study the relationship of IT-mediated interruptions on task performance and the moderating effect of technostress on this relationship. In addition to we will utilize eyetracking (pupillary dilation and gaze duration) to tie the level of IT-mediated interruptions to cognitive resources in low and high technostress individuals.

Keywords: Technostress · Interruptions · Eye tracking · NeuroIS

1 Introduction

Technology has increasingly infringed on the distinction between work and leisure with concepts such as bring your own device (BYOD) and telecommuting [1, 2]. Personal phones with work email, work social media accounts, and work collaborative software are increasingly prevalent in our society. The convenience and availability of work has changed our work-life balance in favor of work. These devices have led to increased interruptions throughout an individual's day which leads to family-to-work conflict [3]. Meanwhile, individuals react to technology in different ways, with some embracing new technologies and others reticent of adoption. Technostress, stress caused by or impacted by technology, is a measure of an individual's ability to cope with technology [4]. Individual responses to coping with technology has further changed the nature of work [5]. Taken together, individual technostress and increasing IT-mediated interruptions may interact, creating a compounding negative effect on work performance.

In fact, a recent call for research in the area of technostress suggested more empirical explanation is needed on indirect variables, as well as, any mediating effects on how technostress is formed [6]. The specific empirical explanations of how and why

This is a U.S. government work and not under copyright protection
in the U.S.; foreign copyright protection may apply 2020
D. D. Schmorrow and C. M. Fidopiastis (Eds.): HCII 2020, LNAI 12197, pp. 291–303, 2020.
https://doi.org/10.1007/978-3-030-50439-7_20

technology creates stress is still being identified [6]. However, technostress has been linked to decreased job satisfaction and job performance [7]. This study aims to show how technostress is itself a mediator of specific task performance.

One such intervening variable could be the increased interruptions imputed by technology and the shift in attention that interruptions require, which increases cognitive load [8]. Interruptions have been shown to increase an individual's workload when they are interrupted mid-workflow [9]. Research has shown perceived IT-mediated interruptions to be inversely related to perceived task accomplishment, meaning the more one feels they are interrupted the less they feel they accomplish [10]. These two studies show that interruptions increase perceived workload while decreasing perceived task accomplishment, but empirical evidence connecting these findings to diminished task performance is limited.

While both technostress and IT-mediated interruptions are linked to lower task performance, scant research exists examining the interaction of these increased interruptions and technostress on work performance. In this study, we examine a potential moderating role of technostress on IT-mediated interruptions and task performance. In this study, we aim to address the following research questions:

1. Do IT-mediated interruptions impact task performance?
2. Does technostress play a moderating role in relationship between IT-mediated interruptions and task performance?
3. How do IT-mediated interruptions of varying complexity affect the attention and cognitive load of individuals working on a task?

The remainder of the paper proceeds as follows. First, we review the theoretical foundation and research surrounding IT-mediated interruptions, technostress, and neuro information systems (NeuroIS). Next, we propose a study to elucidate the relationship between IT-mediated interruptions, technostress, and task performance. Using insights from eyetracking, we propose tying the findings to overall cognitive load and attentional resources. The findings will contribute to the literature in two primary ways. First, we hope to establish the moderating role of technostress in IT-mediated interruptions and work performance. Second, the study will provide insights into the extent to which IT-mediated interruptions disrupt the work process, potentially leading to poor task performance.

2 Theoretical Background

2.1 IT-Mediated Interruptions

IT-mediated interruptions (i.e., interruptions caused by technology) have been found to occur frequently in the workplace, costing managers up to ten minutes of work per hour and creating about 70 suspensions of work per day for office workers [11]. Knowledge workers are particularly susceptible to IT-mediated interruptions, which can last up to 30 min and, in some cases, the individual may never come back to the task once interrupted [12]. IT-mediated interruptions can be problematic outside of work as well.

For example, in a study of after-hours work related IT-mediated interruptions resulted in impacts to both work and home, causing exhaustion and decreased performance [2].

Not all interruptions are the same, some represent a minor distraction while others can completely derail a task. These interruptions require varying amounts of cognitive resources to resolve. A distraction, for example, only briefly interferes with one's task but is easy to come back like signing a paper or answering a question. A total interruption results in a complete break from the primary activity and a shift in mindset, such as meeting with one's supervisor or attending to a new patient [12]. In interruption studies, one found physician's that are interrupted results in an increase in errors for prescriptions and laboratory tests [13], while another showed interruptions of complex decision-making tasks took longer and were less accurate [14]. However, interruptions are not always detrimental to task performance. If an interruption is relevant to the primary task it may in fact improve the task performance [15, 16].

The level of IT-mediated interruption influences how quickly one returns and becomes re-immersed in their primary task. The delay in returning to the primary task is referred to as the switching cost caused by the interruption [17]. In a study of radiologists, a telephone call interruption asking for an urgent parallel image diagnosis led to an increased time on the original task versus the control tasks, but not a decrease in accuracy of diagnosis [18], or more time to complete, but no change in outcome. Last, in another study on IT-mediated interruptions and a creative output task, those that were interrupted with a demanding task (i.e., a task requiring more cognitive resources) performed significantly worse on the primary task when compared to those with a lower-level interruption (i.e., a task requiring less cognitive resources) [19]. So, it shows that both the primary task and the interrupting task have various effects on task performance. The switching cost for more complex interruptions is higher and are proposed to be more harmful to overall task-performance. Therefore, we hypothesize:

H1: The complexity of the IT-mediated interruption is related to overall task performance, such that complex interruptions worsen task performance more greatly than less complex interruptions.

2.2 Technostress

Stress is a complex physiological response to the environment that includes increased affective arousal and is generally associated with negative emotional valence [20]. Technostress is the stress put upon workers while utilizing Information Communication Technologies (ICT) [21]. There are 5 dimensions to technostress: techno-uncertainty, techno-overload, techno-complexity, techno-invasion, and techno-insecurity. Techno-uncertainty is the stress caused by technology changes in an organization. The more changes of workflows, systems, logons, and software add to the stress of an individual. Techno-overload is the stress from too many information channels incurred by utilizing a variety of ICTs. Most have experienced this with notifications from different social media sites, game notifications, text messages, and laptop emails all while attempting to handle all simultaneously. Techno-complexity is the stress caused by one's ability to use and understand technologies to perform one's own job. Some systems are overly complex, do not have a great user interface and add stress and anxiety. Techno-invasion is the stress a technology can incur as it extends work into the home; taking time away from family, friends and leisure activities. Techno-insecurity is the stress caused by one's perception that not having technology skills may cause one to perform poorly or be fired [21].

In the foundational study of technostress, it was found that both gender (men) and age (younger) affect technostress at a higher rate [21]. Tarafdar et al. [7] found certain technostress was negatively correlated with individual productivity and role stress. In the follow-on study, Tarafdar et al. [22] found that technostress was also related to job dissatisfaction, role conflict, decreased innovation in the workplace, less job productivity, reduced commitment to an organization and dissatisfaction of IS. Another study found that compulsive usage of a smartphone also increased technostress [23].

Despite these linkages to task performance, little research has been done in the moderating and conditional effects that technostress is involved in [6]. This study proposes the relationship between IT-mediated interruptions, technostress, and task performance is more complex. We propose that technostress will moderate the relationship between IT-mediated interruptions on task performance similarly to how computer self-efficacy moderates techno-invasion on job anxiety [24]. Higher levels of technostress will exacerbate the effective of IT-mediated interruptions on task performance (Fig. 1). Therefore, we hypothesize:

H2a: Technostress will moderate the influence of IT-mediated interruptions on task performance, such that higher level of technostress will strengthen the negative relationship of IT-mediated interruptions and task-performance

Furthermore, as the complexity of an interruption increases, we expect the moderation effect to increase, having a greater negative impact on task performance. Thus,

H2b: For complex IT-mediated interruptions, technostress will moderate the effect of IT-mediated interruptions on task performance, such that greater levels of technostress will worsen task performance when interruption complexity is high.

We further hypothesize that due to the nature of our interruptions inducing multiple streams of information that may impact technostress' techno-overload, techno-overload may itself directly influence task performance. The hypothesis stated another way is the less an individual perceives the technology is causing an overload the better the performance on the primary task.

H2c: Techno-overload will directly influence task performance, such that the higher techno-overload the lower the task-performance.

2.3 NeuroIS

NeuroIS uses the tools of cognitive neuroscience to investigate how the brain and body respond to information systems [25]. In the context of human-computer interaction, employing this set of methodologies can shed light on how individuals respond to changes in system design. NeuroIS studies have been used to understand locations of IS constructs

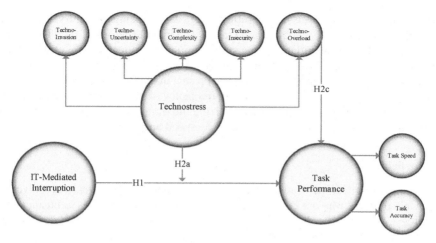

Fig. 1. Research framework of behavioral factors

[26], uncover bias in collaborative decision-making [27], and evaluating the effects of age on web design [28].

In this study, we use eye-tracking to help understand the effect of IT-mediated interruptions on cognitive resources. Previous studies have used gaze duration, pupillary dilation, and saccades to establish visual attention, increased cognitive load, and to record distraction [29].

In the radiologist study, eye tracking was utilized to see how interruptions affect the gaze of the radiologist. They found that immediately after an interruption more time was spent rereading prior dictation notes instead of going back to the radiological image [18]. In a study looking into interruptions and eye-tracking on reading, when individuals were interrupted by a 60 s audio story, they spent more time rereading previously read material and increases overall reading length [30]. Therefore, we hypothesize:

H3a: The post-interruption gaze duration will be positively related to higher complexity interruptions
H3b: Increased pupillary dilation, with a greater delay returning to baseline, will occur in higher-complexity interruptions.
H3c: Individuals with a high-level of technostress will experience greater disruptions in gaze duration and pupillary dilation when interrupted with a high-complexity interruption.

3   Method

The goal of this study is to understand how individuals respond to IT-mediated interruptions while conducting data entry and whether technostress plays a role. The unit of analysis will be the individual. Given this focus, we will set up a study in a controlled environment to avoid many external biases. We will utilize Rissler et al's framework on IT-mediated interruptions as a basis [31]. The boundary conditions will be the person, looking at demographics, traits, and abilities. The task-type being data entry, lower complexity, 6 min in duration, and 5 separate tasks. The manifestations of the IT-mediated interruption will have the initiator be the system initiator, actionable for 3 of the 4 interruptions, irrelevant to the primary task, visual (not auditory/tactile), with user actionable resumption. The manifestation of time will be present in the form of scheduled interruptions, without subject control of it occurring, frequency of 90 s, and duration being time to complete the interruption task. Lastly, utilizing the framework the consequences will manifest itself in task performance measured by speed and accuracy.

3.1 Participants

The participants will come from a student participant pool at a large research university in the United States who are currently enrolled in introduction courses in business and technology. We aim to collect data on 80 subjects in a within-subjects design. The participants will be consented and then randomly assigned to a treatment order.

3.2 Task

Our participants will be required to transcribe five patient encounter notes into a Qualtrics-based survey. The patient encounter notes follow a popular note taking method of evaluation called SOAP. SOAP breaks down a patient encounter into the Subjective assessment, Objective information, final Assessment, and Plan forward. Each participant will be randomized into a random ordering of the types of interruptions that will occur during the data input but will be the same across the 3 interruptions in that data entry. The interruptions will differ in anticipated cognitive load and interruption task (e.g. a low-level interruption would be basic demographic survey questions). Each data entry will be limited to 6 min total time, 90 s of task work followed by an interruption and return to task. In a pilot testing, the average input time for each encounter notes were 8 min and 42 s for the 1035 average character length of the input. One of the data entries will receive no interruptions in order to compare task performance to a control (Fig. 2).

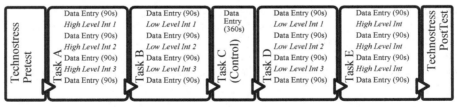

Fig. 2. Example experimental flow (Interruption categories will be counterbalanced between subjects to remove ordering bias)

3.3 Variables

There will be two treatments: high level of interruption, low level of interruption and a control treatment. There are two types of interruptions in each of the treatments. The treatments will be counter-balanced across participants to control for any task order effects.

Independent Variable. The independent variable in this study are the level of interruption and technostress.

The high cognitive interruption will be broken into two types of tasks. One will be a classification task, the second divergent thinking/creativity task. The classification interruption will be a requirement to classify 10 different animals into separate groups. For example, given animals {crocodile, sparrow, gecko, pigeon} and the participant must categorize them into either Bird or Reptile. Each of the interruption will include well-known animals and will not overlap with previous interruptions.

The divergent thinking cognitive interruption will utilize 3 common divergent thinking questions as shown in Table 1. These are adapted from both Guilford and Torrance Test of Creativity in a more recent study on the reliability validity of the measures [32]. The divergent thinking questions are meant to tap into creativity and have shown to have a higher alpha power change in nearly all areas of the brain over more simpler tasks [33]. The tasks will be timed for no more than 3 min but will not show the return to primary task button until after 2 min of the task.

Table 1. High interruption - divergent thinking questions

Divergent thinking	Question excerpt
Unusual Task – Interruption 1	For this task, you should write down all of the original and creative uses for a brick that you can think of
Instances Task – Interruption 2	For this task, you should write down all of the original and creative instances of things that are round that you can think of
Situations Task – Interruption 3	For this task, imagine that people no longer needed to sleep. What would happen as a consequence? Write down all of the original, creative consequences of people no longer needing to sleep

The low cognitive interruption will be a demographic survey and a non-relevant, passive interruption. The demographic survey will be separated into 3 instances of 2 questions in each instance (Table 2). This falls in line with other research utilizing quick information retrieval such as "Which country in the world has the largest number of people? [19].

The second low cognitive interruption will be non-relevant, passive interruptions. This falls in line with a call to research in IT-mediated interruptions regarding informational interruptions gaining little attention and identifying a unique aspect of our study [31]. These interruptions will take the form of a short fact that is interesting, but completely unrelated to the data entry. For example, "A Blue Whale is longer than 3 school busses lined up end to end (Blue Whale 30 m, School Bus 9.1 m)." Upon reading the fact, they can press a button to return to their primary task.

Table 2. Low interruption - demographic questions

Interruption #	Question excerpt
Interruption 1	What is your age?
	What is your gender?
Interruption 2	What is your current college Major?
	Which is your dominant hand?
Interruption 3	What is your primary spoken language?
	Are you Hispanic, Latino/a, or of Spanish Origin?
	What is your race? < select all that apply>

The IV Technostress and its five creators will utilize an adapted scale taken from Ragu-Nathan et al. as shown in Appendix A [21]. These measures need to be adapted slightly due to the inference of jobs in some of the measures. For example, in techno-insecurity a question is "I feel constant threat to my job security due to new technologies." By changing "job security" to "grades" it remains applicable to our student sample. We will conduct confirmatory factor analysis to ensure statistical similarity. The student's will be directed to consider all ICTs used for schoolwork as the context.

Dependent Variables. The dependent variables for this study are task performance and eye-tracking. Task performance will be measured by two factors, task accuracy and task speed.

Task accuracy will be measured using a mathematical concept called Levenshtein's Distance. This concept measures how many substitutions are needed from a given string (e.g. participant's data entry) and a reference string (e.g. original paper copy). The output of this is a ratio that is essentially a percentage and allows for comparisons amongst our sample in terms of task accuracy.

Task speed will be measured by taking the characters typed in the allotted time and dividing by either 6 min or the total completion should it be less.

In addition we will utilize the NASA Task Load Index (TLX) to measure the mental and temporal demands of the primary task, as well as, their perceived performance, effort and frustration [34].

Eye-tracking will be measured utilizing the Gazepoint GP3HD desktop eye-tracker and capture data to include eye gaze over time, heat map, percent of time on screen and pupillary dilation over time. Pupillary dilation will be used as a measure of cognitive load and gaze location and duration will be used to measure visual attention.

Controls and Manipulation Checks. In order to maintain control of our variables, we will control for age [35] as it has been found to be significant between older and younger populations. However, this affect will likely be nonexistent as the undergraduate population is rather similarly aged.

We will also control for technology experience, as that has also been found to impact technostress [36].

3.4 Procedures

Participants will complete the experimental procedure after being provided the informed consent approved by the university's Institutional Review board. The experiment will take place in a laboratory room set up for psycho-physiological analysis. The session will take no longer than 60 min.

Each participant will be provided with paper copies of the patient encounter SOAP notes, clearly labeled and in the order that will be presented to them. They will also be given an ergonomically efficient office chair for comfort and a standard keyboard and mouse for data entry. At this time, we will calibrate the eye tracking software. Should it not calibrate the first time, we will try a second time and if still not successful annotate the results in our laboratory log.

The participant will be shown the Qualtrics start page with their deidentified subject ID inputted. They will be instructed the data entry task has a set length and will be measured on both speed and accuracy. They will also be told that they must complete any task that may interrupt them. This should avoid questions to the experimenter when the survey automatically advances, which would further hinder the task performance and switching cost versus not being told about the interruptions.

Next the participant will conduct the data entry tasks, followed by the surveys. Upon completing both, they will receive a debriefing notice on screen in Qualtrics, as well as, a paper copy for their records and thanked for their time.

3.5 Data Cleaning and Analysis

Eye-tracking data will be separated into each task-interruption combination and compared Data will be aggregated across participants to elucidate average gaze duration on interruptions by treatment.

We will also calculate several other fields to include interruption time length calculated by subtracting the first click from the submission time. Repeated for all 12 interruptions. As well as, averaged for each type of interruption task (4 values), and perceived cognitive load (2 values). Task accuracy and task speed will be calculated as indicated above in section Dependent Variables. Technostress items will be averaged per sub item. Descriptive statistics will be drawn from the demographic survey.

The resulting cleaned data will be inputted into SPSS 25 for analysis utilizing ANCOVA.

4 Potential Implications

This study seeks to elucidate the relationship between, IT-mediated interruptions, technostress and task performance. In addition to identifying the moderating effect of technostress on IT-mediated interruptions and task performance. Lastly, tying the level of IT-mediated interruption to cognitive resources in low and high technostress individuals by utilizing eyetracking. This adds to the body of knowledge in regards to technostress by adding eye-tracking for data entry and IT-mediated interruptions on this specific task. In addition, this study may show how technostress is a moderator. Practice may find that these findings could help identify low performers by targeting technostress over the inevitable interruption, since studies have shown technical support provisions, technology involvement facilitations and innovation support are inhibitors of technostress [22]. The eyetracking data will provide insight into how the level of task interruptions effects overall cognitive and attentional resources, which will further elucidate the relationship between IT-mediated interruptions and technostress.

In our research there may be some way's forward to build a clearer model. This could include using different types of tasks, as there is evidence supporting interrupting a higher cognitive load task is harder to come back to. Another could be adding EEG data that could further isolate psycho-physiological data while conducting the experiment. Another could be to add other measures such as Computer Self-Efficacy or Computer Anxiety as additional factors that influence the task performance in a technology setting or adding burnout for a longer-term indication and impact.

Appendix A: Technostress Survey Items

Construct	Item
Techno-overload adopted from Ragu-Nathan et al. [21]	I am forced by this technology to work much faster
	I am forced by this technology to do more work than I can handle
	I am forced by this technology to work with very tight time schedules
	I am forced to change my work habits to adapt to new technologies
	I have a higher workload because of increased technology complexity
Techno-invasion adapted [21]	I spend less time with my family or friends due to this technology
	I have to be in touch with my school even during my breaks due to this technology
	I have to sacrifice my vacation and weekend time to keep current on new technologies
	I feel my personal life is being invaded by this technology
Techno-complexity adapted [21]	I do not know enough about this technology to handle my school-work satisfactorily
	I need a long time to understand and use new technologies.
	I do not find enough time to study and upgrade my technology skills

(continued)

(*continued*)

Construct	Item
	I find new students know more about computer technology than I do
	I often find it too complex for me to understand and use new technologies
Techno-insecurity adapted [21]	I feel constant threat to my grades due to new technologies
	I have to constantly update my skills to avoid failing
	I am threatened by classmates with newer technology skills
	I feel there is less sharing of knowledge among classmates for fear of failing
Techno-uncertainty adapted [21]	There are always new developments in the technologies we use at school
	There are constant changes in computer **software** in our organization
	There are constant changes in computer **hardware** in our organization
	There are frequent upgrades in computer **networks** in our organization

References

1. Coleman, J., Coleman, J.: Don't take work stress home with you. Harvard Business Review Digital Articles, pp. 2–4. EBSCOhost, July 2016
2. Chen, A., Karahanna, E.: Life interrupted: the effects of technology-mediated work interruptions on work and nonwork outcomes. MIS Q. **42**(4), 1023–1042 (2018)
3. Leung, L., Zhang, R.: Mapping ICT use at home and telecommuting practices: a perspective from work/family border theory. Telemat. Inform. **34**(1), 385–396 (2017)
4. Ayyagari, R., Grover, V., Purvis, R.: Technostress: technological antecedents and implications. MIS Q. **35**(4), 831–858 (2011)
5. WHO: Facing the challenges, building solutions. In: WHO European Ministerial Conference on Mental Health (2005)
6. Tams, S.: Challenges in technostress research: guiding future work. In: Twenty-First Americas Conferences on Information Systems, pp. 1–7 (2015)
7. Tarafdar, M., Tu, Q., Ragu-Nathan, B.S., Ragu-Nathan, T.S.: The impact of technostress on role stress and productivity. J. Manag. Inf. Syst. **24**(1), 301–328 (2007)
8. Borst, J.P., Taatgen, N.A., Van Rijn, H.: The problem state: a cognitive bottleneck in multitasking. J. Exp. Psychol. Learn. Mem. Cogn. **26**(2), 363 (2010)
9. Weigl, M., Müller, A., Vincent, C., Angerer, P., Sevdalis, N.: The association of workflow interruptions and hospital doctors' workload: a prospective observational study. BMJ Qual. Saf. **21**(5), 399–407 (2012)

10. Sonnentag, S., Reinecke, L., Mata, J., Vorderer, P.: Feeling interrupted—being responsive: How online messages relate to affect at work. J. Organ. Behav. **39**(3), 369–383 (2018)
11. Cracker, N.C., et al.: Nursing interruptions in a Trauma intensive care unit: a prospective observational study. J. Nurs. Adm. **47**(4), 205–211 (2017)
12. Healey, A.N., Primus, C.P., Koutantji, M.: Quantifying distraction and interruption in urological surgery. Qual. Saf. Heal. Care **16**(2), 135–139 (2007)
13. Flynn, F., Evanish, J.Q., Fernald, J.M., Hutchinson, D.E., Lefaiver, C.: Progressive care nurses improving patient safety by limiting interruptions during medication administration. Crit. Care Nurse **36**(4), 19–35 (2019)
14. Speier, C., Vessey, I., Valacich, J.S.: The effects of interruptions, task complexity, and information presentation on computer-supported decision-making performance. Decis. Sci. **34**(4), 771–798 (2003)
15. Addas, S., Pinsonneault, A.: E-mail interruptions and individual performance: is there a silver lining? MIS Q. **42**(2), 381–405 (2018)
16. Paul, C.L., Komlodi, A., Lutters, W.: Interruptive notifications in support of task management. Int. J. Hum. Comput. Stud. **79**, 1–15 (2015)
17. Monsell, S.: Task switching. Trends Cogn. Sci. **7**(3), 134–140 (2003)
18. Drew, T., Williams, L.H., Aldred, B., Heilbrun, M.E., Minoshima, S.: Quantifying the costs of interruption during diagnostic radiology interpretation using mobile eye-tracking glasses. J. Med. Imaging **5**(03), 1 (2018)
19. Wang, X., Ye, S., Teo, H.: Effects of interruptions on creative thinking. In: Thirty Fifth International Conference on Information Systems, pp. 1–10 (2014)
20. Shu, Q., Tu, Q.: Wang, K: The impact of computer self-efficacy and technology dependence on computer-related technostress: a social cognitive theory perspective. Int. J. Hum. Comput. Interact. **27**(10), 923–939 (2011)
21. Ragu-Nathan, T.S., Tarafdar, M., Ragu-Nathan, B.S., Tu, Q.: The consequences of technostress for end users in organizations: conceptual development and empirical validation. Inf. Syst. Res. **19**(4), 417–433 (2008)
22. Tarafdar, M., Tu, Q., Ragu-Nathan, T.S., Ragu-Nathan, B.S.: Crossing to the dark side: examining creators, outcomes, and inhibitors of technostress. Commun. ACM **54**(9), 113–120 (2011)
23. Lee, Y.K., Chang, C.T., Lin, Y., Cheng, Z.H.: The dark side of smartphone usage: psychological traits, compulsive behavior and technostress. Comput. Human Behav. **31**(1), 373–383 (2014)
24. Wu, J., Wang, N., Mei, W., Liu, L.: Does techno-invasion trigger job anxiety ? Moderating effects of computer self-efficacy and perceived organizational support (2017)
25. Riedl, R., et al.: On the foundations of NeuroIS: reflections on the gmunden retreat 2009. Commun. Assoc. Inf. Syst. **27**(1), 15 (2010)
26. Dimoka, A.: What does the brain tell us about trust and distrust? Evidence from a functional neuroimaging study. MIS Q. **34**(2), 373–396 (2010)
27. Minas, R.K., Potter, R.F., Dennis, A.R., Bartelt, V., Bae, S.: Putting on the thinking cap: using NeuroIS to understand information processing biases in virtual teams. J. Manag. Inf. Syst. **30**(4), 49–82 (2014)
28. Djamasbi, S., Siegel, M., Tullis, T.: Generation Y, web design, and eye tracking. Int. J. Hum.-Comput. Stud. **68**(5), 307–323 (2010)
29. Schiessl, M., Duda, S., Thölke, A., Fischer, R.: Eye tracking and its application in usability and media research. MMI-interaktiv J. **6**, 41–50 (2003)
30. Cauchard, F., Cane, J.E., Weger, U.W.: Influence of background speech and music in interrupted reading: an eye-tracking study. Appl. Cogn. Psychol. **26**(3), 381–390 (2012)

31. Rissler, R., Nadj, M.T.P., Adam, M., Maedche, A.: Towards an integrative theoretical framework of IT-mediated interruptions. In: ECIS 2017 Proceedings, vol. 2017, pp. 1950–1067 (2017)

32. Silvia, P.J., et al.: Assessing creativity with divergent thinking tasks: exploring the reliability and validity of new subjective scoring methods. Psychol. Aesthetics Creat. Arts **2**(May), 68–85 (2008)

33. Fink, A., et al.: The creative brain: investigation of brain activity during creative problem solving by means of EEG and fMRI, vol. 748, pp. 734–748 (2009)

34. Hart, S.G.: NASA-Task Load Index (NASA-TLX); 20 years later. In: Proceedings of the Human Factors and Ergonomics Society 50th Annual Meeting, pp. 904–908 (2006)

35. Tams, S.: Linking user age and stress in the interruption era: the role of computer experience. In: Proceedings of the 2017 Hawaii International Conference of Systems Science, pp. 5660–5667 (2017)

36. Ayyagari, R., Grover, V., Purvis, R.: Technostress: technological antecedents and implications. MIS Q. **35**(4), 831–858 (2011)

Enhancing Reality: Adaptation Strategies for AR in the Field

Konrad Bielecki[1]([⊠]), Daniel López Hernández[1], Marten Bloch[1], Marcel Baltzer[1], Robin Schmidt[1], Joscha Wasser[1], and Frank Flemisch[1,2]

[1] Fraunhofer FKIE, Fraunhoferstraße 20, 53343 Wachtberg, Germany
`{konrad.bielecki,daniel.lopez.hernandez,marten.bloch,`
`marcel.baltzer,robin.schmidt,joscha.wasser}@fkie.fraunhofer.de`
[2] Institut Für Arbeitswissenschaft (IAW), RWTH Aachen,
Bergdriesch 27, 52062 Aachen, Germany
`flemisch@iaw.rwth-aachen.de`

Abstract. Situational awareness is a crucial aspect for the survivability of combat vehicles and soldiers in the battlefield, as well as for the prevention of fratricide and collateral damage. Optical, radio and acoustic sensors for monitoring the battlefield can significantly increase situational awareness by providing further information to the crew. However, in time-critical situations, a crew can encounter problems when attempting to make meaningful use of this information, as cognitive resources are otherwise occupied during other operations, e.g. controlling a vehicle. Research from various fields clearly indicates that the correct use of Augmented Reality (AR) enables the crew to pay significantly greater attention to the current tasks, while maintaining an overview of the tactical situation. Adaptation strategies for presenting AR symbols as well as a multi modal approach can be used to distribute the workload. This paper deals with the use of AR in the Field, especially in a military context. It describes the practical use of AR, the exploratory methodology for creating new cues and gives an outlook on further possible uses.

Keywords: Augmented Reality · Adaptive systems · Human systems integration

1 Introduction

1.1 Motivation

Even before the 21st century, humans have made use of computer-aided systems that can relieve them of certain tasks. The application ranges from simple cues to automated assistance systems [1]. Especially in the 21st century, Augmented Reality (AR) technology has become increasingly important in the industry, research and military domains [2].

On modern battlefields, AR has the advantage that information of any kind can be presented directly to the crew, which can significantly assist in the understanding of the situation. Using this approach, the crew can keep up to date with the tactical situation while also performing other tasks.

© Springer Nature Switzerland AG 2020
D. D. Schmorrow and C. M. Fidopiastis (Eds.): HCII 2020, LNAI 12197, pp. 304–315, 2020.
https://doi.org/10.1007/978-3-030-50439-7_21

The use of AR can also bring tactical and operational advantages while coordinating multinational efforts, such as within the NATO. Different nations with different languages can share the same repertoire of AR elements and develop a common understanding.

However, the wrong use of AR can also have disadvantages, e.g. if the information is not adequately provided, it can lead to cognitive overload. If the cognitive workload is too high, the crew tends to postpone one or more low priority tasks to a later time (workload debt). If these postponed tasks reach a certain amount, the crew can be more stressed and the failure rate will increase (workload debt cascade) [3]. In addition, if used incorrectly, it can lead to inattentional blindness [4].

In order to ensure the practical use of AR in different scenarios, technologies, humans and AR cues must interact in symbiosis with each other. Therefore, a methodology, Use Cases and an explorative field trial were designed. As the design and use space of human & AR systems is rather large and only partially investigated with experiments, an explorative approach with participation of potential users is more effective to come to working prototypes. An overview of the exploration paradigm can be found in [5], a description of its instantiation as an Exploroscope is provided in [6].

1.2 Methodology and Background

Schwarz and Fuchs [7, 8] present how assistance systems can adapt to the human state using multiple measurements to determine multiple dimensions of the human state, such as situation awareness, attention, fatigue, emotional state, workload and motivation. Depending on the human state, assistance systems can support the human more or less to fulfill their tasks. An option to bi-directionally adapt human and automation behavior are patterns (e.g. [9, 10]) and more specifically interaction patterns (e.g. [11–15]).

Specifically, situational awareness (SA) is a decisive aspect of survivability for combat vehicles and soldiers in the battlefield. The introduction of optical, radio and acoustic sensors for monitoring the battlefield, and in particular the fielding of Battlefield Management Systems (BMS) which provide an overview of the tactical situation, has significantly improved the situational awareness for combat vehicle crews. However, in some cases, the information in the BMS is displayed separately from other sensor feeds, e.g. periscope, leaving the crew to fuse these two key sources of information. In time-critical situations, it is difficult to fully exploit the information provided by the BMS and other information systems, since crew members have to concentrate on the scene where the action is taking place. Ideally, the tactical information in the BMS should be converted and mapped onto the operator's field of view, e.g. by using a heads-up-display, regardless of his role. AR can be used for overcoming these deficiencies by presenting the information from the BMS, directly in the operator's sight, typically in the form of graphical symbols. Visual information may be further augmented by acoustic and haptic cues, e.g. audio alerts or vibrations. Supported by this AR functionality, the crewmembers can pay full attention to what is going on in the vehicle's area of operation, while at the same time stay updated on the tactical situation within the sights' field of view.

The amount of information being displayed to the crew should depend on the current situation (e.g. planning of the route, driving to a certain location, engagement with hostiles, etc.), the current human state (stressed, tired, bored, etc.) and the current quality and availability of information and assistance capabilities. AR can be used to adapt human behavior, e.g. to move the current focus to objects that might be important and to implement interaction patterns which can be helpful to structure the mechanism of behavior adaptation of the human or other assistant systems, e.g. present information of the BMS depending on the human's actions or state.

1.3 bHSI and Interaction Patterns

As can be the case with any changing system, the stability balance present in a system of resting state can quickly become unstable through external factors. This can also happen in an adaptive Human-Machine System if the interaction is not well designed and robust, resulting in a partial state of confusion. E.g. in case of an emergency, the control of a vehicle can be shifted from the less capable actor to the more capable one without understanding of the reasons for such control shift. This results in a subsequent reduction of situation awareness. Interaction patterns can be used to design how interaction happens and therefore, maintain a well-balanced system. In the case of Augmented Reality, interaction patterns are an option to determine which and when a different set of cues should be used in order to adjust to a changing situation. Interaction patterns discussed by [11] and [12] use an escalation scheme based on the severity of the situation. This is similar to a case encountered during military operations, when the crew needs a different type of augmentation depending on patrol or direct combat situations. With the present methodology, it is possible to determine the parameters and extent of the augmentation required in each given scenario. While the trigger for the change of the level of augmentation was not further investigated in these trials, the tools developed during this project can significantly help to identify the appropriate level of augmentation necessary to increase the situational awareness. The tool presented (exploration sandbox) makes it possible to achieve different configurations of each AR cue by changing their properties (e.g. color, brightness, size, etc.) and cooperatively and directly incorporate the user into the design process.

2 Use Cases

Use Cases describe generically what can happen between the actors of a system to achieve a goal of the primary actor. Actors are something or someone that exists outside the technical system under study, but is part of the system under study and under design. Actors take part in a sequence of activities in a dialogue with the technical subsystems to achieve a certain goal. They may be end users, other systems, or hardware devices. Normally, Use Cases describe abstract use situations.

The abstract form of Use Cases facilitates a common understanding between users (commander, gunner, driver, etc.), designers and developers of different faculties (engineers, psychologists, etc.) without going into technical details.

Figure 1 shows a representation of the interaction between two combat vehicle systems (subsystems), each depicted as a human machine system, which includes their respective soldiers interacting with the BMS in a combat vehicle.

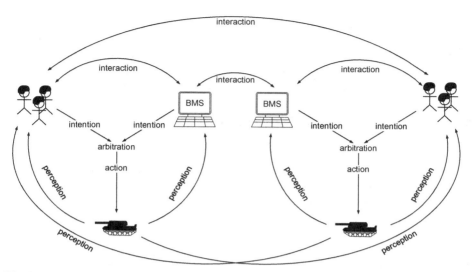

Fig. 1. System diagram of two combat vehicles with their crew and Battle Management System (BMS) [NATO AVT-RTG 290] based on [16].

An urban scenario can have multiple Use Cases. The scenario in Fig. 2 was developed in a workshop together with soldiers from Norway, the Netherlands, the United Kingdom and Germany as part of the NATO Research Task Group 290 "Standardization of Augmented Reality for improved Situational Awareness and Survivability of Combat Vehicles".

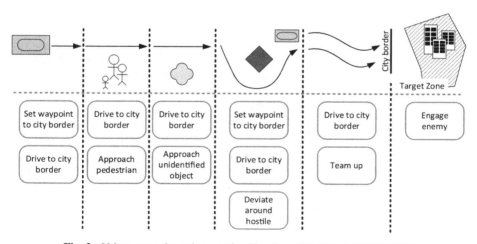

Fig. 2. Urban scenario and respective Use Cases [NATO AVT-RTG 290]

The use case this paper focus is called "Drive to City Border". In this use case, a specific user need of the commander is to know where the blue team is. The respective user story from a commander's point of view is: "As a commander, I want to know where friendly troops are, in order to improve coordinated maneuvers and prevent fratricide". A respective interaction pattern to address blue team tracking in terms of form, color, symbology, etc. was explored, keeping in mind the adaptation of the presented information of the situation, the user state, the quality and information availability.

3   Preparations for Explorative Study

For the exploration of AR cues, two different approaches were used. The first one involves creating a virtual environment to simulate a vehicle, Use Cases and AR cues. The second approach requires a more real world implementation using a test vehicle and positioning sensors. Each approach presents some limitations described more in detail below.

3.1   Laboratory Trials

The virtual approach uses a virtual map where the participant interacts and drives a vehicle through VR glasses (HTC Vive Pro Eye). The map is designed to resemble an urban scenario. It is possible to present to the user not only visual cues, but haptic and auditory cues as well. For the visual cues, the NATO Symbols for Friendly, Neutral, Hostile and Unknown were used. Each cue can be adapted to the user preference in regards to the distance to the object (visibility radius; Fig. 3 left). Additionally, it is possible to mark specific areas with different colors and transparency levels. Other cues can be done through haptic and acoustic modalities. In the acoustic part of the Exploration Sandbox, tones of different frequency, volume and repeatability can be designed. These are played through headphones to the user. Haptic cues can also be implemented through haptic elements. They can reproduce different vibration patterns that vary in frequency, strength and position.

Fig. 3. Left: Map shows the placement of objects and their visibility radiuses; Right: participants point of view in the simulation (Color figure online)

Since the main objective is to increase situation awareness based on the criticality of the situation, a dynamic change of the amount of information presented to the user is necessary. This is achieved by using an interaction pattern with an escalation scheme (Fig. 4). The level of criticality is determined by the distance of the ego-vehicle to an object of interest. Each escalation step uses a specific combination of haptic, visual and acoustic cues. For the laboratory trials, different parameters and cue combinations were used for each object of interest (hostile, friendly, unknown, neutral) depending of their level of threat to the ego-vehicle.

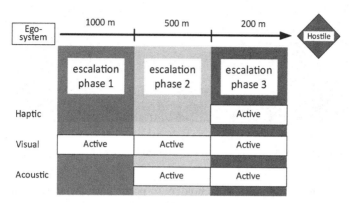

Fig. 4. Example of a distance-based escalation scheme for a hostile object (based on [13])

The activation distances as well as the specific parameters and combination of cues for each escalation step were explored using an Exploration Sandbox (see Sect. 3.3). It is important to mention that participants were involved in the definition of these parameters, based on personal preference and experience.

The first approach takes place in a virtual environment. The second approach, which takes place in the real world, uses a vehicle in combination with AR glasses to drive in the real environment.

3.2   Field Trials

One of the advantages of using a real-world approach is that some of the possible problems encountered when using VR technology can be avoided, e.g. simulation sickness. Nevertheless, this kind of approach requires a more complex setting and in some cases, costly resources which may not be available.

In this field trial, the user sat on a static IFV (Infantry Fighting Vehicle) PUMA, which added realism to the scenario. Symbols, distances and other elements were presented through AR glasses (DreamWorld DreamGlass). Every object of interest was equipped with its own GPS system. This data was used to correctly display the AR symbols in the DreamGlass. Further information on this object, such as the distance to it, can also be displayed (Fig. 5).

3.3   Exploration Sandbox

The exploration sandbox incorporates an extensive range of properties for each modality that can be customized to create many different versions of each AR element (Fig. 6).

Fig. 5. Image from point of view with the AR glasses; Top: NATO symbology and distances; Bottom: alternative view with frame around object (Color figure online)

This allows designers, users and stakeholders to participate dynamically, directly and quickly in the design process and to design behavioral patterns. The created AR elements can be displayed in a combination of visual, haptic and acoustic forms, e.g. through AR or Virtual Reality (VR) glasses, a haptic vest or speakers/headphones, respectively. Since the Exploration Sandbox UI is independent from the rest of the system, it can be used in the laboratory or in field trials. This is possible thanks to the middleware ROS (Robot Operating System) that connects hardware, simulation and UI. Because of this modular approach, new interfaces, e.g. haptic controls or other AR glasses can easily be integrated.

In this context, different modes of augmentation can also be explored to find the right amount of information for the current situation, e.g. training, planning or combat. The elements defined in the process are intended to be used as a standard or guideline in the future design of AR systems for military and partial civil applications.

4   Field Study

A field study was conducted in order to investigate technical feasibility and usability on the one hand, and the understanding and design of the used symbology on the other hand. The study was done with the participation of potential end-users, soldiers of the German Armed Forces at a military base.

The goals of the study was to (1) evaluate the technical and technological properties and challenges of the system in an ecologically valid context and (2) collect feedback and suggestions on the design of the symbology. In total 10 (9 males, 1 female) military drivers took part, the youngest being 30 and the oldest 57 years old (mean $= 44.25$, SD $= 10.91$). The above-mentioned technical setup was placed on site of a military base with direct access to an armored vehicle (IFV PUMA) and availability of the participants.

The study consisted of two parts: first an informal evaluation while being mounted (i.e. in the driver's seat in the armored vehicle), and afterwards a one-on-one participatory design workshop. For the former, the participants were presented with different situations und scenarios in which at least one and at a maximum three elements were shown through the AR-glasses, one of which was a possible waypoint and two were possible friendly symbols. The augmentations could be either static or dynamic, within or out of the field of view, and occluded or open. Of the two friendly symbols, only the first one (Friendly 1) had varying properties, the second one (Friendly 2) was always static, in the field of view and not occluded. For a more detailed description of the scenarios, refer to Table 1.

Note that part of the scenarios were completed with "closed hatch", meaning that the participant was in the inside of the tank as opposed to "open hatch" when the driver's head sticks out of the vehicle. After the situations with varying and possibly dynamic augmentations were completed, situation two (open hatch, two friendly symbols and one way point) was held constant and the properties of the symbol were altered according to the following:

- Change in transparency (0%, 30%, and 70%)
- Fluorescent green vs. NATO color scheme (in this case blue)
- Distance in corner of symbol vs. distance inside the symbol
- Framing of the object and symbol as explanation above vs. symbol directly on object

In order to get as much ad-hoc input as possible, the participants were specifically encouraged to follow the think-aloud protocol [17], especially when they were having issues or concerns. However, since they had no previous experience with this technique,

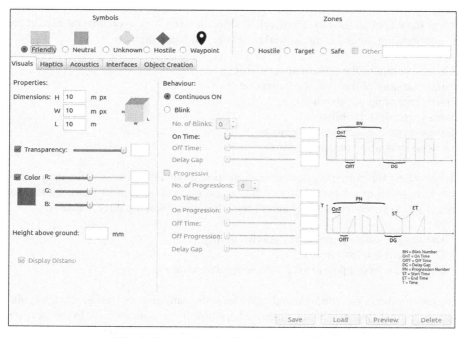

Fig. 6. Exploration Sandbox for visual AR cues

Table 1. Different scenarios of the first study, only the properties of the symbol 'Friendly 1' were altered

Situation Number	Hatch	Friendly 1	Friendly 2	Waypoint	Note
1	Open	Static In field of view Open	–	–	
2	Open	Static Not in field of view Open	Present	Present	
3	Open	Dynamic	Present	Present	Friendly 1 is first not in field of view and occluded, than enters it and becomes visible
4	Closed	Static In field of view Occluded	–	–	
5	Closed	Static In field of view Occluded	Present	Present	

nor was there enough time to extensively instruct them on how to do this, the researcher was interviewing them during the trial about the following themes:

- Understanding of the presented symbology
- Understanding of the distance indication
- Issues regarding understanding
- Issues regarding visibility
- Issues regarding technical equipment
- General evaluation of the system

After finishing the current system state evaluation mounted on the vehicle, the participants were to fill in a 7-point Likert scale indicating their agreement from one (do not agree at all) to seven (completely agree) regarding the following three statements:

1. The system would improve my combat ability.
2. The system is user-friendly.
3. The system would provide me with additional security.

This questionnaire was administered again after the participatory design session evaluating an optimal system with the discussed optimizations implemented. In this session, the participants were shown the different design options that had already been presented to them through the AR glasses and were asked to give their considerations concerning the design of the symbols. It has to be noted that not all had seen the augmentations clearly in the glasses; this made it necessary to present them on a computer screen in order to assess their usefulness and fit.

The way in which the system was implemented and presented to the soldiers while being mounted yielded an average score over all three questions of 4.26 (SD = 1.22), which is very close to the middle of the 7-point Likert scale. For a theoretical ideal implementation this score was 5.44 (SD = 0.82), pointing in a more favorable direction.

Since this was no experimental, validative study, but an explorative one, no inferential statistical analysis was performed. Which also means that possible differences between the two scores are neither proven, nor of particular interest for the goal of this study. The results only indicate that the system in its current state received mixed feedback and it would probably not be advisable to implement it in a vehicle as-is. However, there seems to be potential for use of the technology, even if it was not perceived as overwhelmingly positive. This is indicated by the rather high average score for the theoretical ideal implementation of aforementioned system.

To reach the full potential of the AR system, the collected qualitative data needs to be reviewed in order to identify problems and hindrances, as well as strong points and potential uses. The interviews during and after the trial were transcribed and coded. Topics that were mentioned by more than half of the participants (at least 6) are listed below in descending order:

- NATO green is better visible than blue
- No intuitive understanding of the meaning of the used symbol (NATO symbology)
- Number presented is intuitively understood as distance to target
- Augmentation does not bother while driving with closed hatch
- Augmentations hardly visible (open hatch)
- Augmentations are blurry (open hatch)
- Augmentations are more strongly and clearly visible with closed hatch
- Wrong perception of the color (blue is identified as white, open hatch)
- Double vision of the augmentations

The visibility of augmentations in AR glasses is strongly influenced by the amount of background light [18]. Half of the trials were conducted on a day that could be described as sunny and therefore was relatively bright, as opposed to the following five runs that were completed in cloudy weather, which resulted in the perception of a higher saturation of the augmentations. For this reason, the symbols were visible fundamentally better with a closed hatch because the interior of the vehicle is comparatively dark.

During the participatory design workshops, almost all participants (eight out of ten) determined that the framing design (see Sect. 3) was best suited for the use in an armored vehicle. This was mainly because it had the smallest risk of occluding important information and was least obtrusive. However, there was less consensus about the coloring of the symbol, with some participants preferring the NATO green and others favoring using the corresponding color for the identified object (in this case blue for friendly). This could have been influenced by the weather, respectively light conditions, in which the trials were run. The green symbols were more easily seen in the bright conditions and therefore could have been chosen over the different coloring by participants on this day. Another design consideration that was mentioned by some participants was the display of multiple objects in one location and how to prevent overlapping and therefore ultimately loss of information or information overload. A suggestion by one participant was to extend the frame so that it covered all relevant objects that are in proximity and then displaying a number on how many there were. Solving this issue merits further investigation and needs to be addressed in a follow-up study.

5 Outlook

The results of the exploration provide a discussion baseline for the use of AR cues and propose the use of interaction patterns as an adaptation strategy. The explorative process with the dynamic creation tool can be used to fine-tune the system to find the best combination of cues, properties and activation of each escalation step.

During the field trials, it was clear that there is room for improvement, especially in the GPS communication system, which proved to be inaccurate at times. An alternative for this is to use not only GPS but also other location acquisition systems, e.g. Lidar. This would provide the opportunity to gather extra data about the shape and orientation of objects, and therefore, further classify objects of interest.

Since not all objects represent the same level of threat to the ego-vehicle, even if they are of the same type (e.g. hostile IFV vs. hostile scout soldier), the combat ability of the object of interest in combination to the capabilities of the ego-vehicle can be used in the design of the interaction patterns. Interaction patterns could then be dynamically adjusted depending on the previous object classification, environment and other aspects of the use case. Subsequent changes to the levels of augmentation can also be influenced by this.

When analyzing the interaction between the user and the augmentation and following comments from the explorative workshops, it became clear that the user should remain capable of changing the augmentation or turn it off completely in situations of high stress or according to its personal demand or preference. This could be achieved by incorporating other technologies such as gesture/gaze-based interaction that can add other layer of interaction possibilities by using free resources directly in the line of sight of the user.

References

1. Billings, C.: Human-Centered Aviation Automation: Principles and Guidelines, NASA Technical Memorandum 110381, February 1996
2. Qiao, X., Ren, P., Dustdar, S., Liu, L., Ma, H., Chen, J.: Web AR: a promising future for mobile augmented reality—state of the art, challenges, and insights. Proc. IEEE **107**(4), 651–666 (2019)
3. Wickens, C.D.: Multiple resources and mental workload. Hum. Fact. **50**(3), 449–455 (2008)
4. Dixon, B.J., Daly, M.J., Chan, H.: Surgeons blinded by enhanced navigation: the effect of augmented reality on attention. Surg. Endosc. **27**, 454–461 (2013)
5. Flemisch, F., Baltzer, M., Sadeghian, S., Meyer, R., Lopez Hernandez, D., Baier, R.: Making HSI more intelligent: human systems exploration versus experiment for the integration of humans and artificial cognitive systems. In: 2nd Conference on Intelligent Human Systems Integration IHSI, San Diego (2019)

6. Flemisch, F., et al.: Towards a balanced human systems integration beyond time and space: exploroscopes for a structured exploration of human–machine design spaces. In: HFM-231 Symposium on Beyond Time and Space. NATO-STO Human Factors and Medicin Panel, Orlando (2013)

7. Schwarz, J., Fuchs, S.: Validating a "Real-Time Assessment of Multidimensional User State" (RASMUS) for adaptive human-computer interaction. In: 2018 IEEE International Conference on Systems, Man, and Cybernetics (SMC). IEEE (2018)

8. Fuchs, S., Schwarz, J., Flemisch, F.O.: Two steps back for one step forward: revisiting augmented cognition principles from a perspective of (social) system theory. In: Schmorrow, D.D., Fidopiastis, C.M. (eds.) AC 2014. LNCS (LNAI), vol. 8534, pp. 114–124. Springer, Cham (2014). https://doi.org/10.1007/978-3-319-07527-3_11

9. Alexander, C., Ishikawa, S., Silverstein, M., Jacobson, M., Fiksdahl-King, I., Angel, S.: A Pattern Language: Towns, Buildings, Construction. Oxford University Press, Oxford (1977)

10. Gamma, E., Helm, R., Johnson, R., Vlissides, J.: Design Patterns: Elements of Reusable Object-Oriented Software. Addison-Wesley, USA (1995)

11. Flemisch, F., Onken, R.: Open a window to the cognitive work process! pointillist analysis of man-machine interaction. Cogn. Technol. Work **4**, 60–170 (2002)

12. Baltzer, M.C.A., Lopez, D., Flemisch, F., Wesel, G.: Interaction patterns for cooperative guidance and control: automation mode transition in highly automated truck convoys. In: IEEE International Conference on Systems, Man, and Cybernetics (SMC). IEEE (2017)

13. Baltzer, M.C.A., Lassen, C., López, D., Flemisch, F.: Behaviour adaptation using interaction patterns with augmented reality elements. In: Schmorrow, D.D., Fidopiastis, C.M. (eds.) AC 2018. LNCS (LNAI), vol. 10915, pp. 9–23. Springer, Cham (2018). https://doi.org/10.1007/978-3-319-91470-1_2

14. Baltzer, M.C.A., López, D., Flemisch, F.: Towards an interaction pattern language for human machine cooperation and cooperative movement. Cogn. Technol. Work **21**(4), 593–606 (2019)

15. Hernández, D.L., Baltzer, M.C.A., Bielecki, K., Flemisch, F.: Interaction patterns for arbitration of movement in cooperative human-machine systems: one-dimensional arbitration and beyond. In: Karwowski, W., Ahram, T. (eds.) IHSI 2019. AISC, vol. 903, pp. 116–122. Springer, Cham (2019). https://doi.org/10.1007/978-3-030-11051-2_18

16. Flemisch, F. O., Kelsch, J., Schieben, A., Schindler, J., Löper, C., Schomerus, J.: Prospective engineering of vehicle automation with design metaphors: intermediate report from the H-Mode projects. Zentrum Mensch-Maschine-Systeme **7** (2007)

17. Ericsson, K.A., Simon, H.A.: How to study thinking in everyday life: contrasting think-aloud protocols with descriptions and explanations of thinking. Mind Cult. Act. **5**(3), 178–186 (1998)

18. Kruijff, E., Swan, J. E., Feiner, S.: Perceptual issues in augmented reality revisited. In: 2010 IEEE International Symposium on Mixed and Augmented Reality, pp. 3–12. IEEE (2010)

Confronting Information Security's Elephant, the Unintentional Insider Threat

Matthew Canham[1], Clay Posey[2(✉)], and Patricia S. Bockelman[1]

[1] School of Modeling, Simulation, and Training, University of Central Florida, Orlando, FL, USA
mcanham@ist.ucf.edu
[2] College of Business, University of Central Florida, Orlando, FL, USA
clay.posey@ucf.edu

Abstract. It is well recognized that individuals within organizations represent a significant threat to information security as they are both common targets of external attackers and can be sources of malicious behavior themselves. Notwithstanding these facts, one additional aspect of human influence in the security domain is largely overlooked: the role of unintentional human error. Such lack of emphasis is surprising given relatively recent reports that highlight error's central role in being the root cause for numerous security breaches. Unfortunately, efforts that recognize human error's influence suffer from not employing a commonly accepted error framework and lexicon. We thus take this opportunity to review what the data show regarding error-based breaches across various types of organizations and create a nomenclature and taxonomy rooted in the rich history of safety research that can be applied to the information security domain. Our efforts represent a significant step in an effort to classify, monitor, and compare the myriad aspects of human error in information security in the hopes that more effective security education, training, and awareness (SETA) programs can be devised. Further, we believe our efforts underscore the importance of revisiting the daily demands placed on organizational insiders in the workplace.

Keywords: Unintentional insider threat · Human error · Information security

1 We're Aware Already

1.1 Scoping the Problem

The past decade has been rather bleak for information security. A survey of the ten most notorious data breaches in the U.S. between 2010 and 2019 reveals that over 1.2 billion records were compromised, amounting to an estimated cost of over $2 billion U.S. from those ten events alone [2]. The future appears no less gloomy as it is expected that $5.2 trillion U.S. will be at risk globally to cybercrime activities between 2020 and 2024 [3]. The emergence of widespread ransomware attacks holding cities, hospitals, and other organizations hostage does not instill a sense of optimism for the next decade.

© Springer Nature Switzerland AG 2020
D. D. Schmorrow and C. M. Fidopiastis (Eds.): HCII 2020, LNAI 12197, pp. 316–334, 2020.
https://doi.org/10.1007/978-3-030-50439-7_22

While clandestine organizations filled with cyber threat actors tapping away on keyboards are easily imagined to be the primary threat to information security, the reality is less malicious or dramatic. In fact, data suggests that simple human error is the root cause responsible for a larger portion of data breaches than those due to external threat actors. For example, several studies indicate that human error and human-vector attacks represent the largest source of data breaches both in the present and in the past [4–8]. To help curb these issues, U.S. institutions spent an estimated $370 million U.S. in 2017 on security awareness training for employees [9]. While this expenditure appears to be a significant sum, it represents a mere 0.62% of the estimated $60.4 billion total spent on cyber defense by U.S. institutions during that same period [10]. These figures highlight a strong bias toward relying on technological solutions for what appear to be largely human-centric problems. A nearly universal perspective of security operations personnel is that employees commit these errors because they lack technical or security knowledge, or are simply 'ignorant,' and that the solution is to make employees more security aware [11]. This perspective combined with understaffing and mandated compliance requirements often leads to minimal Security Education, Training, and Awareness (SETA) efforts that appear to be intended to satisfy legal departments rather than address root issues. In other words, organizations place too much emphasis on behavior that is compliant rather than actions that are secure [12]. Unfortunately, these efforts have accomplished little to reduce the role of error-induced data breaches. Better enabling humans with technology and training to cope with cyber-based threats in the upcoming decade will require that we better understand these shortcomings and synthesize a variety of solutions to further enhance security. To that end, we outline the challenges associated with information security, discuss the causes of data breaches from both unintentional errors and intentional actions, provide an overview of human error research, demonstrate how this applies to information security, and finally present a path forward to addressing this serious but sorely underappreciated security threat.

1.2 Cybersecurity vs. Information Security

For the sake of scoping the topic and clarifying the aspects of human behavior on which the present analysis focuses, we distinguish between two important terms. While commonly interchanged, *information security* and *cybersecurity* refer to different types of security concerns [13]. Cybersecurity focuses on securing the systems which maintain information, and these systems are based on information technologies connected to the public Internet infrastructure; information security focuses on protecting information from unauthorized disclosures, manipulations, and restrictions. This distinction is important because a breach of physical records (e.g., printed client list) is a concern of information security even though it might not be considered a 'cyber incident'; conversely, a distributed denial of service (DDoS) attack would be considered both a cyber-incident and a violation of information security through access restriction. The requirements of information security have given rise to a framework referred to as the Information Security Triad. This triad rests upon the three 'pillars' of confidentiality, integrity, and availability [14]. Data breaches that violate the confidentiality pillar expose private information (such as credit card or health information) to unauthorized persons. A real-world example of this type of violation would be the social engineering attack that exposed the confidential emails contained on Hillary Clinton's private email server [15]. The integrity pillar focuses on the prevention of unauthorized modification of data on a system, or the modification of the operating characteristics of a system. Attacks against information integrity modify data in such a way that it becomes untrustworthy, and the attack may not be discovered even after the damage has been done. The Stuxnet worm is often cited as an example of an integrity attack that caused numerous uranium centrifuges to malfunction. Specifically, this malware caused the Siemens centrifuges to spin at rates that would cause them to prematurely wear out and malfunction while simultaneously providing false performance readings to centrifuge operators that left them unaware of an impending problem [16]. The availability pillar protects information assurance, meaning that users can access the data that they are authorized to access. The Dyn DDoS attack that interrupted service across Europe and North America preventing users from accessing Amazon, Netflix, Paypal, Reddit, and other sites is an example of this type of an attack [17]. While not exhaustive of every possible scenario, the Information Security Triad provides a simple, easy-to-use framework for discussing the security of information. Conceptualizing security within these three pillars clarifies the types of security threats posed by errors. For example, an *authorized user* who sends unencrypted sensitive information to an *unauthorized recipient* breaches confidentiality in very much the same way that a victim of a phishing email might. While many in the security industry have acknowledged that human error plays a significant role in security breaches [18], the industry focuses more resources toward malicious threat actors than the error-prone and inattentive employees. This discrepancy suggests that information security might be significantly improved by leveraging lessons learned in safety research. Understanding the causes of industrial accidents, the role human error plays in these events, and reducing their likely causes has saved countless human lives in high-stakes environments and suggests a path forward in applying these lessons to security frameworks. This research, which seeks to understand the underlying causes of human error and reduces these, has saved countless human lives and suggests a path forward. Before suggesting potential paths, we briefly review what several studies indicate about human error within information security contexts.

2 Data Breach Case Examples

Several industry groups, security providers, and insurance companies independently survey, track, and publicly report data breach incidents annually[1]. We review a sample of these efforts that discuss human error in an effort to ground the remainder of our discussion on incidents observed "in the wild." While these surveys provide some interesting insights, they generally lack a uniform nomenclature for describing and discussing human and organizational errors in a way that allows security professionals to make meaningful improvements in their work.

---

[1] Most information security reports focus on breaches of confidentiality rather than integrity and availability; thus, we have focused our efforts on these types of attacks in this section.

2.1 Verizon Data Breach Incident Report (2019)

Since 2008, Verizon has published their annual Verizon Data Breach Incident Report (DBIR), which summarizes the causes of data breaches from the previous year. The 2019 report indicated that "Errors" were responsible for 21% of all breaches in 2018, while "Misuse" by authorized users was responsible for 15%. These statistics represent a 5% increase in Errors since 2013, while Misuse decreased by 2%. Error types according to the DBIR are summarized in Table 1. While the precise statistics for Misuse were not provided in the report, a summary of Misuse types are outlined in Table 2. The distinction between Error and Misuse in the Verizon DBIR seems to be intentionality; where Error is unintentional, Misuse is an action that an individual intentionally carried out [5]. The intentional component of the Misuse category of data breaches will be revisited later.

Table 1. Classification and rates of errors from 2010 to 2018 (Adapted from [5]).

Error type	2010	2018	Difference
Misdelivery	32%	37%	+5
Publishing error	16%	21%	+5
Misconfiguration	0%	21%	+21
Loss	31%	7%	−24
Programming error	0%	5%	+5
Disposal error	5%	14%	−9
Omission	2%	4%	+2
Gaffe	4%	0%	−4

Table 2. Categories of misuse and approximate percentages for 2018 (Adapted from [5])

Misuse type	Approximate percentages
Privilege abuse	70%
Data mishandling	36%
Unapproved workaround	17%
Knowledge abuse	8%
Email misuse	7%
Possession abuse	6%
Unapproved hardware	6%
Unapproved software	4%
Net misuse	2%
Illicit content	1%

2.2 2019 IBM X-Force Threat Intelligence Index

Like the Verizon DBIR, the IBM Threat Intelligence group also provides annual reporting of security incidents from the previous year. The 2019 report was derived from data gathered from 70 billion security events per day, across 130 countries [19]. This report clusters all human-related vulnerabilities under the category of "The Inadvertent Insider," which included incidents related to phishing and social engineering, improper configuration, and failure to use password best practices. It highlights cloud service misconfiguration as the "single biggest risk to cloud security" and noted that misconfiguration resulted in 43% of the total compromised records in 2018. While the report cites human error to be a top concern, it does not elaborate on types of human error beyond misconfiguration [19].

2.3 2018 United Kingdom, Information Commissioner's Office, Data Breach Report

In late 2018, the government of the United Kingdom (UK), Information Commissioner's Office (ICO), contracted with Kroll risk mitigation and investigative service to examine all data breach reports made to the ICO between 2017 and 2018. Of the 2,416 data breach incidents reported to the ICO during this time period, 292 (12.1%) were determined to be deliberate malicious actions, while 2,124 (87.9%) were determined to be caused by human error or through insiders' non-malicious intent [8, 20] (Table 3).

Table 3. Categories of errors report to UK Information Commissioner's Office 2017–2018

Error type	Number reported	Percentage of errors
Data sent by email to incorrect recipient	447	21.05%
Data mailed/faxed to incorrect recipient	441	20.76%
Loss/theft of paperwork	438	20.62%
Failure to redact data	256	12.05%
Data left in insecure location	164	7.72%
Failure to use bcc when sending email	147	6.92%
Loss/theft of unencrypted device	133	6.26%
Verbal disclosure	46	2.17%
Insecure disposal of paperwork	35	1.65%
Loss/theft of only copy of encrypted data	16	0.75%
Insecure disposal of hardware	1	0.05%

2.4 California Data Breach Report (2012–2015)

Beginning in 2012, businesses and government agencies within the state of California are required to notify the California Attorney General's Office of breaches affecting more than 500 California residents. In 2016, the California Department of Justice released a report on 657 reported data breaches occurring between 2012 and 2015, affecting the confidential records of over 49 million Californians [21]. Of these reported breaches, 54% were caused by "Malware and Hacking" affecting 44 million records, 22% were caused by "Physical Theft or Loss" affecting 2.8 million records, 17% were caused by "Errors by Insiders" affecting 2 million records, and finally 7% were caused by "Misuse of Access Privileges by Insiders" affecting 206,000 records. A breakdown of the types of Errors by Insiders is included in Table 4. Unfortunately, an explanation of the different types of Misuse of Access Privileges was not provided in the report.

Table 4. Error types leading to data breach (Adapted from [21])

Error type	Percentage of total
Misdelivery	46%
Web display	35%
Unauthorized employee access	7%
Insecure disposal	4%

2.5 Chubb Cyber Index

As our final example, the Chubb Cyber Index (CCI) provided by Chubb Limited Insurance, a major provider of cyber breach insurance, summarizes cyber incidents between 2009 and 2019 [22]. The CCI cites "Error" as the primary cause of reported incidents, accounting for 21% of the total over this ten-year period. The CCI defines Error as "any unintentional and non-malicious actions, mistakes, or inaction by a natural person" and includes lost devices, misconfiguration, and programming errors as potential examples. The Index also mentions "Misuse" as accounting for 16% cyber incidents. Misuse is defined as occurring when "an internal actor(s) or partner(s) uses a cyber-asset for a purpose other than the original intended use, or the use goes beyond the scope of the original permission."

3 Actions Leading to Data Breaches

Reviewing the incidents leading to data breaches as described in these data breach reports, several patterns emerge. However, a lack of clear error definitions or consistent terminology make comparison across studies difficult. If security professionals are to address the security vulnerabilities posed by human error, a standardized terminology is needed to gain clarity into their underlying causes. Across these five reports, the most frequently reported error involved sending protected information to the incorrect recipient, which we term *transmission errors*. We use the term transmission because the reviewed reports describe multiple channels of transmission that are inclusive of, but not limited to, email. We term the next most common error type *configuration errors*, which refers to errors leading to the misconfiguration of databases, networked systems, web-hosted content, and related situations that lead to the unintentional exposure of data resulting from the configuration setting of a system or component thereof. The inability to locate sensitive items, or items containing sensitive information, is termed *loss errors*. Errors leading to the disclosure of information to unauthorized employees or other individuals, or improper storage of data are termed *leakage errors*. Finally, the insecure disposal of sensitive information, or items containing sensitive information, are termed *disposal errors*. Another point requiring clarification relates to "misuse." Three of the five reports reviewed found that misuse played a significant role in the cause of data breaches; however, the intentionality behind the misuse of privileges, access, or equipment was not clarified. This term needs explanation because someone may intentionally misuse resources with, or without, malicious intent and this distinction is important for security practitioners to understand because they imply different types of prevention strategies [23].

3.1 Understanding Human Error

While few efforts in the information security field have tried to hone in on understanding how human error affects information security, human factors researchers have explored the causes of errors and accidents for decades [1, 24–26]. Emerging from this rich foundation are extensive error taxonomies, descriptions of unsafe policy violations, and their causes. Norman [24] defines human error as <u>any</u> deviation from "appropriate" behavior. Using this definition as a starting point then leads to the *intentionality* of the action as being a centrally defining characteristic in this discussion because a deviation from "appropriate" could be intentionally ignoring a poorly designed policy or unintentionally sending an email to an unauthorized recipient.

Rasmussen [26] proposed that humans operate in three *modes of behavior*: (1) skill-based; (2) rule-based; and (3) knowledge-based. *Skill-based behaviors* are those which involve routine tasks that individuals are usually highly experienced with. *Rule-based behaviors* are those for which a novel situation is encountered, but an existing rule for responding is available for the individual to apply to this new circumstance. These rules are usually in the form of "if encountering X, then do Y"-type of relationship. *Knowledge-based behaviors* occur in unfamiliar situations when neither existent skills nor rules can be applied to address an impediment. These cases require deliberate reasoning to interpret the problem at hand and develop a plan or solution. Distinguishing between these modes of behavior clarifies intentionality because skill-based behaviors are largely unintentional. In other words, this level of behavior is mostly automatic and does not require a significant degree of conscious thought. Knowledge-based behaviors reside on the other extreme of being entirely intentional and deliberate, while rule-based behaviors sit in the middle of being sometimes deliberate and at other times unconscious. Framing human behaviors in this way has led researchers to develop a taxonomy of human error [1, 24] which is illustrated in Fig. 1 and has been adapted to information security contexts [4, 27]. Reviewing these will better provide a context for understanding what is needed moving forward.

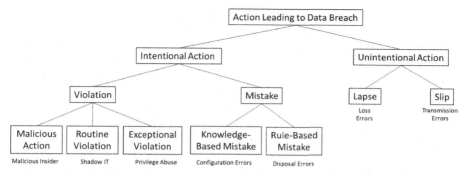

Fig. 1. Taxonomy of actions leading to data breach - Adapted from [1]

3.2 Unintentional Actions

Unintentional actions typically appear in two forms: (1) slips and (2) lapses. *Slips* are errors of commission in which humans correctly understand their objective and know the end-state they wish to achieve, but incorrectly execute the actions to achieve that end state. These almost always occur during skill-based activities. An example from the terrestrial world might be holding car keys in one hand and trash in the other, and then tossing the car keys into the trash receptacle and the trash onto the kitchen counter. Within information security, the majority *transmission errors* are very likely slips. The auto-complete function present in most email client software inserts suggested usernames when only a portion of the recipient's username has been entered. Different systems use different algorithms to determine which suggested recipient will appear at the top of the suggested list. Some email clients will suggest a recipient most recently communicated with, while another might suggest the one most communicated with. One of the most commonly reported reasons for mis-directed emails is the insertion of a username by the auto-complete function that is similar to the one that the sender intended but is not the intended recipient [28].

In contrast to a slip, a *lapse* is an error of omission in which an intended action was not carried out, typically because of a memory failure on the part of an individual. Such lapses can be the result of performing an action sequence out of the typical order, or it could be due to distraction. These errors usually occur during rule-based activities when something disrupts the if-then rule application. A non-security example of a lapse might be getting milk out of the refrigerator, but then forgetting to put it back in the refrigerator because a distraction like an upset child interrupts the normal cycle of pouring milk into cereal and then returning the container to the refrigerator. Lapses are likely culprits for *loss errors* that occur when a user misplaces sensitive information or a storage device containing sensitive data.

3.3 Intentional Actions - Mistakes

Distinct from lapse and slip errors, *mistakes* are committed with intentionality and typically manifest in one of two forms: (1) *rule-based mistakes* or (2) *knowledge-based mistakes*. Rule-based mistakes may occur when an individual correctly assesses a situation but applies the incorrect rule, or by incorrectly applying the correct rule. As the term implies, these issues occur during rule-based behavioral actions. A non-security example of a mistake might be adjusting the thermostat in a room above the desired temperature with the intention that it will warm up faster, when in fact this action results in the furnace overshooting the originally intended temperature. An example of a rule-based mistake within a security environment might be a user forwarding a questionable email to their work email address and opening it on the work network because they believe that the security of the network at work "is better than at home". In the latter example, the user is intentionally applying a rule incorrectly because of a faulty mental model.

Knowledge-based mistakes occur in unfamiliar situations when an individual lacks pre-existing skills or rules upon which to rely. There are many reasons that errors occur in this category, many related to a lack of knowledge or awareness of actionable environmental cues or correct responses. An everyday example of a knowledge-based mistake might be observed when individuals attempt to assemble a piece of Ikea furniture without the instructions. A large proportion of *configuration errors* (i.e., a misconfiguration of an information resource that leads to the unintentional exposure of data) are certainly knowledge-based mistakes because of the scale and complexity of the types of systems that need configuring.

3.4 Intentional Actions – Deliberate Violations

Contrary to error, employees intentionally choose to violate policies and rules for a host of reasons, some for malicious purposes but most are not [23, 29]. The categories of deliberate violations of rules and policies most relevant to the previously reviewed data breach reports are *malicious actions*, *routine violations*, and *exceptional violations*. Though not the focus of our discussion, a brief example of a malicious misuse is provided on the Chubb Cyber Index: "Employee A is given access to the company's financial information to prepare for an audit. Employee A is only supposed to review this information for potential problems or issues that may arise during the audit. Instead, Employee A copies the information and sells it to a competitor" [30].

Unintentional insider threats, on the other hand, represent issues that stem from nonmalicious activity. Specifically, we define an *unintentional insider threat* as an organizational insider who does not harbor malicious intentions for their actions but who still actively chooses to go against established norms and policies to achieve some other goal despite knowing that harm *could* be caused. This definition aligns with that of *organizational deviance*: "Employee deviance is…the voluntary behavior that violates significant organizational norms and in doing so threatens the well-being of an organization, its members, or both" [31]. For example, an individual who violates a security policy because s/he simply wishes to complete an assigned task in a timely manner is an unintentional rather than intentional insider threat. Thus, *intention whether to commit harm or illegal activity* (e.g., theft, sabotage) is the differentiating factor, not *the volitional decision to go against policy or standard procedures*. An analogy here is that not everyone who exceeds the prescribed speed limit while driving intends to commit vehicular manslaughter, but a very small portion do. Within this scope, deliberate violations fall within two categories: (1) routine violations and (2) exceptional violations [24].

Routine violations are the everyday violations of rules and policies that many commit and are rarely enforced. Examples include exceeding the speed limit while driving, walking across the street against the signal, or reusing passwords on multiple sites. The primary cause of these violations are rules and policies that invite violation by competing with, or causing impedance to, achieving higher priority objectives. With few exceptions, most organizations prioritize completing work objectives over security policy adherence. In this vein, managers are more likely to discipline employees who fail to achieve work goals than they are to discipline high-performing employees who regularly violate security policies. Imagine a top sales performer who outsells everyone else in their division by 300%, being terminated from their position for not using passwords of sufficient complexity. Companies thrive by selling products and services not by adhering to security policies. This situation creates competing interests, and it is through this lens that security policies should be framed. Every security policy that requires an employee to exert some amount of effort causes *bureaucratic friction* against which that employee must operate to complete their assigned tasks. When this friction exceeds the perceived benefit, that employee will likely violate that policy, and can be rewarded for doing so [29]. One of the most frequent examples of routine violations is termed shadow IT [32]. *Shadow IT* occurs when a user (or group of users) use unapproved work arounds or rely on unauthorized information technology resources for operation. Examples include external cloud services, unauthorized devices, and unauthorized wireless network access points that are connected to organizational networks.

While routine violations are widespread and generally ignored, *exceptional violations* occur infrequently and are punished with uneven regularity depending upon organizational culture. As the name implies, these violations are usually committed in an exception to the norm, usually in response to an urgent situation or emergency requiring rule suspension. A non-security example would be the permission for emergency response vehicles to exceed speed limits when utilizing emergency lights and audible siren in responding to a sufficiently critical situation. Likewise, southbound roadways may be converted for northbound traffic (or vice versa) by law enforcement and transportation departments when individuals attempt to flee the path of a hurricane. In information security environments, exceptional violations often occur when workloads are high, or deadlines are critical. Under these circumstances, users are very unlikely to comply with policies that compete with the completion of 'traditional' work demands [33].

4   SETA is not a Panacea

While our discussion thus far demonstrates that information security risks stem from employees across various levels of technical expertise and demand various cognitive, behavioral, and affective responses, the current organizational approaches to employee training often create self-imposed limitations on both content and scope of learning/training opportunity. We assert that the focus on and even fascination with awareness, above all else, is at the core of the problem for modern organizations.

4.1   Why Awareness Alone Won't Help the Masses

NIST Special Publication 800-50 [34] reiterates statements from preceding publication 800-18, stating that: "Awareness is not training. The purpose of awareness presentations is simply to focus attention on security. Awareness presentations are intended to allow individuals to recognize IT security concerns and respond accordingly". In NIST's framework (see Fig. 2), education is ultimately reserved for the highest tier of IT professionals. Education is, essentially a degree. Training is establishing some skills (usually validated by a certificate), and it is the level of exposure that the "middle" specialists need to do their jobs in IT. Awareness then lets the common employees have what they need to recognize risks and act safely. We strongly advise that such a well-defined demarcation is dangerous.

This arbitrary progression set forth in the "awareness-focused" approach not only disregards well-established research in human learning and performance, but it has set a precedence and norm in IT for what "good" training paradigms are. In essence it recommends that "good" training excludes training and performance expertise and fails to call for any actual behavioral changes in most of the target participants.

Researchers have likewise proposed awareness-focused approaches. For example, a recent review of social engineering training programs suggested that information security awareness training is the key component for more resilience against socially engineered cyber threats [35], but the review lacked data on the overall efficacy of the awareness approach. Such criticism of similar studies has been noted by others [e.g., 36]. While important, this line of research appears to build a case that the main consideration in establishing an efficacious security awareness training program is primarily one of instructional methodology or delivery, in that the way that the courses or sessions are designed and delivered is the main aspect. Aligned with NIST, researchers who advocate awareness often recommend a diverse portfolio of communication and delivery mechanisms including but not limited to: e-mail broadcasting, social media advertising, and blogging; game-based delivery methods; video-based and independent, self-paced delivery methods; and simulation-based delivery methods like email phishing campaigns to measure employees' awareness levels [35].

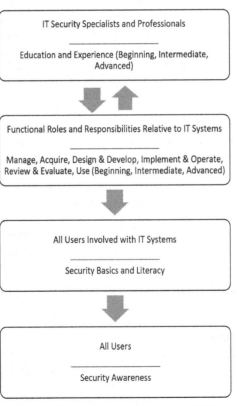

Fig. 2. IT security learning continuum, NIST Special Publica-tion 800-50

There at least two major problems with this position that looks to rid organizations of human-related security issues in such broad strokes. The critique is not with authors, such as those cited who report on these approaches advocated by the NIST recommendations, but challenge security experts to reconsider assumptions. To the contrary, research suggests that the approach of awareness, which essentially primes and warns, may increase certain types of risks.

First, the notion that a method for optimal learning/training delivery can be identified or declared based on anything other than concise learning/training objectives is flawed. "Security awareness" is, at best, a problematic learning objective (if it is a learning objective at all as most learning/training experts would not consider it such[2]). When constructing learning/training objectives, human performance, cognition, and education experts avoid metrics that require instructors to make assumptions about participants' mental states. For example, the objectives "participants will be aware of social engineering techniques" and "students will know about common security threats" would be too vague for expert learning designers.

This is not to assert that awareness is trivial. To the contrary, in the tradition of Harrold Bloom, and subsequentially taxonomical variants [e.g., 37], humans need knowledge to progress to more abstract cognitive states of analysis and synthesis. If higher-order cognitive skills are required [38] (which they arguably are in anticipation of or response to the security cases described above), then it makes sense that foundational awareness is necessary yet glaringly insufficient. The second problem, then, is that security awareness approaches cannot prepare participants to meet the higher-order cognitive demands required for real-world information security needs. Whereas the first flaw focuses on delivery methods failing to address objectives, the second points to key objectives being missed altogether.

In summary, most information security threats faced by organizational employees will require a response that draws from a triadic set of knowledge, skills, and affect that transcends awareness. Of course, variability in the threats' complexities will likely dictate which facets of the triad are most important in a given response, but in general, employees will first need to apply *what* they know about a risk. They will also need to apply that knowledge in a certain way, using a tool or technique—that is, *how* to respond. And they need to have a motivation, a driving force behind a response (e.g., moral or ethical framework, perceived need to share a company value system, need to receive a paycheck)—the affective *why*. We note, however, that differences exist among motivational aspects and their impact on employees' secure intentions and behaviors (e.g., intention to comply with policy versus intention to protect organizational informational resources) [39]. Educators know that introductory learners receive instruction in this what-how-why triadic form; novice learners also demonstrate learning in this form [40]. Further, such efforts should not be relegated to one type of employee or professional role. Because issues of information security extend beyond any one person or single organizational department, it is imperative that organizations provide 'normal' employees the opportunity to experience all SETA components. It is imperative, though, that the goals be clearly articulated.

---

[2] While beyond the scope of this paper to provide a tutorial on writing learning or training objectives, most experts in these fields agree that these objectives must at least declare an expectation of observable participant/learner behavior that demonstrates measurable change under a given condition (often time framed) [cf. Mayer 38].

4.2   Beyond Awareness to Achievement

There are three major imperatives to be taken from our discussion on SETA:

1. Behavioral, cognitive, and affective aspects of SETA efforts must be incorporated at appropriate levels for all, not just the highest-level decision makers and operational experts;
2. Identification of behavioral, cognitive, and affective aspects of information security will require collaboration with human performance and learning experts, not just information security, technology, and organization experts;
3. Efficacy must be determined by objective measures at multiple levels, meaning that SETA programs' evaluations must shift from impossible-to-measure goals (e.g., "employees will be aware") to metrics that can be tracked over time.

These steps will not only impact the present challenge of improved learning/training paradigms but will also provide more powerful data for tracking shifts in demographic preparation and response across sectors. These data can become inputs for improved reporting systems, feedback for intuitive design, and support for more resilient AI security support at individual, team, and organizational levels.

5   Confronting the Elephant in the Room

We have offered three major points of discussion in this paper, which we believe represent a significant opportunity for improving organizational information initiatives. First, one of the significant threats to organizational information security is due to employees' unintentional error. The current data show this trend increasing, yet little is being done about it. In fact, it is particularly troubling that this trend was known much earlier, with roughly 50% of data breaches from 1993 and 2005 known to be caused by human error [4]. Second, the few research organizations that are tracking and reporting these error-based trends fail to utilize a common error framework and terminology to categorize and discuss these important issues. And third, current organizational SETA efforts largely focus on awareness-level knowledge. Founded largely on NIST recommendations, the training and education components are often left to the IT and security professionals.

While error in its various forms lacks the allure of more exotic threats like industrial espionage and employee sabotage, it remains an ever-present issue that must begin receiving the meaningful attention it deserves. In addition, organizational leaders must realize that relying on awareness-heavy programs to complete compliance checklists will amount to little more than that: being compliant rather than secure. In our minds, this strategy is very much aligned with the problem of rewarding A while hoping for B [41]. In that vein, we propose three recommendations that have the possibility to fundamentally change the way that organizations handle unintentional insider threats and the SETA functions.

5.1   Creating a Confidential Error Reporting System

One of the primary obstacles to addressing human error as a threat to information security is the lack of a uniform classification and reporting system. While all the reviewed data breach reports mentioned earlier roughly agree with each other, they lack consistency in the terminology used, and specific definitions are mostly missing. For human error to be properly addressed, we need a standardized reporting system with clarification of the specific factors contributing to the error. A system that might be used as a referent is the Common Vulnerabilities and Exposures (CVE) repository hosted and managed by the MITRE Corporation. The CVE provides a central point of reference for identified information security vulnerabilities [42]. A similar reporting system that catalogs Human Vulnerabilities and Exposures (HVE) would need to be designed somewhat differently than the CVE system, however. One difference would be the need to protect the privacy of the individuals involved in the incident, and by extension, those reporting the incident (if they are part of the same organization). Such confidential incident reporting systems that might be utilized as a guide exist outside of information security. For example, the Aviation Safety Reporting System (ASRS) designed and operated by the National Air and Space Administration (NASA) allows pilots and other aviation employees to voluntarily and confidentially report "close call" incidents to alert the aviation community of critical issues and reduce the likelihood of similar incidents. These reports are eventually made available to the public through a database and monthly newsletters. Reports of special concern may be used to generate alerts to specific airlines, industry segments, or issue safety recalls. Two key aspects of the ASRS that should be considered for a security error reporting system are (1) confidentiality and (2) immunity. An individual may submit a report anonymously or choose to include contact information. Individuals and companies who voluntarily submit reports are considered to be acting in "good faith" and are allowed a certain amount of immunity by the Federal Aviation Administration (FAA) if the incident is later discovered by an investigator.

5.2 Creating and Cultivating Cyber-Resilient Organizations

The second recommendation that we believe is vital for a greater and fruitful focus on human error in information security is for organizational leaders to move from an avoidance mentality to one of tolerance. An *avoidance* mentality acknowledges the numerous threats in existence and actively attempts to limit exposure to each one in the hopes that those efforts will be enough to stem the tide of potential attacks or executed vulnerabilities. A *tolerance* mentality acknowledges known threats but also accepts the high likelihood that new issues will emerge, and negative events will occur [4]. Thus, avoidance is likened to putting out fires as they occur, and tolerance is more aligned with constant learning cycles no matter what new issues can and will emerge.

The tolerance approach is evident within organizations that consistently function in high-risk, complex environments. Such organizations are termed high reliability organizations (HROs), and they are intentionally designed to handle unexpected events through active anticipation and resilience [43]. Organizations that fit the HRO mold are air traffic control centers, aircraft flight decks, nuclear facilities, and hospital emergency rooms, among others. These organizations continue to operate and offer services despite "a million accidents waiting to happen that didn't" [43]. While a complete explanation of HROs is not appropriate here, we assert that many organizational environments are sufficiently complex for most employees where a single error can cause substantial harm. Further, though the outcome might not result in loss of human life as in instances of emergency room and aircraft traffic control center failures, significant harm can be caused digitally with the press of a keyboard button.

Research has shown that HROs are successful despite their challenges because they embrace five general philosophies. These approaches include: (1) preoccupation with failure; (2) reluctance to simplify; (3) sensitivity to operations; (4) commitment to resilience; and (5) deference to expertise [44, 45]. Briefly, organizations whose cultures give attention to even the seemingly smallest of issues, resist unwarranted simplifications (e.g., "that's just people being people"), provide opportunities for all employees to understand the inner workings and interdependencies of the organizations' systems, promote the importance of approaching perfection though it can never be attained, and actively search for and listen to expertise wherever it exists within the organization when issues arise are consistent performers in high-risk environments. Thus, it "is not that [HROs are] error-free but that errors don't disable [them]" [43].

By applying the important lessons learned from HROs into modern organizations, we have the possibility to further limit human error's influence on information security. Unfortunately, some of these aspects are in direct opposition to the current operations within organizations. For example, issues are waved off at times because they are said to be due to "that's just employees being stupid and ignorant," [46, 47] and many employees are deterred if not prohibited from engaging with the department formally charged with information security when issues arise [48, 49] though the employees likely have much more knowledge of the affected operational processes than the security professionals themselves. Therefore, to achieve cyber resiliency, organizational leaders will need to actively promote substantial cultural changes, which will take much time and effort.

5.3   Creating and Implementing Human-Centric Security Systems

Finally, organizational security systems must become human-centric and designed around the ways that humans *actually* behave rather than how they are *supposed* to behave. The HVE described above would provide much data and insight in determining how to design secure systems around actual behavior. Established principles of user interface design could provide tremendous insights into how to design security around the human [50]. "The more secure you make something, the less secure it becomes" [51]. This quote by Donald Norman in context makes the argument that adding security controls to a system (virtual or physical) creates second order effects that lead to an overall reduction in the security of that system. This occurs because the security controls introduce impediments to the user in accomplishing their goals. An idea that prevails throughout the security industry is that usability is necessarily at odds with security. We challenge this notion, because of Norman's argument that people will find work arounds. Systems that may seem superficially less secure, may ultimately be more so. Passwords serve as excellent case studies in this regard. For example, consider a 22-character password consisting of random alphanumeric and symbols in comparison to a 22-character passphrase made from only lower-case letters. The 22-character complex password will be significantly more resilient to dictionary attacks; however, considering the two examples from a human-centric perspective it becomes obvious that the English-based passphrase is more secure. Here is the 22-character complex password, **3963%0TF32(gqe7oo3N*!2)**. And here is the 22-character password, **lily had a little**

lamb. Two weeks from now, when you are expected to log back into the systems for which you have just created these two passwords, which will you be able to remember, and which will you need to write down? This password example demonstrates that sometimes less security is more actually more secure.

Another candidate area that could better leverage usability is Shadow IT usage. When users are resorting to alternative informational technology resources instead of the approved or authorized resources, this should be a strong signal to security staff that something in the existing system needs to be fixed. Users do not decide to use unauthorized IT services because they are inferior to the authorized services. They use unauthorized services because the authorized services DO NOT WORK as well as the alternatives. Much in the same way that muddy tracks across grass provide evidence of poor landscape design [1], Shadow IT provides evidence of poorly designed IT. Sometimes less security is more secure.

6 Summary

Errors and the misuse of data systems are responsible for a majority of data breaches according to several surveys [5, 19, 21, 22]. Comparing studies and analyzing trends to identify root causes is nearly impossible because of a lack of standardized terminology and uniform classification. Human error results from many causes, and security awareness training only addresses a subset of these. To effectively address this exposure, we recommend the creation of a Human Vulnerabilities and Exposures (HVE) database similar to the established Common Vulnerabilities and Exposures (CVE) database, creating and cultivating Cyber-Resilient Organizations (CROs), and implementing human-centric security systems that are complimentary to human cognitive processing rather than contradictory to it. Until steps such as these steps are taken, the significant security exposure to human error will persist.

Acknowledgement. This research was in part sponsored by the U.S. Army CCDC Soldier Center and was accomplished under Cooperative Agreement Number W911NF-15-2-0100. The views and conclusions contained in this document are those of the authors and should not be interpreted as representing the official policies, either expressed or implied, of U.S. Army CCDC Soldier Center or the U.S. Government.

References

1. Reason, J.: Human Error. Cambridge University Press, Cambridge (1990)
2. Goldberg, M.: 10 of the biggest data breaches over the last decade (2019). https://www.bankrate.com/finance/banking/us-data-breaches-1.aspx#slide=1. Accessed 30 Jan 2020
3. Bissell, K., LaSalle, R., Dal Cin, P.: The Cost of Cybercrime: Ninth Annual Cost of Cybercrime Study. Accenture (2019)
4. Im, G.P., Baskerville, R.L.: A longitudinal study of information system threat categories: the enduring problem of human error. Database Adv. Inf. Syst. **36**(4), 68–79 (2005)
5. Verizon: Data Breach Investigations Report (2019)

6. Baskerville, R.: A taxonomy for analyzing hazards to information systems. In: Katsikas, S.K., Gritzalis, D. (eds.) SEC 1996. IAICT, pp. 167–176. Springer, Boston, MA (1996). https://doi.org/10.1007/978-1-5041-2919-0_14

7. Reilly, R.B.: 95% of successful security attacks are the result of human error (2014). https://venturebeat.com/2014/06/19/95-of-successful-security-attacks-are-the-result-of-human-error/. Accessed 30 Jan 2020

8. Targett, E.: Revealed: human error, not hackers, to blame for vast majority of data breaches (2018). https://www.cbronline.com/news/kroll-foi-ico. Accessed 30 Jan 2020

9. Metinko, C.: Cybersecurity training sees flood of M&A (2018). https://www.forbes.com/sites/mergermarket/2018/08/17/cybersecurity-training-sees-flood-of-ma/#5d8e709d2266. Accessed 30 Jan 2020

10. Statista: Spending on cybersecurity in the United States from 2010 to 2018 (2019). https://www.statista.com/statistics/615450/cybersecurity-spending-in-the-us/. Accessed 30 Jan 2020

11. Carpenter, P.: Transformational Security Awareness: What Neuroscientists, Storytellers, and Marketers Can Teach Us About Driving Secure Behaviors. Wiley, Indianapolis (2019)

12. Cram, W.A., D'Arcy, J., Proudfoot, J.G.: Seeing the forest and the trees: a meta-analysis of the antecedents to information security policy compliance. MIS Q. **43**(2), 525–554 (2019)

13. von Solms, B., von Solms, R.: Cybersecurity and information security–what goes where? Inf. Comput. Secur. **26**(1), 2–9 (2018)

14. Conrad, E., Misenar, S., Feldman, J.: CISSP Study Guide, 2nd edn. Syngress, Waltham (2012)

15. Debenedetti, G. The email headache that won't go away (2016). https://www.politico.com/story/2016/07/hillary-clinton-email-fbi-fallout-225113. Accessed 30 Jan 2020

16. Response, S.S.: W32.Duqu: the precursor to the next Stuxnet (2011). https://www.symantec.com/connect/w32_duqu_precursor_next_stuxnet. Accessed 30 Jan 2020

17. Graff, G.M.: How a dorm room minecraft scam brought down the Internet (2017). https://www.wired.com/story/mirai-botnet-minecraft-scam-brought-down-the-internet/. Accessed 30 Jan 2020

18. Spadafora, A.: 90 percent of data breaches are caused by human error (2019). https://www.techradar.com/news/90-percent-of-data-breaches-are-caused-by-human-error. Accessed 30 Jan 2020

19. IBM: X-Force Threat Intelligence Index (2019)

20. Targett, E.: Personal Communication with M. Canham (2020)

21. Justice, C.D.O.: California Data Breach Report, 2012–2015 (2016)

22. Chubb: Chubb cyber index: providing data driven insight on cyber threat trends (2020). https://chubbcyberindex.com/#/incident-growth. Accessed 30 Jan 2020

23. Willison, R., Warkentin, M.: Beyond deterrence: an expanded view of employee computer abuse. MIS Q. **37**(1), 1–20 (2013)

24. Norman, D.: The Design of Everyday Things, Revised and Expanded edn. Basic Books, New York (2013)

25. Perrow, C.: Normal Accidents: Living with High Risk Technologies, Updated edn. Princeton University Press, Princeton (2011)

26. Rasmussen, J.: Skills, rules, and knowledge; signals, signs, and symbols, and other distinctions in human performance models. IEEE Trans. Syst. Man Cybern. **SMC-13**(3), 257–266 (1983)

27. SKYbrary: Human error types (2016). https://www.skybrary.aero/index.php/Human_Error_Types. Accessed 30 Jan 2020

28. Rader, E., Munasinghe, A.: "Wait, do I know this person?" Understanding misdirected Email. In: Proceedings of the 2019 CHI Conference on Human Factors in Computing Systems, Glasgow, Scotland (2019)

29. Posey, C., et al.: Bridging the divide: a qualitative comparison of information security thought patterns between information security professionals and ordinary organizational insiders. Inf. Manag. **51**(5), 551–567 (2014)

30. Chubb: Chubb Cyber Library (2020). https://chubbcyberindex.com/#/cyber-library. Accessed 30 Jan 2020

31. Robinson, S.L., Bennett, R.J.: A typology of deviant workplace behaviors: a multidimensional scaling study. Acad. Manag. J. **38**(2), 555–572 (1995)

32. Silic, M., Back, A.: Shadow IT–a view from behind the curtain. Comput. Secur. **45**, 274–283 (2014)

33. Posey, C., Canham, M.: A computational social science approach to examine the duality between productivity and cybersecurity policy compliance within organizations. In: International Conference on Social Computing, Behavioral-Cultural Modeling & Prediction and Behavior Representation in Modeling and Simulation (SBP-BRiMS), Washington D.C. (2018)

34. Wilson, M., Hash, J.: SP 800-50: Building an Information Technology Security Awareness and Training Program, NIST, Gaithersburg (2003)

35. Aldawood, H., Skinner, G.: Educating and raising awareness on cyber security social engineering: a literature review. In: 2018 IEEE International Conference on Teaching, Assessment, and Learning for Engineering (TALE), Wollongong, NSW, Australia. IEEE (2018)

36. Bulgurcu, B., Cavusoglu, H., Benbasat, I.: Information security policy compliance: an empirical study of rationality-based beliefs and information security awareness. MIS Q. **34**(3), 523–548 (2010)

37. Cannon, H.M., Feinstein, A.H.: Bloom beyond Bloom: Using the revised taxonomy to develop experiential learning strategies. In: Developments in Business Simulation and Experiential Learning: Proceedings of the Annual ABSEL Conference, Orlando, FL (2005)

38. Mayer, R.E.: Applying the Science of Learning. Pearson/Allyn & Bacon, Boston (2011)

39. Burns, A., et al.: Intentions to comply versus intentions to protect: a VIE theory approach to understanding the influence of insiders' awareness of organizational SETA efforts. Decis. Sci. **49**(6), 1187–1228 (2018)

40. Kennedy, D.: Writing and Using Learning Outcomes: A Practical Guide. University College Cork (2006)

41. Kerr, S.: On the folly of rewarding A, while hoping for B. Acad. Manag. J. **18**(4), 769–783 (1975)

42. MITRE: Common vulnerabilities and exposures (2020). https://cve.mitre.org/. Accessed 30 Jan 2020

43. Weick, K.E., Sutcliffe, K.M.: Managing the unexpected: sustained performance in a complex world. Wiley, Hoboken (2015)

44. Weick, K.E.: Organizational culture as a source of high reliability. Calif. Manag. Rev. **29**(2), 112–127 (1987)

45. Roberts, K.H.: Some characteristics of one type of high reliability organization. Org. Sci. **1**(2), 160–176 (1990)

46. Field, T.: Insider threat: 'you can't stop stupid' (2010). https://www.bankinfosecurity.com/insider-threat-you-cant-stop-stupid-a-2789. Accessed 30 Jan 2020

47. Matyszczyk, C.: IT and security professionals think normal people are just the worst (2019). https://www.zdnet.com/article/it-professionals-think-normal-people-are-stupid/. Accessed 31 Jan 2020

48. Kraemer, S., Carayon, P., Clem, J.: Human and organizational factors in computer and information security: pathways to vulnerabilities. Comput. Secur. **28**(7), 509–520 (2009)

49. Safa, N.S., Von Solms, R., Furnell, S.: Information security policy compliance model in organizations. Comput. Secur. **56**(February), 70–82 (2016)

50. Adams, A., Sasse, M.A.: Users are not the enemy. Commun. ACM **42**(12), 40–46 (1999)
51. Norman, D.A.: The way I see it when security gets in the way. Interactions **16**(6), 60–63 (2009)

Adapting Interaction to Address Critical User States of High Workload and Incorrect Attentional Focus – An Evaluation of Five Adaptation Strategies

Sven Fuchs(✉) , Stephanie Hochgeschurz , Alina Schmitz-Hübsch ,
and Lerke Thiele

Fraunhofer Institute for Communication, Information Processing and Ergonomics FKIE,
53343 Wachtberg, Germany
sven.fuchs@fkie.fraunhofer.de

Abstract. We have developed an approach to Dynamic Adaptation Management in Augmented Cognition systems that processes task and operator state indicators to dynamically select and configure context-sensitive adaptation strategies in real time. This dynamic approach is expected to avoid much of the potential cognitive cost associated with adaptations. Following an overview of our conceptual approach and the description of a proof-of-concept implementation in the anti-air warfare domain, this paper describes the conceptualization and operationalization of five adaptation strategies: Context-Sensitive Help, Automation, Scheduling, Visual Cueing, and Decluttering. These strategies were designed to address two critical user state diagnoses – high workload and incorrect attentional focus. We then report an experiment that evaluated each strategy's impact on those two critical states. Twenty-four participants (18m/6f) took part in the lab-based experiment and performed a naval air surveillance task in six different conditions (five adaptation conditions and one control condition). Two adaptation strategies significantly reduced the average duration of critical state episodes. Two other adaptation strategies also showed promising trends for being effective in addressing the cognitive state problems.

Keywords: Adaptive systems · Adaptation strategies · Augmented Cognition · User state assessment · Workload · Attention · Physiological monitoring

1 Introduction

Most Augmented Cognition systems use physiological measures to detect critical cognitive states and trigger adaptation strategies to address the problem state and restore or augment performance. However, the mere presence of a critical state reveals little about the type of adaptation that would be appropriate with respect to the given situation. Without accounting for context, it is likely that adaptations are inappropriate for the current situation, triggered or withdrawn at inopportune moments, potentially interrupting or

© Springer Nature Switzerland AG 2020
D. D. Schmorrow and C. M. Fidopiastis (Eds.): HCII 2020, LNAI 12197, pp. 335–352, 2020.
https://doi.org/10.1007/978-3-030-50439-7_23

confusing the user. Fuchs, Schwarz and Flemisch [1] stated that many adaptation frameworks demonstrated to date may be successful in diagnosing and addressing isolated problem states but lack the flexibility necessary to be effective in complex systems with heterogeneous elements and rich interactions, concluding that more flexible frameworks are needed.

This paper describes an approach to dynamically adapting system interactions to address critical operator states. Task and operator state indicators are processed to dynamically select and configure context-sensitive adaptation strategies in real time. Following an overview of the conceptual approach and a proof-of-concept implementation, we describe the design of five adaptation strategies meant to address two critical user state diagnoses – high workload and incorrect attentional focus.

1.1 User State Diagnostics

In previous work, Schwarz, Fuchs, and Flemisch [2] have proposed a multidimensional assessment of user state as a more holistic approach to user state analysis in adaptive system design. This approach has resulted in a diagnostic component named RASMUS ('Real-Time Assessment of Multidimensional User State') that detects performance decrements of the user and analyzes which critical user states are likely to have caused the performance decrement [3]. RASMUS diagnostics enable technology to detect not only when the human operator's performance declines and which cognitive states are critical, but also identifies contextual factors (system state, task state, user state as indicated by physiological and behavioral metrics) that may have contributed to the situation. RASMUS forwards detections of potentially critical user states and associated contextual indicators to the adaptation management component, thereby enabling dynamic adaptive systems to not only determine when the user needs support, but also to infer what kind of support is most appropriate to restore performance.

For the purposes of the work reported in this paper, RASMUS diagnostics were configured to detect two critical user states – high workload and incorrect attentional focus. The physiological and contextual indicators used to diagnose these states are listed in Table 1. Diagnostic outcomes for these user states were validated in a prior experimental study [4].

Table 1. Contextual indicators used to diagnose cognitive problem states.

High workload	Incorrect attentional focus
Number of tasks high Click frequency high Heart rate variability low Pupil size high Respiration rate high	>1 tasks present highest priority task not selected

1.2   Dynamic Adaptation Management

The goal of our Dynamic Adaptation Management concept is to determine when the user needs support and to infer what kind of support is most appropriate to restore user performance in a given situation. To that end, Fuchs and Schwarz [5] have developed ADAM, an "Advanced Dynamic Adaptation Management" component that is summarized herein for easy reference. ADAM processes RASMUS diagnostics, determines an appropriate adaptation goal and selects the adaptation strategy that is best suited to mitigate the detected critical state, and thus, to restore the user's performance.

ADAM assumes that declines in performance are symptoms caused by underlying cognitive problem states. For example, task omission may be caused by excessive workload, incorrect attentional focus, or even motivational issues. Accordingly, ADAM assumes that task performance can be restored by adapting the interaction with strategies that address the diagnosed cognitive state problem. According to Breton and Bossé [6], humans should receive support when their "cognitive capabilities are not sufficient to adequately perform the task" (p. 1–4). Thus, ADAM assumes a need for adaptation when a performance decrement coincides with at least one critical cognitive state. A performance decrement was included as a prerequisite for adaptation to allow operators to self-adapt to the problem state, considering that "having adaptive system working together with an adaptive operator will likely be unsuccessful. An adaptive system is more likely to work successfully when it starts reallocating tasks as soon as the operator is no longer able to adapt properly to changing task demands" ([7], p. 10). Also, intervening too early may favor complacency ("based on an unjustified assumption of satisfactory system state," [8], p. 23) and hinder development of resilience and coping strategies.

To trigger adaptation, ADAM also requires that the pool of adaptation strategies include a strategy that is (a) capable of addressing the diagnosed cognitive state problem and (b) suitable in the given context. For example, automating tasks to support the operator may only make sense if a sufficient number of tasks is present. Once a need for adaptation is indicated, ADAM dynamically selects the best strategy from a pool of candidate strategies based on RASMUS diagnostic output. The approach to dynamic adaptation management involves five steps (Fig. 1) that are detailed below.

First, ADAM selects an adequate adaptation objective based on the diagnosed cognitive state problem. Adaptation objectives are abstract descriptions of cognitive manipulations that describe how a certain critical cognitive state can be addressed so that it is no longer critical (e.g. shift user attention to critical task). Each objective may be achieved through several strategies (e.g. cueing, decluttering, etc.). The second step is therefore to select, from a pool of available adaptations, a strategy that is linked to address the adaptation objective and that is suitable under current conditions. To that end, all adaptations able to achieve the adaptation objective become candidate strategies. It is possible that certain conditions must be met in order to trigger a strategy. ADAM evaluates known conditional prerequisites for all candidate strategies to determine the best strategy under current circumstances. If prerequisites are not fulfilled, the strategy is deemed unsuitable in the given context and further strategies from the pool will be evaluated for their suitability. An adaptation strategy may also comprise contextual parameters which can be used to further tailor it to current task and user state (Step 3). For example, task urgency

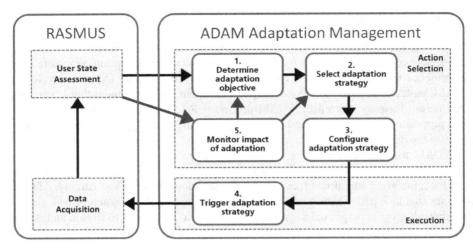

Fig. 1. Five steps of ADAM dynamic adaptation management.

could inform cue salience. Once selected and configured, the adaptation strategy is activated (Step 4) in the information display, altering human-machine interaction in a way that affects the cognitive state problem and serves the adaptation objective. Monitoring the effects of adaptation (Step 5) with respect to task performance and cognitive state changes will determine whether and how adaptation is continued. If the adaptation objective was accomplished and the underlying problem states are no longer present, it is important to withdraw this context-specific adaptation, as inappropriate continuation could have negative effects on the operator and task performance.

We expect this dynamic approach to avoid much of the potential cognitive cost associated with adaptations. To understand how each adaption strategy effects cognitive user states in stand-alone, adaptation strategies were analyzed individually, before examining the effect of the dynamic approach as a whole.

2   Method

Goal of the reported experiment was to examine each adaptation strategy's capacity to mitigate the two critical user states of high workload and incorrect attentional focus.

2.1   Task Environment

Given the generic nature of the adaptation management approach, it can be applied to various operational and instructional settings. The experimental testbed for the work reported herein was a naval air surveillance task (Fig. 2), implemented in Java and connected to an existing AAW simulation driven by Presagis Stage [9]. The simulation was simplified for learnability but designed to maintain the essential cognitive demands of the real-world task. The operator is to perform all tasks with a focus on keeping the Identification Safety Range (ISR) around the own ship clear of threats. The simulation comprises four simplified AAW-tasks (cf. Table 2 for task descriptions): identification of contacts, creation of new contacts, warn, and engage contacts. Figure 2 shows a screenshot of the task environment. The blue dot in the center of the map represents the own ship. Identified radar contacts are visualized in green (neutral), blue (friendly), or red (hostile). New, unidentified contacts (yellow) must be identified as neutral, friendly, or hostile according to certain criteria, such as location, direction, and IFF code. Hostile contacts that enter perimeters marked by blue or red circles around the own ship must be warned or engaged, respectively.

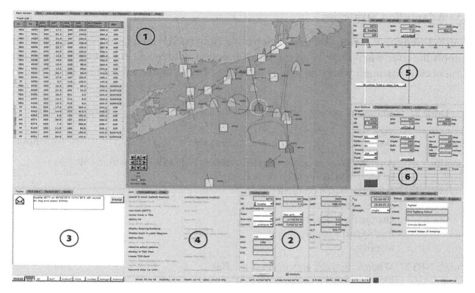

Fig. 2. Screenshot of the task environment. (Color figure online)

Tasks occur at scripted times throughout the scenario and may occur simultaneously. In this case, the task with the highest priority must be performed first. Each task is associated with a time limit for task completion. Time limits were adopted from outcomes of an earlier study that employed the same tasks and simulation software [10]. If a task is not completed within the defined time limit or completed incorrectly, RASMUS detects a performance decrement. Table 2 indicates the priorities (with 500 being highest and 100 being lowest) and the respective time limits of each task.

2.2 Adaptation Framework

The adaptation objective is determined based on cognitive state diagnoses reported at the time of critical performance. In order to demonstrate dynamic selection of an adaptation based on diagnosed cognitive state problems, context-sensitive adaptation objectives were defined for each cognitive problem state (Fig. 3). For proof-of-concept purposes, five adaptation strategies were designed and implemented, so that each objective was linked to only one adaptation strategy designed to address the diagnosed cognitive state problem (although the conceptual framework allows selection from multiple strategies per objective). The strategies are detailed in Sect. 2.3.

As performance decrements are detected and the system determines a need for adaptation, the appropriate strategy is invoked based on the selected adaptation objective (Table 3). While adaptation is active, performance and cognitive state criteria are continuously monitored to examine whether a strategy was successful in achieving the associated adaptation objective and must be withdrawn.

2.3   Adaptation Strategies

Context-Sensitive Help. A Context-Sensitive Help (CSH) adaptation was designed to support the user when there is an executive function bottleneck. An executive function bottleneck is assumed when the user is in a state of high workload but task execution is

Table 2. Description, priority, and time limit for each subtask in the experimental task.

Task	Description	Priority	Time limit
Identify	Any unidentified contacts must be identified as friendly, neutral, or hostile based on predefined criteria. Identified contacts may change their behavior in a way that requires reassigning their identity	100 outside ISR, 300 within ISR	30 s
Create NRTT	When a message appears in the message panel a contact (NRTT) must be added manually to the TDA. Information required to create the NRTT is displayed in the message	200	30 s
Warn	Contacts identified as hostile must be warned as soon as they enter the Identification Safety Range (ISR; indicated on the TDA by a blue circle around the own ship)	400	20 s
Engage	Contacts identified as hostile that have been warned must be engaged as soon as they enter the Weapon Range (WR; indicated on the TDA by a red circle around the own ship)	500	10 s

Abbreviations: ISR – Identification Safety Range; TDA – Tactical Display Area; NRTT – Non-real-time track; WR – Weapon Range

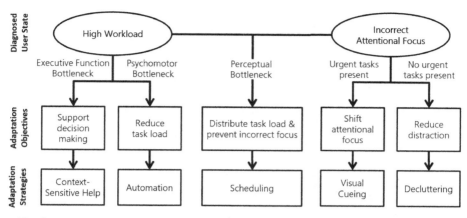

Fig. 3. Context-sensitive selection of adaptation strategies in the proof-of-concept system.

Table 3. Trigger rules for adaptation strategies.

Adaptation strategy	Diagnosed user state	Context parameters to determine objective
Context-Sensitive Help	High workload	Low mouse click frequency
Automation	High workload	High mouse click frequency, priority of attended task > priority of automated task
Scheduling	High workload + incorrect attentional focus	High mouse click frequency
Visual Cueing	Incorrect attentional focus	High priority task present
Decluttering	Incorrect attentional focus	Only low priority tasks present

delayed. Delayed task execution is operationalized by a low frequency of mouse clicks despite high task load. CSH points out the next step of action and places cognitive affordances where the user needs support [11]. To overcome executive function bottlenecks or mental blockades, this adaptation adds a yellow tag to the track with the highest-priority task that is labelled with the task to be performed ("Identify", "Warn", "Engage"). In case an NRTT message is the highest-priority task, its envelope is highlighted in yellow (Fig. 4). To activate CSH, three criteria must be met at the same time: A performance decrement, high workload, and a low number of mouse clicks. The adaptation is withdrawn as soon as the tasks that caused the performance decrement are complete or workload is no longer critically high.

Automation. Automation changes the extent of human involvement in the task by taking over tasks previously performed by a human operator. To address psychomotor bottlenecks during task execution, the dual-mode Automation strategy described in [5] was employed in a slightly redesigned form. A psychomotor bottleneck may arise when the

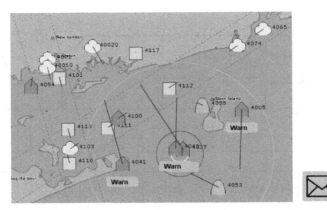

Fig. 4. Context-Sensitive Help for contacts (left) and NRTT envelope (right). (Color figure online)

user knows what to do but is incapable of keeping up with the task load. Following the Adaptive Automation approach (i.e. [12–14]), the user should be kept "in the loop" and Automation is only to be activated on demand. Activation of the Automation strategy requires three criteria to be met: A performance decrement, high workload, and a high mouse click frequency, the latter being an indicator of high psychomotor workload. If all three criteria are met, identification of contacts in uncritical positions is automated to free up operator resources for higher priority tasks. To prevent automation-related errors and complacency effects (e.g. [15]), all automated identifications are marked with a "?" (Figure 5). Suspect identifications can then be verified manually later when the user is less taxed.

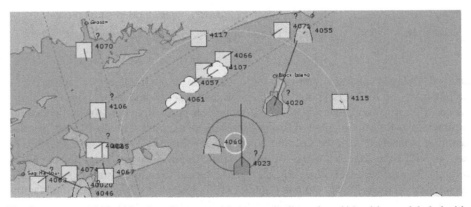

Fig. 5. Automated identification; Contacts with automatically assigned identities are labeled with a question mark.

Scheduling. In case of a perceptual bottleneck, sensory inputs may be processed incorrectly or not at all. Hence, a perceptual bottleneck is assumed when high workload and incorrect attentional focus occur at the same time. To address this problem, a Scheduling strategy was designed that suppresses the presentation of non-critical new tasks (identification of tracks outside of ISR, NRTT tasks) when high-priority tasks are present. Our Scheduling strategy thus converts simultaneously occurring events into sequential form (cf. [11]). The strategy assists in distributing workload more evenly and prevents distraction from uncritical new tasks. Four criteria must be met to trigger the scheduling strategy: A performance decrement, high workload, incorrect attentional focus, and presence of high-priority tasks (identification inside of ISR, warn, engage). Suppressed tasks are presented as soon as any of the four criteria is no longer met.

Visual Cueing. Operator attentional focus can be shifted by increasing the saliency of relevant information objects [16]. A Visual Cueing strategy was designed to immediately refocus the operator's attention to the highest-priority task in case there are urgent tasks that the user's attention is not focused on. The strategy is operationalized as a red arrow pointing from the cursor to the task with highest priority (Fig. 6). Three conditions must be met to trigger this strategy: A performance decrement, incorrect focus of attention, and presence of tasks with a high priority (identification in ISR, warn, or engage). The arrow disappears as soon as the tasks that caused a performance decrement are completed or when the focus of attention is no longer critical.

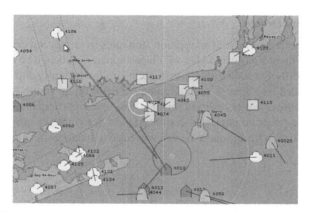

Fig. 6. Visual Cueing; a red arrow points at the highest-priority contact (4010). (Color figure online)

Decluttering. When the operator's attentional focus is incorrect but there are no urgent tasks present, less intrusive means of manipulating user's attention are sufficient, as to not cause task interruption problems. Decluttering refers to reducing the amount or complexity of information to be displayed [17] and is similar to cueing in that it provides attentional filtering by increasing the salience of some display objects compared with others [11].

A Decluttering strategy was designed to reduce distraction caused by currently irrelevant information/objects in non-time-critical settings. In order to facilitate task detection, contacts that do not require action are faded to 30% opacity to reduce their salience

(Fig. 7). The Decluttering strategy is triggered when the following criteria are met: A performance decrement, incorrect focus of attention, and presence of low-priority tasks (identification outside of ISR). Decluttered contacts fade in again when all tasks that caused the performance decrement are complete or attentional focus is no longer inadequate.

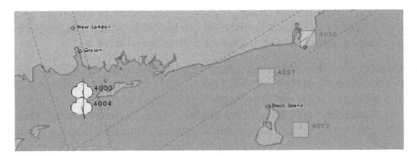

Fig. 7. Decluttering; opacity of currently irrelevant contacts is reduced.

2.4   Experimental Design

To investigate the effectiveness of adaptation strategies in mitigating critical user states, a lab-based experiment was designed in which participants worked on a naval air surveillance task in two different scenarios (24 min each; similar difficulty and task load). Each scenario was divided into three 8-min long phases, resulting in six phases (Fig. 8). One phase was used as the control condition. In all other phases, participants were assisted by one of the five adaptation strategies whenever ADAM determined a need for adaptation. To avoid order effects, a Latin square design was used to balance the order of conditions.

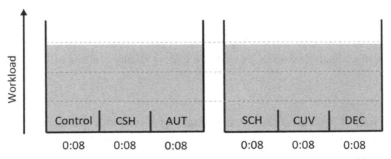

Fig. 8. Two scenarios were divided into six 8-min phases (five adaptation conditions and one control condition). CSH = Context-Sensitive Help, AUT = Automation, SCH = Scheduling, CUV = Visual Cueing, DEC = Decluttering.

We hypothesized that adaptation strategies mitigate cognitive state problems by either reducing high workload, redirecting attention to the highest-priority task at the moment, or both. Changes in cognitive state problems were investigated using three dependent variables: (1) quantity, (2) average, and (3) cumulative duration of diagnosed critical user states (cf. Table 1).

2.5 Participants

Twenty-four employees of the Fraunhofer Institute for Communication, Information Processing and Ergonomics (18 males, 6 females) took part in the lab-based experiment. Participants were between 19 and 48 years old ($M_{age} = 31.96$, $SD_{age} = 7.18$) and 62.5% reported having very good or good knowledge of computer games, while 37.5% stated having little or no knowledge of computer games.

2.6 Apparatus

Figure 9 shows the research testbed with the sensors utilized for user state assessment: a Tobii X120 eye tracker underneath the monitor, a Zephyr BioHarness3 multisensor chest strap on the left, and a webcam positioned on top of the monitor. A 24-inch monitor is used to display the user interface of the Anti-Air-Warfare simulation. The Tactical Display Area (TDA) located in the center displays virtual contacts in the surroundings of the own ship.

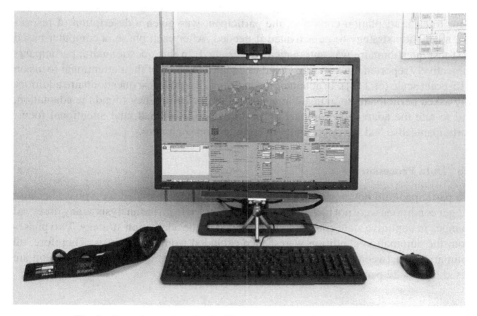

Fig. 9. Experimental testbed with user screen and sensor equipment.

The experimental testbed provides information on tasks and task state, behavioral activity, and RASMUS user states diagnostics for high workload and incorrect attentional focus. These cognitive states were chosen because of their particular relevance to the task and because RASMUS diagnostics were successfully validated for these states in a previous experiment [4]. Along with every critical cognitive state diagnostics, RASMUS reports the status of the contextual indicators that contributed to the state diagnosis (cf. Table 1).

2.7 Procedures

At the beginning of each session, participants gave informed consent. Participants donned the Zephyr BioHarness3 multisensor chest strap after its contact surfaces were moistened to improve conductivity. A computer-based pre-test questionnaire was then administered to assess demographic data, experience with computer games and the current cognitive, motivational, and emotional states of participants. Afterwards, participants read instructions for the experimental task. Next, the eye tracker was calibrated. Participants then completed a 13-min training scenario with moderate workload. During this training phase, participants were able to ask questions and received guidance from the experimenter to ensure full comprehension of all tasks. Towards the end of the training scenario, workload was reduced to ensure that participants had enough time to finish all tasks before recording the physiological baseline. In a 2-min baseline phase, physiological indicators were measured to represent an uncritical state used by RASMUS diagnostics to detect cognitive state problems during the two experimental scenarios.

Finally, the two scenarios with three phases each were run in counterbalanced order. Prior to each adaptation condition, the participant was given a description of respective adaptation strategy to be activated if needed. After each phase, a computer-based post-test questionnaire was automatically presented. In this questionnaire, participants subjectively reported their workload and rated the adequacy of their attentional focus on a 15-point scale (cf. [18]). In adaptation conditions, the post-test questionnaire additionally asked participants to indicate how helpful or disturbing they found the adaptation, and to rate the adaptation's impact on performance, workload, and attentional focus. Participants also had the opportunity to comment on each item.

2.8 Data Processing

If a participant did not experience the adaptation strategy intended for a phase because trigger conditions were not met, the phase was excluded from the analysis as any observed changes in cognitive state could not be attributed to the adaptation strategy. Two phases from the automation condition were hence excluded from the analysis. Therefore, all comparisons between the adaptation strategy automation and the control condition are based on $n = 22$ participants.

2.9 Analysis

Statistical analyses were performed with SPSS 25. To analyze the experimental hypothesis, quantity, average duration, and cumulative duration of critical state episodes for both high workload and incorrect attentional focus were compared between adaptation conditions and the control condition. Dependent t-Tests for paired samples were computed when normality assumptions were met. Otherwise, Wilcoxon signed-rank tests were used. For all comparisons the acceptable type one error rate was set to $\alpha = .01$ based on a Bonferroni correction due to multiple tests. Additionally, effect sizes (Pearson's r) were calculated. The effect was considered small when $r = .10$, medium when $r = .30$, and large when $r = .50$ [19].

3   Results

Table 4 shows means and standard deviations of how often and how long each adaptation strategy was activated. Decluttering was by far activated most often, whereas Context-Sensitive Help remained active the longest on average. Pronounced interindividual differences were present as suggested by the large standard deviations. While some participants rarely met the criteria for activating adaptation strategies, others did quite often and for long intervals.

Table 4. Occurrences, average, and cumulative durations of adaptation strategy activations.

Adaptation strategy	Occurrences M (SD)	Average duration (s) M (SD)	Cumulative duration (s) M (SD)
Context-Sensitive Help	5.63 (3.20)	31.28 (30.79)	154.42 (128.35)
Automation	3.38 (2.60)	1.79 (1.48)	7.08 (7.64)
Scheduling	7.38 (4.49)	13.35 (8.91)	95.46 (78.85)
Visual Cueing	8.88 (5.81)	3.17 (1.15)	30.71 (24.69)
Decluttering	18.08 (7.57)	10.47 (4.73)	188.67 (106.68)

3.1   Critical Workload

It was tested whether adaptation strategies were effective in mitigating diagnosed critical workload episodes by either reducing quantity, average duration, or cumulative duration of diagnosed cognitive user states compared to the control condition. Results of these comparisons and descriptive statistics of critical workload episodes are shown in Table 5.

Quantity. No adaptation strategy significantly reduced the quantity of diagnosed critical workload episodes (all $ps > .302$, all |r|s $< .22$).

Average Duration. On average, critical workload episodes were significantly shorter with Visual Cueing than in the control condition ($z = -2.77, p = .006, N = 24, r = -.40$). Scheduling showed a statistical trend in the same direction ($z = -1.83, p = .067, N = 24, r = -.26$). Other comparisons were not statistically significant (all $ps > .28$, all |r|s $< .17$).

Table 5. Quantity, average, and cumulative duration of high workload episodes in comparison to the control condition.

	Quantity		Average duration (in seconds)		Cumulative duration (in seconds)	
	M (SD)	p (r)	M (SD)	p (r)	M (SD)	p (r)
CSH	8.25 (3.76)	.519 (.14)	35.25 (29.48)	.819 (.03)	231.25 (117.81)	.773 (.06)
AUT	8.91 (3.54)	.613 (.08)	26.71 (20.06)	.570 (.09)	212.00 (123.58)	.535 (.14)
SCH	10.04 (5.41)	.507 (.14)	25.92 (22.98)	.067 (.26)	206.54 (112.54)	.247 (.24)
CUV	10.29 (4.03)	.303 (.21)	22.66 (24.49)	.006 (.40)	193.83 (121.62)	.088 (.35)
DEC	9.38 (4.68)	.798 (.11)	49.71 (94.96)	.278 (.16)	228.17 (113.61)	.633 (.10)
CTL	9.04 (4.24)	–	37.06 (32.72)	–	240.17 (91.09)	–

Notes. CSH = Context-Sensitive Help, AUT = Automation, SCH = Scheduling, CUV = Visual Cueing, DEC = Decluttering. $N = 24$, except for AUT-CTL comparison ($n = 22$).

Cumulative Duration. No adaptation strategy significantly reduced the cumulative duration of diagnosed critical workload episodes (all $ps > .087$, all |r|s $< .36$), but the comparison between Visual Cueing and the control condition was marginally significant ($t(23) = 1.78, p = .088, r = .35$). With Visual Cueing, episodes of critical workload tended to be shorter overall.

3.2 Incorrect Attentional Focus

It was further investigated, whether adaptation strategies effectively mitigated episodes of incorrect attentional focus by either decreasing quantity, average duration, or cumulative duration of incorrect attentional focus in comparison to a control condition. Results of these comparisons and descriptive statistics of episodes of incorrect attentional focus are shown in Table 6.

Quantity. Episodes of incorrect attentional focus occurred significantly more often with Decluttering, compared to the control condition ($z = -3.23, p = .001, N = 24, r = -.47$). A similar statistical trend was found for Context-Sensitive-Help ($t(23) = -2.08$, $p = .049, r = .40$) and Visual Cueing ($t(23) = -2.04, p = .053, r = .39$). Other differences were not statistically significant (all $ps > .492$, all |r|s $< .16$).

Average Duration. Episodes of incorrect attentional focus were significantly shorter with Decluttering compared to the control condition ($z = -3.57, p < .001, N = 24$, $r = -.52$). Additionally, there were marginally significant differences with Context-Sensitive Help ($z = -1.97, p = .049, N = 24, r = -.28$) and Visual Cueing ($z = -2.49$, $p = .013, N = 24, r = -.36$). Other comparisons were not statistically significant (all $ps > .252$, all |r|s $< .17$).

Cumulative Duration. No adaptation strategy significantly reduced the cumulative duration of episodes of incorrect attentional focus (all $ps > .103$, all |r|s $< .34$).

Table 6. Quantity, average, and cumulative duration of incorrect attentional focus episodes in comparison to the control condition.

	Quantity		Average duration (in seconds)		Cumulative duration (in seconds)	
	M (SD)	p (r)	M (SD)	p (r)	M (SD)	p (r)
CSH	21.54 (6.55)	.049 (.40)	11.63 (5.71)	.049 (.28)	236.05 (96.58)	.226 (.25)
AUT	17.50 (4.15)	.493 (.15)	15.08 (7.26)	.671 (.09)	258.00 (97.53)	.750 (.07)
SCH	18.92 (5.21)	.749 (.07)	13.50 (5.23)	.253 (.16)	248.13 (92.08)	.717 (.08)
CUV	21.67 (6.97)	.053 (.39)	10.92 (3.29)	.013 (.36)	227.79 (79.98)	.104 (.33)
DEC	23.63 (6.06)	.001 (.47)	10.67 (4.29)	<.001 (.52)	242.00 (83.70)	.404 (.17)
CTL	18.38 (5.45)	–	14.64 (6.73)	–	254.88 (94.70)	–

Notes. CSH = Context-Sensitive Help, AUT = Automation, SCH = Scheduling, CUV = Visual Cueing, DEC = Decluttering. $N = 24$, except for AUT-CTL comparison ($n = 22$).

4 Discussion

In summary, two adaptation strategies turned out to be effective in addressing cognitive state problems: Visual Cueing and Decluttering effectively reduced the average duration of critical state episodes in comparison to a control condition. Two more strategies, Scheduling and Context-Sensitive Help, showed similar statistical trends in the same direction with moderate effect sizes. Both Scheduling and Context-Sensitive-Help were associated with a marginally significant lower average duration of cognitive state problems in comparison to the control condition. The moderate sizes of these non-significant effects suggest that the study was most likely underpowered. Therefore, a replication with a larger sample size is necessary to further test the effectiveness of the adaptation strategies.

Decluttering significantly reduced the average duration of diagnosed incorrect attentional focus episodes, confirming its intended function; however, at the same time, the strategy significantly increased the number of such problem state episodes. Context-Sensitive Help und Visual Cueing showed trends in the same direction. This seemingly contradictory observation can be explained with the trigger mechanism for adaptations. An adaptation strategy is activated when a performance decrement coincides with at least one critical cognitive state. Thus, a critical state episode is required for an adaptation strategy to be triggered and can thus not be proactively avoided. This may explain the general ineffectiveness of adaptation strategies in reducing the quantity of cognitive state problems. Also, as two adaptation strategies effectively reduced the average duration of critical state episodes, the problem states were able to reoccur more often.

Visual Cueing was designed to address incorrect attentional focus. While it did not significantly reduce the average duration of such critical state episodes, it showed a promising statistical trend in the intended direction with a moderate effect size. In addition, although not designed for this purpose, Visual Cueing significantly reduced the average duration of high workload episodes. This effect can be easily explained, as Visual Cueing frees the participant from searching for and identifying the highest-priority task, thereby reducing workload. Also, the arrow cue may have led to conclusions about the overall task state. For example, a cue pointing to an ID task would indicate that this was currently the highest-priority task, thereby implying that no (higher-priority) warnings or engagements were necessary at that time. Finally, the effect may be explained with outcomes of a previous experiment using the same task environment [20], in which we tried to induce workload and incorrect attentional focus independently in separate scenarios. Results indicated that occurrences of the two states overlapped considerably. In fact, incorrect attentional focus was more successfully induced by increasing workload than by adding distracting elements to the scenario, suggesting that the two constructs are tightly coupled and may generally coincide.

Contrary to our expectations, the strategy of automating low-priority ID tasks did not significantly reduce critical workload. The specific reason is unknown but the result is consistent with other findings in automation research (e.g. [21]). It is possible that automated changes to the situation require users to invest additional workload for change detection and mental model updates.

Generally speaking, the effectiveness of individual adaptations may have been limited by their highly specific respective purposes. In order to limit the cognitive costs associated with each strategy, adaptation strategies for the Dynamic Adaptation Management approach are designed to temporarily address temporary states, limiting their individual long-term effectiveness. A single adaptation strategy may support certain aspects of a situation when needed but can negatively affect other aspects of the same situation. For instance, Decluttering reduced the salience of a hostile track until it actually entered the identification safety range, which made it more difficult to anticipate such tasks. In a dynamic interplay of strategies, a subsequent visual cue on the task, once it has occurred, could counteract that negative effect of Decluttering. In a Dynamic Adaptation Management setting, different adaptation strategies may complement each other and may thus lead to a higher combined effectiveness in addressing cognitive state problems.

5 Conclusions

Visual Cueing and Decluttering significantly reduced the average duration of critical user state episodes, Scheduling and Context-Sensitive Help showed promising statistical trends in the same direction. However, neither the quantity nor the cumulative duration of critical user states was significantly reduced by any of the five adaptation strategies. It was suggested that this was partly due to the configuration of adaptation strategies as a reaction to a user experiencing a cognitive state problem.

Additional analyses will investigate the effect of these five adaptation strategies on task performance. As a next step, a study is underway to examine the effectiveness of a dynamic interplay between all adaptation strategies rather than evaluating each adaptation strategy individually. It is expected that adaptation strategies only reveal their full potential when complementing each other within a Dynamic Adaptation Management approach.

References

1. Fuchs, S., Schwarz, J., Flemisch, F.O.: Two steps back for one step forward: revisiting augmented cognition principles from a perspective of (social) system theory. In: Schmorrow, D.D., Fidopiastis, C.M. (eds.) AC 2014. LNCS (LNAI), vol. 8534, pp. 114–124. Springer, Cham (2014). https://doi.org/10.1007/978-3-319-07527-3_11

2. Schwarz, J., Fuchs, S., Flemisch, F.: Towards a more holistic view on user state assessment in adaptive human-computer interaction. In: Proceedings of the 2014 IEEE International Conference on Systems, Man, and Cybernetics (SMC), pp. 1228–1234. IEEE (2014)

3. Schwarz, J., Fuchs, S.: Multidimensional real-time assessment of user state and performance to trigger dynamic system adaptation. In: Schmorrow, D.D., Fidopiastis, C.M. (eds.) AC 2017. LNCS (LNAI), vol. 10284, pp. 383–398. Springer, Cham (2017). https://doi.org/10.1007/978-3-319-58628-1_30

4. Schwarz, J., Fuchs, S.: Validating a "real-time assessment of multidimensional user state" (rasmus) for adaptive human-computer interaction. In: Proceedings of the 2018 IEEE International Conference on Systems, Man, and Cybernetics (SMC), pp. 704–709. IEEE (2018)

5. Fuchs, S., Schwarz, J.: Towards a dynamic selection and configuration of adaptation strategies in augmented cognition. In: Schmorrow, Dylan D., Fidopiastis, Cali M. (eds.) AC 2017. LNCS (LNAI), vol. 10285, pp. 101–115. Springer, Cham (2017). https://doi.org/10.1007/978-3-319-58625-0_7

6. Breton R., Bossé, É.: The cognitive costs and benefits of automation. In: The Role of Humans in Intelligent and Automated Systems. Proceedings of the RTO Human Factors and Medicine Panel (HFM) Symposium (RTO-MP-088). NATO RTO, Neuilly-sur-Seine, France (2003)

7. Veltman, H.J.A., Jansen, C.: The adaptive operator. In: Vincenzi, D.A., Mouloua, M., Hancock, P. (eds.) Human Performance, Situation Awareness, and Automation: Current Research and Trends, vol. 2, pp. 7–10. Lawrence Erlbaum Associates, Mahwah (2004)

8. Billings, C.E., Lauber, J.K., Funkhouser, H., Lyman, G., Huff, E.M.: NASA aviation safety reporting system (Technical report. TM-X-3445). NASA Ames Research Center, Moffett Field, CA (1976)

9. Presagis Stage Homepage. https://www.presagis.com/en/product/stage/. Accessed 21 Jan 2020

10. Schwarz, J.: Benutzerzustandserfassung zur Regelung Kognitiver Assistenz an Bord von Marineschiffen. In: Söffker, D. (ed.) 2. Interdisziplinärer Workshop Kognitive Systeme: Mensch, Teams, Systeme und Automaten, DuEPublico, Duisburg-Essen (2013)

11. Fuchs, S., Hale, K.S., Stanney, K.M., Juhnke, J., Schmorrow, D.D.: Enhancing mitigation in augmented cognition. J. Cogn. Eng. Decis. Making 1(3), 309–326 (2007)

12. Rouse, W.B.: Adaptive aiding for human/computer control. Hum. Factors 30(4), 431–443 (1988)

13. Scerbo, M.W.: Theoretical perspectives on adaptive automation. In: Parasuraman, R., Mouloua, M. (eds.) Automation and Human Performance: Theory and Applications, pp. 37–63. Lawrence Erlbaum Associates Inc., Hillsdale (1996)

14. Scerbo, M.W.: Adaptive automation. In: Karwowski, W. (ed.) International Encyclopedia of Ergonomics and Human Factors, pp. 1077–1079. Taylor & Francis, London (2001)

15. Parasuraman, R., Molloy, R., Singh, I.L.: Performance consequences of automation-induced 'complacency'. Int. J. Aviat. Psychol. 3(1), 1–23 (1993)

16. Posner, M.I.: Orienting of attention. Q. J. Exp. Psychol. 32(1), 3–25 (1980)

17. Kroft, P., Wickens, C.D.: The display of multiple geographical data bases: Implications of visual attention (Report No. ARL-01-2/NASA-01-2). University of Illinois, Aviation Research Laboratory, Savoy, IL (2001)

18. Fuchs, S., Schwarz, J., Werger A.: Adaptive Mensch-Maschine-Interaktion: Ganzheitliche Onlinediagnose und Systemadaptierung (AMIGOS). Final project report (grant no. E/E4BX/EA192/CF215). Fraunhofer FKIE, Wachtberg, Germany (2016)
19. Cohen, J.: Statistical Power Analysis for the Behavioral Sciences. Lawrence Erlbaum Associates, Hillsdale (1988)
20. Fuchs, S., et al.: Anwendungsorientierte Realisierung adaptiver Mensch-Maschine-Interaktion für sicherheitskritische Systeme (ARAMIS). Final project report (grant no. E/E4BX/HA031/CF215). Fraunhofer FKIE, Wachtberg, Germany (2019)
21. Kaber, D.B., Endsley, M.R.: The effects of level of automation and adaptive automation on human performance, situation awareness and workload in a dynamic control task. Theor. Issues Ergon. Sci. **5**(2), 113–153 (2004)

Information-Theoretic Methods Applied to Dispatch of Emergency Services Data

Monte Hancock[1], Katy Hancock[2(✉)], Marie Tree[3], Mitchell Kirshner[4], and Benjamin Bowles[3]

[1] 4Digital, Gloucester, USA
[2] Murray State University, Murray, USA
khancock11@murraystate.edu
[3] Sirius 20, New York, USA
[4] University of Arizona, Tucson, USA

Abstract. The dispatch of emergency services is a complex cognitive task. Current decision support methods rely heavily on manual analysis of maps and map overlays. This paper aims to use call for service/emergency (CFS) dispatch data from various cities to look for patterns not usually amenable to visual analysis that could be used to create decision support tools or methods for dispatchers who must allocate first responder resources under emergency conditions. The authors have collected from the Police Data Initiative, a publicly available government repository that contains millions of annotated 911 dispatch records. The authors have selected three major American cities (Hartford, CT; Lincoln, NE; and Orlando, FL). Three experiments are performed to assess possible benefits of augmenting conventional manual methods with automated analysis derived using methods of data science. In particular, high-dimensional and non-linearly coded information not amenable to manual analysis are considered.

Keywords: 911 system · Crime mapping · Crime informatics

1 Introduction

Responding to emergencies is a problem that exists worldwide. Increasing the speed and efficiency of such responses can result in far more favorable outcomes such as lives being saved and criminal being apprehended. A variety of factors can affect the response to emergency calls, for example, time of day, jurisdiction, the type of crime, the location of the crime, the availability of resources, and dispatcher characteristics.

There is a small but growing body of research that looks at optimizing dispatch services. This research mainly focuses on temporal and spatial characteristics of the call event and on dispatcher characteristics. The current paper looks to compare visual pattern recognition with machine learning and data science that can identify high-dimensional information not readily visible to either the naked eye or to common data analysis techniques. Should the latter technique prove to provide more information, they can be further developed to improve emergency dispatch services.

© Springer Nature Switzerland AG 2020
D. D. Schmorrow and C. M. Fidopiastis (Eds.): HCII 2020, LNAI 12197, pp. 353–370, 2020.
https://doi.org/10.1007/978-3-030-50439-7_24

2 Dispatch

In the United States, calls for emergency services (Call-for-Service, CFS), such as police, fire, and medical services, are handled by 911 dispatch, who may handle one or more types of emergency calls. Dispatchers must make quick decisions, often with only partial information and with limited resources (Shively, 1995). Dispatchers may receive information from a variety of sources, including police officers, fire departments, ambulance dispatch, calls from citizens (both emergency and non-emergency), hazardous materials crews, and from other dispatchers (Shively, 1995; Terrell, McNeese, & Jefferson, 2004). Dispatchers must take this information and decide what priority level to assign, who to dispatch, and whether more information must be gathered. Filtering, analyzing, and acting on all of this information is a highly complex task that is inherently stressful and difficult; as such, software has been created to help optimize these tasks.

The most common group of dispatch technologies is the Computer Aided Dispatch (CAD) system, which is "an interactive, visual based technology used by members of dispatch teams to enter information regarding an emergency situation, receive recommendations for appropriate emergency response, and to share information about a given emergency with fellow colleagues" (Terrell, 2006, p. 61). CAD can retrieve information about the caller, such as phone number and location, record information about the type of incident, suggest call priority and resource allocation, produce photos of the call location, search for warrants and citations, and store all this information (Terrell, 2006). CAD systems are now also integrated with geographic information systems (GIS) technology, making locating callers much easier and more accurate (Neitzel, 2019).

In addition, the United States is seeing massive overhauls to the basic 911 dispatch technology [Police Executive Research Forum (PERF), 2017]. For example, Next-Generation 911 (NG911) systems are being developed and tested (in partial deployment) to include capabilities related to text, photo, and video messaging, information sharing across agencies, and recording response times (Neitzel, 2019). In addition, FirstNet, created by Congress in 2012 and operational in 2017, is a nationwide wireless communication network that can allow police agencies to share information (text, photos, and videos) with other agencies and the public (PERF, 2017). These functions can revolutionize how people communicate with emergency services, alter dispatch decision-making, and change the ability to gather and analyze evidence (PERF, 2017). About 33 states have opted-in to the FirstNet service (PERF, 2017) and 33 states, the District of Columbia, and 2 tribal jurisdictions have plans in place to implement NG911 and in total have received over $100 million USD in federal grants to assist in implementing the system (National Telecommunications and Information Administration, 2019).

Other software exists to help do data analysis of spatial data, the most notable being ArcGIS. Such software can analyze spatial patterns, look at aggregate information, include information about groups living within the mapped area, look for routes, and similar analyses. Such software, while extremely useful, does not perform the direct non-linear mixing of features that are described in this paper.

With all the new developments in dispatch technology creating vast amounts of new information, the integration of these technologies with advanced machine learning and data science techniques can further improve the outcomes of 911 dispatch. The current paper will look at such techniques using software developed by the researchers that has many of CADs basic features as well as software developed that uses advanced machine learning techniques to analyze the data.

2.1 Three Data Experiments

The researchers performed a series of experiments. In the first experiment (Experiment 1), two "analysts" used mapping software created by the researchers. This software created annotated visual overlays of 911-Call-for-Service data on a city map, allowing analysts to visually look for different types of patterns in the CFS data (Fig. 1). It should be noted that research indicates that there are patterns in crime with regard to location and time. For example, crime tends to peak in summer months and drop in winter months. In addition, urban areas, lower income areas, and racially heterogeneous areas are expected to have higher levels of crime generally. Robberies are most common in public places while domestic violence and sexual assault would be more likely in residences. Theory also suggests that an intersection of residential and business zones would have higher levels of crime. As such, the researchers expected some patterns to be easily observable. Analysts were told they could look for whatever patterns they liked, but were told that expected patterns might include demographic and zoning factors. It was also suggested they might look for patterns in the Hartford dataset [1] related to the police shooting of Michael Brown, Jr. in August of 2014-this and subsequent, related events were the impetus for protests, rallies, and riots across the United States in the days and months that followed. (The Hartford data set was the only one used that includes data for 2014.)

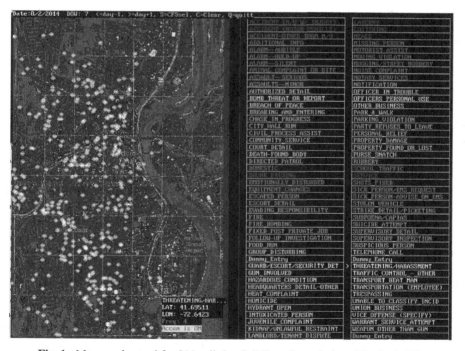

Fig. 1. Map overlay tool for 911 calls by Call Type through time (Hartford, CT)

In the second experiment (Experiment 2), pattern processing software was used to quantify non-linear relationships in the CFS data. These are relationships that would not be readily observable to the naked eye, or to simple analytic techniques (linear correlation, statistical normalization), but which could augment such techniques.

In the third experiment, (Experiment 3), a data imputation technique was used to model annual CFS activity for three cities. Each model captures patterns of CFS activity from relationships implicit in a year and city (the reference set). These models were then applied to other years than those of the reference set, and to other cities than the reference city. The purpose was to create and demonstrate a methodology for comparing CFS data from one time and place to others, at the individual call level, without having to choose a specific pairing of calls. The comparison is objective, and quantifiable at the individual call level (e.g., we could identify anomalous calls... calls that do not fit the model). Experiment Three shows that, with proper information encoding, different cities can be directly compared in an objective, numeric manner.

3 The Data

3.1 Data Collection

Data for this study were gathered from the Police Data Initiative (https://www.policedat ainitiative.org/datasets/calls-for-service/Additional), a site that has publicly available datasets of compiled 911 dispatch calls from a variety of cities. The researchers selected 3 cities due to the differences in their demographic, geographic, and political makeup and due to the types of information in the datasets: Hartford, Connecticut; Lincoln, Nebraska; and Orlando, Florida. All three datasets covered, at the least, an entire calendar year, and included latitude, longitude, and incident information for each call.

3.2 Features, Feature Vectors, and Target Variables

A feature is an attribute of an entity being analyzed. Features are typically represented by numeric values. A feature vector is an ordered array of features. A set of feature vectors is a Feature Set.

Let N be a positive integer; when each feature vector in a feature set consists of N real numbers, the feature set can be regarded as a collection of points in the N-Dimensional Euclidean Space R^N. This geometrizes the associated data analysis problem, facilitating the application of methods from Linear Algebra and Numeric Regression.

For Experiments Two and Three, seven basic features were used. These were chosen because they are known to be correlated with call-for-service activity, and were available for all three cities in the study. They are (Table 1):

The Ground Truth (target variable for analysis and regression experiments) was chosen to be the Call Priority, which is assigned by Dispatch to each call for service at the time the call is received. A Call Priority value of 3 indicates Low Priority; a value of 2 is Medium Priority; and a value of 1 indicates High Priority.

Call type-call type was operationalized as criminal, bad non-criminal, service, and administrative. Criminal calls involved those with putative criminal intent, such as robbery, assault, or theft. Bad non-criminal were calls that involved "bad" but likely non-criminal behavior, such as loitering and suspicious person calls. Service calls were calls for service such as traffic problems, roadway obstructions, etc. Administrative calls were calls that did not fit into the other categories, such as "food run" and comfort stops.

Table 1. The seven basic features for Call-for-Service data

FEA 1	Call_Type (1 = criminal, 2 = bad-non-criminal, 3 = service call, 4 = admin)
FEA 2	DOW (day of week, 1 = Sunday)
FEA 3	MOY (Month of Year, 1 = January, 2 = February, ... 12 = December)
FEA 4	DOM (Day of Month, 1–31)
FEA 5	Longitude (decimal degrees, West indicated by negative values)
FEA 6	Latitude (decimal degrees, North indicated by positive values)
FEA 7	HOD (Hour of Day, 0–24, decimal value)

4 Demographics of the Subject Cities

4.1 Hartford, CT

Hartford is a New England city in the Northeastern United States about 100 miles inland from Boston, Massachusetts. It has a population of about 122,587 people, and it has the unfortunate distinction of being one of the poorest cities in the U.S. Another distinction of Hartford is its racial makeup: Caucasians, Blacks/African Americans, and Hispanics are almost equally distributed (33.1%, 36.9%, and 44.5%, respectively). The median household income ($34,338) is significantly below the national average and the poverty level is a staggering 30.1%. The percentage of high school graduates only comes in at 74.2% of the population, and, correspondingly, only 16.8% of residents hold advanced degrees. Despite the low number of graduates, the area does offer a number of higher education choices, including the University of Hartford, the University of Connecticut, and Trinity College.

4.2 Lincoln, NE

Nebraska is a state in the Midwestern United States, and Lincoln is the state's capitol. According to world population review, of the 200 largest cities in the U.S., Lincoln comes in 72nd with about 287,401 people. An interesting fact about this area is it is not suburbanized like many other cities. Lincoln's population is predominately Caucasian (85.2%). The median household income is reflective of a middle class city ($55,224); however, the poverty level (14.2%) is slightly above the national average. The University of Nebraska-Lincoln is by far the largest higher learning institution in the area, with over 25,000 students. Over 90% of the population finished high school, and close to 40% of people hold degrees at the bachelor's level or higher.

4.3 Orlando, FL

Orlando, Florida is located in the Southeastern most part of the United States, approximately 200 miles north of Miami. Similar to Lincoln, Orlando ranks 77th out of 200 large cities with about 285,700 people, and the metropolitan area ranks as one of the fastest growing populations in the nation. In contrast to Lincoln, Orlando is more racially diverse, with Caucasians (60.7%) and Blacks/African Americans (25.4%) making up a majority of the city, with a significant percentage of Hispanics among them (31.1%). The median household income ($48,511) is below the national average with the city having a poverty level (18.2%) above the national average. Orlando is similar to Lincoln in regards to both high school and higher education graduate percentages, and the University of Central Florida enrolls the most students in the U.S. at over 60,000 students.

5   Experiments

5.1   Experiment 1: Visual Data Analysis Using Map Overlays

Hartford, Connecticut. Upon visual analysis, Analyst 1 found that both serious assault and homicide were clustered in their own specific blocks of the Hartford map. In addition, robberies seemed to map along major roads. When mapping larceny and muggings (as these crimes are similar to robbery), a similar, albeit less apparent, pattern of mapping along roadways occurred. Roads in the north central area were problematic for theft; this area seems to be zoned as industrial and neighborhood mix. Larceny and robbery seemed more prominent in the southern part of the map; this area has mixed zoning, is racially diverse, and has a low income level. Analyst 1 also observed that one area of serious crime occurred in an area of the city with a clear income divide. A mapping of alarm hold-ups indicated mapping along roads that also mapped for robbery as well as in blocks that mapped for serious crimes; Analyst 1 theorized that cluster analysis would reveal a mixture of serious crime types and theft-as if there was a link somehow between the serious crimes and the theft.

In terms of the impact of the Ferguson, Missouri shooting event (August 9, 2014), Analyst 1 indicated that there seemed to have more shots fired in 2014 and 2015 (as compared to 2012 and 2013), but fewer group disturbances. The Analyst also found clear evidence for an increase in picketing, especially on the anniversary of the Ferguson event (August 9, 2015). Looking for minor assault from the date of the shooting through the end of the month, the Analyst found the most calls for minor assault on August 16, a day that had a rally scheduled (thestruggle.org). The Analyst also discovered that on the day of the rally, while it was tied with another day for the highest calls for loitering and mugging, the city had the least robbery and directed patrol calls in August. While acknowledging she has no way of knowing the statistical significance of the findings, the Analyst's assessment was "Overall, [I] would say the level of law and order typical of Hartford, CT was maintained in August of 2014. Some people chose to exercise their freedom of speech without obtaining proper permits, possibly blocking businesses, etc. When proper process was followed for the 8/16 rally, a few chose the action of minor assault and used it as an opportunity for street crimes. Probably more interesting is that robberies and people requesting patrols went down as people were occupied with other events."

Analyst 2 made some similar observations in Hartford to Analyst 1. For example, he noted that robberies were frequent and traced the structure of the roads. Similar to the findings of Analyst 1 with mugging and larceny, Analyst 2 observed lines in data being quite common and indicative of high-crime streets, especially Russ Street and

Main Street. Uniquely, he noted a square almost free of robbery that was located near a park, cemetery, and a middle school; he also noted that this was almost free of juvenile complaints. Analyst 2 considered this curious as it was near the middle school. Analyst 2 did not look at the Ferguson event.

Lincoln, Nebraska. Analyst 1 first looked at indecent exposure from 11/1-3/31 and 4/1-10/31 for all the years in the dataset. As she expected, indecent exposure was down in the colder months versus the warmer months. Indecent exposure incidents in the inner city seemed tied to roads while incidents in the outskirts did not seem to be tied to roads, although the Analyst suggested the map might not show smaller roads. Continuing with sex crimes, the Analyst noted that most prostitution incidents occurred in the outskirts of the city. Deciding to look at what other crimes happened on the outskirts of town, the Analyst discovered that kidnapping and, to a lesser extent, bomb threats and "theft-coin operated." The Analyst described the outskirts of town as being a diverse zoning area, with business, agriculture, residential, development areas, and more. There are also many roads into and out of town as well as the airport.

For the Lincoln map, Analyst 2 observed that burglaries are far more common north of the city's center. When looking at DUI calls, he noted that these were not as correlated with major roads as expected-indeed, many of these calls were located within neighborhoods. Another surprising finding was that drug calls were not linking with population centers; these calls were common on the outskirts of town. Finally, Analyst 2 noted that the southern exclave near 40.7349, -96.7727 reports some crimes (health, fire, drugs), but not others (DUI). The presence of drugs but not DUI was reported as surprising.

Orlando, Florida. For Orlando, Analyst 1 randomly selected 6 weeks (Sunday–Monday) from each of the 3 years available (2015, 2016, and 2017) and randomly selected call types (prostitution, missing persons, hit and run, arson). It would appear that the influence of prostitution was spreading and becoming more common in fringe areas over time. To investigate this theory, the analyst mapped the cumulative plots of each year for this type of call. Although the cumulative plots were not as convincing of the theory of spreading as the random week data, it still seemed that incidents in the fringes are occurring more often over time. Juvenile kidnapping were observed in neighborhoods and major roads but seemed to have an inherent randomness. Similarly, hit and run seemed to have no location pattern, occurring both in neighborhoods and along major roads. For arson, the Analyst concluded there was a pattern but not enough information to make any other conclusions.

In contrast, Analyst 2 first noted that drownings in Orlando are clustered around population centers rather than lakes. In addition, areas with significant housing make many more suspicious persons calls, whereas areas with larger traffic but few houses (e.g. the Orlando airport, Disney, etc.) make much fewer suspicious persons calls. Another observation noted by Analyst 2 is that many kidnappings are in unusually straight lines-for instance, one long line is almost parallel to 17, running north to south (north of I4 and south of 423). Finally, Analyst 2 observed that prostitution calls were not present at all in some highly populated areas, suggesting to him that prevention efforts might have been successful in those areas.

Algorithm for Assessing Feature Performance. To assess the performance of a member of a sports team, statistics are computed from the game play. However, it could be that these are biased by the presence of other players on the team. Sometimes synergistic affects can hide the true contribution of a single player. So, it is best to assemble statistics on the player as they perform with a range of other players. If teams generally perform better when a player is present, and worse when they are not, we are justified in attributing the difference to their presence/absence.

Features can be evaluated in the same way. Selectively include/suppress features, and accumulate performance statistics by attributing outcomes to each feature depending upon whether that feature was used or not. In this way, we identify, not just the best features, but the best set of features.

For Experiment two, this feature evaluation technique was applied to the seven basic features, and the larger FLINKED set (described below) of features.

Algorithm:

'-Phase A:

'Step 1: read in the data file

'Step 2: uniform randomly segment into equal size calibration, training, validation files

'Step 3: compute class centroids and standard deviations for each of the segments

'Step 4: Z-Score each feature using its class mean and standard deviation)

'

'-Phase B:

'Step 1: uniform randomly select a subset of the features to test

'Step 2: use the training set as the reference for an N-Nearest Neighbor classifier for the vectors in the validation set

'Step 3: for each feature, aggregate performance statistics (used, and not used)

'

'-Repeat Phase B for L epochs

'

' Compute performance metrics for the "best" feature clique using the validation file.

Figure 2 and Fig. 3 show the results of this assessment applied to the Hartford, CT CFS data.

5.2   Experiment Two: FLINKING Feature Sets to Expose Non-linear Patterns

The Transform. For this experiment, the non-linear FLINKING transform was applied to expose non-linear patterns not easily discernable by manual inspection. This is a polynomial transform, F, defined here:

Let a feature set consist of N real features, $a_{j1}, a_{j2}, ..., a_{jN}$, so that the feature space is R^N.

The j-th feature vector is then a vector of attributes which we denote here as

$$\{A_j\} = \{(a_{j1}, a_{j2}, a_{j3}, ..., a_{jN})\}$$

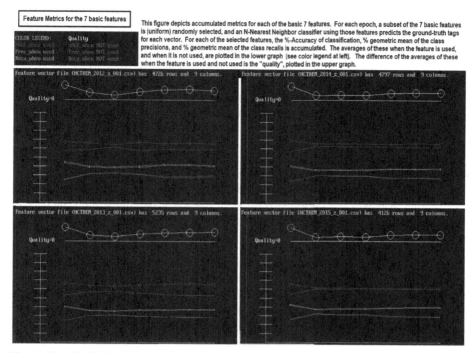

Fig. 2. Plots for 2012, 2013, 2014, and 2015, showing the relative contribution of the seven basic features to prediction of the Call Priority Shown are the respective contributions when each feature was used/not-used to classification accuracy, precision, recall, and overall quality.

Feature File: HCTREM_2012_z_001.csv
Norm: 1
Trials: 3210

Metric	Fea 1	Fea 2	Fea 3	Fea 4	Fea 5	Fea 6	Fea 7
%ACC_when used	67.5	66.4	64.3	67.2	65.9	66.2	65.9
Prec_when used	45.1	40.1	39.3	40.6	42.2	42.3	41.5
Reca_when used	43.9	38.9	38.4	39.2	41.5	41.5	40.7
%ACC_when NOT used	57.4	58.6	61.6	57.8	59.2	59.0	59.3
Prec_when NOT used	25.5	31.9	33.1	31.5	29.0	29.0	30.0
Reca_when NOT used	24.9	31.3	32.2	31.1	27.5	27.8	28.8
Rel_Quality	16.2	7.8	5.0	8.9	11.3	11.4	10.0

Feature File: HCTREM_2014_z_001.csv
Norm: 1
Trials: 3655

Metric	Fea 1	Fea 2	Fea 3	Fea 4	Fea 5	Fea 6	Fea 7
%ACC_when used	69.7	65.7	65.7	65.8	65.9	66.1	66.0
Prec_when used	41.3	36.5	36.9	36.9	38.0	38.2	38.5
Reca_when used	40.4	35.8	36.0	36.2	37.6	37.7	37.8
%ACC_when NOT used	57.4	62.7	62.7	62.6	62.4	62.2	62.3
Prec_when NOT used	24.0	30.3	29.9	29.9	28.2	28.1	27.5
Reca_when NOT used	23.9	30.0	29.9	29.7	27.5	27.5	27.1
Rel_Quality	15.3	5.0	5.4	5.6	7.8	8.1	8.5

Feature File: HCTREM_2013_z_001.csv
Norm: 1
Trials: 3076

Metric	Fea 1	Fea 2	Fea 3	Fea 4	Fea 5	Fea 6	Fea 7
%ACC_when used	66.9	64.0	63.6	64.8	65.8	66.2	66.0
Prec_when used	40.9	36.3	35.9	37.1	37.2	37.4	37.7
Reca_when used	40.1	35.6	35.3	36.3	36.8	36.9	37.1
%ACC_when NOT used	52.9	56.7	57.4	55.9	54.0	53.8	53.8
Prec_when NOT used	24.3	30.2	30.9	29.4	28.9	28.7	28.3
Reca_when NOT used	24.2	30.2	30.7	29.3	28.4	28.4	28.0
Rel_Quality	15.5	6.2	5.2	7.9	9.5	9.9	10.3

Feature File: HCTREM_2015_z_001.csv
Norm: 1
Trials: 4941

Metric	Fea 1	Fea 2	Fea 3	Fea 4	Fea 5	Fea 6	Fea 7
%ACC_when used	65.8	61.6	61.6	61.3	61.3	61.5	61.2
Prec_when used	42.9	37.8	38.7	38.4	39.7	40.0	40.0
Reca_when used	42.2	37.1	37.8	37.8	39.1	39.4	39.4
%ACC_when NOT used	54.7	60.2	60.3	60.7	60.6	60.3	60.8
Prec_when NOT used	25.8	32.5	31.3	31.7	29.9	29.6	29.3
Reca_when NOT used	25.2	31.8	30.8	31.0	29.0	28.8	28.4
Rel_Quality	15.1	4.0	5.3	4.7	6.9	7.4	7.4

Fig. 3. Table showing the results depicted in Fig. 2 in numeric form.

Then define:

$$F(A_j) = ((a_{j1}, a_{j2}, a_{j3}, \ldots, a_{jN})) = (\prod_{k=1}^{N} a_{j1}a_{jk}, \prod_{k=2}^{N} a_{j2}a_{jk}, \ldots, \prod_{k=N}^{N} a_{jN}a_{jk})$$

That is, each transformed feature is replaced with a collection of products of the original features of vector A_j. This corresponds to a polynomial kernel (vis. the Vapnik "Kernel Trick"). Notice that it increases the dimensionality of the problem from N to N(N + 3)/2.

An early instance and description of this transform for data preparation can be found in Pao, where it is referred to as the Functional Link. This name has led us to use the term "FLink" as a transitive verb to describe this data operation, which has valuable properties. For example, it makes the Parity Two problem linearly separable without increasing the dimension of the feature space; and, supports a wide variety of straightforward generalizations.

FLINKING Results. FLINKING the seven basic features generates 28 additional features that are pairwise products of the seven basic features (giving a total of 35 features). Ground Truth is again Call Priority (priority assigned by Dispatch to call, 1–3, 1 highest). Figure 4 below shows the progression from the original seven basic features (upper left box), to the intermediate form of z-scored columns (upper right box), to the final 35 FLINKED features (bottom box):

Incident	CallType	DOW	MOY	DOM	Longitude	Latitude	HOD	Priority
A	2	1	1	1	-72.6862	41.7453	22.48333	1
B	4	1	1	1	-72.67854	41.76833	1.4	3
C	4	6	1	6	-72.68168	41.79466	2.083333	2
D	2	5	1	5	-72.71104	41.76634	18.26667	3
E	2	1	1	1	-72.66802	41.79485	0.75	2
F	2	5	1	5	-72.68929	41.75811	17.03333	2
G	1	3	1	3	-72.66071	41.79799	22.25	2
H	2	5	1	5	-72.71246	41.74244	23.75	2
I	4	4	1	4	-72.65443	41.78671	23.2	3
J	4	1	1	1	-72.68065	41.7801	0.25	1

Raw form: Feature Data for 10 randomly selected Incidents
first week of January, 2012, Hartford, Connecticut

Incident	CallType	DOW	MOY	DOM	Longitude	Latitude	HOD	Priority
A	-1.047199	-1.11874	-1.61222	-1.66254	0.9077764	0.01356193	1.40864	1
B	0.8083974	-1.11874	-1.61222	-1.66254	1.037091	0.8084489	-1.6774	3
C	0.8083974	1.431631	-1.61222	-1.09542	0.9841543	1.71723	-1.57737	2
D	-1.047199	0.921557	-1.61222	-1.20884	0.4885352	0.7397175	0.791434	3
E	-1.047199	-1.11874	-1.61222	-1.66254	1.214705	1.723813	-1.77254	2
F	-1.047199	0.921557	-1.61222	-1.20884	0.8556128	0.4557071	0.610906	2
G	-1.974997	-0.09859	-1.61222	-1.43569	1.338094	1.832177	1.374487	2
H	-1.047199	0.921557	-1.61222	-1.20884	0.4644497	-0.08519	1.594047	2
I	0.8083974	0.411482	-1.61222	-1.32227	1.444225	1.442831	1.513542	3
J	0.8083974	-1.11874	-1.61222	-1.66254	1.001542	1.21478	-1.84573	1

Z-Scored: Feature Data for 10 randomly selected Incidents
First week of January, 2012, Hartford, Connecticut

Incident	CallType	DOW	MOY	DOM	Longitude	Latitude	HOD	FEA8	FEA9
A	-1.0472	-1.11874	-1.61222	-1.66254	0.9077764	0.0136	1.40864	1.096626	1.171545
B	0.808397	-1.11874	-1.61222	-1.66254	1.037091	0.8084489	-1.6774	0.653506	-0.90439
C	0.808397	1.431631	-1.61222	-1.09542	0.9841543	1.71723	-1.57737	0.653506	1.157327
D	-1.0472	0.921557	-1.61222	-1.20884	0.4885352	0.7397175	0.791434	1.096626	-0.96505
E	-1.0472	-1.11874	-1.61222	-1.66254	1.214705	1.723813	-1.77254	1.096626	1.171545
F	-1.0472	0.921557	-1.61222	-1.20884	0.8556128	0.4557071	0.610906	1.096626	-0.96505
G	-1.975	-0.0986	-1.61222	-1.43569	1.338094	1.832177	1.374487	3.900613	0.194719
H	-1.0472	0.921557	-1.61222	-1.20884	0.4644497	-0.0852	1.594047	1.096626	-0.96505
I	0.808397	0.411482	-1.61222	-1.32227	1.444225	1.442831	1.513542	0.653506	0.332641
J	0.808397	-1.11874	-1.61222	-1.66254	1.001542	1.21478	-1.84573	0.653506	-0.90439

Incident	FEA10	FEA11	FEA12	FEA13	FEA14	FEA15	FEA16	FEA17	FEA18	FEA19
A	1.688315	1.741011	-0.95062	-0.0142	-1.47513	1.251582	1.803657	1.859953	-1.01557	-0.0152
B	-1.303315	-1.343994	0.838382	0.653548	-1.356	1.251582	1.803657	1.859953	-1.16024	-0.90445
C	-1.303315	-0.8855347	0.795588	1.388204	-1.27515	2.049567	-2.308104	-1.56824	1.408546	2.45844
D	1.688315	1.2659	-0.51159	-0.77463	-0.82879	0.8492668	-1.485752	-1.11402	0.450213	0.681692
E	1.688315	1.741011	-1.27204	-1.80518	1.856201	1.251582	1.803657	1.859953	-1.35894	-1.9285
F	1.688315	1.2659	-0.896	-0.47722	-0.63974	0.8492668	-1.485752	-1.11402	0.788496	0.41996
G	3.18413	2.83549	-2.64273	-3.61854	-2.71461	0.00972	0.1589523	0.141548	-0.13193	-0.18064
H	1.688315	1.2659	-0.48637	0.0892	-1.66928	0.8492668	-1.485752	-1.11402	0.428017	-0.0785
I	-1.303315	-1.068919	1.167508	1.166381	1.223544	0.1693177	-0.6634	-0.54409	0.594273	0.5937
J	-1.303315	-1.343994	0.809644	0.982025	-1.49208	1.251582	1.803657	1.859953	-1.12047	-1.35902

Incident	FEA20	FEA21	FEA22	FEA23	FEA24	FEA25	FEA26	FEA27
A	-1.5759	2.599253	2.680382	-1.463535	-0.0219	-2.27104	2.764043	-1.509216
B	1.876572	2.599253	2.680382	-1.672019	-1.3034	2.704332	2.764043	-1.724206
C	-2.25822	2.599253	1.766058	-1.586673	-2.76855	2.543074	1.199945	-1.078062
D	0.729351	2.599253	1.948922	-0.7876263	-1.19259	-1.27597	1.461304	-0.590563
E	1.983012	2.599253	2.680382	-1.958372	-2.77917	2.857723	2.764043	-2.019497
F	0.562985	2.599253	1.948922	-1.379436	-0.7347	-0.98492	1.461304	-1.034302
G	-0.13551	2.599253	2.314653	-2.157302	-2.95387	-2.21598	2.061214	-1.921092
H	1.469005	2.599253	1.948922	-0.7487952	0.137345	-2.56995	1.461304	-0.561447
I	0.622796	2.599253	2.131788	-2.328408	-2.32616	-2.44016	1.748395	-1.909654
J	2.064888	2.599253	2.680382	-1.614706	-1.95849	2.975715	2.764043	-1.665105

Incident	FEA28	FEA29	FEA30	FEA31	FEA32	FEA33	FEA34	FEA35	Priority
A	-0.0225	-2.34192	0.824058	0.0123	1.27873	0.000184	0.0191	1.984267	1
B	-1.344079	2.78874	1.075558	0.838435	-1.739612	0.6535897	-1.35609	2.813658	3
C	-1.881088	1.727887	0.96856	1.690019	-1.552379	2.948879	-2.70871	2.488109	2
D	-0.894203	-0.95672	0.238667	0.361378	0.3866434	0.547182	0.585438	0.626368	3
E	-2.86591	2.946919	1.475508	2.093924	-2.153112	2.971531	-3.05553	3.141895	2
F	-0.5508788	-0.73849	0.732073	0.389909	0.5226991	0.207669	0.278394	0.373206	2
G	-2.630444	-1.97334	1.790496	2.451625	1.839193	3.356873	2.518304	1.889215	2
H	0.1029814	-1.92695	0.215714	-0.0396	0.7403547	0.00726	-0.1358	2.540986	2
I	-1.907811	-2.00131	2.085786	2.083773	2.185895	2.081761	2.183785	2.29081	3
J	-2.019622	3.068594	1.003086	1.216653	-1.848571	1.47569	-2.24215	3.406701	1

Z-Scored and FLINKED: Feature Data for 10 randomly selected Incidents
First week of January, 2012, Hartford, Connecticut

Fig. 4. The two-step progression from the seven basic features to FLINKED data

These product features are not amenable to visual analysis, as their presentation varies in a non-linear way across maps and map overlays.

The Information Assessment process was applied to the FLINKED feature set. Results are shown in Fig. 5. Results indicated that there were some non-linear combinations of features that yielded better results than the basic seven features. In particular, some of these bilinear features show improved precision (specificity), which equates to reduced false alarm rates when the goal is event detection.

The most informative combination was call type and the day of the week. This would theoretically make sense, as the call type reflects the level of seriousness of what is going on and the day of the week might reflect scheduling. If more resources (such as officers) are scheduled for a particular day, then dispatchers may be more willing to assign a higher priority.

Day of the week can also reflect scheduling of dispatchers. While CAD can recommend a priority level, dispatchers always have the power to override that recommendation, so day of the week might reflect the decisions of different dispatchers.

The second most powerful feature combination was call type by itself-again, reasonable as call type indicates the level of severity of what is happening. Finally, the third most powerful combination was call type and latitude and fourth was call type and longitude. Indeed, location would potentially have a strong influence on priority of the call. If certain locations are perceived as more "dangerous" or are known to have more crimes happen there, calls about events in those places could receive a higher priority.

The graphs and table below show that the added features (particularly FEA8 through FEA14) contribute significantly to the performance of the classifier in all metrics.

5.3 Experiment 3: Data Conformation for Year-to-Year and City-to-City Modeling

Experiment three creates a number of city-to-self across years, and city-to-city models for the purpose of assessing the stability of the call-for-service information binding over time, and across space. Two questions about stationarity of call-for-service data are addressed:

1. If a pattern models is derived from call-for-service data for a city, how well can this model reconstruct call-for-service data for the city for a different year?
2. If a pattern model is derived from call-for-service data for a city, how well can this model reconstruct call-for-service data for a different city?

Conforming Call Type Codes. Data representation schemes vary from jurisdiction to jurisdiction, since each police force establishes a data representation schemes that suits its jurisdiction's patterns of activity. In particular, the numeric codes representing Call Type are unique to each city.

To make comparison of cities possible, it was necessary to create a common a mapping from the disparate call-type codes used in multiple cities to a uniform set of call-type codes that could be used for all cities. This was accomplished by creating a set of 27 general call-type codes, and assigning each of the city-specific codes to one of these (tables x, y, and z). The assignment was performed manually according to our intuitive interpretation of the meaning of the original codes.

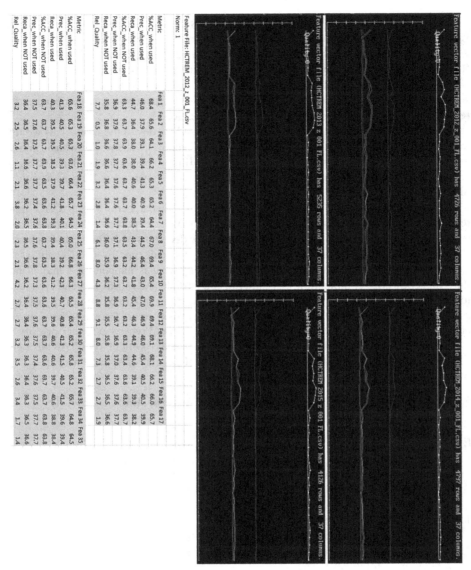

Fig. 5. The relative information content of the FLINKED features. Notice the elevated performance of features 8 – 14.

Uniform Recoding of Call Type. Tables were created for Orlando, Florida and Hartford, Connecticut that mapped the Call Type codes each city used to a common set of 27 codes. The table of common codes is Table 2; the corresponding Orlando table is Table 3; the corresponding Hartford table is Table 4. Conformation is perform by replacing the Call Type codes for each city with the corresponding value in the common code table.

Table 2. Commom Call Type Codes

Money	1
Sex	2
Suspicious	3
Accident	4
Theft	5
Arson_bomb	6
Alarm	7
Assault	8
Fire_expl	9
robbery	10
Police Business	11
Trespass/Vandalism/Disturbance/Alcohol	12
Medical	13
Missing	14
DUI	15
Kidnap	16
Weapon	17
Domestic	18
Drugs	19
Police Business Escalated	20
Suicide	21
Child	22
Other	23
Body	24
Homicide/Murder	25
burglary	26
mentally ill	27

Table 3. Orlando Call Type Codes

Row Labels Orlando	Call_Type	marie code
911_hang_up	1	11
Abandoned_vehicle	2	11
Accident_(general_disturbance)	3	4
Accident_(injuries)	4	4
Accident_(minor)	5	4
Accident_with_road_blockage	6	4
Arson_fire	7	6
Attempted_rape	8	2
Attempted_suicide	9	21
Bad_check_passed	10	11
Bank_alarm	11	7
Bank_robbery	12	10
Battery	13	8
Battery_-_cutting	14	8
Battery_-_fight_in_progress	15	8
Battery_-_shooting	16	8
Bomb_explosion	17	6
Bomb_threat	18	6
Car_jacking	19	10
Check_the_well_being_of	20	11
Child_abuse	21	22
Child_neglect	22	22
Commercial_alarm	23	7
Commercial_B&E	24	12
Commercial_robbery	25	10
Community_Orientated_Policing_detail	26	11
Dead_person	27	24
Deviate_sexual_activities	28	2
Direct_traffic	29	11
Disabled_occupied_vehicle	30	11
Domestic_disturbanc	31	18
Door_alarm	32	7
Drowning	33	24
Drug_violation	34	19
Drug_violation_-_armed	35	17
Drunk_driver	36	15
Escaped_prisoner	37	20
Fire	38	9
Fraud-counterfeit	39	11
Fugitive_from_justice	40	20
General_disturbance	41	12
General_disturbance_-_armed	42	17
General_investigation	43	11
Hit_and_run_(injuries)	44	4
Hit_and_run_(minor)	45	4
Hit_and_run_with_road_blockage	46	4
Hold-up_alarm	47	7
Home_Invasion	48	12
House/business_check	49	11
Immediate_backup	50	20
Industrial_accident	51	4
Kidnapping	52	16
Law_Enforcement_Officer_escort	53	11
Liquor_law_violation	54	12
Man_down	55	20
Mentally-ill_person_(non-violent)	56	27
Mentally-ill_person_(violent)	57	20
Missing_person_-_adult	58	14
Missing_person_-_juvenile	59	14
Murder	60	25
Noise_ordinance_violation	61	12
Non-emergency_assistance	62	11
Obstruction_on_highway	63	11
Other_sex_crimes	64	2
Person_robbery	65	10
Prostitution	66	2
Prowler	67	12
Rape	68	2
Reckless_vehicle	69	4
Residential_alarm	70	7
Residential_B&E	71	12
Rush-Officer_needs_help	72	20
Security_checkpoint_alarm	73	7
Sick_or_injured_person	74	13
Stalking	75	8
Stolen_vehicle	76	5
Suicide	77	21
Suspicious_car/occupant_armed	78	17
Suspicious_incident	79	3
Suspicious_person	80	3
Suspicious_vehicle	81	3
Theft	82	5
Theft_by_shoplifting	83	5
Threats/assault	84	8
Threats/assaults_-_armed	85	8
Trespasser	86	12
Trespasser_-_armed	87	17
Unknown_trouble	88	23
Vandalism/criminal_mischief	89	12
Vehicle_alarm	90	7
Vehicle_B&E	91	12
Weapons/armed	92	17

Table 4. Hatfrod Call Type Codes

Row Labels Hartford	Call_Type	marie code
ACCIDENT_(M/V_W/_INJURY)	1	4
ACCIDENT_(MOTOR_VEHICLE)	2	4
ACCIDENT-OTHER_THAN_M/V	3	4
ADDITIONAL_INFO	4	11
ALARM--AUDIBLE	5	7
ALARM--HOLD-UP	6	7
ALARM--SILENT	7	7
ANIMAL_COMPLAINT_OR_BITE	8	12
ASSAULT--SERIOUS	9	8
ASSAULTS--MINOR	10	8
AUTHORIZED_DETAIL	11	11
BOMB_THREAT_OR_REPORT	12	6
BREACH_OF_PEACE	13	12
BREAKING_AND_ENTERING	14	12
CHASE_IN_PROGRESS	15	20
CITY_HALL_RUN	16	11
CIVIL_PROCESS_ASSIST	17	11
COMMUNITY_SERVICE	18	11
COURT_DETAIL	19	11
DEATH-FOUND_BODY	20	24
DIRECTED_PATROL	21	11
DOMESTIC	22	18
DRUNK_DRIVING	23	15
EMOTIONALLY_DISTURBED	24	27
EQUIPMENT_CHANGES	25	11
ESCAPED_PERSON	26	20
ESCORT_DETAIL	27	11
EVADING_RESPONSIBILITY	28	20
FIRE	29	9
FIRE_BOMBING	30	6
FIXED_POST_PRIVATE_JOB	31	11
FOLLOW-UP_INVESTIGATION	32	11
FOOD_RUN	33	11
GROUP_DISTURBING	34	12
GUARD/ESCORT/SECURITY_DET	35	11
GUARD/ESCORT/SECURITY_DET	36	11
GUN_INVOLVED	37	17
HAZARDOUS_CONDITION	38	11
HEADQUARTERS_DETAIL-OTHER	39	11
HEAT_COMPLAINT	40	11
HOMICIDE	41	25
HYDRANT_OPEN	42	11
INTOXICATED_PERSON	43	12
JUVENILE_COMPLAINT	44	22
KIDNAP/UNLAWFUL_RESTRAINT	45	16
LANDLORD/TENANT_DISPUTE	46	11
LARCENY	47	5
LOITERING	48	12
MEALS	49	11
MISSING_PERSON	50	14
MOTORIST_ASSIST	51	11
MOVING_VIOLATION	52	4
MUGGING/STREET_ROBBERY	53	10
NOISE_COMPLAINT	54	12
NOTARY_SERVICES	55	11
NOTIFICATION	56	11
OFFICER_IN_TROUBLE	57	20
OFFICERS_PERSONAL_USE	58	11
OTHER_BUSINESS	59	11
PARK_&_WALK	60	11
PARKING_VIOLATION	61	13
PARTY_REFUSES_TO_LEAVE	62	12
PERSONAL_RELIEF	63	11
PROPERTY_DAMAGE	64	12
PROPERTY_FOUND_OR_LOST	65	11
PURSE_SNATCH	66	5
ROBBERY	67	10
SCHOOL_TRAFFIC	68	11
SERVICE	69	11
SHOTS_FIRED	70	17
SICK_PERSON/EMS_REQUEST	71	13
SICK_PERSON-ADVISE_ON_EMS	72	13
STOLEN_VEHICLE	73	5
STRIKE_DETAIL/PICKETING	74	12
SUBPOENA/CAPIAS	75	11
SUICIDE_ATTEMPT	76	21
SUPERVISORY_DETAIL	77	11
SUPERVISORY_INSPECTION	78	11
SUSPICIOUS_PERSON	79	3
SUSPICIOUS_VEHICLE	80	3
TELEPHONE_CALL	81	11
THREATENING/HARASSMENT	82	8
TRAFFIC_CONTROL_-_OTHER	83	11
TRANSPORT_BEAT_MAN	84	11
TRANSPORTATION_(EMPLOYEE)	85	11
TRESPASSING	86	12
UNABLE_TO_CLASSIFY_INCID	87	23
UNION_BUSINESS	88	11
VICE_OFFENSE_(SPECIFY)	89	2
WARRANT_SERVICE_ATTEMPT	90	11
WEAPON_OTHER_THAN_GUN	91	17

5.4 Information Model for Year-to-Year Comparison

The Lincoln, NE CFS data set consists of 337,730 calls for service records. Of these, 115,871 are for 2017 calls; 117,120 are for 2018 calls; and 104,739 are for 2019 calls (but the 2019 data ends at 11/14/19).

In practical applications it is not uncommon for some collected records to have "missing" values: values that could not be collected, were collected incorrectly, or were corrupted or lost after collection. Rather than just discard such partial records (which can comprise a significant proportion of the entire corpus) mathematical methods called <u>data imputation techniques</u> are sometimes used to fill missing fields with consistent estimated values derived from the contextual patterns present in the data.

The weakest imputation technique is to replace missing values with some fixed agnostic "fill value" chosen for that purpose. This completely ignores the data context, and can lead to confusion later in the analysis ("Is this a REAL zero… or a FILL zero?"). A better technique, which attempts to retain some of the information context of the data, is to replace missing values with their population means ($O(n)$ in the number of records). This approach is fast and simple; but, because it is single-field oriented, it ignores the intra-record information-context of the data. This readily produces inconsistent and anomalous records.

For numeric data, a context sensitive method is the nearest neighbor normalization technique. This can be applied reasonably efficiently (at worst $O(n^2)$ in the number of records) even to large data sets having many dimensions.

The following describes the nearest neighbor normalization imputation method (from [5]).

This technique proceeds in the following manner for each feature to be imputed in a given vector, V_1:

1. From a reference set of feature vectors, find the one, V_2, which:

   a. Shares a sufficient number of populated fields with the vector to be imputed (this increases the likelihood that the nearest vector is representative of the vector being processed).
   b. Has a value for the feature being imputed, F_m.
   c. Is nearest the vector to be imputed (possibly a weighted distance).

2. Compute the weighted norms of the vector being imputed, V_1, and the matching vector found in step 1, V_2, in just those features present in both.
3. Form the normalization ratio $Rn = |V_1|/|V_2|$.
4. Create a preliminary fill value $P = R_n*F_m$.
5. Apply a consistency test to P to obtain F'_m, the final fill value.
6. Fill the gap in V_1 with the value F'_m.

In this approach, the reference set establishes an implicit model of intra-vector relationships among the features in each feature vector: certain combinations of feature values tend to co-occur, indicating multi-factor correlations, both linear and non-linear.

In this way, the imputation scheme just described allows a reference file to serve as an information model for the problem space.

To assess the information-theoretic commensurability/consistency of two data sets, one is used as reference to completely reconstruct the other, feature-by-feature. If the files encode similar patterns, the reconstruction of the target file will be similar to the original. High reconstruction error indicates information-theoretic differences, quantifies them, and indicates in which vectors and for which features they occur.

It is important to note that the imputation approach is an information theoretic, not a statistical approach. The vectors for reconstruction are chosen based upon their proximity to the target, not parametrics. This makes the imputation approach a valuable and distinct adjunct to conventional coding techniques.

Reconstruction error is measured here as the RMS Error of the reconstruction based upon the z-scores of the original features.

First, we reconstruct data for a single city for different years, and sort the RMS z-score reconstruction error in ascending order:

Figure 6 shows the sorted (ascending) imputation errors for a randomly selected subset of the Lincoln, NE 2018 Call-for-Service data imputed from Lincoln, NE 2017 Call-for-Service data.

Figure 7 shows the sorted (ascending) imputation errors for a randomly selected subset of the Lincoln, NE 2019 Call-for-Service data imputed from Lincoln, NE 2017 Call-for-Service data.

The reconstruction is fairly good (RMS z-score error < 0.5) for about 95% of the data. This suggests that the information model provided by the 2017 CFS data is stationary for Lincoln, NE.

Manual inspection of the imputation data showed that almost all of the error was attributable to difficulty in recovering the latitude and longitude of CFS calls. Figure 8 shows the scatterplot of reconstruction error for Latitude vs. Longitude for the Lincoln, NE imputation of 2018 CFS data from Lincoln, NE 2017 CFS data. Figure 9 shows the same view after removal of the high-error vectors. In both cases, the error is zero mean, and normally distributed.

5.5 Information Model for City-to-City Comparison

Figure 10 shows that city-to-city imputation might not have the same fidelity as within-city imputation. Approximately 10% of the CFS vectors of Lincoln, NE could not be accurately imputed from the Hartford, CT CFS data. Once again, manual analysis showed that the majority of the error was in the location data. This is to be expected, since different cities will have different numbers and arrangements of CFS "hot spots", and these will not be represented in a foreign reference set.

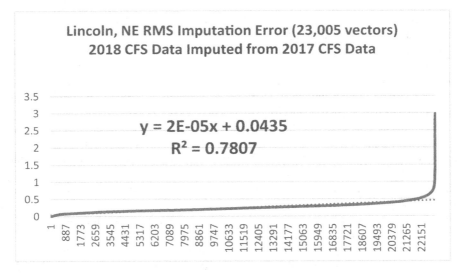

Fig. 6. RMS imputation error 2018 CFS data from 2017 CFS data

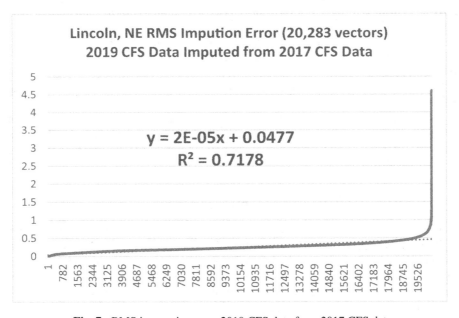

Fig. 7. RMS imputation error 2019 CFS data from 2017 CFS data

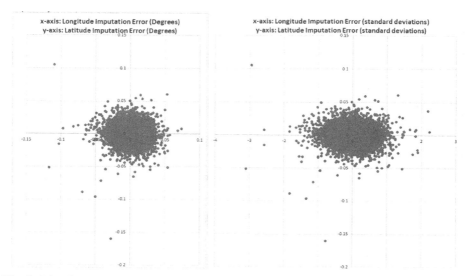

Fig. 8. Lincoln, NE (Longitude, Latitude) imputation error scatterplot, imputing 2018 CFS data from 2017 CFS data, no excision

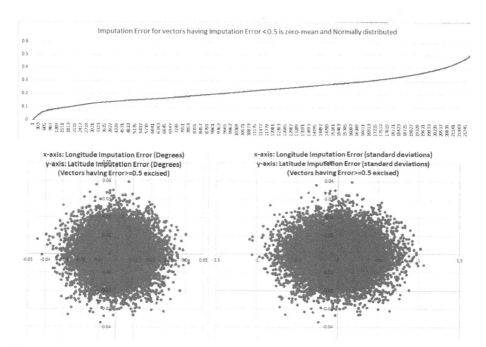

Fig. 9. Lincoln, NE (Longitude, Latitude) imputation error scatterplot, imputing 2018 CFS data from 2017 CFS data, vectors having imputation error > 0.5 excised

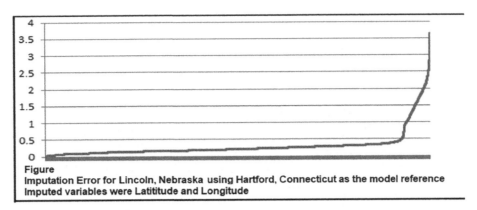

Figure
Imputation Error for Lincoln, Nebraska using Hartford, Connecticut as the model reference
Imputed variables were Latititude and Longitude

Fig. 10. Impute Lincoln, NE from Hartford, CT as reference

6 Conclusions

This empirical investigation has shown that certain data science methods (the use of non-linear transforms, and vector imputation) can be applied to CFS data to evaluate the information content of CFS data, characterize non-linear information that is not visually apparent, and provide a mechanism for information-theoretic comparison of jurisdictions through time and across space.

References

1. Theodoridis, S.: Pattern Recognition. Elsevier, Amsterdam (2008). ISBN 9780080949123
2. Pao, Y.: Adaptive Pattern Recognition and Neural Networks. Addison-Wesley, Boston (1989). ISBN 0-201-12584-6
3. Hanlon, et al.: Feedback control for optimizing human wellness. In: Proceedings of 22nd
4. International Human-Computer Interaction Conference, Copenhagen, Denmark, July 2020
5. https://www.policedatainitiative.org/datasets/calls-for-service/Additional

Experimental Evaluation of an Adaptive Planning Assistance System in Manned Unmanned Teaming Missions

Felix Heilemann[✉] and Axel Schulte

Universität der Bundeswehr, Werner-Heisenberg-Weg 39, 85579 Neubiberg, Germany
{felix.heilemann,axel.schulte}@unibw.de
https://www.unibw.de/fmff

Abstract. The task-based guidance of multiple unmanned aircraft (UAV) from aboard a manned aircraft increases the mission performance and reduces the potential risk for the crew. In time-critical situations an adaptive assistance system can simplify or take-over the UAV to avoid mishaps. This article describes and evaluates the effects of such planning assistance with different intervention levels in a human-in-the-loop experiment with German Air Force pilots. For this purpose, we present three different intervention levels (hint, simplification, take-over). The three intervention levels are then examined in four different threat situations to determine their appropriateness. The results show that too high intervention is rated negatively in low threat situations. In the case of a threat to the manned fighter and the unmanned systems, the simplification and take-over intervention were evaluated very positively and the time between the occurrence of the threat and the delegation of the countermeasures was drastically reduced.

Keywords: Manned-unmanned teaming · Planning assistance system

1 Introduction

Current developments in automation, planning and artificial intelligence allow future UAV systems to perform increasingly complex tasks. At the moment these systems heavily depend on a connection to the command center on the ground, which delegates the corresponding tasks and makes further decisions if necessary. Communication delays or disturbances, e.g. by hostile jamming, to this command center can eliminate these advantages [9]. This problem is tackled by the concept of manned-unmanned teaming (MUM-T). In MUM-T, manned and unmanned mobile assets (air, land, sea, space) interoperate to pursue a common mission objective. The unmanned platforms as well as their mission payloads are commanded by the manned asset. The required mission planning and management capabilities for a single operator are a highly relevant field of research [1,3]. The high work demands, arising from the multi-platform mission management

D. D. Schmorrow and C. M. Fidopiastis (Eds.): HCII 2020, LNAI 12197, pp. 371–382, 2020.
https://doi.org/10.1007/978-3-030-50439-7_25

and tasks execution, besides the usual pilot tasks, necessitates a certain degree of automation. Although highly automatic planners are feasible to solve such multi-vehicle planning problems in real time, they increase the risk for automation-induced errors such as the loss of situational awareness, complacency, or opacity [19]. The Institute of Flight Systems (IFS) at the Universität der Bundeswehr München addresses these problems by developing adaptive assistance, mission management and guidance systems. In previous research we studied the team-based guidance of three UCAVs (unmanned combat aerial vehicle) from aboard a single-seat fighter aircraft [4]. Even though all missions were successfully completed, the intentionally chosen high degree of automation temporarily led to mental under load of the pilots and was lacking in adaptability to balance the operator's activity and work demands in the sense of degrading situational awareness and complacency over the course of the mission. The experimental subjects further expressed the desire to be able to assign dedicated tasks to the UCAVs during mission execution, especially in less demanding situations [4]. In the helicopter domain, we investigated multi-UAV guidance on a task-based level [16]. During the mission execution the crew was supported by an adaptive assistance system, that recognized the currently performed tasks of the crew, determined the workload and proactively avoided phases of excessive stress [2]. The evaluation of the concept with German helicopter pilots showed the advantages of the concept such as reduced workload and increased performance [14–16]. On this background we developed interaction concepts [7,8] and implemented a mixed-initiative mission planner for the guidance of multiple UCAVs from aboard a manned fighter cockpit [5,6]. This work experimentally evaluates the intervention possibilities of this system with German Air Force pilots and is structured as follows: First the different actors, their roles and relations are described with the help of a work system analysis. Based on this, we present the human machine interface for the task delegation and the intervention possibilities in this process. The different interventions are then experimentally examined and evaluated in realistic mission situations.

2   Approach

The initial step in the development of a MUM-T system for the cockpit-based cooperative UCAV guidance is a top-down analysis of the individual system participants and their relationships. The Work System notation, described in [17], provides a semantical and graphical language for such a top-level system design with strong focus on human-automation work share and is used in the following. Within the Work System there exist two roles, defined as follows:

- *Worker*: The worker knows, understands and pursues the Work Objective by own initiative. There has to be a human taking the role of the worker, in any case.
- *Tools*: The tools receive the orders from the worker and will only execute them when commanded. Usually conventional automation (e.g. FMS, auto pilot) takes the role of a tool.

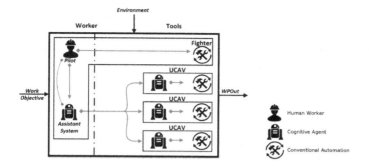

Fig. 1. Work system of the mixed-initiative planner, as part of the Assistance System (Color figure online)

Three different types of actors, i.e. humans, artificial cognitive agents, and conventional automation, are now integrated into the Work System. These actors are assigned in accordance with their capabilities and given requirements, to the Worker or Tool role. The entities of the Work System can stand in a hierarchical (green connector) or heterarchical (blue connector) relationship to each other. Figure 1 depicts the current work system for the cooperative UCAV guidance in the fighter domain at the IFS. The system contains cognitive agents installed aboard the UCAVs and the manned fighter aircraft. The cognitive agents aboard the UCAVs are responsible for the execution of assigned tasks in the role of a tool. In case of errors or situational changes, these agents are capable to pursue their tasks independently as long as the plan is not affected, however they cannot pursue the Work Objective independently. The cognitive agent (Assistant System) aboard the manned fighter supports the pilot in mission planning and execution and therefore adopts the role of a worker in this system. The pilot (i.e. the Human Worker) stands in a hierarchical as well as a heterarchical relationship to the assistance system. The hierarchical relationship enables the pilot to delegate tasks to the UCAVs through the assistance system [7]. A mixed initiative mission planner in the assistance system integrates the delegated task into the mission plan of the UCAVs, considering resources, constraints and timings [6], and then delegates the task to the UCAVs through the hierarchical relationship between the AS and the UCAV Agent. The heterarchical relationship between the pilot and the assistance system enables the system to support in the task assignment processes, the resolution of planning conflicts and the identification and improvement of sub-optimal plans. The intervention concept is based on the basic requirements for assistance according to Onken and Schulte [13], which are defined as follows:

1. Draw the attention of the assisted human operator(s) with priority on the objectively most urgent task or subtask.
2. If the person is overtaxed, transfer the task situation into a manageable one for him.
3. Only take-over tasks that the human is principally not capable to accomplish, or which are of a too high risk or likely a cause of too high costs.

These two distinct modes of cognitive automation (i.e. the hierarchical and heterarchical relationship) [17] allow the mission planning to be initiated and executed by both parties. Regardless of who initiated the planning, the following steps have always to be performed:

1. Selection of the desired task
2. Delegating the task
3. Integrating the task into the mission plan of the UCAV

According to the assistance levels presented above, steps 1–3 of this process can be partially or completely taken over by the assistance system. These different types of the cockpit-based UCAV guidance (with/without assistance) and the corresponding human machine interface are presented below. Afterwards, the appropriateness of the different interventions is evaluated in different mission scenarios with German Air Force pilots.

3 Human Machine Interface

This chapter describes the Human Machine Interface (HMI) for the cockpit-based cooperative UCAV guidance with different levels of intervention. First, the hierarchical relation of the HMI, the task creation and delegation, is presented. Then the adaptation of the HMI to the heterarchical relation, i.e. the different intervention levels, of the assistance system are presented.

3.1 Delegation

The hierarchical relationship between the pilot and the assistance system enables the task delegation to the UCAVs. This interaction takes place via the multifunctional display of our experimental fighter cockpit, shown in Fig. 2. The pilot first selects a target on the tactical map. According to the selected target [10] the

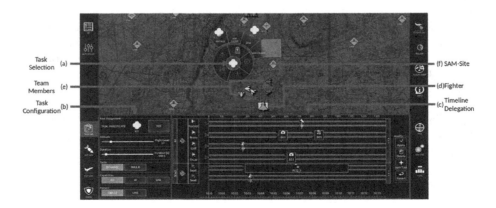

Fig. 2. Human machine interface

pilot can then create a task using a radial context menu at the target location, see Fig. 2a. After the task creation the pilot can adapt the task parameters, (Fig. 2b) and delegate it to the team members with the delegation interface, shown in Fig. 2c. The task creation and delegation process is described in detail in [8]. The position of the own fighter aircraft (grey symbol) and the team members are indicated in Fig. 2d&e. The red circles, Fig. 2f, mark enemy radar or missile defense positions which shall be suppressed or, if possible, circumvented for safe mission execution.

3.2 Assistance

The heterarchical relationship between the assistance system and the pilot allows the system to support the pilot in the task creation and planning process, compare Fig. 3. These intervention levels are analogous to the to the basic requirements, as follows:

1. Hint: Pop up dialog box (PUD) at the target with description of the missing task, the pilot has to create the task himself and assign it to a team member via the delegation interface as described before, see Fig. 3.1.
2. Simplification: PUD at the target with description of the missing task, as well as the most suitable team member for the task, is shown to the pilot. Additionally, the position for the new task is visualized in yellow on the time line. For the delegation of the task and the replanning, the pilot can accept the proposal directly in the PUD, Fig. 3.2.
3. Take-over: PUD on the target with description of the automatically delegated task and the corresponding team member. The pilot can revise this decision by clicking decline, restoring the old plan, Fig. 3.3.

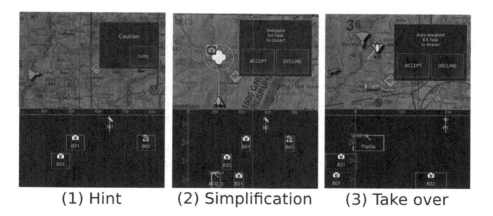

(1) Hint   (2) Simplification   (3) Take over

Fig. 3. Different types of intervention, hint simplification, take-over

4   Experimental Setup

The evaluation of the different interventions is carried out in a human-in-the-loop experiment. In this context, the influence of incorrect interventions (propose engagement of SAM position that can be circumvented) on the mental workload (MWL), and system acceptance are examined. Another focus is set on the experimental determination of the adequacy of the intervention level, i.e. to high/low intervention, in the missions. Therefore, we first develop hypotheses and define missions and situations based on these hypotheses. These missions are then integrated into the simulator, shown in Fig. 4, and carried out by German Air Force pilots. Retrospectively the different situations in the missions are replayed and evaluated with the help of questionnaires.

Fig. 4. The MUM-T fighter simulator at the IFS.

4.1   Hypotheses

The following aspects are to be examined with regard to the mental workload and appropriateness of the intervention:

H1: False intervention increases the mental workload (i.e. propose/take-over engagement of SAM position that can be circumvented).
H2: Simplification/take-over intervention reduces the MWL if the target, a UCAV or the fighter is threatened.
H3: Intervention is undesired if there is enough time to solve the problem.
H4: The automatic delegation of tasks is preferred when the own fighter is threatened.

4.2   Missions

The hypotheses are examined with three missions containing the desired types of situations. For briefing technical reasons and to avoid automation surprises,

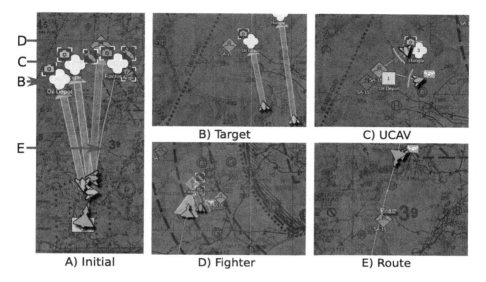

Fig. 5. Example mission (A) with the different mission phases: Endangering of high value target (B), UCAV (C), Fighter (D) and Route (E).

a fixed intervention level was defined for each mission, i.e. all interventions in a mission were on the same level. Each mission contained the reconnaissance, engagement and battle hit assessment of a high value target and the reconnaissance of two secondary targets. The missions were pre-planned and the pilot was responsible for the target engagement and target verification. The other tasks were performed by the UCAVs. Throughout the mission area pop up threats, i.e. enemy surface to air missile sites (SAM), had to be expected. The rules of engagement in these missions stated that those threats should only be engaged if an aircraft or high-value target is endangered. An exemplary mission with the individual mission phases is shown in Fig. 5A. The letters on the left side indicate the positions of the different threats in the mission scenario. In Fig. 5B-E the situations occurring during the mission execution are shown in detail. The first situation in this mission (B) is the threat to a target, in the sense that no aircraft is within range of the SAM site and there is sufficient time to complete the task, i.e. delegate a HARM task to suppress the enemy SAM site. The second situation (C) describes the pop-up of a threat with one or more UCAVs in range, represented with a red circle. In this case, an immediate reaction of the pilot is necessary. An escalation of this situation is shown in Fig. 5D, with an additional threat to the manned fighter aircraft. The last situation (E) shows a threat to the route. In accordance with the rules of engagement, the threat has to be circumvented. Each of the missions covered these four different situations and to eliminate possible spill-over effects from previous pop-up SAMs, the occurrence of two threats in the missions were at least 60 s apart.

4.3 Data Acquisition

After the mission execution, the four situations, shown in Fig. 5, were replayed and the adequacy of the intervention was determined. Additionally, the impact of the intervention on the pilots' mental workload for the situation was assessed. For this purpose, the mental workload with and without assistance was determined using NASA-TLX questionnaires. As performance measure we evaluated the interaction time between the pop up and the elimination of the threat.

5 Results

The missions were carried out with eight active German Air Force pilots. First, the effect of the intervention on the mental workload is discussed, followed by an evaluation of the performance and questionnaires.

5.1 Mental Workload

The effects of the different intervention stages on the mental workload without an acute threat are shown in Fig. 6. The interventions in the case of an endangerment of the route, Fig. 6 left side, showed a slight workload increase for all types of interventions. In case of the hint and simplification, the pilots had to reject the dialogue message. For the take-over intervention the pilot had to contradict the faulty intervention, i.e. the engagement of the SAM site, otherwise the rules of engagement would have been violated and therefore the mental workload slightly increases. However this increase was too small to support the hypothesis H1. The workload results for an active threat to an UCAV or the manned fighter are shown in Fig. 7. When the UCAV and the fighter are threatened, the hints lead to a higher workload. This can be explained by the fact that in addition to the delegation of the engagement of the threat the hint had to be rejected. For the simplification and take-over interventions little or no effect on the mental workload can be observed and therefore the hypothesis H2 has to be rejected.

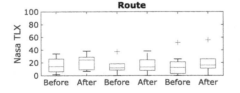 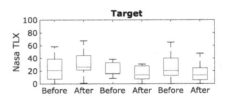

Fig. 6. Nasa TLX scores of the different intervention levels for the threatening of the route and a target.

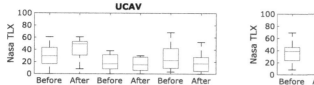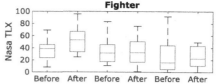

Fig. 7. Nasa TLX scores of the different intervention levels for the threatening of the UCAV and fighter.

5.2 Performance

The evaluation of the interaction times is divided into the different intervention levels (i.e. hint, simplification, take-over). For the hint, only the dialog had to be removed for the route, thus showing a very short reaction time. In case a target, UCAV, or the fighter was threatened, a HARM task had to be created and delegated to a UCAV. There is a huge time difference if a target or a UCAV is threatened in contrast to a manned fighter. This can be explained by the fact that in the case of a threat to the own fighter aircraft, evasive maneuvers were immediately carried out and then the delegation of the task was addressed. For the intervention with the simplification it showed up that the proposals for the route, the UCAV and the fighter were processed equally fast. The elimination of the threat to the target was somewhat faster, which could be interpreted as a higher situational awareness of the area. The evaluation of the take-over intervention only shows a valid time for the route, as the wrong decision of the system had to be counteracted. For the target, the UCAV, and the fighter, the time when the dialogue was closed is shown here, but the threat is eliminated from the system intervention.

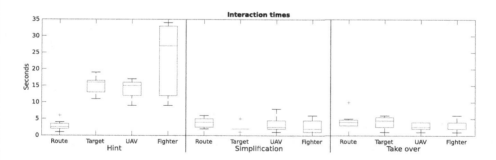

Fig. 8. Interaction time of the individual intervention stages for the elimination of a threat to the route, target, UAV or fighter.

5.3   Questionnaires

The evaluation of the questionnaires is shown in Fig. 9 and the results are referenced as follows (Question number, Intervention, Situation), the intervention and
situation are abbreviated with the initial letter. The hint was evaluated rather
negatively in the situations, since the pilots did not feel supported (Fig. 9.1H).
Hypothesis H3 is supported by the fact that the majority of pilots did not wish
more intervention if a route was threatened (Fig. 9.4HR). On the other hand, if
the targets are threatened, the pilots desired more intervention although there
was enough time to solve the problem (Fig. 9.2HT & 4HT). In case of a threat to
the UCAV and the fighter (Fig. 9.2HU & 2HF), the pilots did not have enough
time to solve the problem with the hint, which also correlates with the interaction times (Fig. 8). In case of a threat to the target, the UCAV, and the fighter,
the pilots also desired more support by the system (Fig. 9.4H) than a simple
hint. One problem here was that in addition to the dialog message, a sound
notification is triggered when a threat occurred. The pilots remarked that such
an intervention would be useful if they had no situation awareness for the task.
In contrast to the hints, the simplification of the task was evaluated very positively. However, some test persons expressed the desire for more support of the
system, especially in case of a threat to the fighter (Fig. 9.1SF & 4SF). For the
automatic take-over of tasks it was shown that the faulty interventions when
threatening a route (Fig. 9.1TR & 3HR) were evaluated too positively by the
pilots, because here a false system decision was made. A reason for the positive

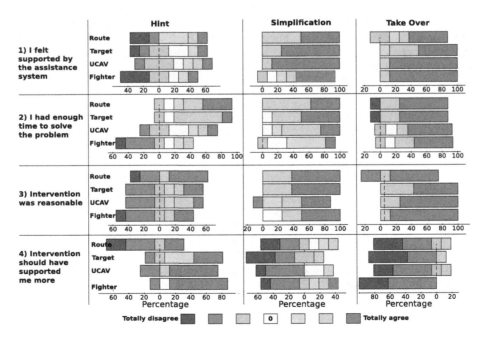

Fig. 9. Evaluation of the questionnaires after the mission.

evaluation could be the simple correction of the incorrect intervention by the decline button (compare Fig. 3). In the case of a threat to the manned fighter, the intervention level was found to be more appropriate than the simplification (Fig. 9.4SF vs. Fig. 9.4TF). This supports the hypothesis H4, however significance cannot be shown.

6   Conclusion and Future Work

In this work we experimentally evaluated the effects of different intervention levels in MUM-T mission. Starting with a work system analysis we derived two modes of automation, delegation and assistance, required in such MUM-T missions and presented their realization in the HMI. The assistance functions of this system were then systematically evaluated with German Air Force pilots in four different threat situations (route, target, UCAV and fighter). For each of the three intervention levels (hint, simplification and take-over) the impact of the intervention on mental workload, performance and system satisfaction was evaluated. The results showed that simplification and take-over intervention are desirable in time-critical situations. However, the automatic take-over of tasks in non mission critical situations (threat of route, target and UCAV) was criticized. This corresponds with the third basic requirement of [13] to only take-over tasks that the human is principally not capable to accomplish. A similar effect was observed for the hint intervention, here the intervention was unnecessary, because in case of a threat to the route or the target, the pilot directly realised the situational change and in case of a threat to the UCAVs or manned fighter aircraft, a higher intervention was desired. In order to achieve an adaptive intervention, more extensive analyses of the current situation, the pilot's situational awareness [18] and his current mental workload [11] must be incorporated into the decision-making process of the assistance system [12].

References

1. Behymer, K., et al.: Initial evaluation of the intelligent multi-UXV planner with adaptive collaborative/control technologies (impact). Technical report, Infoscitex Corp. Beavercreek (2017)
2. Brand, Y., Schulte, A.: Design and evaluation of a workload-adaptive associate system for cockpit crews. In: Harris, D. (ed.) EPCE 2018. LNCS (LNAI), vol. 10906, pp. 3–18. Springer, Cham (2018). https://doi.org/10.1007/978-3-319-91122-9_1
3. Chen, J.Y.C., Barnes, M.J.: Human-agent teaming for multirobot control: a review of human factors issues. IEEE Trans. Hum.-Mach. Syst. 44(1), 13–29 (2014)
4. Gangl, S., Lettl, B., Schulte, A.: Management of multiple unmanned combat aerial vehicles from a single-seat fighter cockpit in manned-unmanned fighter missions. In: AIAA Infotech@ Aerospace (I@ A) Conference, p. 4899 (2013)
5. Heilemann, F., Hollatz, F., Schulte, A.: Integration of mental resources in the planning of manned-unmanned teaming missions: concept, implementation and evaluation. In: AIAA Scitech 2020 Forum (2020)
6. Heilemann, F., Schmitt, F., Schulte, A.: Mixed-initiative mission planning of multiple UCAVs from aboard a single seat fighter aircraft. In: AIAA Scitech 2019 Forum, p. 2205 (2019)

7. Heilemann, F., Schulte, A.: Interaction concept for mixed-initiative mission planning on multiple delegation levels in multi-UCAV fighter missions. In: Karwowski, W., Ahram, T. (eds.) IHSI 2019. AISC, vol. 903, pp. 699–705. Springer, Cham (2019). https://doi.org/10.1007/978-3-030-11051-2_106

8. Heilemann, F., Schulte, A.: Time line based tasking concept for MUM-T mission planning with multiple delegation levels. In: Ahram, T., Karwowski, W., Vergnano, A., Leali, F., Taiar, R. (eds.) IHSI 2020. AISC, vol. 1131, pp. 1014–1020. Springer, Cham (2020). https://doi.org/10.1007/978-3-030-39512-4_154

9. IJtsma, M., Lassiter, W., Feigh, K.M., Savelsbergh, M., Pritchett, A.R.: An integrated system for mixed-initiative planning of manned spaceflight operations. In: 2019 IEEE Aerospace Conference, pp. 1–8. IEEE (2019)

10. Lindner, S., Schwerd, S., Schulte, A.: Defining generic tasks to guide UAVs in a MUM-T aerial combat environment. In: Karwowski, W., Ahram, T. (eds.) IHSI 2019. AISC, vol. 903, pp. 777–782. Springer, Cham (2019). https://doi.org/10.1007/978-3-030-11051-2_118

11. Mund, D., Schulte, A.: Model- and observation-based workload assessment and activity determination in manned-unmanned teaming missions. In: 33rd EAAP Conference (European Association for Aviation Psychology): Dubrovnik, Croatia, 24–28 September 2018 (2018)

12. Müller, J., Schulte, A.: Concept of an adaptive cockpit to maintain the workflow of the cockpit crew. In: Ahram, T., Karwowski, W., Vergnano, A., Leali, F., Taiar, R. (eds.) IHSI 2020. AISC, vol. 1131, pp. 952–958. Springer, Cham (2020). https://doi.org/10.1007/978-3-030-39512-4_145

13. Onken, R., Schulte, A.: System-ergonomic Design of Cognitive Automation - Dual-Mode Cognitive Design of Vehicle Guidance and Control Work Systems, 1st edn. Springer, Heidelberg (2010). https://doi.org/10.1007/978-3-642-03135-9

14. Schmitt, F., Roth, G., Barber, D., Chen, J., Schulte, A.: Experimental validation of pilot situation awareness enhancement through transparency design of a scalable mixed-initiative mission planner. In: Karwowski, W., Ahram, T. (eds.) IHSI 2018. AISC, vol. 722, pp. 209–215. Springer, Cham (2018). https://doi.org/10.1007/978-3-319-73888-8_33

15. Schmitt, F., Roth, G., Schulte, A.: Design and evaluation of a mixed-initiative planner for multi-vehicle missions. In: Harris, D. (ed.) EPCE 2017. LNCS (LNAI), vol. 10276, pp. 375–392. Springer, Cham (2017). https://doi.org/10.1007/978-3-319-58475-1_28

16. Schmitt, F., Schulte, A.: Experimental evaluation of a scalable mixed-initiative planning associate for future military helicopter missions. In: Harris, D. (ed.) EPCE 2018. LNCS (LNAI), vol. 10906, pp. 649–663. Springer, Cham (2018). https://doi.org/10.1007/978-3-319-91122-9_52

17. Schulte, A., Donath, D.: A design and description method for human-autonomy teaming systems. In: Karwowski, W., Ahram, T. (eds.) IHSI 2018. AISC, vol. 722, pp. 3–9. Springer, Cham (2018). https://doi.org/10.1007/978-3-319-73888-8_1

18. Schwerd, S., Schulte, A.: Mental state estimation to enable adaptive assistance in manned-unmanned teaming. In: 8th Interdisziplinärer Workshop Kognitive Systeme: Mensch, Teams, Systeme und Automaten. Verstehen, Beschreiben und Gestalten Kognitiver (Technischer) Systeme. Duisburg, 26–28 März 2019 (2019)

19. Wiener, E.L., Curry, R.E.: Flight-deck automation: promises and problems. Ergonomics **23**(10), 995–1011 (1980)

Cognitive Variability Factors and Passphrase Selection

Lila A. Loos[✉], Michael-Brian C. Ogawa[✉], and Martha E. Crosby[✉]

University of Hawai'i at Mānoa, Honolulu, HI 96822, USA
{lila7194,ogawa,crosby}@hawaii.edu

Abstract. Security policies require a secret code to access electronic information. Challenges exist between the usability and memorability of passwords. This study spotlights individualistic behavioral assimilation of passphrase styles for design insight and recall abilities. Data captured categorical authentication behavior toward enhanced usability outcomes. Validated locus of control personality and memory associative instruments demonstrated the internal and external personality types and cognitive response types that contribute to the systematic quest toward a more memorable passphrase scheme. Personalized criteria contributed to practical evaluation employing a repeated measures structure. This study tested 58 participants who successfully completed a passphrase survey consisting of four rulesets applied to imposed and user created passphrases designed for repeated measures. Although electrophysiological data was collected, it was not analyzed in time for this publication. Results indicate that memory associative factors of cognition represent a significant factor in the recall of 75% of imposed passphrase category types. The locus of control and memory associative variables are significant at the .05 level. Internally controlled participants preferred the created room objects and created no vowel passphrases. Additionally, the created room objects and animal association passphrases ranked the highest among the externally controlled subjects. The imposed passphrases constructed without vowels and associated with animals received the least recall. This descriptive study informs passphrase usability identifying cognitive demands that impact memory.

Keywords: Passphrase authentication · Locus of control personality · Cognition · Memorability

1 Introduction

Password objectives spanning more than 40 years have positioned the importance of security and convenience in human computer interaction. Password security continues to dominate computer authentication in a non-standard system [20]. Although, Herley, Van Oorschot, and Patrick [12] question if passwords will be universally used by the year 2019, the secret keyword remains ubiquitous. Varying password selection criteria persists with inconsistent conditions among websites [6] and usability problems continue to exist with passwords [22].

D. D. Schmorrow and C. M. Fidopiastis (Eds.): HCII 2020, LNAI 12197, pp. 383–394, 2020.
https://doi.org/10.1007/978-3-030-50439-7_26

Bonneau, Herley, Van Oorschot, and Stajano [5] report the difficulty to replace passwords given security and human interaction weaknesses. Usability, deployability, and security advantages framed their focus on evaluation principles. Results from their study suggest passwords are not expected to be a displaced authentication technology. Therefore, the application of a criteria based methodology furthered their discussion on computer authentication.

Establishing a memorable password is expected to lessen the burden of password creation and recollection. Addas, Thorpe, and Salehi-Abari [2] suggest the use of a geographic cue to assist user memory during passphrase recall. However, recollection data indicated that while users were not constrained to a passphrase format, allowing a free form passphrase suggests to impede the memory advantage of utilizing the Geo-Hint authentication system. Another study conducted by Yıldırım and Mackie [28] supports memorability proposed password composition guidelines that suggest to augment memorability and alleviate the difficulty of managing multiple accounts with unique requirements.

Grassi, Garcia, and Fenton [11] define successful digital authentication of a memorized secret to a system when a subscriber provides secret information for that identity. Unlike symmetric keys that are controlled by the verifier, memorized passwords are constructed by the user and are expected to be successfully recalled. Therefore, similar passwords may be composed and used in other logins. The reuse of password credentials support memorability by lessening the burden of recall errors. Herley, Van Oorschot, and Patrick [12] offer insight beyond passwords. The current memorability challenge is offset by password resets. However, security implications and user management of numerous passwords are concerns that impact the progress of enhanced user authentication that addresses security and usability.

Passphrase structures simulate natural language and therefore suggest to increase usability [22]. A password memory study devised a way for users to create memorable passphrase consisting of four words. Jeong, Vallat, Csikszentmihalyi, Park, and Pacheco [14] propose that electing four random words is more secure than a standard password consisting of a combination of characters and numbers. Besides collecting passwords for security reasons, Bonneau and Preibusch's [6] study indicate that not all websites necessitate security passwords. In fact, passwords are collected by websites to establish a database of user information targeted for other reasons such as marketing purposes that add weight to usability efforts. Mitigating password enrollment, resets, sharing, and reuse are motives that contribute to this passphrase study. The research examines memorable communication between the user and computer with emphasis on individual selection and recollection of various passphrase types for discrete locus of control personalities and behavioral preferences.

2 Research Design

The passphrase study is designed to expose all university respondents, N = 58, to all levels of the independent variable predictors that result in regulating variance and alleviating the random assignment of the sample [23]. Repeated measures identify within subject differences in passphrase selection and memorability and it is expected to result in participant similarities that differ among others [8]. Subjects were scheduled for 90 min appointments at the University of Hawai'i at Mānoa's Hawaii Interdisciplinary Neuroscience Technology Laboratory (HINT) lab. The survey was administered using an anonymized Qualtrics platform licensed by the University of Hawai'i at Mānoa. Physiological data was collected measuring heart rate, skin conductance, and facial corrugator muscle using the BIOPAC Systems, Inc. equipment. Electronic sensors, amplifiers, and transducers complied with the criteria of the International Organization for Standardization that detail laboratory methods for quality management principles [21]. The results from the electrophysical data collection will be reported in the future.

This study intends to predict associations based on theory, clearly defined research questions, hypotheses testing and regression analysis. The survey was processed through a systematic progression of pretesting design principles. Three pilot studies were conducted to arrive at the final survey. The completed design is governed by four rule sets for each imposed and user constructed category that are defined by four words containing five characters.

The survey data collected was password secured and manually scored by the researcher. Data was cleaned by comparing the responses to the passphrase requirements and categorizing individual records for the associated variable scores. All conditions will be tested using IBM's statistical package for the social sciences (SPSS) that assess the significance and effects of interactions [10]. Data analysis is expected to reveal relationships and patterns to predict the effects of cognition and locus of control personality on passphrase construction and memorability behavior.

3 Passphrase Instruments

Fifty-eight participants were presented with a survey consisting of imposed and user constructed passphrase recall based on four rulesets. Data was collected using existing validated instruments as well as a passphrase survey that was tested using three pilot studies. In addition to the various passphrase memory types, the survey contained Rotter's [24] locus of control (LOC) personality test and Ekstrom, Dermen, and Harman's [9] memory associative factor cognitive test. During passphrase recall, the Stroop [27] word and color test was randomly distributed throughout the survey to interrupt immediate retrieval from working memory. Table 1 illustrates the characteristics of the survey questions.

3.1 Locus of Control Personality

The passphrase survey was designed using self-reports of the locus control personality test. Locus of control deciphers a situation to inform individual approaches to one's surroundings [16]. Locus of control is interpreted as internal or external personality. Responses are construed to attribute behavior based on self-perception that are internally determined. Situations contingent on the external or to be controlled by destiny are revealed in the validated generalized reinforcement questions rooted in social, academic, and attitudes of living [24].

Table 1. Passphrase survey composition

Question Type	Characteristics	Instrument	Usefulness
Self-report	Expression of attitude, belief, feeling	LOC internal external test	Indicative of self-knowledge
	Identity	Demographics	Indicative of self-knowledge
Closed ended	Multiple choice format	LOC internal external test	Indicative of self-knowledge
	Memory recall of numbers and names	Memory associative tests	Indicative of working memory processing
	Physiological responses	EDA, EKG, fEMG	Psychophysiological inferences of emotion
Open ended	Memory recall of passphrases	Passphrase recall tests	Indicative of working memory processing
	Passphrase Construction	Passphrase construction rule sets	Indicative of cognitive processing

Source: Adapted from Spector ([26], p. 253)

The locus of control personality scale is indicative of internal traits that have a positive response to perceived regulation of behavior expectancies that have been modified and applied to predict latent behavior [15]. Factors representing internally controlled personalities are reliant on personal unique characteristics. Externally controlled individuals distinguish events with unpredictability allowing environmental factors to add complexity to distinctive contexts [24].

Results of the measurements surrounding decision making are implied to indicate personal assessments of the presented stimuli. Responses are expected to amplify behavioral responses to cognitive stressors that impact this design studying execution during passphrase construction and recall.

3.2   The Stroop Effect on Cognitive Load

Passphrase decisions were captured during cognitive load. A rise in passwords coupled with varying usage and non-standard requirements produce usability difficulties [1]. To interfere with cognitive load, the Stroop test randomly appeared throughout the survey as a distractor mechanism. It prompted the participant to distinguish the word from the ink color. Therefore, if the word BLUE is displayed in yellow ink, the correct answer is blue. The obstruction interrupts processes allocated to attention and is most effective when cognitive load is high [17].

3.3   Memory Associated Factor-Referenced Cognitive Tests

To assess working memory retrieval, this study employed two memory associative factor tests. The assessment of immediate recovery was supported by three pilot tests. Results are expected to lighten the burden of memorability and authentication by examining cognitive load and informing research on passphrases targeted to support recollection. Respondents were timed to memorize, recall, and match 15 numbers with objects and 15 first names to last names [9]. Results are expected to provide perception to immediate recall abilities. Cognitive processing that effectively organizes associations to arrive at a meaningful command of grouping words advances the ability to manage elements of mental effort attributed to load. Therefore, this study approaches the ability of working memory to advert the diminishing effects over time.

3.4 Passphrase Rulesets

The passphrase survey was designed with four rulesets that required 16 repeated recalls for each category. The imposed passphrase and participant created passphrases were used to classify individual responses to these classifications. Each passphrase consisted of four words requiring five characters per word. Clear instructions for each question supported cognitive load. To create a passphrase, descriptions for each ruleset were defined by the following: (a) vowels removed from all words; (b) select a word for each category of food, nature, sports, and transportation; (c) select visual objects located in the test room; (d) select words associated with an animal. The same rulesets applied to passphrases that were imposed on the participants.

3.5 Psychophysiological Inferences of Emotion

Each participant was measured for psychophysiological factors including heart rate, skin conductance, and facial electromyography. Electronic sensors captured signals from the electrocardiogram (EKG), electrodermal activity (EDA), and facial electromyography (fEMG). Visible occurrences in these physiological states are expected to generate quantifiable dimensions of emotion to passphrase events [7].

Physiological responses demonstrated by reactions to cognitive processing are targeted on behaviors from the autonomic nervous system that traverse to the sympathetic nervous (SNS) and parasympathetic nervous (PNS) systems. Therefore, arousal in electrical signals were captured by the heart rate and skin conductance to indicate homeodynamic effects regulated by the demand of cognitive stimuli [3]. Whereas the corrugator muscle exhibits a non-reflective affect of emotion [25], such activity will be differentiated through activation measurements.

Variability in physiological responses are anticipated to reveal participant experiences during the processing of passphrase selection and recall. Reactions from the passphrase survey stimuli were designed to provide physiological understanding of working memory processing during the encoding and decoding of passphrases. As of this writing, measurements collected were not analyzed for heart rate pulses, facial electromyography muscle activation, and skin conductance arousal significances.

3.6 Passphrase Survey

The passphrase survey consisted of 12 variables measuring 187 items including locus of control, memory associations, and passphrase recall for imposed and created passphrases based on four rulesets shown in Table 2.

Table 2. Passphrase survey items

Locus of Control Personality	
Internal locus of control	29 items
External locus of control	29 items
Memory Associative Factors	
Object number associations	15 items
First name last name associations	15 items
Imposed and User Created Passphrase Rulesets Four Words, 5 Characters per Word 16 Recalls per Category	
No vowels	32 items
Four categories: food, nature, sports, transportation	32 items
Room objects	32 items
Animal associations	32 items
Stroop Word and Color Test	
Randomized distractor used during passphrase recall	98 items
Demographics	
Number of logins, age, gender, study focus/major, employment status	5 items

4 Passphrase Survey Descriptive Results

This study consisted of 58 university participants who were exposed to all levels of the independent variables alleviating the essential random designation of subjects [23]. The repeated measures design categorizes discrete differences in passphrase recall and selection among the sample [8]. An anonymized Qualtrics survey was used to present the questionnaire that was accompanied by instructions and examples to facilitate working memory [19].

4.1 Demographics

Figure 1 displays the sample (N = 58), comprised of subjects who manage an average of 14 active online accounts. Fifty-nine percent are currently employed. The average age is 21 years old.

4.2 Locus of Control and Memory Associative Measures

The repeated measures within subjects effects between the memory associative and locus of control variables resulted in the significance of (p = 0) and the rejection of the null hypotheses, which is below the level of (p ≤ .05). The partial Eta Squared reported that 51.5% of the sample is attributed to variability. Tukey's test resulted in the significance (p = 0) in a pairwise comparison. Therefore, the mean difference is significant at the .05 level.

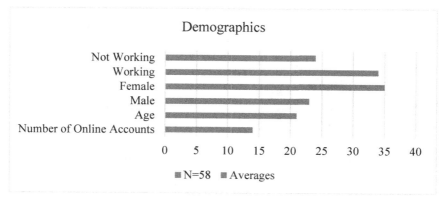

Fig. 1. Demographics

4.3 Locus of Control and Memory Associative Factors on Passphrase Types

The sample results indicated that 53% of participants were internally controlled while 62% scored below the mean for working memory recall of numbers to objects and first names to last names cognitive factor analysis (Fig. 2).

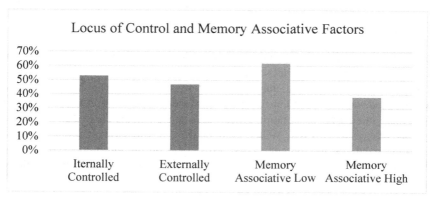

Fig. 2. Locus of control and memory associative factors sample

Specifically, the distribution for internal locus of control resulted in 33% who scored below the mean for cognitive processing compared with 21% who achieved scores indicative of high memory association ability. The majority of external controlled participants scored low on the factor analysis instrument. The outcomes are shown below in the Fig. 3.

4.4 Locus of Control and Passphrase Types

The repeated measures within subjects effects between the locus of control variable and imposed passphrases resulted in the significance of ($p = 0$) and the rejection of the null

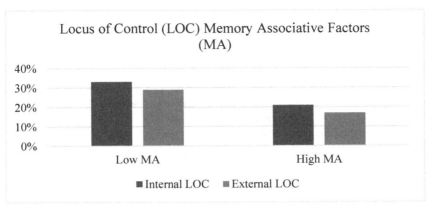

Fig. 3. Locus of control and memory associative factors descriptives

hypotheses, which is below the level of (p ≤ .05). The partial Eta Squared reported that 38.6% of the sample is attributed to variability. In comparison, the locus of control variable and the created passphrases resulted in the significance of (p = .036) and the rejection of the null hypotheses, which is below the level of (p ≤ .05). The partial Eta Squared reported that 4.40% of the sample is attributed to variability.

4.5 Memory Associative Correlation and Passphrase Types

Results displayed in Table 3 depict the memory associative variable was significant with the recall of three forms of imposed passphrases: (a) four categories (p = .023), (b) room objects (p = .006), and (c) animal associations (p = .001). Therefore, the memory associative test was significant at the (p ≤ .01) level for correlation with the imposed room objects and animal association passphrases as well as the created four categories passphrase rule (p = .003). The Wilks' Lambda (p = .05) statistic identified differences between the means of the memory associative effect on the imposed passphrase sets. Furthermore, unlike the memory associative effect on the created passphrase rule sets, the imposed passphrase rule sets reported a significance of (p = .002) with a Partial Eta ratio of variance that suggested 41.73 percent of variance can be explained by the memory associative variable. Regarding the effect of memory associations and the created passphrase rule sets, Roy's Largest Root multivariate hypotheses reported significance at (p = .002). A complete list of positive correlations among passphrase variables relating to the cognitive test for immediate retrieval is presented below.

The majority of respondents were internally controlled. These subjects scored highest in all eight passphrase types. The imposed passphrase types are: no vowels (INV), four categories (I4C), room objects (IRO), and animal associations (IAA). The created passphrase types are: no vowels (CNV), four categories (C4C), room objects (CRO), and animal associations (CAA). As shown in Fig. 4, internal control subjects scored the highest recalling the created room objects passphrase.

The created no vowels passphrase and created room objects were ranked among the highest recall categories by the internally controlled. Additionally, 69% of the sample

Table 3. Correlations for memory associative and passphrase types (N = 58)

	I4C	IRO	IAA	C4C	CRO
Memory Associative (MA)	.023*	.006**	.001**	.003**	
Imposed Four Categories (I4C)					
Imposed Room Objects (IRO)	.033*				
Imposed Animal Associations (IAA)	.046*	.000**		.022*	
Created Four Categories (C4C)					
Created Animal Associations (CAA)				.022*	.003**
Created Room Objects (CRO)					

*Correlation is significant at the 0.05 level (2-tailed)
**Correlation is significant at the 0.01 level (2-tailed)

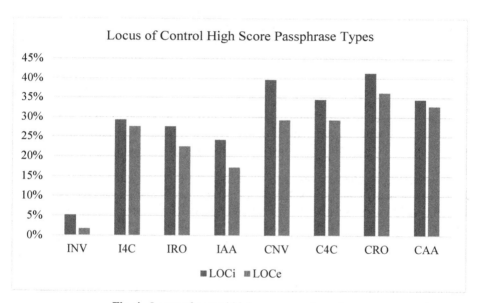

Fig. 4. Locus of control high score passphrase types

were effective in creating and recalling a passphrase without vowels. External controlled subjects favored to create passphrases based on room objects and animal associations. Respectively, these subjects accounted for 36% of the total 78% and 33% of the total 67%.

The imposed four category passphrases followed by imposed room objects received the highest scores for both internal and external controlled subjects. Internals preferred the room objects while externals were inclined to the four categories which accounted for a 28% success rate by each group.

In contrast to the favored created no vowels passphrase, the imposed no vowel version received the lowest success for both internal and external locus of control respondents. Subsequently, five percent of the internal controlled and two percent of the external controlled sample recalled this passphrase type. The overall approach to recalling a complex passphrase resulted in partial success of the four word passphrase. Individual encoding data will be analyzed in the future.

5 Concluding Remarks

Responses to passphrase recall measured short term memory using associative factors. The locus of control personality test reflected dichotomous answers that represent the underlying performance of internal and external traits. This study was designed to predict relationships in non-experimental research that utilize causal inference based on statistical results of hypotheses testing of the research questions. The predictive qualities of cognition were measured for processing of passphrase types that were presented to the subjects for recall. Presenting passphrases that were constructed of categories, room objects, and animal associatives resulted in a significant relationship between memorability factors of the passphrase structures. This finding is indicative of a systematic process to aid cognitive recall and avert its decrease over time.

The locus of control antecedent trait is perceived to influence decision making during passphrase selection and recall. Behavior moderated by the environment or external forces are predicted to reject unique conditions that are expected to be innate to those encompassing situational command [16]. Consequently, controlled individuals are perceived to embrace predictable abilities compared with peripheral based contingencies [18].

The created passphrase room objects received the highest recall ability from internally and externally controlled subjects. It can be inferred that visual objects assist in recalling passphrases. Subsequently, internals selected the created no vowels category whereas externals preferred the created animal associations.

Unlike the successes from the created rulesets, the imposed no vowel passphrase received the lowest recall from both locus of control personalities. This may be attributed to the difficulty with the four words that require a method for positive encoding. The imposed animal associations followed as the ruleset that was also problematic for memorability.

6 Future

The physiological data collected will be processed using BIOPAC's AcqKnowledge system software for electrodermal arousal patterns, heart rate frequency patterns, and activation of the facial corrugator muscle. These psychophysiological inferences are expected to reveal physical responses to the eight passphrase rulesets. Individual physiological predictors will be analyzed for correlation to cognitive processing to reflect the decision making process of constructing and selecting computer passphrases. The locus of control personality antecedent suggests to stimulate neural motion as behaviors governed by the control features are evident by pulse patterns during cognitive though processing [13]. Physiology and cognition are anticipated to establish perceptive processing to inform memorable computer authentication.

The combination of cognitive processing, locus of control personality, and psychophysiological factor outcomes are expected to provide performance insight during the processing of working memory recollection of computer passphrases. The amalgamation of diverse factors will copiously characterize individual perceptions and behavior of passphrase types that strengthen the support of cognitive retrieval.

References

1. Adams, A., Sasse, M.A.: Users are not the enemy. Commun. ACM **42**(12), 40–46 (1999)
2. Addas, A., Thorpe, J., Salehi-Abari, A.: Geographic hints for passphrase authentication. In: 2019 17th International Conference on Privacy, Security and Trust (PST), pp. 1–9. IEEE (2019)
3. Berntson, G.G., Cacioppo, J.T., Tassinary, L.G. (eds.): Handbook of Psychophysiology. Cambridge University Press, Cambridge (2017)
4. Biddle, S.J.: Motivation and perceptions of control: tracing its development and plotting its future in exercise and sport psychology. J. Sport Exerc. Psychol. **21**(1), 1–23 (1999)
5. Bonneau, J., Herley, C., Van Oorschot, P.C., Stajano, F.: The quest to replace passwords: a framework for comparative evaluation of web authentication schemes. In: 2012 IEEE Symposium on Security and Privacy, pp. 553–567. IEEE (2012)
6. Bonneau, J., Preibusch, S.: The password thicket: technical and market failures in human authentication on the web. In: WEIS (2010)
7. Cacioppo, J.T., Tassinary, L.G.: Inferring psychological significance from physiological signals. Am. Psychol. **45**(1), 16 (1990)
8. Cacioppo, J.T., Tassinary, L.G., Berntson, G.G.: Psychophysiological science: interdisciplinary approaches to classic questions about the mind. In: Handbook of psychophysiology, vol. 3, pp. 1–16 (2007)
9. Ekstrom, R.B., Dermen, D., Harman, H.H.: Manual for Kit of Factor-Referenced Cognitive Tests, vol. 102. Educational testing service, Princeton (1976)
10. Gaur, A.S., Gaur, S.S.: Statistical Methods for Practice and Research: A Guide to Data Analysis Using SPSS. Sage, Thousand Oaks (2006)
11. Grassi, P., Garcia, M., Fenton, J.: DRAFT NIST Special Publication 800-63-3 Digital Identity Guidelines. National Institute of Standards and Technology, Los Altos, CA (2017)
12. Herley, C., van Oorschot, P.C., Patrick, Andrew S.: Passwords: if we're so smart, why are we still using them? In: Dingledine, R., Golle, P. (eds.) FC 2009. LNCS, vol. 5628, pp. 230–237. Springer, Heidelberg (2009). https://doi.org/10.1007/978-3-642-03549-4_14
13. Hu, Q., West, R., Smarandescu, L.: The role of self-control in information security violations: insights from a cognitive neuroscience perspective. J. Manag. Inf. Syst. **31**(4), 6–48 (2015)
14. Jeong, B., Vallat, A., Csikszentmihalyi, C., Park, J., Pacheco, D.: MementoKey: keeping passwords in mind. In: Extended Abstracts of the 2019 CHI Conference on Human Factors in Computing Systems, pp. 1–6 (2019)
15. Loos, L.A., Crosby, M.E.: Cognition and predictors of password selection and usability. In: Schmorrow, D.D., Fidopiastis, C.M. (eds.) AC 2018. LNCS (LNAI), vol. 10916, pp. 117–132. Springer, Cham (2018). https://doi.org/10.1007/978-3-319-91467-1_10
16. Lowes, S., Lin, P.: Learning to learn online: using locus of control to help students become successful online learners. J. Online Learn. Res. **1**(1), 17–48 (2015)
17. MacLeod, C.M.: Half a century of research on the Stroop effect: an integrative review. Psychol. Bull. **109**(2), 163 (1991)

18. Marks, L.I.: Deconstructing locus of control: implications for practitioners. J. Couns. Dev. **76**(3), 251–260 (1998)
19. Mavilidi, M.F., Zhong, L.: Exploring the development and research focus of cognitive load theory, as described by its founders: interviewing John Sweller, Fred Paas, and Jeroen van Merriënboer. Educ. Psychol. Rev. 1–10 (2019). https://doi.org/10.1007/s10648-019-09463-7
20. Morris, R., Thompson, K.: Password security: a case history. Commun. ACM **22**(11), 594–597 (1979)
21. Naden, C.: International Organization for Standardization. https://www.iso.org/home.html. Accessed 17 May 2019
22. Nielsen, G., Vedel, M., Jensen, C.D.: Improving usability of passphrase authentication. In: 2014 Twelfth Annual International Conference on Privacy, Security and Trust, pp. 189–198. IEEE (2014)
23. Pelham, B.W., Blanton, H.: Conducting Research in Psychology: Measuring the Weight of Smoke, 4th edn. Cengage, Boston (2013)
24. Rotter, J.B.: Generalized expectancies for internal versus external control of reinforcement. Psychol. Monogr. Gen. Appl. **80**(1), 1 (1966)
25. Russell, J.A.: Core affect and the psychological construction of emotion. Psychol. Rev. **110**(1), 145 (2003)
26. Spector, P.E.: Sage encyclopedia of social science research methods. In: The SAGE Encyclopedia of Social Science Research Methods, Sage Publications Inc. (2004)
27. Stroop, J.R.: Studies of interference in serial verbal reactions. J. Exp. Psychol. **18**, 643–662 (1935)
28. Yıldırım, M., Mackie, I.: Encouraging users to improve password security and memorability. Int. J. Inf. Secur. **18**(6), 741–759 (2019)

Variable Self-Efficacy as a Measurement for Behaviors in Cyber Security Operations

Ricardo G. Lugo[1(✉)], Benjamin J. Knox[2], Øyvind Josøk[1,3], and Stefan Sütterlin[4,5]

[1] Institute of Psychology, Inland Norway University of Applied Sciences, Lillehammer, Norway
Ricardo.Lugo@inn.no
[2] Department of Information Security and Communication Technology,
Norwegian University of Science and Technology, Gjøvik, Norway
[3] Child and Youth Participation and Competence Development (BUK),
Inland Norway University of Applied Sciences, Lillehammer, Norway
[4] Faculty of Health and Welfare Sciences, Østfold University College, Halden, Norway
[5] Centre for Digital Forensics and Cyber Security, Tallinn University of Technology,
Tallinn, Estonia

Abstract. Training and development of skills gives individuals the knowledge and procedures required to perform their tasks. Training also requires individuals to develop reflective skills that help consolidate and transfer knowledge to long-term memory, where it then can be recalled in appropriate situations. But to develop reflective skills, one must first engage with the task's demands and then analyse how one's skills were appropriate or not for the situational demands, and how one could have also approached the situation differently. This process leads to an adaptable self-efficacy. It is still unclear how personality trait measures of self-efficacy can predict performance in cyber security operations. We developed an adaptable self-efficacy measure which was collected via an app. Findings support previous research that specific self-efficacy has better predictive validity, but also finds that an increased number of measurements could yield better effect sizes. This research shows how multiple measurements of specific self-efficacy can be done with minimized invasiveness while providing better measurement reliability and predictability. General findings and implications for further research are also discussed.

Keywords: Cyber security · Self-efficacy · Assessment

1 Cyber Security

1.1 Cyberspace

With the rapid technological advances in the twentieth century, the term 'cyberspace' emerged as an additional domain alongside the physical domains of land, sea, air and aerospace and is defined as 'the interdependent network of information technology infrastructures, and includes the internet, telecommunications networks, computer systems, and embedded processors and controllers in critical industries' [1]. This domain

D. D. Schmorrow and C. M. Fidopiastis (Eds.): HCII 2020, LNAI 12197, pp. 395–404, 2020.
https://doi.org/10.1007/978-3-030-50439-7_27

is more than just information-communication technologies, it also involves the creation, storage, and use of information that includes ICT devices, networks, software, and data [1, 2]. Today's cyberspace connects people to various systems and other people through various technologies and networks, such as mobile devices and personal computers.

While cyber incidents raise increased media attention, the role of the human factor in cyber defense still lacks a comprehensive scientific framework [3]. Interdisciplinary approaches to investigate human factor in cyber defense operations include human interaction, the physical and social operating environment, decision-making processes and psychological determinants of performance in cyber officers [4]. The rapid technological developments and definition of the cyber domain as a battlefield changed the cognitive demand profiles for cyber defense officers and challenged the traditional military organizational structures. Decisions on tactical levels can have large geostrategic implications and are often highly complex, based on insufficient or unreliable information, and under time pressure.

1.2 Self-Efficacy in Cyberspace

Training and development of skills gives individuals the knowledge and procedures required to perform their tasks. Training also requires individuals to develop reflective skills that help consolidate and transfer knowledge to long-term memory, where it then can be recalled in appropriate situations when needed [5]. But to develop this reflection, one must first engage with their environment and then analyze how one's skills were appropriate or not for the situation, and how one could have also approached the situation differently [6]. This process leads to an adaptable self-efficacy. Perceived self-efficacy is defined as the 'beliefs in one's capabilities to organize and execute the courses of action required to produce given attainments' (p. 3) [5] and is divided into a specific and a global component. General self-efficacy relates to the overall belief that one is in control over one's own life, actions, and decisions that shape one's life, while specific self-efficacy is the belief into one's performance in a certain task or described situation. Self-efficacy is also contingent on outcome expectancies, since one has to consider the desired outcome and judge if one possesses the skills necessary to reach those outcomes [5, 7].

Self-efficacy can be strong and weak all within one person, as being confident in one's skills in one area of functioning does not automatically generalize to other areas. Self-efficacy is realized through four separate efficacy-activated processes [5]: (1) Cognitive processes, including goal setting, self-appraisals, anticipatory scenarios, and analytic thinking; (2) Motivational processes, which include causal attributions, self-regulatory processes, outcome expectancies, and cognized goal/reinforcements; (3) Affective processes, affected by anxiety arousal, vigilance, rumination, and situations; and (4) Selection processes, by choosing of environments. These processes work in conjunction with each other, are dynamic and can be influenced in different ways. Bandura [5] identified four influencing factors for perceived self-efficacy: a) mastery experience, b) vicarious experience, c) social persuasion, and d) physiological and emotional state feedback. Self-efficacy in task-specific contexts is the result of previous learning-, success-, and failure experiences and can be represented as both a top-down (experience; intuition) and bottom-up (analytical approach) process.

Embracing the full complexity of human computer interaction in hybrid contexts reveals the need to explore and measure emergent properties [8]. One can assume that for cyber defense officers, having highly trained skills that are adaptable to the cyber domain would give the foundation for expertise. High and resilient self-efficacy is a marker for expertise. It could be argued that being familiar for the ambiguity that the cyber domain encompasses would then require adaptable coping mechanisms which would will lead to better decision-making strategies that are not reliant on intuition, but are the foundation of what is deemed as expert knowledge as described by [9].

1.3 Performance in Cyber Operations

Measuring Self-Efficacy. There is a debate on how trait measures can predict performance. Research [10, 11] indicates that situational SE predicts performance better than trait SE. Nevertheless, human judgment of other people's competence is very often biased by and based on generalized trait SE.

Bandura [11] has also stated that specific and situational aspects of self-efficacy have more predictive power than trait measurements and suggests to use state measurements. Trait measurements are usually considered more distal factors while state measurements are more proximal. With recent technological advances, opportunities to gather data during observation have become more efficient and less invasive. There has been recent debates about the reliability of measurements collected under observation due to homogeneity of the participants [12]. While Hedge et al. were referring to retrospective verbal accounts (cognitive task analysis), this might also be relevant in quantitative approaches also [12]. This may be due to the participant having time to reflect over one's performance before expressing their experience. Within cyber security, specific self-efficacy was shown to predict task performance, while general self-efficacy better predicted general experiences [13]. These findings, although significant, had small effect sizes ($r < .3$) and were taken prior to the task.

Adaptive Self-Efficacy. The Root Mean Square of the Successive Differences (RMSSD) is one of a few time-domain tools used originally to assess heart rate variability, the successive differences being neighboring RR intervals. Using the RMSSD formula and applying it to multiple measurements of self-efficacy would give an indication of variability of one's self-efficacy, similar to the assessment of affective variability [14]. Originally derived from psychophysiology, the root mean square of successive differences between normal heartbeats (RMSSD) is obtained by first calculating each successive time difference between heartbeats in milliseconds [15]. Then, each of the values is squared and the result is averaged before the square root of the total is obtained. RMSSD has been adopted to other behavioral measurements explain intra-individual differences. Koval et al. [14] used RMSSD to validate the measure of affective instability where high RMSSD represents high moment-to-moment variability. A higher RMSSD reflects more intra-individual variability (IIV), which is defined as "relatively short-term changes that are construed as more or less reversible and that occur more rapidly" [16]. Higher affective IIV predicts higher neuroticism [17], declining mental and physical health [18]. Dawood and Pincus [19] used IIV measurements (RMSSD) to explain pathological narcissism and depression. Higher IIV in life satisfaction and in perceived control predicted earlier mortality [20, 21]. Using the IIV approach reflect "the new person-specific paradigm in psychology" that emphasizes people as dynamic systems where trait nor state measurements are able to access changes during specific situations [22] (p. 112).

1.4 Purpose of the Study

We argue that multiple measurements of self-efficacy taken during an exercise would give better indices than either trait or situational measurements taken prior to an exercise. Being able to measure an adaptive form of self-efficacy while engaging with one's environment would give a better understanding of how self-efficacy processes are influenced by situational demands. Such demands help develop coping strategies, both psychological and affective that are the basis for the self-efficacy construct.

2   Methods

2.1   Participants

Participants: Twenty-five cyber cadet officers (M_{age} = 22.7 years, SD = 0.71) were assessed during a cyber defence exercise.

2.2   Cyber Defense Exercise

Data was collected during the Norwegian Defense Cyber Academy's (NDCA) annual Cyber Defense Exercise (CDX). This is an arena that facilitates the opportunity for fourth year cyber engineer students to train in tactics, techniques and procedures for handling various types of cyber attacks. The exercise contributes to improving appreciation for the human and technical competences necessary to establish, manage and defend a military infrastructure under simulated operational conditions. The students worked in four teams of 9 or 10 members, took decisions and acted in order to strengthen operational freedom, mission assurance and control in the cyber domain. The four teams participating in the exercise worked independently from each other but not against each other.

2.3   Self-Efficacy

For the CDX, both general, situational, and specific self-efficacy were measured.

Trait Self-Efficacy (SE). Trait self-efficacy was measured prior to the exercise and the General Self-Efficacy Scale [23] was used. The scale is comprised of ten Likert-scale items from 1–4, with higher scores indicating higher trait self-efficacy. The scale has shown validity in several domains and across cultures [24] and has acceptable internal validity ($\alpha = .75 - .91$).

Situational Self-Efficacy (SSE). Situational self-efficacy was measured using three scales for positivity, arousal, and confidence. Participants before starting the CDX had to answer three questions pertaining to it while looking at a visual analogue scale that was later coded on a scale of 0 to 10 (none - very). The three questions were as follows: 'how positive are you to the task', 'how aroused do you feel about the task', and 'how confident are you right now'. To compute a state self-efficacy score, the three scales were aggregated to a total score where higher scores indicated higher self-efficacy. This scale followed the guidelines established by Bandura [11] for creating self-efficacy measurements. The scale created was tailored to the unknown task with unipolar formulations of their current beliefs.

Situation Specific Self-Efficacy (SSSE). Situation specific self-efficacy was measured 10 times per day over the four day exercise with the Hybrid Space App (HSA). This was developed for collecting and analyzing individual cognitive processes when engaging in hybrid contexts [25]. Figure 1 shows an example of the data collected, both HS measurements (see below) and SSSE measurements. Previous research has shown that both individual [26] and macrocognitive factors [27] influence performance in such hybrid contexts. Previous studies in similar show that general and specific self-efficacy measurements yielded significant but small effects [13] and this might be due to the nature of the measurement (prior to task). The HSA was designed to record behavioral data during tasks instantly to minimize reflective processes. The HSA and an adaptable self-efficacy score was computed based on the formula

$$RMSSD = \sqrt{\frac{1}{(N-1)} \sum_{j=0}^{n} \left(RR_{j+1} - \overline{RR} \right)}$$

where N is the total number of R peaks, RR_j is the jth RR interval, RR is the mean of the RR intervals, $\overline{RR_j}$ denotes the average of the RR intervals up to the jth.

2.4 Hybrid Space

The Hybrid Space is mapped on a Cartesian plane visualizing the cyber-physical and tactical-strategic dimensions. Participants were asked to simultaneously mark their cognitive location within the Hybrid Space [28] (see Fig. 1) each hour from 08:00 to 20:00. Simultaneously, they were asked to mark their self-efficacy and effort, through sliding scales. In addition, students noted their current task at each position, to provide context for further analysis.

Hybrid Space operationalization:

Movement in the Hybrid Space is operationalized through four constructs and represents the dependent variables in the study. Four dependent variables were created (see Fig. 2):

- Total HS Travel: distance traveled in the Cartesian Plane measured by Euclidian distance
- X-Axis: Movement along the cyber-physical domain (x-axis)

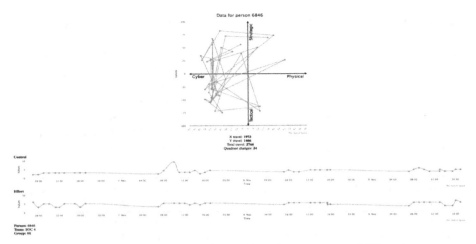

Fig. 1. Sample visualization of a participant's movements and SE scoring

- Y-axis: Movement along the strategic-tactical domain (y-axis)
- Quadrant Changes: Number of quadrant changes

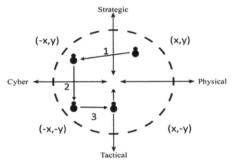

Movement in the Hybrid Space: (1) operator reporting quadrant change (x,y) to (-x,y); (2) operator moving along the y-axis; (3) an operator reporting movement to an axis but not crossing to other quadrant

Fig. 2. Description of Hybrid Space movements

2.5   Ethical Considerations

Participants gave oral informed consent prior to the study and were debriefed about the study's purpose after completing the data collection. Participants were informed that they could withdraw from participation at any time and without any consequences throughout and after the session. The study has been approved by the Norwegian Social Science

Data Services (NSD; project number 43901) and was conform to the NDCA's ethical guidelines for experimental studies.

2.6 Data Reduction and Analysis

Statistical analysis was done with JASP [29]. All variables were checked for normality. Pearson's bivariate correlations were calculated and all variables were entered in the calculation. For the regression analysis, self-efficacy was entered as the independent variable if was correlated with the dependent variable, and the Hybrid Space variables were used as dependent variables. Significance level was set to .05 (2-tailed).

3 Results

Descriptive statistics and correlations can be found in Table 1.

Table 1. Descriptive statistics and correlations

	M	SD	1	2	3	4	5	6
1. Trait SE	3.65	.46	1					
2. Situational SE	91.68	54.40	−.082	1				
3. SSSE	219.04	75.50	.252	−.167	1			
4. X axis	1447.76	777.41	.057	−.011	.557**	1		
5. Y axis	1207.00	573.38	−.090	−.11	.515*	.727**	1	
6. Total HS travel	2097.76	997.88	−.005	−.058	.583**	.955**	.895**	1
7. Quadrant changes	16.48	7.34	−.011	−.173	.379	.806**	.840**	.884**

SE: Self-Efficacy; SSSE: Specific Situational SE; $*p < .05$; $**p < .01$

Neither trait nor situational self-efficacy correlated with any of the Hybrid Space movements. Since these did not correlate with any of the dependent variables, they were not used in the regression analysis. However, SSSE did predict three of the four HS measurements (X-axis, Y-axis, Total travel) and showed tendencies for quadrant changes (see Table 2 for regression statistics), supporting the hypothesis.

4 Discussion

This study aimed at establishing an adaptable self-efficacy measurement that can be measured multiple times while being unobtrusive as opposed to more used approaches, such as trait or situational (i.e. immediately before) measurements. Results showed that the adaptable self-efficacy measure (SSSE) predicted three of the four performance variables of the Hybrid Space, while showing tendencies for the fourth. The trait and situational self-efficacy assessment (both one-time assessments) had no association with

Table 2. Regressions

	R^2	F	P	β
X-axis	27.7	9.50	.006	.557
Y-axis	26.5	7.57	.012	.515
HS travel	33.9	10.78	.004	.583
Quadrant changes	14.4	3.53	.074	.379

Predictor: Situation Specific Self-Efficacy (SSSE)

the dependent variables, in line with Bandura's (2006) guidelines. The adaptable self-efficacy showed also moderate effect sizes (.515 − .583).

The findings also support Zuckerman's [10] stipulation that state measurements are better at predicting situational aspects while trait measurements are better at describing more general aspects. Previous research [13] could only find small effects from self-efficacy, but these were also taken prior to exercises. The Hybrid Space app makes it possible to measure self-efficacy in very unobtrusive ways. It can be incorporated into devices (i.e., laptops, mobile phones) that are part of the environment, and measurements are collected immediately. By applying measurements of intra-individual variability to self-efficacy, one can increase the sensitivity of the factor in order to explain behavior.

Self-efficacy is both a global and specific psychological measure. It has been a key psychological factor in domains across human functioning, but it can be difficult to measure accurately. Most research uses either global trait measurements, as the General Self-efficacy Scale [23], or situational measurements as defined by Bandura [11]. Both such measurements are one-time measurements and thus may omit a person using different psychological strategies that may help them perform and influence their own self-efficacy. One-time measurements do not incorporate psychological and affective processes while interacting with one's environment. This may lead to small reported effects sizes since measurements are one-time and not really reflecting how self-efficacy is influencing performance as seen in the Wee et al. study [13]. Koval et al. [14] showed that intra-individual variability measurements of affect could predict health outcomes. Other research has shown that applying intra-individual variability measures to behavioral factors has shown good effects in predicting depression [30] and stress [31]. The moderate effects found in this study are similar to the effects found by Bos et al. [30] and Dawood and Pincus [19].

4.1 Limitations

Most studies using inter-individual variability measurements have only examined linear associations as did this experiment. The optimal level of moment-to-moment variability depends on the context. Also, this experiment did not consider an event that required confidence had occurred or not, as well as what was the nature of that event was not considered in our analysis, which is a major limitation. The participants in this study are both limited (N = 25) and cannot be generalized to other groups due to the limited sample of cyber security cadets. Another limitation is that the evening and the morning measurement may give different values of self-efficacy as fatigue was not accounted for and some variation of scores between the last measurements of one day differed from initial measurements on the next day.

4.2 Conclusion

These findings support previous research that variable self-efficacy has better predictive validity, but also finds that an increased number of measurements could yield better effect sizes. This research shows how multiple measurements of specific self-efficacy can be done with minimized invasiveness while providing better measurement reliability and predictability. Future research should include variable self-efficacy measurements as they are more sensitive and have better explanation power than the other trait or state measurements.

Acknowledgments. We would like to thank the Norwegian Defence Cyber Academy and the officer cadets for their cooperation in this study.

References

1. Kuehl, D.T.: From cyberspace to cyberpower: defining the problem. Cyberpower Natl. Secur. **30**, 1–17 (2009)
2. Mayer, M., et al.: How would you define Cyberspace. First Draft Pisa **19**, 2014 (2014)
3. Gutzwiller, R.S., et al.: The human factors of cyber network defense. In: Proceedings of the Human Factors and Ergonomics Society Annual Meeting. SAGE Publications, Los Angeles (2015)
4. Mancuso, V.F., et al.: Human factors in cyber warfare II: emerging perspectives. In: Proceedings of the Human Factors and Ergonomics Society Annual Meeting. SAGE Publications, Los Angeles (2014)
5. Bandura, A.: Self-Efficacy: The Exercise of Control. WH Freeman, New York (1997)
6. Zimmerman, M.A.: Taking aim on empowerment research: on the distinction between individual and psychological conceptions. Am. J. Commun. Psychol. **18**(1), 169–177 (1990)
7. Pajares, F., Schunk, D.H.: Self-beliefs and school success: self-efficacy, self-concept, and school achievement. Perception **11**, 239–266 (2001)
8. Hoffman, R.R., Hancock, P.A.: Measuring resilience. Hum. Factors **59**(4), 564–581 (2017)
9. Kahneman, D., Klein, G.: Conditions for intuitive expertise: a failure to disagree. Am. Psychol. **64**(6), 515 (2009)
10. Zuckerman, M.: The distinction between trait and state scales is not arbitrary: comment on Allen and Potkay's "On the arbitrary distinction between traits and states" (1983)
11. Bandura, A.: Guide for constructing self-efficacy scales. Self-efficacy Beliefs Adolesc. **5**(1), 307–337 (2006)
12. Hedge, C., Powell, G., Sumner, P.: The reliability paradox: why robust cognitive tasks do not produce reliable individual differences. Behav. Res. Methods **50**(3), 1166–1186 (2018)
13. Wee, J.M.C., Bashir, M., Memon, N.: Self-efficacy in cybersecurity tasks and its relationship with cybersecurity competition and work-related outcomes. In: 2016 USENIX Workshop on Advances in Security Education (ASE 2016) (2016)

14. Koval, P., et al.: Affective instability in daily life is predicted by resting heart rate variability. PloS One **8**(11), 1–10 (2013)
15. Shaffer, F., Ginsberg, J.: An overview of heart rate variability metrics and norms. Front. Public Health **5**, 258 (2017)
16. Nesselroade, J.R.: Interindividual differences in intraindividual change (1991)
17. Timmermans, T., Van Mechelen, I., Kuppens, P.: The relationship between individual differences in intraindividual variability in core affect and interpersonal behaviour. Eur. J. Pers. **24**(8), 623–638 (2010)
18. Segerstrom, S.C., et al.: Briefly assessing repetitive thought dimensions: valence, purpose, and total. Assessment **23**(5), 614–623 (2016)
19. Dawood, S., Pincus, A.L.: Pathological narcissism and the severity, variability, and instability of depressive symptoms. Pers. Disord. Theory Res. Treatment **9**(2), 144 (2018)
20. Boehm, J.K., et al.: Variability modifies life satisfaction's association with mortality risk in older adults. Psychol. Sci. **26**(7), 1063–1070 (2015)
21. Eizenman, D.R., et al.: Intraindividual variability in perceived control in a older sample: the MacArthur successful aging studies. Psychol. Aging **12**(3), 489 (1997)
22. Molenaar, P.C., Campbell, C.G.: The new person-specific paradigm in psychology. Curr. Dir. Psychol. Sci. **18**(2), 112–117 (2009)
23. Scholz, U., et al.: Is general self-efficacy a universal construct? Psychometric findings from 25 countries. Eur. J. Psychol. Assess. **18**(3), 242 (2002)
24. Luszczynska, A., Scholz, U., Schwarzer, R.: The general self-efficacy scale: multicultural validation studies. J. Psychol. **139**(5), 439–457 (2005)
25. Jøsok, Ø., Hedberg, M., Knox, B.J., Helkala, K., Sütterlin, S., Lugo, R.G.: Development and application of the hybrid space app for measuring cognitive focus in hybrid contexts. In: Schmorrow, D.D., Fidopiastis, C.M. (eds.) AC 2018. LNCS (LNAI), vol. 10915, pp. 369–382. Springer, Cham (2018). https://doi.org/10.1007/978-3-319-91470-1_30
26. Knox, B.J., Lugo, R.G., Jøsok, Ø., Helkala, K., Sütterlin, S.: Towards a cognitive agility index: the role of metacognition in human computer interaction. In: Stephanidis, C. (ed.) HCI 2017. CCIS, vol. 713, pp. 330–338. Springer, Cham (2017). https://doi.org/10.1007/978-3-319-58750-9_46
27. Lugo, R., et al.: Team workload demands influence on cyber detection performance. In: Proceedings of 13th International Conference on Naturalistic Decision Making (2017)
28. Jøsok, Ø., Knox, B.J., Helkala, K., Lugo, R.G., Sütterlin, S., Ward, P.: Exploring the hybrid space. In: Schmorrow, D.D.D., Fidopiastis, C.M.M. (eds.) AC 2016. LNCS (LNAI), vol. 9744, pp. 178–188. Springer, Cham (2016). https://doi.org/10.1007/978-3-319-39952-2_18
29. JASP Team: JASP (version 0.10. 2) [computer software] (2019). https://jasp-stats.org
30. Bos, E.H., de Jonge, P., Cox, R.F.: Affective variability in depression: revisiting the inertia–instability paradox. Br. J. Psychol. **110**(4), 814–827 (2019)
31. Segerstrom, S.C., Sephton, S.E., Westgate, P.M.: Intraindividual variability in cortisol: approaches, illustrations, and recommendations. Psychoneuroendocrinology **78**, 114–124 (2017)

Probing for Psycho-Physiological Correlates of Cognitive Interaction with Cybersecurity Events

Nancy Mogire[1]($\boxtimes$), Randall K. Minas[2], and Martha E. Crosby[1]

[1] Information and Computer Sciences, University of Hawaii at Manoa,
Post 317 1680, East-West Road, Honolulu, HI 96822, USA
{nmogire,crosby}@hawaii.edu
[2] Shidler College of Business, University of Hawaii at Manoa,
2404 Maile Way Suite E601f, Honolulu, HI 96822, USA
rminas@hawaii.edu

Abstract. Changes in psychophysiological signals of the human body are highly revealing of cognitive and emotional responses to stimuli, capturing even subtle and transient events. Based on these properties, changes in recorded psychophysiological signal could reflect changes occurring in the computing network while the human factor is a part of it. This paper outlines a methodology for exploring the psychophysiological correlates of cognitive interaction with cybersecurity events. This continues the discussion on a dissertation research project exploring what neurological and physiological signal changes might reveal in the context of digital interactions from a cybersecurity standpoint.

Keywords: Human factor · Human-Computer interaction artifacts · Tokens of interaction · Digital evidence · Digital incidents · Digital events · Psycho-physiological signals · Signal changes · Cognition · Cognitive responsiveness · Cybersecurity applications

1 Introduction

1.1 Research Overview

The proliferation of devices that measure and record psycho-physiological signal devices in user space provides an opportunity to harness human cognitive functioning for potential cybersecurity applications.

This research investigates how the electrical signals generated from the functioning of the body, respond to human interaction with digital incidents [1]. If we can find that response-related signal changes are consistently notable, and we can retrieve these changes from recorded signal with an accuracy that is greater than chance, then we can claim that psycho-physiological signals contain markers of digital incidents.

Potential applications of these markers include: in digital investigations for triangulation of other evidence, in cybersafety management as input to tools for regulating immersive digital experiences of locked-in individuals.

© Springer Nature Switzerland AG 2020
D. D. Schmorrow and C. M. Fidopiastis (Eds.): HCII 2020, LNAI 12197, pp. 405–415, 2020.
https://doi.org/10.1007/978-3-030-50439-7_28

1.2 Cognition and Psychophysiological Artifacts

Cognition processes fall under one of two systems: conscious cognition—also known as system 1 cognition—and automatic (system 2) cognition [2].

During a digital task, conscious cognition is mainly dedicated to the task while automatic cognition attends to a broader scope of elements of human functioning [2, 3] including the task processing, evaluation of its presentation features and assessment of other components of the general environment. Cognition may include appraisal activities such as fetching or forming various heuristics and making of non-conscious judgements.

Cognition is a necessary part of the human functioning that is involved in completing digital tasks [4] (Fig. 1).

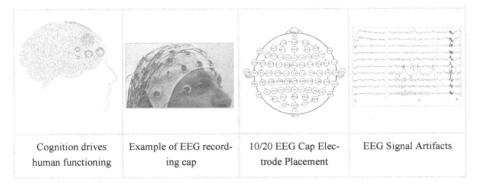

| Cognition drives human functioning | Example of EEG recording cap | 10/20 EEG Cap Electrode Placement | EEG Signal Artifacts |

Fig. 1. Cognition, EEG correlates and measurement

Cognitive functioning activates various body systems such as the brain, facial brow muscles, heart and electrodermal systems. These systems generate electrical signals during their functioning [5].

As such, cognition can be said to create psycho-physiological signals artifacts. Examples of such signals include: Electroencephalograms (EEG), Electromyograms (EMG), Electrocardiograms(ECG) and Electrodermograms (EDR). These signals have known structural forms and follow predictable change patterns [6–9].

Various relationships between cognition and psycho-physiological signal change have been studied and documented. For example, some signals have been found to reflect such cognitive experiences as variation in mental workload [10], shift in attentive focus, and experiences of emotional affect such as disgust [11].

1.3 Cybersecurity Context and Thresholds

We assume that cybersecurity events inherently feature, some response-evoking characteristics e.g. salience, aversiveness. What counts as an event is dependent on context, and as such security-relatedness is a wide and varying spectrum.

We differentiate events from incidents by taking events to be those that have potential security implications, while incidents are the events that have links to actual breaches.

Events: can potentially cause security breaches. Incidents: events where occurrence can be directly linked to a breach that occured (Fig. 2).

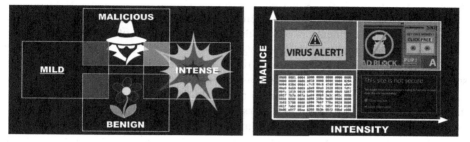

Fig. 2. Security-relatedness is a wide spectrum of variability [1]

1.4 Research Approach Overview

This study involves presentation of cybersecurity related events as stimuli, to participants within a lab setting while they undertake generalized computing tasks that mimic regular device usage.

Throughout each lab session, as the participant works on the tasks and experiences the stimuli, we record psychophysiological signals (EEG, EMG, EDA, ECG) using various body sensors together with relevant body signal acquisition software. We assume that response-evoking properties e.g. salience and aversiveness, are inherent in cybersecurity events.

The study takes a non-blind followed by blind study methodology, and utilizes relevant signal analysis methods including event related potential analysis and wavelet transformation.

In the non-blind phase we match signal changes to event timings using a separately tracked record of event stimuli display timings. In the blind phase, event timings are initially unknown to the primary researcher but held by another researcher. We attempt to identify event timings via signal analysis and then verify against the actual timings. If we can retrieve event timings from signal with accuracy greater than chance, then we can make the claim that psycho-physiological signals contain markers of digital events and incidents [1].

2 Research Design and Process

2.1 Stimuli Design

To be able to probe psychophysiological signals for markers of interaction with cybersecurity events, it was necessary to create conditions mimicking real scenarios that occur in relation to cybersecurity breaches. Towards this goal, we designed a study so that we could present security related material such as pop-ups, page crash reports and error warnings to the study participants while recording their psychophysiological signal data was recorded.

The key assumption is that these security related events would inherently have response evoking properties. Such properties may include dubiousness, aversiveness, salience or other characteristics that elicit cognitive responses [3, 12–14]. These response-provoking elements would be the basis for selection of stimuli events to be displayed in the experimental study setting.

The presentation of these events as stimuli needs be indirect rather than as though such events are the primary computing tasks in order to mirror how such events might occur within a regular computing environment. In this case the study is designed so that they can be presented within the workflow of a computer-based activity serving as a distractor. The activity is packaged in a web-based application that ties together the study introductory and closing material, the assignments a given participants needs to complete for their lab session, and the events and triggers serving as the cybersecurity stimuli.

In summary, the stimuli setup components are as follows:

Distractor Exercise: In this case the exercise involves a set of research tasks that lead to creation of computer workstation specifications for a specified group of professionals. The activity workflow takes participants through a path allowing for the presentation of intended stimuli.

Security Events: These are web elements and content intended as the stimuli to be interacted with in the study. They are selected because they feature various properties associated with security breach scenarios. The properties are initially assumed to be response-evoking and then tested later via a survey during the stimuli validation stage. There are two key selection criteria used to include stimuli events in the setup:

i.   **Security-relatedness** i.e. elements should have properties that cause the element to be judged as having security implications e.g. pop-up dialog that prompts a decision on an unsolicited file downloading. Non-security-related events will be utilized as controls.

ii. **Intensity of material** (e.g. visual noise, decision or no decision). Both the security event set and the control event set of elements are further considered on their visual and cognitive demand intensity. Hence the final selections are intended to fit in either into either the intense or the mild category. These selections are later tested via a survey to ensure they elicit the meanings we intended. The purpose of this criteria is to help in the evaluation of the role played by an event's intensity in determining responses evoked in the psychophysiological signal.

Control Events: These are elements selected because they appear to be decidedly non-security related. The goal is to utilize them as control stimuli for comparative study of responses evoked during interactions with them against those evoked during interactions with security-related stimuli. It is supposed that the responses elicited in the two stimuli groups would be distinct from each other, with the former being less aversive in nature than the latter.

2.2 Stimuli Validation Surveys

To ensure that stimuli is perceived to have the implications we intend, we conducted evaluation surveys to ask a subset of people from the sample population – from which the participant pool for the lab study will be drawn – to rate the elements on various metrics to let us know how these elements are perceived with respect to potential effect on security of a device.

The surveys focused on two goals as discussed next,

Survey 1: Evaluation of User Perceived Significance and Risk Level of Material:
The perception of risk may influence the level of responsiveness of a person interacting with digital material. In this survey the goal was to ascertain that the material selected for presentation had a high chance of being interpreted as security-related and potential to elicit responses mimicking those typical in security events (Figs. 3 and 4).

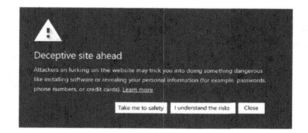

Fig. 3. Example of material presented for evaluation: a warning web page

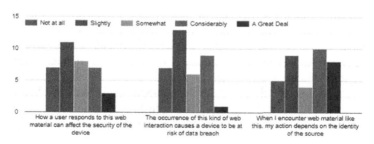

Fig. 4. Sample of summarized results for a subset of the evaluation criteria

Based on the outcomes of these surveys we narrowed down on material that was sound to use as stimuli in as far as perception of security-relatedness was intended. Similarly we identified sound control stimuli against the security-related property.

The selection was limited to materials with evaluation that was not more than 2 SD from the mean, hence abstracting the security relatedness property down to a narrow field where assumptions on general perception could be reasonably made.

Survey 2: Categorization of Material by Security-Relatedness and Intensity: The second survey asked participants to rate material on a likert scale from low to high perceived intensity based on various criteria including visual noisiness and cognitive demand presented by a decision. Similarly, the survey asked participants to rate the material on perceived relevance to security.

Based on the survey, we grouped the elements into four categories based on security relatedness and intensity for use in the analysis phase. for comparative analysis of response signals. The categories used were: Malicious-Intense, Malicious-mild, Benign-Intense, Benign-Mild.

2.3   Other Data Validation Manipulations

- **Single clock setup for timing synchronization:** The same cpu is used to run the stimuli display application and the signal acquisition process. This is one way to achieve millisecond accuracy in the matching of stimuli presentation timing with the timing of response-related changes in recorded signal.
- **Repeated measurement:** Stimuli is presented multiple times with minor variations, to allow for verification of observations by comparing responses to similar stimuli.
- **Handedness questionnaire:** An evaluation of hand dominance versus preference is conducted to account for hand-preference-related variance in speed of response to stimuli.

2.4   Data Collection

Data collection will be conducted in a specialized lab setting with body sensors and data acquisition software that will make the measurement of psychophysiological signals possible.

The setup will include a participant area walled into a semi-private space within the lab, where the stimuli display computer will be available for the participant to undertake the computer-based tasks. The researcher will have a workspace away from the participant's work area from where they will manage the data acquisition and storage during the study sessions (Fig. 5).

In the lab, participants will undertake the computer-based exercise hence allowing us to present the security event stimuli. Throughout the session, we will record the participant's body signals using relevant sensors, and corresponding signal acquisition software.

2.5   Data Cleaning and Processing

During data collection, the signal files will be initially output in the default formats e.g. EDF for the EEG. To process the data, the files will first be converted into spreadsheets with numerical data.

Once the data format is changed, data is manually inspected for integrity. For example in EEG data collection, the baseline is a fixed default gyroscope setting e.g. 4000 mv and therefore none of the recorded values should not be below this (Fig. 6).

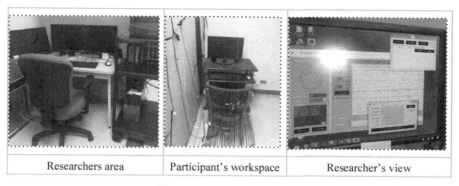

Researchers area	Participant's workspace	Researcher's view

Fig. 5. Data collection setting

s337data.xlsx

	A	B	C	D	E	
1	P3	P4	POZ	PO7	PO8	O1
2	-0.9079993926	-4.30796478	-2.236424876	2.650015757	2.857742157	-1.2
3	-0.4057702449	-3.437510874	-1.305111146	2.408334562	3.179946729	-0.14
4	0.3146382045	-1.616029309	0.04293533846	1.292611653	2.400555732	1.1
5	1.053819037	0.6924996851	1.451605749	-0.3832231668	0.7478738304	2
6	1.579789384	2.868993696	2.522256872	-2.131011596	-1.293498002	2.8
7	1.697283176	4.307687184	2.926687808	-3.428607784	-3.116887161	2.6
8	1.312918196	4.587711691	2.507248041	-3.867743376	-4.159580805	1.6
9	0.4753842766	3.598706396	1.333586605	-3.268640803	-4.057620163	0.088
10	-0.6241460705	1.582733656	-0.3033871871	-1.731083809	-2.749752631	-1.7
11	-1.69499956	-0.9224526637	-1.9622227	0.3901385873	-0.5037923026	-3
12	-2.423862194	-3.225860447	-3.169177571	2.576988654	2.142217057	-4.1
13	-2.564925742	-4.673490534	-3.551984612	4.281443018	4.51064195	-3.5
14	-2.01728103	-4.833604818	-2.95109253	5.064835445	5.955285226	-2.7
15	-0.8657004819	-3.624280379	-1.475304546	4.707692493	6.030601123	-0.68
16	0.6295154914	-1.345638469	0.5163800445	3.263339462	4.616127543	1.7

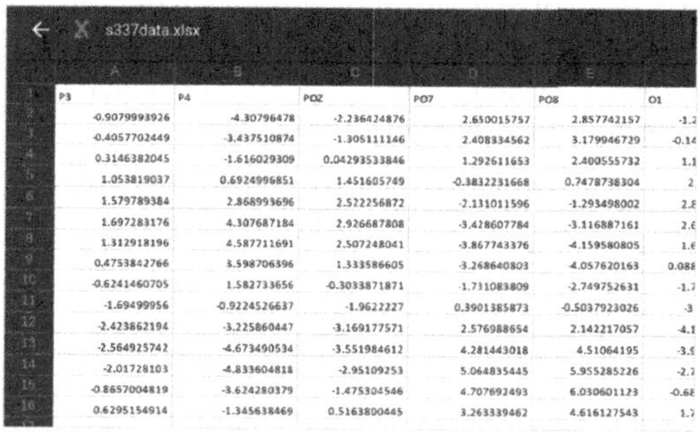

Fig. 6. Section of spreadsheet depicting data collected with emotiv EEG headset

After inspection, the next steps include: trimming out extraneous signal recordings i.e. before task start and after task end, labelling sensor fields in the data as needed, preparing data for analysis by formatting into matrices and storing as variables. Further processing includes removing baselines, extracting epochs and rejecting extraneous epochs, leaving the cleaned, formatted and epoched data ready for analysis.

2.6 Data Analysis

Analysis will be done using two techniques as follows,

Event Related Potential (ERP) Analysis: This will entail running an Independent Component Analysis algorithm on the EEG signal to isolate the ERP event-related potential(ERP) components contained within it. These components reveal the neurological activity underlying various signal changes.

For this analysis, we will be searching for specific ERP components as per the study hypotheses drawn from theory. If anticipated ERP components are present and corresponding to stimuli presentations, they will be utilized in evaluating the study hypotheses (Fig. 7).

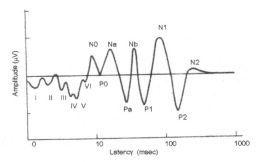

Fig. 7. Graph depicting some of the major ERP Components [16]

Wavelet Transforms: These will be used to analyze and classify the signal change patterns. This process will involve removal of noise from the signal i.e. signal denoising [15], wavelet transformation, feature extraction and classification, determining thresholds and finally locating any structural break points for matching with stimuli display timings to see if there is any correspondence (Fig. 8).

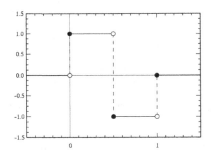

Fig. 8. Example wavelet transform plot: Haar wavelet [17]

2.7 Evaluating Outcomes

The outcomes will be evaluated in two phases:

Non-blind Phase: In this phase, changes in signal will be matched to corresponding stimuli displays if any, by referencing the record of stimuli event display timings. This will facilitate recognition and evaluation of changes in the signal, that relate to security events in particular. Findings from this stage will be used for guidance in the blind phase of the analysis (Fig. 9).

Blind Phase: The blind evaluation will run opposite of the one in the non-blind phase. Starting from psychophysiological signals, we will attempt to locate changes that suggest responses to security events. Based on any changes located, we will hypothesize a timeline of when security-event stimuli was presented (Fig. 10).

This process will not include any attempt to specify the event that was displayed, but rather that a security event was interacted with.

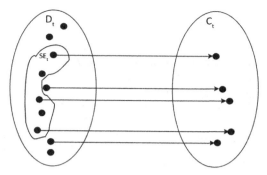

Fig. 9. Finding event markers in signal by mapping security event display timings to changes noted in recorded signal

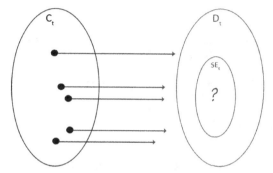

Fig. 10. Changes noted in signal mapped to hypotheses of timings that the corresponding event stimuli was displayed

Afterwards, the hypothesized timings will be verified against the actual event stimuli display timings on record. If we find that we can retrieve the event stimuli timings via the signal, with an accuracy that is greater than chance, then we can claim that the signals contain markers of interaction with those events.

Finally we will work on determining other relevant measures such as error rates and thresholds, which would mark the end of the study.

Fig. 11. Human Computer Interaction (HCI) Outputs = Inputs to Cybersecurity Applications

3 Conclusion

The potential impact of this project is that it highlights and models outputs of human-computer interaction as inputs to cybersecurity applications. If we can harness human cognition in the computing environment for any markers of appraisal, such markers can be useful for various applications (Fig. 11).

For example in digital investigations, they can be used to triangulate other evidence and hence support the justice function. Another envisioned application would be in immersive digital experiences of locked-in individuals, where such markers can be used as input to cybersafety management tools to regulate the experiences of those users. This would contribute towards making cybersafety accessible.

References

1. Mogire, N.: Tokens of interaction: psycho-physiological signals, a potential source of evidence of digital incidents. In: HotSoS 2020, 7–8 April 2020, Lawrence, KS, USA. ACM (2020). ISBN 978-1-4503-7561-0/20/04, https://doi.org/10.1145/3384217.3384226
2. Kahneman, D.: Thinking. Fast and Slow. Macmillan, Basingstoke (2011)
3. Posner, M.I., Snyder, C.R.R.: Attention and cognitive control. In: Solso, R.L. (ed.) Information Processing and Cognition: The Loyola Symposium, pp. 55–85. NJ: Lawrence Erlbaum Associates, Hillsdale (1975). [Ref list]
4. Karray, F., Alemzadeh, M., Saleh, J.A., Arab, M.N.: Human-computer interaction: overview on state of the art (2008)
5. Waller, A.D.: Eight Lectures on the Signs of Life from their Electrical Aspect. EP Dutton, New York City (1903)
6. Loeb, G.E., Loeb, G., Gans, C.: Electromyography for Experimentalists. University of Chicago Press, Chicago (1986)
7. Camm, A.J., et al.: Heart rate variability: standards of measurement, physiological interpretation and clinical use. Task force of the European society of cardiology and the North American society of pacing and electrophysiology. Circulation 93(5), 1043–1065 (1996)
8. Boucseinm, W.: Electrodermal Activity. Springer, Heidelberg (2012). https://doi.org/10.1007/978-1-4614-1126-0
9. Luck, S.J., Kappenman, E.S.: Electroencephalography and event-related brain potentials. In: Cacioppo, J.T., Tassinary, L.G., Berntson, G.G. (eds.) Handbook of Psychophysiology, pp. 74–100, 4th edn. Cambridge University Press, Cambridge (2016). (Cambridge Handbooks in Psychology)
10. Porges, S.W., Campbell, B.A., Hayne, H., Richardson, R.: Attention and information processing in infants and adults. In: Autonomic Regulation and Attention. Lawrence Erlbaum Associates, Hillsdale, pp. 201–223 (1992)
11. Kreibig, S.D., Samson, A.C., Gross, J.J.: The psychophysiology of mixed emotional states. Psychophysiology 50(8), 799–811 (2013)
12. Lang, P., Bradley, M.M.: The international affective picture system (IAPS) in the study of emotion and attention. In: Handbook of Emotion Elicitation and Assessment, 19–29 April 2007
13. Porges, S.W.: Autonomic regulation and attention. Attention and information processing in infants and adults, pp. 201–223 (1992)
14. Sawaki, R., Luck, S.J.: Capture versus suppression of attention by salient singletons: electrophysiological evidence for an automatic attend-to-me signal. Attention Percept. Psychophys. 72(6), 1455–1470 (2010)

15. Denoising and Compression (n.d.). https://www.mathworks.com/help/wavelet/denoising-and-compression.html
16. Kuok, A.: Brainstem potentials, 7 June 2007. https://zh.wikipedia.org/wiki/File:Brainstem.PNG. Accessed 2 March 2020
17. Haar wavelet 3 June 2006. https://commons.wikimedia.org/wiki/File:Haar_wavelet.svg. Accessed 2 March 2020

Tracking and Improving Strategy Adaptivity in a Complex Task

Jarrod Moss[(✉)], Gary Bradshaw, Aaron Wong, Jaymes Durriseau, Philip Newlin, and Kevin Barnes

Mississippi State University, Mississippi State, MS 39762, USA
jarrod.moss@msstate.edu

Abstract. Strategies are a major component of increasing expertise and performance in complex tasks. High performers often have better strategies than low performers even with similar practice. Relatively little research has examined how people form and modify strategies in tasks that permit a large set of possible strategies. One challenge with such research is determining strategies based on behavior. We have developed an algorithm that accurately identifies the strategies that people employ in a complex decision-making task based on task behavior. In this paper, we report different methods to identify strategies that human participants are using in a complex decision-making task and document the efficacy of our methods. Participants have difficulty applying strategies consistently and thereby fail to gain useful performance-related feedback about the strategy's effectiveness. A further challenge is to identify the optimal strategy to use as the properties of the task change over time. We refer to these challenges as strategy consistency and strategy adaptivity. These analyses have led to the construction of a strategy coaching module that enhances the task interface to support consistent application of a strategy and identification of suboptimal strategies. We report initial data on the effectiveness of this strategy coaching module.

Keywords: Decision-making strategies · Training · Adaptivity · Skill acquisition · Human performance

1 Introduction

In complex tasks, one major component of increasing expertise and performance is the strategies that experts use to perform a task [1]. A strategy is a sequence of actions performed to solve a problem or accomplish a task. The process of selecting a strategy for a task can be based on a competition between familiar and available strategies [2]. Strategies can be modeled as production rules with utility values and are selected based on their relative utilities according to a softmax function [3]. A strategy's utility rate increases based on its base-rate of success [2, 3].

In order for strategies to compete in this manner, they must already exist as production rules in the system. One theory which has addressed how strategies might develop and change is the RCCL theory. It includes four components: Represent the task, Consider

© Springer Nature Switzerland AG 2020
D. D. Schmorrow and C. M. Fidopiastis (Eds.): HCII 2020, LNAI 12197, pp. 416–433, 2020.
https://doi.org/10.1007/978-3-030-50439-7_29

sets of task features to include in a strategy, Choose a strategy, and Learn the strategy's success rate [4]. The last two components of choosing a strategy and learning its success rate correspond to the strategy competition component of ACT-R mentioned above [2]. However, if the success rates of strategies are low enough, then the task representation may be altered to include new features that can be utilized in forming new strategies. In simple tasks like those reported in the RCCL paper, this is a relatively straightforward task. However, in more complex tasks, the act of identifying new features and forming strategies based on those features is a significant problem that has received little attention. One possibility is that searching through the space of possible strategies is itself a problem-solving activity where search for a possible strategy operates in a secondary space to the original task. These kinds of dual space searches have been proposed to account for rule induction and scientific reasoning [5, 6].

In a recent category induction study, Prezenski, Brechmann, Wolff, and Russwinkel [7] demonstrated that participants appeared to use a heuristic to generate simpler one-feature rules before more complex multi-feature rules, indicating that the search for a category rule was systematic; however the space of category rules in this research was relatively small. Many experimental tasks permit only a few possible strategies and thereby lack the potential to investigate how large spaces of possible strategies can be explored. One reason that strategy formation has not often been studied in more complex task environments is that it is difficult to identify the strategies that people are using and to determine when they are using them.

The most common approach to examining strategies in studies is to identify one simple measure of task performance that can discriminate between two strategies that researchers have identified and, in some cases, taught to participants. For example, in the Building Sticks Task, the first move made characterizes which of two strategies the person is using [2, 8]. In a study on how participants adapted their strategy in Space Fortress, researchers first taught and trained one flight control strategy before modifying the environment to examine how it impacted the flight control strategy [9]. In this case, the proportion of the time spent on a certain region of the screen could be used to see if participants continued to use the original strategy or a modified strategy. The limited space of possible strategies in these tasks is too restricted to garner evidence for a heuristically-guided strategy search. Moving to complex tasks that allow for a large space of strategies requires a method for tracking the features that people are using in their strategies. This paper presents the results of using machine learning classifiers to track a person's strategy over time as they perform a task.

The ability to track strategy development and change based on behavior in a task enables the examination of strategy development in more complex tasks than have previously been possible. In addition, it offers the potential to augment cognition and training in complex task environments via monitoring and intervening when strategies are not effective. The goal of the research reported here was to develop and test a method for automatically tracking strategy development and change in a complex task and then to design interventions to improve strategy use and adaptation as the task environment changes.

We use a modified version of the Abstract Decision Making (ADM) task, a computerized multitasking environment that is predictive of performance on air traffic control and emergency dispatch tasks [10]. A modified version of the task has also been used to examine individual differences in a multitasking situation with interruptions [11]. The task used in the current research has been further modified to increase the space of strategies that participants can employ to select the next subtask to work on, and we refer to this task variant as the strategic ADM (sADM).

Key task properties that make it both challenging and similar to many real-world decision-making tasks. People must frequently choose which subtask to work on next. The choice of one subtask means that other subtasks must be postponed, which has implications for the person's score because the subtasks have a deadline and penalty for missing that deadline. It is analogous to the decisions that might occur in emergency dispatch where the dispatcher has to make a choice about which of many emergency calls to handle first. Each decision necessarily means that the handling of other calls will be delayed. Therefore, one of the main sources of variance in performance in the sADM is the selection strategy that people use to select the next subtask to work on.

Following an overview of the sADM task and the strategy classification algorithm, a study validating the performance of the algorithm is reported. The strategy detection algorithm was validated via simulation of strategies in a computational cognitive model that performs the sADM task. In this manner, the underlying ground truth is known and the algorithm can be assessed on how well it can recover the underlying strategy driving behavior. Next we describe using this algorithm to track strategies in a set of data on humans performing the sADM task. We close with a discussion of a strategy coaching module that we have developed that uses this strategy tracking algorithm to improve strategy use and adaptation of one's strategy to changing task demands.

2   Description of the sADM Task

The sADM task consists of two interleaved activities: *selecting an object* to work on and *processing that object* by sorting it into a bin based on its attributes. This structure mimics real-world tasks such as emergency dispatch where there are multiple tasks one could work on and each requires some set of actions to process it before moving on to the next task [10]. Processing an object involves querying its attributes and then sorting it into one of four bins based on those attributes. Additional objects appear in the queue either after an object has been sorted or during the sorting process (i.e., interrupting the flow of the sorting process). The basic structure of the task is shown in Fig. 1.

The sADM task requires users to sort objects into one of four different bins depending on the object's properties. Prior to beginning work on the task, participants memorize the attributes of the bins so that they can correctly sort objects into appropriate bins. The interface is text-based and controlled via four keys on the keyboard. Each object is only identified by its name and the participant must execute a series of keystrokes to query the properties of the object. For example, a participant might select the object SOF from the example queue of objects shown in Fig. 2. After selecting SOF, the person then presses keys to query the interface to report that the object is yellow, then executes another query to identify it as large, and finally a third query to identify that the object is shaped like a triangle. Based on this object attribute information, held in working memory, the participant now knows that the object belongs in bin 4 and can execute a series

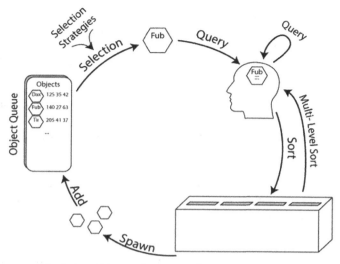

Fig. 1. A conceptual depiction of the routine sequence of actions in the sADM task. An object is selected for processing from a queue of objects that could be processed. Each object has features that affect the user's performance score. Object attributes must be queried and the results held in working memory. The object must be sorted into the bin that matches its attributes. Some objects require multiple rounds of querying and sorting. Finally, new objects may appear after sorting and the process repeats until the time limit is reached.

of key presses to sort it. Because each bin only accepts certain attribute combinations (e.g., large, orange triangles), the user must query objects to see their attributes and then sort them into the appropriate bins. Some objects require querying and sorting on one set of attributes (visual-based attributes of size, color, and shape) and others require querying and sorting on two sets of attributes (sound-based attributes in addition to visual attributes). Objects that require two levels of querying/sorting therefore require about twice as long to process.

Object Name	Time/ Deadline	Points	Penalty	Queue Position	Selection Distance
SOF	001/060	203	-45	1	2
ZIR	047/120	402	-100	2	1
TIW	029/030	103	-38	3	0
WOH	015/020	153	-10	4	1
JIQ	100/110	237	-57	5	2

Fig. 2. Sample queue showing all object features available in the sADM task. The last two columns do not appear explicitly in the queue that participants see, they are implicit in their position in the queue.

A task block is presented for six minutes, and the goal is to score as many points as possible. Objects have a range of point values, penalties, and deadlines as shown in Fig. 2. Sorting the object correctly awards the participant the point value of the object. Sorting the object into the wrong bin subtracts the object's point value from the current score. Every object has a deadline and timer that counts down from that deadline to zero. Every time the counter reaches zero, the counter resets to the deadline and the penalty value for that object is deducted from the score. This penalty for elapsed deadlines occurs for all objects in the queue, including the object that is currently being processed by the participant. Therefore, performance depends on utilizing strategies that take into account multiple, dynamically changing factors (e.g., points, deadline, time until the object's next deadline).

The queue of objects initially starts with 1–3 objects, and each time an object is successfully sorted one to three new objects appear probabilistically. The task adjusts the probability of new objects arriving so that the queue contains approximately 7–12 objects at any given time. Objects can also occasionally arrive while the participant is processing an object as a form of interruption. When this interruption occurs, the participant must choose whether to continue processing the current object or switch to the interrupting object.

The primary way to maximize performance is to sort as many high point objects before their deadline and to prevent high penalty objects from remaining in the queue and accumulating penalties. Selection strategies used to select objects to work on are therefore critical to performance. In the queue, the participant can see the object's name, deadline, point value, penalty value, how many seconds remain before the deadline elapses again, and the number of levels that an object must be queried/sorted on. These object features, along with the position of the object in the queue, are the possible features that can be used to guide one's selection strategy.

3   Strategy Classification in the sADM Task

The sADM selection strategy classification algorithm takes as input a list of the objects on each queue presented to a participant, including all of the object features, and whether each object was selected from the queue. Each sADM task block yields a list of objects that appeared in the queue for each selection that a participant made and the nine features of each object: points, deadline, time until deadline, penalty value, the number of sorting levels required, whether work on the object was interrupted, position in the queue, selection distance, and the selection number. When a person begins object selection, the middle object in the queue is initially highlighted and they move the selector up or down using key presses before selecting an object. The selection distance is simply how many times the selector had to be moved to reach an object. The selection number is a feature that can be used to group together all objects that appeared on the same queue. For example, there may have been nine objects present on the queue when a participant made their fourth object selection. These nine objects would all have a selection number of four. The category label that the classification algorithms are being trained on is a binary value (selected/unselected). Both a linear support vector machine (SVM) and a decision tree (DT) classifier were implemented using the scikit-learn library [12] and trained to predict which object a person would select from this data.

After each of the DT and SVM classifiers were trained to predict which objects a participant selected, then the classifiers were analyzed to determine the features present in that participant's strategy. A strategy is represented as a series of features ordered by importance and valence. Valence refers to whether higher or lower values on a feature were more likely to be selected. For example, a strategy represented as [Points+, Deadline−] means that higher point value was the most important feature, but the participant also preferred lower deadline values.

The SVM and DT classifiers use different techniques to learn to categorize instances of selected and unselected objects from sADM data. Each classifier gets information about a number of features for each selected and unselected object (e.g., points, deadline, time in queue). The SVM classifier divides up this multidimensional feature space by separating the selected and unselected objects with a hyperplane. The DT classifier builds up a hierarchical set of rules to classify objects as selected or unselected (e.g., if the object has the highest point value then it is selected, if not then if it has the lowest deadline then it is selected, otherwise it is not selected). Therefore, the DT classifier represents a participant's strategy as a sequence of binary decisions while the SVM classifier represents the strategy as a hyperplane. The components of the DT or the location of the hyperplane in the feature space can be analyzed to determine the important features of a participant's strategy.

Both the SVM and DT methods take a common approach to preprocessing feature data. All features within one selection number are scaled to a 0 to 1 scale. Features are scaled within each selection number to reflect the structure of the decision. For example, an object in a queue with the highest point value will have a 1 after scaling. This method of scaling is important because these classifiers treat each object as a separate piece of data to be trained on. They do not make a selection of one object from the queue of objects. Instead each object from each queue is a separate instance labeled as either selected or unselected. The classifier's job is to learn to predict whether each object is selected or unselected. There are more unselected than selected objects, and so both classifiers are trained with weights to balance the contribution of both selected and unselected objects on the resulting trained classifier. Following feature scaling, the data are split into three cross-validation folds such that the classifier is trained on two-thirds of the data and produces a prediction accuracy on the other third.

All objects from a given selection queue are semi-randomly placed into one cross-validation fold so that objects from multiple selection queues are not spread across folds. The process is semi-random because the selection queues from a given block are divided up into those that occur in the first, second, and third third of the block. Each of the three cross-validation folds will contain one third of the data from each third of the block. This constraint was included so that any strategy differences that occurred over time in the block would be equally represented in each cross-validation fold.

Traditional machine learning applications of these classifiers have the primary goal of maximizing prediction accuracy on novel data. However, our goal is to best represent the decision-making strategy that a participant was using. Therefore the cross-validation process was not used to maximize accuracy, but it was instead use to tune hyperparameters of the classifiers. These hyperparameters control how complex the classifier is allowed to be, which translates roughly into the number of object features that were used to classify the data. More features indicate a more complex strategy, but classifiers can also use a plethora of features to over-fit the training data, generating high performance on training sets but limiting the generality for new data.

A grid search was performed for each hyperparameter and the primary measure obtained was to assess the agreement on the feature extracted from each cross-validation fold. If perfect agreement was found for multiple hyperparameter values, then the hyperparameter with the highest cross-validation prediction accuracy was selected. The purpose of this approach is to allow the complexity of the strategy to vary so long as all three folds yielded the same strategy. This approach allows for the most complex strategy supported by the data (including accuracy on untrained data) to be extracted.

Both machine learning classifiers learn to classify a single object as either selected or unselected. However, a good representation of a participant's strategy should lead to a prediction about which object that participant will select from a queue of objects. In order to test the classifiers' ability to predict this behavior, testing accuracy was calculated on predicting the object that would be selected from a queue as opposed to simply allowing the classifier to classify each object as selected or not selected. Both the DT and SVM classes in the scikit-learn library have a method that allows for a probability to be generated instead of a binary classification. For all of the objects in a queue, the classifiers rank ordered the objects that would be selected according to their predicted probability of being selected. A rank order accuracy score was calculated by this formula: (queue_size-rank)/(queue_size-1). Here queue_size is simply the number of objects in the queue and rank is the rank order assigned to the object that the participant actually picked. This rank accuracy score has a maximum value of 1 when the person's selection matches the classifier's top ranked object and has a minimum of 0 when the person picks the object ranked last by the classifier. The expected value of this rank accuracy score if the classifier assigned ranks randomly would be 0.5.

This rank accuracy score was used instead of a binary accuracy score so that the accuracy score would retain some sensitivity to a person's underlying strategy even if the person did not always pick the optimal object under a strategy. For example, a person picks the second highest point value because of an error or it was simply close enough to the maximum point value object (satisficing behavior). The final strategy reported was based on which of the DT and SVM classifiers reported the highest rank accuracy score on the test data when there was agreement on the strategy across the cross validation folds. This approach using both classifiers was used since some strategies, such as a weighted additive strategy for two features, would be better matched by the structure of an SVM, while other strategies would be better matched by the structure of a DT.

3.1 Validating the Strategy Classification Algorithm

Without knowing the strategy a person was actually using, it is difficult to establish the accuracy of strategy detection. People reported using a number of object features in complex ways, but this self-report data did not predict their actual selections very well. To circumvent this validation conundrum, an ACT-R cognitive model was created that interacts directly with the same sADM task that human participants perform. ACT-R is a cognitive architecture that incorporates a number of constraints consistent with theories of human cognition [3] and was used to simulate behavior across a range of selection strategies in the sADM task. The ACT-R model consistently employs the strategy every time it selects a subtask in the sADM, providing an ideal set of data with a known ground truth for comparison with the output of the strategy classification algorithm. The complexity of the model's strategy can be controlled, and it is also possible to insert a controlled amount of noise into the model's behavior to examine the performance of the strategy tracking algorithms in the presence of errors or satisficing behavior.

Human participants face a number of choices at various times in the sADM task such as how many object properties to query before sorting, and how to handle an error in sorting. An idealized participant was implemented in the ACT-R model: other than selecting the next object to work on, the model always performed the same sequence of actions. It queried the minimum number of object properties to determine the correct bin, and it never made a sorting error. Different ACT-R models only varied in the selection strategy used. After initial construction of the model, and using default parameter values, the ACT-R model's performance was compared to human participants in terms of the number of objects that it could sort in a 6-min sADM task block. Compared to human participants that spent about 1 h practicing the task, ACT-R's performance fell in the 60th percentile of the human data. Therefore, the data presented to the strategy classifier algorithm being tested is similar to a human with moderate task experience.

The results from an ACT-R model of eight object selection strategies are presented here. Each model employed only one of these strategies. The two simplest strategies each used a single feature that was directly presented to participants in the queue: pick the object with the highest point value and pick the object with the lowest deadline. The third and fourth strategies were combinations of using points and deadline. A points-deadline weighted strategy selected the object with the highest weighted sum of points and deadline. A points-deadline threshold strategy selected objects based on points if there were objects with point values greater than 400 in the queue and otherwise it picked the object with the lowest deadline.

The remaining four strategies were noisy versions of the simple single-feature strategies. For example, the model would choose randomly 10% of the time and the other 90% it would follow the points strategy (i.e. a points-noise or a deadline-noise strategy). Finally, a model that incorporated satisficing behavior for both the points and deadline strategies was implemented: points-satisfice and deadline-satisfice. The points-satisficing strategy was to pick the first object within 50 points of the highest point value in the queue. These variants were created to simulate noise expected to be in human data. People are known to satisfice in decision making and occasionally make errors.

Table 1 shows the results of applying the algorithm to each of 60 simulated 6-min blocks of sADM data. The rank accuracy score is the rank accuracy on the testing data from the cross-validation process. Rank accuracy scores can be converted into an approximate rank order of the object selected by the person by assuming a given queue size. In this case, the mean queue size (7.13) was used to produce the rank order shown in the table. The final column is the accuracy of the strategy reported by the algorithm. For a strategy to be scored as correct, all features and their correct valences had to be reported by the algorithm.

The algorithm does well with one and two feature strategies. It also performs reasonably well with single feature strategies in the presence of noise or satisficing. However,

Table 1. Results of recovering strategy in simulated data

Strategy	Rank score	Rank order	Strategy accuracy
Points	0.997	1.02	100%
Deadline	0.999	1.00	100%
Points-Deadline-weighted	0.951	1.3	95%
Points-Deadline-threshold	0.959	1.25	90%
Points-noise	0.944	1.35	95%
Points-Deadline-weighted-noise	0.877	1.75	70%
Points-satisfice	0.957	1.26	88%
Points-Deadline-weighted-satisfice	0.929	1.44	63%

the accuracy of the algorithm drops with two feature strategies in the presence of noise or satisficing. In these cases, the strategy accuracy declines because only one of the two features (either points or deadline) is identified but not both. This drop in accuracy with noise and more complex strategies is most likely due to the limited amount of data with which to train the algorithm. When pairs of blocks are used instead of single blocks, so the number of selections in the data is roughly doubled, the accuracy of the algorithm on two feature strategies improves to the same level as without noise or satisficing.

In summary, the strategy classification algorithm does well on this simulated data, but there are limits to its ability to detect more complex strategies with noise in the data. People have been shown to simplify their decision-making procedures, often reducing a multi-attribute decision to a single feature under time pressure [13], so the algorithm may perform reasonably well in the time-pressured sADM task with human data.

4   sADM Strategy Classification with Human Data

Given the success of the strategy classifier algorithm on simulated data from the sADM task, the next step was to test the ability to track strategies on people performing the task. The ability to identify strategies automatically based on behavior in a task that has a sufficiently complex space of strategies opens up the possibility to address a number of questions that have received little attention in the literature. The primary question to be addressed is to what degree does the strategy algorithm identify stable strategies in peoples' behavior. Of course, in this case, the actual strategy that participants are using is not known, but the predictive accuracy assessed with the rank accuracy score is highly correlated with the strategies identified in the simulated sADM data. Therefore, this measure can be used to address the question of how well the strategy classifier algorithms work on human data.

The sADM task was designed with three different strategy scenarios, or blocks of the task where one particular selection strategy is optimal. By examining performance on the sADM task in which a number of these strategy scenarios are encountered, we can address the question of how often people adapt to the scenario and select the best strategy for that scenario. Finally, the data will provide an opportunity to examine points where people fail to adapt their strategies or where there is high variability across individuals in performance. These points of failure or high variability are targets for training augmented with the ability to know what strategy (if any) people are using. So the third purpose of this study is to identify these points where our strategy-based coaching system can have the most impact in improving strategy-based performance in the task.

4.1 Method

Participants. The participants were 64 students or staff from Mississippi State University who participated in exchange for compensation. All participants stated that they were native-English speakers with normal hearing and color vision.

Design. The experiment consisted of two sessions in which a total of 20 blocks of the sADM task were completed. The values of object features in each block were designed so that a particular selection strategy would yield the best performance. For example, a selection strategy that picks the object with the shortest deadline will lead to the best score when the variability of the deadline feature is high across objects, but the variabilities of the points and penalty features are low. In this manner, blocks of the sADM task can be designed to have an optimal strategy.

Four consecutive blocks all had the same optimal selection strategy, which was shifted for the next four blocks. These groups with the same optimal strategy are referred to as strategy scenarios. Three different strategy scenarios were used: points, penalty, and deadline, reflecting the best strategy to maximize overall score.

Participants saw a total of four of these strategy scenarios in the following order: points, deadline, points, penalty. They were not informed about the scenarios or when the switches occurred. These four scenarios consisted of four 6-min blocks for a total of 16 blocks. This strategy scenario design was used so that the impact of the scenarios on the strategies that participants used could be examined. A final set of four blocks were then presented in which the optimal strategy shifted within each block. The results presented in this paper focus on only the first 16 blocks.

Procedure. For the first 1.5-h session, participants completed a tutorial on how to complete the sADM task which also included memorizing the attributes of each of the bins that objects were sorted into. Following the tutorial, the participants completed a practice block in which they simply had to successfully sort three objects to finish the block. Object features in this block were designed to have identical values so that no selection strategy was better than any other. After the practice block, participants completed eight six-minute blocks which constituted a points strategy scenario followed by a deadline strategy scenario.

In the second session, participants completed 12 more blocks of the sADM task after a brief refresher on the attributes for each bin. The first four blocks of this session were again a points strategy scenario followed by four blocks of a penalty strategy scenario. The last four blocks were the ones in which the optimal strategy changed midway through the block.

Participants were told that they would receive a performance incentive based on the average of their top three sADM block scores. The incentive was to encourage high performance throughout the entire study because there was alway an incentive to improve on one's high score. Participants' top three scores were shown after each block.

Participants answered two questions at the end of each block. The first asked participants to identify the features they were basing their object selections on. Participants simply selected all of the features they had used during the prior block. The second question asked participants to rank order the importance of all of the selected features. These questions assessed how participants thought they were selecting objects and were used to determine to whether this was consistent with the task selections they made.

sADM Performance Measures. The primary sADM performance measure is simply the score earned in each block. In addition, the number of sorting errors, sorting time, and query efficiency were also used as measures of performance for the sorting component of the sADM. A sorting error is placing an object in the wrong bin. Note that a sorting error results in the object's point value being deducted from the total score. Sorting time is simply how long it took a person to sort the object from the time they selected it from the queue until it was successfully sorted (i.e., the total time they were actively sorting the object). Query efficiency is a measure of how many excess queries were made on an object. To correctly sort an object requires knowledge of only two of its attributes for each level of sorting. Query efficiency is simply the number of attribute queries made divided by the minimum number required to sort it and has a value of 1 if a person queries only the minimum number of features necessary to ensure error-free sorting.

4.2 Results

Selection Strategies Used in the sADM. The strategy classification algorithm described earlier was used to determine the strategy for each participant for each block of the sADM task. Table 2 shows the number of blocks in which a given selection strategy was identified for each of the three strategy scenarios used. A strategy is identified by the constituent object features and a valence. For example, "Points+" means that the strategy selects the highest point value object.

The fourth column of the table shows the mean rank accuracy score of that strategy, which ranges from 0 to 1 with an expected value of 0.5 for random object selections. The fifth column shows the rank order of the object predicted by the classifier assuming a queue size equal to the mean of all queue sizes in the data, 7.13 objects. A value of 1 in this column would mean that the object selected by the participant was the same as the top object predicted by the strategy classifier. A value of 2 in this column would mean that the object selected by the participant was the object the strategy classifier ranked as the second most likely object, and so on.

There are a few properties of this data worth noting. First, the optimal strategy appears in this list for each strategy scenario. It is the most likely strategy under the points scenario, and it is the second most likely strategy under the penalty scenario. For the deadline scenario, participants preferred a time until deadline strategy over the deadline strategy. These object features are correlated, and the time until deadline strategy performs nearly as well as the deadline strategy in our ACT-R simulations of the sADM.

Table 2. A sample of strategies used in different sADM strategy scenarios

Strategy scenario	Strategy	N	Mean rank score	Mean rank
Deadline	SelectionDistance−	82	0.92	1.47
Deadline	TimeTillDeadline−	60	0.85	1.93
Deadline	No Strategy	24	-	−
Deadline	Deadline−	19	0.80	2.21
Deadline	FinalLevel−	19	0.78	2.35
Deadline	Points+	14	0.87	1.78
Deadline	QueuePosition−	13	0.81	2.14
PenaltyPoints	TimeTillDeadline-	61	0.86	1.86
PenaltyPoints	PenaltyPoints−	51	0.85	1.90
PenaltyPoints	SelectionDistance−	45	0.94	1.34
PenaltyPoints	No Strategy	27	−	−
PenaltyPoints	PenaltyPoints−, TimeTillDeadline−	10	0.81	2.18
Points	Points+	120	0.89	1.65
Points	No Strategy	101	−	−
Points	TimeTillDeadline−	89	0.86	1.85
Points	SelectionDistance−	73	0.92	1.48

Second, the strategy classifier could not identify a strategy a substantial portion of the time. This is reported as "No Strategy" in Table 2. The histograms of rank accuracy score in Fig. 3 show these "No Strategy" values at the level of chance (i.e., 0.5). This figure shows that the proportion of these failures to identify a strategy occur predominantly in the first half of the task. However, they continue to be present in a substantial number of blocks in the second half of the task.

Third, there is substantial variability in how well the strategy identified explains the selections that people make. There are a few possible explanations for the strategy not correctly predicting the object selected by a participant. First, there can be ties in the features (e.g., two objects in the queue with the same highest point value). However, these ties occur rarely and based on our ACT-R data, they can be expected to lead to an average rank order of 1.04. The second explanation is that the strategy classifier has only captured a portion of the strategy. However, our ACT-R data shows that the classifiers do very well at capturing strategies that utilize up to two features without noise or satisficing. The third explanation is that participants did not always select the optimal object for a given strategy. Participants could be satisficing and picking near optimal objects, simply making errors, or exploring the strategy space without settling on a specific strategy. Our ACT-R simulations incorporating minor amounts of satisficing behavior and random selections (i.e., errors) are in line with the range of data in Fig. 3.

Based on these comparisons with the ACT-R simulation data, our interpretation of these data is that there is substantial variability in how consistently participants are

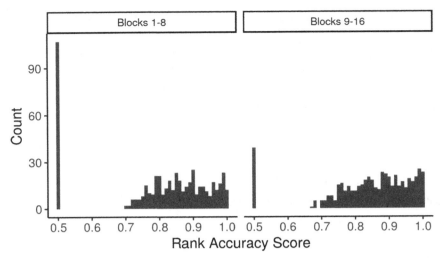

Fig. 3. Histogram of rank accuracy scores per block for the first and second half of the 16 blocks of the sADM task.

carrying out their strategies. In analyses reported below, we will use the rank score as a measure of strategy consistency in order to examine whether the consistency with which one can apply a strategy impacts performance on the task.

Predictors of Task Performance. If participants use the optimal selection strategy for a given strategy scenario in the sADM task, then performance should be higher than if they used any other strategy. Figure 4 displays the mean score on the sADM task across participants for each block for participants who used the optimal selection strategy and those who used some other strategy. Use of the optimal strategy was determined by examining the object features that were important for each participant's selection decisions for each block. If the most important feature matched the optimal strategy, then the participant was considered to be using the optimal strategy. This measure does allow for participants to be using additional object features as long as the most important feature was optimal for a given block within a strategy scenario. Figure 4 shows that using the optimal strategy generally led to better performance.

In order to examine this relationship between strategy use and total score while also accounting for other performance-related measures, a set of measures from the sorting portion of the sADM task were all used to predict total score in a linear mixed effects model. The measures from the sorting portion of the sADM were the number of sorting errors, sorting time, and query efficiency. A binary predictor was included that was set to 1 if the optimal strategy was used or 0 for other strategies. Finally, the rank score from the strategy classifier was included as a predictor as a measure of the consistency with which participants applied the strategy they were using in a given block.

The resulting model fit is shown in Table 3. All performance measures from the sorting portion of the task had an impact on score. Better query efficiency was always a significant predictor of scoring more points. Sorting errors and the time taken to sort an object were also significant but their effect decreased with practice (as these

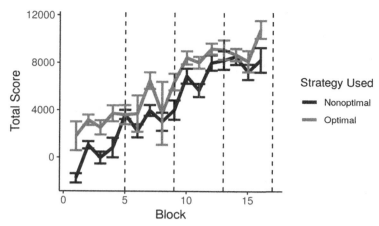

Fig. 4. Mean score on the sADM task for each block for participants who used the optimal selection strategy and those who used a nonoptimal strategy.

measures generally asymptote with practice). Importantly, using the optimal strategy was a significant predictor of higher scores, consistent with the data shown in Fig. 4. In addition, strategy consistency was a predictor that increased in importance in later blocks. Figure 5 depicts predictions from this model for the effects of strategy consistency and optimal strategy use. Consistent strategy users achieve higher scores especially in later blocks, and optimal strategy users outperform non-optimal strategy users.

Table 3. Model fit for model predicting total score from sADM performance and strategy use

Term	Estimate	SE	df	t	p
(Intercept)	6162.84	2323.21	734.76	2.653	0.01*
Block	365.01	205.81	636.79	1.774	0.08
Sort errors	−309.96	781.19	652.29	−0.397	0.69
Sort time	−162.93	40.49	215.55	−4.024	<.001*
Rank accuracy score	39.91	2400.71	825.88	0.017	0.99
Optimal strategy used	638.44	196.22	809.05	3.254	0.001*
Query efficiency	−1949.96	864.55	330.02	−2.255	0.02*
Block x sort errors	−846.84	96.68	756.62	−8.759	<.001*
Block x sort time	−57.2	5.63	696	−10.16	<.001*
Block x rank accuracy score	507.71	233.7	636.86	2.173	0.03*

* indicates significance at $\alpha = .05$

The left half of Fig. 6 shows the proportion of participants who used the optimal strategy in each block. The points scenario was used from blocks 1 through 4 and for blocks 5 through 8 (at the beginning of both sessions), and the figure shows that

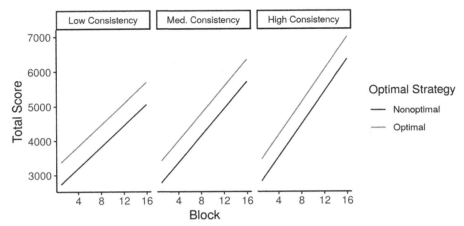

Fig. 5. Depiction of the effects of using the optimal strategy and increasing strategy consistency as measured with the rank accuracy score.

the proportion of participants using a points-based strategy steadily increased in these blocks. However, participants struggled to identify the optimal strategy in the deadline scenario presumably because the time until deadline strategy appeared better or at least more salient. Finally, the proportion of optimal strategy users in the penalty scenario was intermediate between these other scenarios.

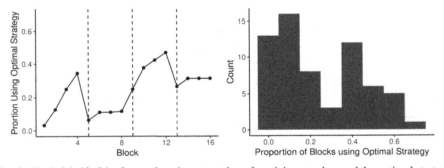

Fig. 6. The left half of the figure plots the proportion of participants who used the optimal strategy on each block of the sADM. The right half is a histogram showing the frequency with which individual participants used the optimal strategy.

4.3   Discussion

One of the primary questions to be addressed in this study was the extent to which the strategy classifier algorithms could be used on human data. The results strongly support that these algorithms work well. In the second half of the task, the algorithms identified strategies at above chance levels of prediction accuracy over 92% of the time. The rate was a bit lower in the first half of the task, 79%, likely due to people still learning the task. In support of this, we also find that sorting errors, query efficiency, and the amount of time taken to sort an object also asymptote after the first half of the task. These results could mean that simply learning how to execute the task well interferes with the ability to employ a consistent selection strategy.

The second question addressed in this study was to what extent people adapt their strategy to match the strategy scenario's optimal strategy. The data shown in the left half of Fig. 6 show that the proportion of participants using the optimal strategy never reaches 50% on any block. The right half of Fig. 6 shows that there are substantial individual differences in the rate at which individuals adapt their strategy to the current scenario. These results raise questions about how people search for an effective strategy or whether other sources of individual differences (e.g., working memory, reasoning ability) may underlie these differences in optimal strategy use. These questions are beyond the scope of the current paper, but this variability in optimal strategy use provides a point of where intervention may be introduced by a strategy coaching system.

Two possible points of intervention for strategy coaching arise from this study. First, there is the variability in optimal strategy use, suggesting participants do not know how to identify the optimal strategy. Second, there is substantial variability in the predictive rank accuracy score shown in Fig. 3, suggesting that, behavior was so inconsistent that a consistent selection strategy could not be found. However, for many more cases, the algorithm struggled to predict exactly which object was selected. When people select the object the algorithm ranks as second or third most likely to be selected, then it indicates that either the strategy identified is not complete (i.e., there are additional object features that people are using) or people are not picking the optimal object for their strategy. Our simulations of various levels of random noise and satisficing behavior in the ACT-R sADM data do appear similar to the range of predictive accuracy values seen in Fig. 3. It is impossible to know exactly which of these two possibilities (or both) are true, but this variability in consistently applying a strategy is another potential target for the strategy coaching system. In summary, the results support the use of these algorithms in tracking strategies and support two potential targets for a strategy coaching system: optimal strategy use and consistent strategy use.

5 Strategy Coaching and Broader Implications

The strategy classifier algorithm and the results from the study of strategy use in the sADM task lay the groundwork for utilizing knowledge of the strategies that people are employing to improve strategy use. We have developed a strategy coaching system for the sADM task in order to examine the potential for training people on strategy use. The coaching system includes three components: a declarative knowledge training phase, an online strategy consistency monitoring system, and an optimal strategy monitoring system.

The declarative knowledge training phase establishes a foundation of knowledge that ensures that all participants recognize that selection strategies have performance implications and establishes a common vocabulary for discussing them. First, this phase explains what a selection strategy is and how someone might use three simple selection strategies to choose an object from the queue. It presents a points, a deadline, and a penalty strategy. It also explains that in the sADM, a strategy's effectiveness can only be assessed over a matter of minutes because the impact of penalty occurs over time as unselected objects accumulate penalties by remaining in the queue. It then shows a video of the computer performing the task over a few minutes with these three different strategies. The video provides an opportunity to observe the impact of different strategies without having the additional load imposed by completing the task. Finally, there is explicit instruction about the variability in object features that may be present in the queue and their relationship to strategy effectiveness. A test consisting of having to identify which strategies should be employed for a given set of object queues is given both at the start and end of this phase to measure knowledge. This training has been piloted on a small group of participants ($N = 19$) with significant improvements seen from pretest to posttest, $t(18) = 4.3, p < .001$. In addition, there is an improvement in consistency of strategy use from sADM blocks before the training phase to those after the training phase, M $= .61$ to M $= .76, t(18) = 4.7, p < .001$.

The strategy consistency and optimal strategy monitoring components utilize the strategy classification algorithm to track whether people are employing strategies consistently and if so whether the strategy is the optimal one for the strategy scenario being presented. For consistency monitoring, if low consistency is detected, then the participant is prompted to identify a feature they would like to base their selection on and the interface highlights the object consistent with selecting based on that feature. This scaffolding persists for 10 selections and then the objects stop being highlighted. This approach is intended to help temporarily minimize the load of identifying the best object and thereby encourage consistent strategy application. The optimal strategy intervention is triggered if participants continue to use a non-optimal strategy after the first two blocks of a strategy scenario. The task is paused early in the third block and participants are asked to identify the object feature in the queue with the highest variability. This intervention is intended to prompt individuals to shift out of their current strategy to the optimal strategy, perhaps because they have not noticed the change in the queue with the shift in the strategy scenario. This type of scaffolding has been used successfully in intelligent tutoring systems [14].

One of the key questions to be addressed in future work with this coaching module is whether the interventions lead to higher strategy consistency and more optimal strategy use in future strategy scenario shifts that occur post-intervention. More broadly, a future research topic is to identify the cognitive processes involved in effective strategy development and adaptation. In particular, we expect that individual differences might lead someone to be better at developing effective strategies in a dynamic task. Identifying these processes and individual differences will provide a basis for training applications, and it opens up the possibility to investigate transfer of strategy training across tasks.

Even if transfer of strategy training is confined to the task it is trained in, there may be substantial benefits to identifying those individuals most likely to benefit from strategy training within a task. Given the importance of strategies in expert performance [1], an individual who can flexibly shift strategies within a task as the task environment changes should exhibit high levels of performance even if the skills acquired are limited to the task at hand.

The strategy classifier algorithm that has been developed for the sADM has enabled these questions to be addressed within this task designed to capture some key elements of other time-pressured real-world tasks. However, there is nothing specific to the sADM task in the algorithm. Any task where strategic decisions are made based on features present in the task environment should be able to be analyzed with the same approach.

References

1. Schunn, C.D., McGregor, M.U., Saner, L.D.: Expertise in ill-defined problem solving domains as effective strategy use. Mem. Cogn. **33**, 1377–1387 (2005). https://doi.org/10.3758/BF0319 3370
2. Lovett, M.C., Anderson, J.R.: History of success and current context in problem solving. Cogn. Psychol. **31**, 168–217 (1996)
3. Anderson, J.R., Bothell, D., Byrne, M.D., Douglass, S., Lebiere, C., Qin, Y.: An integrated theory of the mind. Psychol. Rev. **111**, 1036–1060 (2004)
4. Lovett, M.C., Schunn, C.D.: Task representations, strategy variability, and base-rate neglect. J. Exp. Psychol. Gen. **128**, 107–130 (1999)
5. Klahr, D., Dunbar, K.: Dual space search during scientific reasoning. Cogn. Sci. **12**, 1–48 (1988)
6. Simon, H.A., Lea, G.: Problem solving and rule induction: a unified view (1974)
7. Prezenski, S., Brechmann, A., Wolff, S., Russwinkel, N.: A cognitive modeling approach to strategy formation in dynamic decision making. Front. Psychol. **8**, 1335 (2017)
8. Schunn, C.D., Lovett, M.C., Reder, L.M.: Awareness and working memory in strategy adaptivity. Mem. Cogn. **29**, 254–266 (2001). https://doi.org/10.3758/BF03194919
9. Moon, J., Betts, S., Anderson, J.R.: Individual differences and workload effects on strategy adoption in a dynamic task. Acta Psychol. (Amst.) **144**, 154–165 (2013)
10. Joslyn, S., Hunt, E.: Evaluating individual differences in response to time-pressure situations. J. Exp. Psychol. Appl. **4**, 16–43 (1998)
11. Bai, H., Jones, W.E., Moss, J., Doane, S.M.: Relating individual differences in cognitive ability and strategy consistency to interruption recovery during multitasking. Learn. Individ. Differ. **35**, 22–33 (2014)
12. Pedregosa, F., et al.: Scikit-learn: machine learning in python. J. Mach. Learn. Res. **12**, 2825–2830 (2011)
13. Oh, H., Beck, J.M., Zhu, P., Sommer, M.A., Ferrari, S., Egner, T.: Satisficing in split-second decision making is characterized by strategic cue discounting. J. Exp. Psychol. Learn. Mem. Cogn. **42**, 1937–1956 (2016)
14. Azevedo, R., Hadwin, A.F.: Scaffolding self-regulated learning and metacognition - implications for the design of computer-based scaffolds. Instr. Sci. **33**, 367–379 (2005)

Computing with Words in Maritime Piracy and Attack Detection Systems

Jelena Tešić[1], Dan Tamir[1(✉)], Shai Neumann[2], Naphtali Rishe[3], and Abe Kandel[4]

[1] Texas State University, San Marcos, TX 78666, USA
{jtesic,dan.tamir}@txstate.edu
[2] Eastern Florida State College, Cocoa, FL 32922, USA
[3] Florida International University, Miami, FL 33199, USA
[4] University of South Florida, Tampa, FL 33620, USA

Abstract. In this paper, we propose to apply recent advances in deep learning to design and train algorithms to localize, identify, and track small maritime objects under varying conditions (e.g., a snowstorm, high glare, night), and in computing-with-words to identify threatening activities where lack of training data precludes the use of deep learning. The recent rise of maritime piracy and attacks on transportation ships has cost the global economy several billion dollars. To counter the threat, researchers have proposed agent-driven modeling to capture the dynamics of the maritime transportation system, and to score the potential of a range of piracy countermeasures. Combining information from onboard sensors and cameras with intelligence from external sources for early piracy threat detection has shown promising results but lacks real-time updates for situational context. Such systems can benefit from early warnings, such as "a boat is approaching the ship and accelerating," "a boat is circling the ship," or "two boats are diverging close to the ship." Existing onboard cameras capture these activities, but there are no automated processing procedures of this type of patterns to inform the early warning system. Visual data feed is used by crew only after they have been alerted of a possible attack. Camera sensors are inexpensive but transforming the incoming video data streams into actionable items still requires expensive human processing.

Keywords: Deep learning · Artificial neural networks · Convolutional neural networks · Fuzzy logic · Computing with Words · Autonomous vehicle · Usability · Human computer interaction

1 Introduction

The rise of maritime piracy and attacks on transportation ships has posed a significant burden on the global economy [1]. Researchers have proposed agent-driven modeling to counter the threats, facilitating the potential of piracy countermeasures but lacking real-time update capability for situational context [2]. Onboard sensor information combined with intelligence from external sources proved valuable for early piracy threat detection [3]. In all scenarios, visual data feed is used by crew only after they have been alerted to

© Springer Nature Switzerland AG 2020
D. D. Schmorrow and C. M. Fidopiastis (Eds.): HCII 2020, LNAI 12197, pp. 434–444, 2020.
https://doi.org/10.1007/978-3-030-50439-7_30

a possible attack: camera sensors are inexpensive but transforming the incoming video data streams into actionable items still requires expensive human processing.

In this paper, we propose to apply recent advances in deep learning to design and train algorithms to localize, identify, and track small maritime objects under varying conditions (e.g., a snowstorm, high glare, night). The crew can benefit from an early automated warning, e.g., "a boat is approaching the ship and accelerating" and decide on the countermeasures based on the developing scenario. Existing onboard cameras capture surrounding activities, but there is no automated processing of threatening patterns to inform early warning systems. Automated warning such as "a boat is circling the ship," or "two boats are diverging close to the ship" can help direct piracy countermeasures more effectively. The state-of-art deep learning activity detection and recognition systems depend on large corpora of training data, which is infeasible to use in this scenario. We propose a computation-with-words approach, where any activity, e.g., "a boat is circling the ship," is analyzed using natural language-based human reasoning.

Humans have a remarkable capability to reason, compute, and make rational decisions in the environment of imprecise, uncertain, and incomplete information. In so doing, humans employ modes of reasoning that are approximate rather than exact.

For example, consider a scenario where two friends are each renting a personal wave-runner (small ski-jet boat) and exploring the local lake together while maintaining a "*safe distance*[1]" from each other and other boats on the lake. To this end, a computer-based system might deploy numerous high-resolution sensors for accurately measuring location, speed, velocity, and distance from other marine vehicles. The sensor data and information are generally represented via numerical data. Even this task of maintaining "safe distance" from lake traffic, which is relatively simple for most humans, requires the autonomous system to obtain large amounts of data, maintain a relatively large database/knowledgebase of general knowledge, and perform complex real-time inference. The approach of employing Computing with Numbers (CWN) is well established and is highly effective in many applications. Yet, due to the efficiency and effectiveness of the human mind, there are applications where the human approach, which involves computation using natural language reasoning and "calculations," that is, Computing with Words (CWW), is more simple, efficient, and effective. This raises an intriguing research question concerning the computational and human-computer interaction benefits of using CWW for tasks involving Artificial Intelligence (AI).

The goal of the research reported here is to explore the utility of CWW as a tool for identifying threatening activities where lack of training data precludes the use of deep learning. The main contribution of this paper is that it should serve as an initiator of theoretical and applied research in the field of CWW and its applications in areas such as autonomous vehicles and vehicle activity detection. To the best of our knowledge, there is no published research concentrating on the disruptive concepts of using CWW for efficient computation and Human Computer Interaction (HCI) and Artificial Intelligence (AI) applications.

The rest of the paper is organized as follows. Background concepts are defined in Sect. 2. Section 3 reviews state of the art. Section 4 introduces the CWW-based Maritime

---

[1] The term "safe distance" has a fuzzy connotation. Nevertheless, a CWN engineer might attempt to provide a crisp and accurate definition for this term.

Activity Detection System (CWWMAD). Section 5 provides a qualitative analysis of the CWWMAD effectiveness and efficiency, as well as its usability. Finally, Sect. 6 concludes and proposes further research.

2 Background

In this section, we provide background concerning Maritime Activity Detection. The CWW theoretical and applied concepts are detailed in a previously published paper [4].

2.1 Maritime Object Localization, Identification, and Tracking Using Deep Learning

Maritime sensor imagery quality varies greatly with the weather conditions, sea surface movements, size, and distance of maritime vessels. This degrades the performance of state-of-art Deep Convolutional Neural Networks (DCNN) for identifying maritime vessels. On the other hand, compared to the consumer domain, the maritime domain lacks the abundance of alternative targets that could be incorrectly identified as maritime objects: this allows relaxing the parameter constraints learned on urban natural scenes in consumer photos, adjusting parameters of the model inference, and achieving robust performance and high average precision (AP) measure for transfer learning scenarios [5].

DCNNs trained on large corpora of labeled consumer images provide robust generalized modeling for starting and initializing a network with transferred features from almost any number of layers and boosts generalization. In our previous work [4], we have relied on this finding and expanded the consumer dataset to the maritime domain for adapting the DCNNs to reflect the target domain. We have utilized domain characteristics to refine the deep learning framework and have shown that our transfer learning strategy produces models that reliably and accurately identify sea-surface objects in overhead imagery data. Furthermore, we have demonstrated successful single-source domain adaptation from consumer and maritime data sources to maritime object recognition [5]. In this paper, we use CenterNet as a baseline [6], as it has emerged as a fast and lean deep network that produces the same quality of recognition results with reduced model size and inference time. CenterNet object detector builds on successful key-point estimation networks, finds object centers, and regresses to their size. The algorithm is simple, fast, accurate, and end-to-end differentiable without expensive non-maximum suppression (NMS) post-processing step. Every object is modeled as a single point, centered at its bounding box, and the approach skips the expensive step of an exhaustive search of the possible object locations.

Frame-to-frame target tracking can greatly improve the accuracy of the system and reduce the number of false positives in the dynamic learning schema. Towards this end, we have adapted the DeepSort algorithm for real-time object tracking. The system optimizes cosine similarity through a simple re-parametrization of the conventional Softmax classification regime [7]. At inference time, the final classification layer facilitates nearest neighbor queries on unseen individuals using the cosine similarity metric [7]. Figure 1 depicts the CenterNet domain adaptation to maritime data. The adaptation produces high, real time, recall localization inference at a fraction of the model size. As illustrated in the figure, the system, trained on maritime data, can identify and track small objects through longer periods of occlusions.

Fig. 1. CenterNet domain adaptation to maritime data

2.2 The IPATCH Dataset

The IPATCH project data collection focuses on non-military protection measures for merchant shipping against piracy [3]. The goal of the project is to develop an on-board automated surveillance and decision support system providing early detection and classification of piracy threats and supporting the captain and crew in selecting appropriate countermeasures against piracy threats. The data collection was carried out employing the vessel 'VN Partisan,' where the ship was traveling at a constant speed, while fishing boats and "pirate" boats re-enacted scenarios. A subset of the IPATCH dataset collected in 2015 was released for public use as the PETS dataset [3]. The recordings, which represents a series of realistic maritime piracy scenarios, presents several challenges of object detection and tracking including the fusion of data from sensors with different modalities and sensor handover, tracking objects passing from one field of view (FoV) to another with minimal overlapping FoVs, event detection, and threat recognition. Piracy attacks on a vessel typically fall into one of five scenarios [3]. The PETS database contains only a small sample of each of the listed scenarios.

2.3 Activity Recognition

Recent two-stream DCNNs have made significant progress in recognizing human actions in videos. Despite their success, methods extending the basic two-stream DCNN have not systematically explored possible network architectures to utilize spatiotemporal dynamics within video sequences further. Furthermore, activity recognition in maritime videos lacks sufficient annotations. The annotated activities are categorized into seven groups:

- **Speeding:** Sudden acceleration of the mobile object.
- **Loitering:** The detected object stands still or moves slowly in the same area.
- **Group formation:** A mobile object comes close to another and holds an interaction.

- **Group Separation:** A mobile object departs from a group.
- **Moving Around (Circling):** A boat is moving and has appeared in two or more sides of the ship.
- **Sudden Direction Change:** sudden change of the trajectory.

2.4   Computing with Words, Human-Computer Interaction, and Usability

The theoretical and to some degree the applied background of the present paper is described in detail in [4] "Computing with Words—a Framework for Human-Computer Interaction," authored by a subset of the present authors with a contribution by Lotfi Zadeh. A further in-depth review of the background can be found reference [4] citations.

3   Related Work

The DCNN framework is designed so that the individual neurons respond to overlapping regions in the visual field [8]. DCNNs for object recognition in images consist of multiple layers of small neuron collections, inspecting small portions of the input image, called receptive fields. The results of these collections are then tiled so that they overlap to obtain a better representation of the original image; this is repeated for every layer. One major advantage of DCNNs is the use of shared weights in convolutional layers, which means that the same filter (weights bank) is used for each pixel in the layer; this reduces the required memory size and improves performance. The performance of the DCNNs in the ImageNet Large Scale recognition challenge has approached the capabilities of human recognition [9]. When it comes to noisy imagery, however, the processes humans use to identify a specific target are largely unknown. Nevertheless, recent advancements in DCNN research have changed the expectations from an image and video understanding system, significantly raising the bar. DCNNs, however, continue to exhibit shortcomings, which has spurred great activity in the research community but with limited its effectiveness in real-life situations. Due to a large number of network parameters that have to be trained on, every DCNN system requires a significant number of labeled training samples to perform well. Pascal VOC, COCO, and ImageNet [9], and benchmarks motivated a breakthroughs in the field as training samples were collected via a well-executed and expensive crowd sourcing endeavor to label millions of object instances in imagery created by consumers using their hand-held devices. To achieve similar performance in other domains, one has to consider the replication of these process at comparable scale, and this is prohibitive when it comes to the periscope imagery domain, where crowd sourcing effort or labeling uniformity to achieve comparable benchmark at such a large scale are not available. Recent advances in DCNN development for object recognition have demonstrated that one can apply high-capacity DCNNs to bottom-up region identification in order to localize and segment objects, and when labeled training data is scarce, supervised pre-training for an auxiliary task, followed by domain-specific fine-tuning, yields a significant performance boost [8, 10–12].

Transfer learning focuses on storing knowledge gained while solving one problem and applying it to a different but related problem and has gained traction in domain adaption problem in computer vision. In the domain adaption problem, we focus on utilizing multiple existing source data to build a model that performs well on different but related dataset. For tasks where sufficient number of training samples is not available, a DCNN trained on a large dataset for a different task is tuned to the current task by making necessary modifications to the network and retraining it with the available data [13, 14]. Lately, multiple groups proposed a one-shot learning approach for deep learning setup, and demonstrated that it is consistent with 'normal' methods for training deep networks on large data [13]. Domain adaptation of DCNN systems has been used to produce segmentation maps and to improve category identification when applied to satellite imagery and remote sensing [15]. It should be noted that generating synthetic data can serve as an intermediate stage for DCNN training.

Numerous publications address the topic of CWW from the theoretical point of view as well as relevant applications [16–20]. Several of these papers and patents, e.g., [16, 17] allude to the possibility of using CWW for Maritime navigation, activities, and Piracy Alert systems. Nevertheless, a thorough search for literature that is using the approach presented in this paper did not yield any relevant publications.

Several papers, cited in [4] address the topic of HCI in CWW-based system and conclude that the affinity between the way that the CWW-based system operates and natural language oriented HCI significantly improve these systems' usability. These observations are in line with our expectations. Nevertheless, except for reference [4], we could not identify papers that specifically address the HCI of CWW-based applications such as Maritime Activity Detection, and Piracy Alert systems.

4   CWW Procedure

4.1   Background

Assume that a boat that is monitoring maritime activity is the "friendly boat," referred to as Boat Φ, and the boats that are monitored are the "adversary boats," referred to as the Ψ boats. In the present paper, we study the problem of identifying whether a single Ψ-boat is "circling" Boat Φ. A byproduct of our procedure is identifying whether any Ψ is "too close" to Φ.

The DCNN is fed by onboard video cameras placed according to the schematics of Fig. 2. These cameras provide incomplete and overlapping coverage of the area around Φ. Hence, at certain times, the location of an adversary boat in the image obtained by a specific camera is known, and at other times it is unknown.

"Four AXIS P1427-E Network cameras were added to the ship; three of them at the side and one at stern. The camera technical characteristics are the following: 5-megapixel resolution; 35°–109° FoV – Autofocus; Frame rate 30 fps; Digital PTZ" [3].

We are fixing a computer graphics frame (affine coordinate system and a point of reference - origin), referred to as the Cameras View Frame (CVF), where Boat Φ is assumed to be in its origin. The CVF is time dependent as is generated by applying affine transformations on the location of the adversary boat as obtained from onboard cameras using DCNN. Next, we mark the subregions of the CVF, as depicted in Fig. 3. The regions {A, B, C, D} denote proximity of the adversary boat to Boat Φ. The virtual region 'INV' denotes that none of the cameras identifies an adversary boat at the given CVF. To distinguish between individual frames of each camera and the view frameset

Additional visual cameras
Three at the side
1 at stern
Additional Thermal cameras
Two at the side
1 at stern

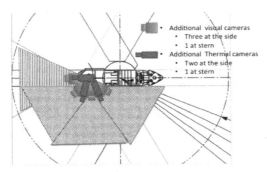

Fig. 2. Schematic representation of added cameras in the VN Partisan [3]

by the monitoring system, we refer to the camera view frames as the CVFs, while the video frames of the individual cameras are referred to as Image Frames (IFs). We assume that the cameras mounted on the boat, along with the DCNN, can provide the following information, as illustrated in the Fig. 3, at a rate of 30 CVFs per second.

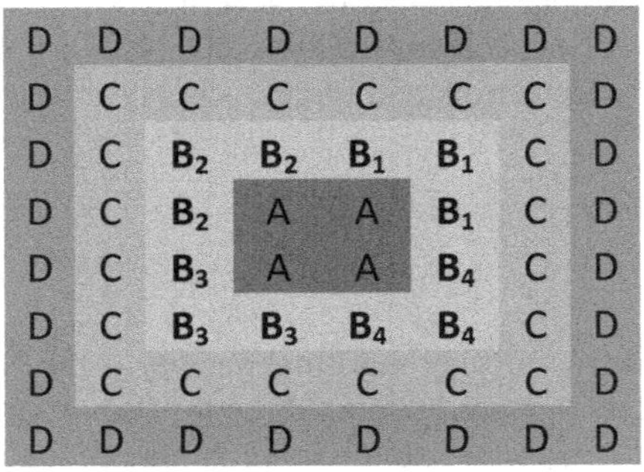

D D D D D D D D
D C C C C C C D
D C B2 B2 B1 B1 C D
D C B2 A A B1 C D
D C B3 A A B4 C D
D C B3 B3 B4 B4 C D
D C C C C C C D
D D D D D D D D

Fig. 3. Cameras view frame subregions

We further assume that at every time unit (of 1/30 of a second or longer) the system provides the following information: (1) either (x, y, 0) – meaning that an adversary boat is invisible in the current frame, or (2) (x, y, 1) meaning that an adversary boat is visible and is located in the affine point $(x,\ y,\ 1)^T$. In this system, (x, y, 0) means a vector to the direction of (x, y), which represents the direction at which the adversary boat was visible in the last frame. In this case, (0,0,0) is the 0 vector – meaning that there is no knowledge about an adversary boat. Given the above, we can construct the following list of directions (Table 1):

Table 1. List of directions

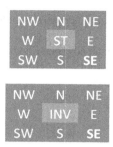

For example, the direction NW means that the adversary boat is "currently" moving in the Northwest direction in the CVF, ST means that the boat is loitering, and INV means that the boat is invisible at the current CVF. Note that the smallest time unit for 'currently' is 1/30 of a second (the frame rate of the cameras). The largest 'currently' unit is arbitrary and is determined by monitoring considerations.

A "circling" activity in the context of the present paper is defined to be:

Circling: An adversary boat is moving and has appeared at two or more sides of Boat Φ. It should be noted that this activity can be detected by the dynamics of the boat location change. For example: consecutive movement through regions B_1, B_2, and B_3 constitutes a circling event.

The situation assessment unit, referred to as CWWMAD (MAD stands for Maritime Activity Detection), is capable of providing the crew of Boat Φ with several indications of alert levels, which can be considered as flags, where "Alarm" is the "Red Flag." The alert levels are denoted as {invisible, watch, inspect, suspect, and Alarm}. The alert levels are determined by the current location, history of directions, and history of activities of the adversary boat. Additionally, the direction is utilized to visualize the adversary boat's trajectory and to produce synthetic training data.

4.2 CWWMAD Inference Procedure

The CWWMAD procedure for identifying whether an adversary boat is too close Boat Φ, or is circling Φ, can be implemented using the DCNN readings denoted in Fig. 3. It is described using the following notation:

1) No adversary- If the current and the previous n readings of the adversary boats' locations are 'INV', then it is assumed that no adversary boat is in proximity to Boat Φ.

2) Watch - If the current reading of an adversary boat's location is 'D,' it is assumed that the adversary boat is in the 'D' region and might become a threat.

3) Inspect - If the current reading of an adversary boat's location is 'C,' it is assumed that the adversary boat is in the 'C' region and requires an elevated level of monitoring.

4) Suspect - If the current reading of and an adversary boat's location is 'B$_1$', 'B$_2$', 'B$_3$', or 'B$_4$', it is assumed that the adversary boat is in one of the 'B' regions and might become a reportable threat if it crosses into the 'A' region completes a circling operation, as defined above.

5) Alarm - If the current reading of an adversary boat's location is 'A' or if the adversary boat has completed a circling activity (as defined above) then the crew must receive an alarm and undertake a response.

5 Analysis

5.1 System Verification

A simplification of the problem of detecting circling and several other maritime activities via a CWW enables system implementation using a relatively simple rule-based system. We have constructed a rule-based system RB-CWWMAD for the detection of "circling" and of "getting too close to our boat".

Due to the simplicity of the CWW problem statement, we have been able to manually verify the soundness and completeness of CWWMAD. This is significant since currently there is not enough realistic video data that can be used for automatic verification.

5.2 Usability

The following is quoted from [4]. Internal citation numbers are removed to fit the present paper.

"Tamir et al. have developed an effort-based theory and practice of measuring usability [-]. Under this theory, learnability, operability, and understandability are assumed to be inversely proportional to the effort required from the user in accomplishing an interactive task. A simple and useful measure of effort can be the time on the task. Said theory can be used to determine usability requirements, evaluate the usability of systems (including comparative evaluation of "system A vs. System B"), verify their compliance with usability requirements and standards, pinpoint usability issues, and improve usability of system versions. It is quite obvious, and supported by research literature, that an interface that uses a natural language or a formal language that are "close" to natural languages reduces the operator effort and can improve system usability [-].

A natural language interface can accompany a CWN-based as well as a CWW-based system. A CWW-based system, however, provides the advantage that the system itself operates and is being controlled in a way that is closer to human reasoning, decision making, and operation. This increases the coherence between the system and its user interface and, thus, it simplifies the system design and human-controlled operation of the system."

5.3 Additional Activities

The CWWMAD system for circling detection can be expanded to include additional activities, such as "divergence of boats" and "acceleration towards our boat." These activities would be detected via simple extensions of the current CWW rule-based inference engine. Another activity of interest, which might be more challenging, is detecting that an adversary boat is following Boat Φ. This system would require a more complex inference engine.

6   Conclusion and Further Research

We have demonstrated that the integration of a CWW-based inference engine with a DCNN is a viable tool for detecting and providing alerts concerning several maritime activities, including boat circling. The inference can be implemented with a relatively simple rule-based system. Additionally, we have discussed the utility of a CWW-based human-computer interface (HCI) to enable ergonomic interaction with the CWW system. Analysis of the system's performance and its usability shows that the approach has a potential for the facilitation of mitigation of maritime piracy, with an ergonomic and efficient system.

In the future, we plan to extend the activity detection capability of the system and include other maritime activities such as "following," "diverging," and "accelerating towards our boat." We also plan to examine the utility of more complex inference engines that include game-theory considerations and space search-based inferencing.

We plan to obtain and annotate additional realistic video data, which would enable automatic testing of CWW, as well as training a deep-learning artificial neural network (ANN) and comparing the performance of the CWW-based inference engines to the performance of ANN-based approaches. We also plan to construct a simulation system to generate synthetic data for training deep-learning ANNs. With the addition of computer graphics capabilities to the simulation, stakeholders will be able to assess the utility and usability of maritime activity inference engines.

References

1. Marchione, E., Johnson, S.D.: Spatial, temporal and spatio-temporal patterns of maritime piracy. J. Res. Crime Delinquency **50**(4), 504–524 (2013)
2. Vaněk, O., Jakob, M., Hrstka, O., Pěchouček, M.: Agent-based model of maritime traffic in piracy-affected waters. Transp. Res. Part C: Emerg. Technol. **36**, 157–176 (2013)
3. Patino, L., Cane, T., Vallee, A., Ferryman, J.: PETS 2016: dataset and challenge. In: The IEEE Conference on Computer Vision and Pattern Recognition (CVPR) Workshops, pp. 1–8 (2016)
4. Tamir, D., Newman, S., Rishe, N., Kandel, A., Zadeh, L.: Computing with words a framework for human computer interaction. In: The International Conference on Human Computer Interaction HCII-2019, pp. 356–372 (2019)
5. Russakovsky, O., et al.: Imagenet large scale visual recognition challenge. Int. J. Comput. Vision **115**(3), 211–252 (2015). https://doi.org/10.1007/s11263-015-0816-y

6. Warren, N., Garrard, B., Staudt, E., Tesic, J.: Transfer learning of deep neural networks for visual collaborative maritime asset identification. In: 2018 IEEE 4th International Conference on Collaboration and Internet Computing (CIC), Philadelphia, PA, pp. 246–255 (2018)

7. Wojke, N., Bewley, A.: Deep cosine metric learning for person re-identification. In: IEEE Winter Conference on Applications of Computer Vision (WACV) (2018)

8. Pailla, D.R., Kollerathu, V., Chennamsetty, S.S.: Object detection on aerial imagery using CenterNet. arXiv:1908.08244, August 2019

9. Hou, X., Wang, Y., Chau, L.: Vehicle tracking using deep SORT with low confidence track filtering. In: 2019 16th IEEE International Conference on Advanced Video and Signal Based Surveillance (AVSS), Taipei, Taiwan, pp. 1–6 (2019)

10. Feichtenhofer, C., Fan, H., Malik, J., He, K.: SlowFast networks for video recognition. In: The IEEE International Conference on Computer Vision (ICCV), pp. 6202–6211 (2019)

11. He, K., Gkioxari, G., Dollár, P., Girshick, R.: Mask R-CNN, International Conference on Computer Vision (ICCV), 2017, p. 5274 (2017)

12. Zhou, X., Wang, D., Krähenbühl, P.: Objects as Points, arXiv, April 2019

13. Hara, K., Vemulapalli, R., Chellappa, R.: Designing Deep Convolutional Neural Networks for Continuous Object Orientation Estimation, arXiv (2017)

14. Çevikalp, H., Dordinejad, G.G., Elmas, M.: Feature extraction with convolutional neural networks for aerial image retrieval. In: 25th IEEE Signal Processing and Communications Applications Conference (SIU) (2017)

15. Maggiori, E., Tarabalka, Y., Charpiat, G., Alliez, P.: Convolutional neural networks for large-scale remote-sensing image classification. IEEE Trans. Geosci. Remote Sensing 55(2), 645–657 (2017)

16. Zadeh et al.: Method and System For Feature Detection, United States, Patent Office, US9,916,538B2 (2018)

17. Liu, J., Martínez, L., Wang, H., Rodríguez, R.M., Novozhilov, V.: Computing with words in risk assessment. Int. J. Comput. Intell. Syst. 3(4), 396–419 (2010)

18. Zhou, C., Yang, Y., Jia, X.: Incorporating perception-based information in reinforcement learning using computing with words. In: Mira, J., Prieto, A. (eds.) IWANN 2001. LNCS, vol. 2085, pp. 476–483. Springer, Heidelberg (2001). https://doi.org/10.1007/3-540-45723-2_57

19. Wójcik, A., Hatłas, P., Pietrzykowski, Z.: Modeling communication processes in maritime transport using computing with words. Arch. Transp. Syst. Telematics 9(4), 47–51 (2016)

20. Le Pors, T., Devogele, T., Chauvin, C.: Multi agent system integrating naturalistic decision roles: application to maritime traffic. In: IADIS International Conference Intelligent Systems and Agents, pp. 196–210 (2009)

Modeling User Information Needs to Enable Successful Human-Machine Teams: Designing Transparency for Autonomous Systems

Eric S. Vorm[1] and Andrew D. Miller[2]([⊠])

[1] U.S. Naval Research Laboratory, Washington, DC, USA
eric.vorm@nrl.navy.mil
[2] Indiana University Purdue University Indianapolis, Indianapolis, IN, USA
andrewm@iupui.edu

Abstract. Intelligent autonomous systems are quickly becoming part of everyday life. Efforts to design systems whose behaviors are transparent and explainable to users are stymied by models that are increasingly complex and interdependent, and compounded by an ever-increasing scope in autonomy, allowing for more autonomous system decision making and actions than ever before. Previous efforts toward designing transparency in autonomous systems have focused largely on explanations of algorithms for the benefit of programmers and back-end debugging. Less emphasis has been applied to model the information needs of end-users, or to evaluate what features most impact end-user trust and influence positive user engagements in the context of human-machine teaming. This study investigated user information preferences and priorities directly by presenting users with an interaction scenario that depicted ambiguous, unexpected, and potentially unsafe system behaviors. We then elicited what features these users desired most from the system to resolve these interaction conflicts (i.e., what information is most necessary for users to trust the system and continue using it in our described scenario). Using factor analysis, we built detailed user typologies that arranged and prioritized user information needs and communication strategies. This typology can be adapted as a user model for autonomous system designs in order to guide design decisions. This mixed methods approach to modeling user interactions with complex sociotechnical systems revealed design strategies which have the potential to increase user understanding of system behaviors, which may in turn improve user trust in complex autonomous systems.

Keywords: Human-computer interaction · Human factors · Human-machine teaming · Autonomous systems · Human-autonomy interaction

D. D. Schmorrow and C. M. Fidopiastis (Eds.): HCII 2020, LNAI 12197, pp. 445–465, 2020.
https://doi.org/10.1007/978-3-030-50439-7_31

1   Introduction

In this paper, we sought to capture the questions users ask when the autonomous systems they are interacting with behave unexpectedly or produce unexpected results. Through this modeling process, we argue that we can derive design strategies that better support user information needs (i.e., what information could an interface provide that would answer user questions in such a scenario). Our theoretical position is informed by a wealth of findings from both laboratory studies and real world mishap investigations [1–3,10,13] which outline numerous examples of how these off-nominal, unexpected system behaviors can have very serious consequences on human decision making. Thus, we argue that by improving how autonomous systems communicate in a way that aligns with user expectations and priorities, we can improve user trust, and strengthen the overall human-machine team environment. To accomplish our modeling task, we used a mixed method called Q methodology, a technique in which participants assess a bank of questions and rank them in a normalized distribution. These rankings can then be quantitatively analyzed using factor analysis in order to find commonalities amongst the factor groups. We ran a human-in-the-loop study with 110 participants in the US and UK, using the same interaction scenario. Our findings resulted in a detailed analysis of a range of user information needs and priorities. With this information we built a detailed user typology, which corresponds to the four distinct user types found in our data. In subsequent sections, we detail and describe each user type, discuss the ramifications and uses of building such a typology, and discuss how utilizing this approach to user interaction modeling can help guide the design of intelligent systems which are transparent and intelligible to their users. This work will ultimately help designers tailor intelligent systems based on the desires and priorities of users.

2   Background and Motivation

Generating explanations that are meaningful and relevant to lay users is complicated by a variety of factors, both psychological and technological [4]. Models that are able to be explained and understood by humans are said to be "intelligible" [5]. Intelligibility is a major component of the umbrella term of "transparency," which has lately come to refer to both the degree to which a system's inner workings can be seen and understood by the user, as well as other factors such as fairness, accountability, and privacy. For this work, we focus on the intelligibility component of intelligent system transparency, particularly in the role it plays in helping end users understand and trust intelligent autonomous systems.

A good deal of research has been done towards developing methods to make models more intelligible [6–8]. Gregor and Benbasat [1] presented a detailed review of explanation types, and identified a set of useful constructs used to generate explanations to users of early intelligent systems. These include: trace or line of reasoning (explaining why certain decisions were or were not made by reference to the underlying data and rule base), justification or support (linking "deep" domain knowledge to portions of a procedure, such as providing a textbook reference or hyperlink to explore deeper), control or strategic (explaining the system's behavior by providing its problem solving strategies and reasoning rules), and terminology (providing users with term definitions to aid in their comprehension).

These constructs have been used broadly to enhance user understanding and trust in intelligent systems with some success. Studies have found that users who consider systems to be "intelligible" tend to perform better on system tasks, demonstrate more appropriate trust (generally defined as knowing when and when not to use the system, depending on the circumstances), and often report higher levels of usability and satisfaction during interaction [3,9–11]. Exploring intelligibility in context-aware systems, Lim, et al. [7] examined how different explanation types (Why, Why Not, What If, and How To) had an impact on user trust, performance, and comprehension of system functions. Participants interacted with different context-aware systems and were shown basic input-output cycles, along with different reasoning traces in the form of the questions above. Participants were then measured on their understanding of how the system functioned. They found that the Why and Why-Not explanations improved participants' task performance and understanding, as well as increased their trust in the system, while neither the How-To nor What-If explanations showed any improvements over no explanations. Interestingly, there were no significant differences between Why and Why-Not intelligibility types. These findings suggest that most users have a basic desire to understand system behaviors and inner workings, and that answering these user questions in the correct format plays a central role in determining the interaction outcome.

Despite these findings, however, much of the work in the realm of explainability and intelligibility has been done to produce what some might call "back end" explanations, or explanations that would be considered intelligible by programmers and experts, but few others. This is not an indictment against these efforts - they are vital to the development of safe and effective algorithms, and methods to enable programmers to better validate models and detect potential unsafe deviations from operating parameters are critical. It is important to note, however, that the human user comprises at least 50% of the human-machine system, at least in principal, and so should receive commiserate levels of modelling and consideration towards system design. In reality, it could be argued that humans comprise a considerably higher portion of the variability in system performance since it is their interactions with systems that ultimately determine much of the system's output. Yet a cursory review of the scientific literature on explainability and transparency reveals that user-centered studies focusing on the development of design strategies for autonomous systems are squarely in the minority, with most attention given to algorithm development, visualization techniques, or debugging efforts [8,23,27].

This problem is not unique to today's advanced autonomous systems. Early intelligent systems such as expert systems were only able to provide the most basic of explanations. These tended to be focused on verbalizing internal states, goals, and plans. These explanations were interpreted from a knowledge base, which limited their ability to answer questions from a static dataset and often bore little resemblance to human language [14]. As intelligent systems matured further, more sophisticated attempts to provide explanations emerged, and began to incorporate some degree of justification, offering not only the *what*, but also the *why* [15]. These systems offered explanations that were both understandable and satisfying, although only in limited scope. The newest generation of intelligent systems now strive to consider a variety of factors in their explanation capabilities, including the decision context, knowledge of the user, knowledge of the history of system performance, including reliability; modeling and knowledge of the goals of the user, and awareness of the domain [9].

Determining which of these features to include in an explanation, at what time, and in what format is still the subject of much investigation. Our own recent work has sought to explore methods of providing explanations by category (system parameters and logic, different qualities of data, how user personalization plays a role, providing some justification of why one option was recommended over another, and what other users have done in similar circumstances before) as a means of improving transparency and intelligibility of intelligent system recommendations [16]. While much work remains focused on developing methods to make models explain themselves to users, we argue that much of this model-based information is superfluous to most end-users, and that by prioritizing information that users care about most, system designs can achieve a higher net effect in terms of user comprehension, trust, and perceived usability. In many cases, these data already exist in the underlying system architecture, which means making them available to users through an interface is often relatively inexpensive and simple. Tracing decisions made by systems, especially in examples such as recommender systems, is also relatively easily accomplished. Most design decisions such as these, however, tend to be made out of pragmatic considerations. In other words, most designs of autonomous intelligent systems are driven by an effort to reduce clutter and streamline interface layout. In many cases, these decisions are made according to the priorities of the designers, rather than the end users, a phenomenon Cooper termed "the inmates running the asylum" [12].

This means that in order to truly design an autonomous system interface that supports user trust and decision making, we must first determine what information users think is most important, especially in the context of highly complex autonomous, distributed systems that have the capability of choosing behaviors with little to no user input. Our purpose for this research is to assess the relative value or importance of various bits of information that would be potentially available in an interface with an autonomous system, and to quantify these priorities in a manner that supports future system designs.

To do this, we employed a method known as Q-methodology. Q-methodology is a mixed method, often referred to as the "scientific study of subjectivity" [17] and has been successfully used in previous HCI work [18] to elicit design feedback from stakeholders. Q-methodology asks users to sort statements or questions according to their preferences or priorities into a fixed matrix that represents a normalized distribution. Using a factor analytic approach, Q-methodology can identify patterns of subjectivity and thought in the data, which is used to identify groups, or clusters of people who share similar opinions and ways of thinking about a given issue. Interpretation and classification of these clusters is then made using traditional factor analysis techniques, effectively combining the strengths of both qualitative and quantitative research. The results from this method are data that is deep in texture and nuance that could otherwise be passed over by a purely statistical or survey sampling approach. We describe the steps of our methodology in the following section.

3   Methods

Since we are interested in understanding what features users find more or less valuable to help them understand and trust intelligent autonomous systems, we needed a way for them to consider and prioritize a large number of design elements. To do this, we created a bank of questions that users might ask of a system when it behaves unexpectedly or uncharacteristically. The motivation behind this decision was that by asking participants to identify what questions *they* would ask, we could more accurately model what information they considered vital to their decision making. This approach, we argued, would minimize issues common to survey or ethnographic methods (i.e., bias, response interpretation), while still allowing for detailed user priorities to be captured in a quantifiable and reliable manner.

These questions were deliberately developed to represent a variety of approaches to providing users critical information that could help them resolve conflicts with interactions in intelligent autonomous systems. In order to narrow down the potential list of conflicts, we chose to focus our study on interactions with intelligent recommender systems, or systems that provide recommendations to users (i.e., decision support algorithms). For example, in a basic recommender system scenario, if a user was presented with a restaurant recommendation which seemed out of place for their tastes, the user would probably want to know *how* that recommendation was made. To answer *how*, however, the user could be given a variety of information. For example, they may care to learn what data was used to create that recommendation, and in doing so better understand the recommendation. Or perhaps they might want to know whether or not the system actually has a model of themselves and their tastes, or whether the recommendation was made randomly. Because there are a variety of potential answers to "how was this recommendation made?" where some of those answers would be more valuable and satisfying than others to individual users, it is important that we try to model these so that our interface designs provide answers that are meaningful to the intended users of the system, rather than system architects.

In our approach, each question we developed is mapped to a potential design feature that could be achieved through an interface. Our purpose was to help determine which potential design features would most help users understand and trust intelligent autonomous systems, and which would be considered a nuisance or irrelevant.

3.1   Question Bank Development

Because the question bank we developed from an earlier project [16] was not specific to any one type of system, we first had to develop questions that would be most appropriate for interactions with our intended system in this project. To do this we started with Ram's taxonomy of question types as an initial starting point to ensure that we used a variety of question types [19]. Ram's taxonomy is useful because it describes a wide breadth of questioning strategies, and was developed explicitly to enhance the explainability of intelligent systems to end users. We refined these questions using Silveira et al's taxonomy of user's frequent doubts [20]. After iterative evaluation and consultation within the project team and with experts in intelligent system design in the US and UK, we arrived at an initial bank of 36 questions.

Fig. 1. Our interactive testbed for this project was the Deep Securities and Accounting Management (DSAM) system. This system emulated an intelligent autonomous system that provides recommendations to its users.

Once the set of questions was developed, we presented participants with the Deep Securities and Accounting Management (D-SAM) system (Fig. 1). D-SAM is a research testbed, and was developed by reviewing recent submissions to the United States Patent and Trademark Office's Patent Full-Text and Image Database (PatFT). By exploring recent patent submissions, and combining these with our knowledge of intelligent autonomous systems research, we developed a near-future, plausibly relevant financial management system that embodies many of the most advanced efforts in intelligent systems today, and assumes their success in the near future. Our users were asked to interact with D-SAM, which resulted in a system-generated recommendation the user had to determine whether to accept or reject. This interaction deliberately introduced ambiguity and uncertainty into the scenario in the form of an unexpected or seemingly inappropriate recommendation. This ambiguity and unexpected system behavior is the most common combination found to result in significant user conflicts with intelligent systems [10], and thus served to create the need for users to seek additional information from the system in order to determine whether or not it could be trusted, or if its recommendation should be disregarded. This ambiguity and unexpected system behavior is one in which the concept of transparency is theoretically most critical, hence we used it to frame our study.

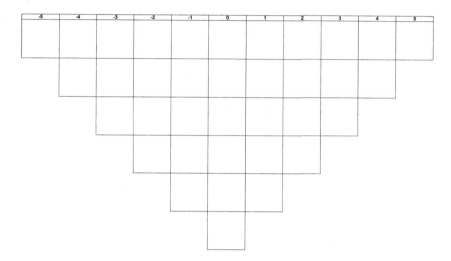

Fig. 2. The forced distribution matrix. Cards are arranged from right (most important to me, +5) to left (least important to me, −5).

Once presented with the interaction scenario, participants were then given a stack of 36 numbered index cards, each containing a different question. Each question on the card was meant to represent a different information seeking strategy. The concept behind this is that a user could potentially ask the system a wide variety of questions about all manners of different things. Our goal was to better understand what questions were more or less frequently prioritized as important to users, as a proxy of inferring their information priorities and preferences. Some example questions were "How current is the data used in making this recommendation?", "Precisely what information about me does the system know?", and "What have other people like me done in response to this recommendation?" Participants were then given time to sort these 36 cards into the fixed distribution matrix described in Fig. 2 above. Participants were encouraged to consider each question as if it were something they would ask themselves, and then to determine which questions, if answered, would have the greatest impact on trust and their willingness to act on recommendations generated by the system.

Once cards were sorted, participants recorded their arrangements on a paper form, and answered two additional questions: "Briefly describe why you chose this question as your most/least important question to ask." Paper forms were then collected and prepared for analysis and interpretation.

Each column in the matrix in Fig. 2 is given a number value, corresponding to the degree of preference expressed by each participant - +5 for the rightmost column corresponding to "most important to me," −5 for the leftmost column corresponding to "least important to me," and everything in between. Each participant's sort then represents a full arrangement of their preferences in a forced and normalized distribution. Once completed, each participant sort was arranged in a correlation matrix with each other participant sort. This matrix was then submitted to factor analysis.

Using principal components analysis (PCA) for factor extraction [21], we extracted 8 initial factors. We tested several possible solutions, ranging from two to eight factor groups, and ultimately settled on a four factor solution because together they explained the majority of variance (61%), and divided the majority of respondents into a relatively small number of groups that were distinct from one another, yet large enough to permit statistical analysis. We then used the VARIMAX method to obtain optimal rotation [22]. 11 of the participant's arrangements were confounded because they loaded on more than one factor, and 18 participants failed to load on any of the four factors we extracted. This resulted in four distinct viewpoints of information priorities and preferences of the remaining 89 individuals.

3.2 Factor Interpretation

Once factor extraction and rotation was complete, we next set about analyzing how each factor group arranged their questions in order to intuit and interpret their reasoning and prioritization strategy. To accomplish this, we produced a weighted average of each participant's arrangement of cards from within their factor group, and combined those arrangements into one exemplar composite arrangement per group, also known as a "factor array." We then compared each group factor array to one another in order to derive a statistical basis of comparison. By examining the placement of each question within each factor array and comparing those arrangements to each other factor array, we can begin to detect patterns, which can be used to infer how and why these clusters of individuals prioritize and value information differently.

To do this, we examined each factor array's **distinguishing questions**. A distinguishing question is found when the participants in a factor group place a question in a significantly different position from all participants loading on other factors. For example, the highest ranked question from factor group two was "What is the history of reliability for this system?" (composite score 5, $Z = 1.85$, $p < 0.01$). This question was placed significantly higher than any other factor group, thus partially defining factor group two. By examining distinguishing questions for each factor group, we began to uncover unique differences amongst the groups, and to describe how each group prioritized information differently.

Finally, in order to fully appreciate our findings, we examined participants' qualitative feedback to contextualize and verify our analyses. This feedback was solicited from participants in the form of two questions which they answered in open comments on the data collection sheet, "Briefly describe why you chose this question as the MOST/LEAST important question to you."

The result of this factor analysis is a detailed user typology that sorts participants into four cohesive, like-minded groups based on their shared priorities, reasoning strategies, and patterns of thought. In the following section we describe our findings from each of the four user typologies, and later discuss the potential implications of these findings in the greater design space of intelligent system transparency and intelligibility in Sect. 5.

4 Results

We identified four distinct user typologies for intelligent system transparency. In this section, we describe each group based on its quantitative features, and then provide an analytical interpretation of the characteristics associated with people in the group. A summary table of findings is available in Table 2, while a detailed table of all findings by question type is available in the Addendum.

4.1 Factor Group 1 - "Interested and Independent"

Factor group one was defined by 24 participants and explained 14% of the total study variance with an eigenvalue of 20. 71% reported they had little to no working knowledge of intelligent systems. Roughly 60% of factor group one were less than 40 years old. Individuals in this group most want to know **"why was this recommendation the BEST option,"** indicating a desire for some sort of justification for why a recommendation was made, above and beyond a basic explanation (composite score 5, $Z = 1.42$, $p < 0.05$). Individuals in this group also demonstrated an interest in some of the underlying components of how systems function, and would like to know **"What if I decline? How will that decision be used in future recommendations by this system?"** (composite score 4, $Z = 1.29$, $p < 0.01$) and **"Can I influence the system? Will it consider my input?"** (composite score 3, $Z = 1.06$, $p < 0.01$).

Individuals in factor group one were least interested in the opinions or behaviors of others when considering what to do with a computer-generated recommendation. They ranked questions like **"Is there anyone in my social network that has received a similar recommendation"** (composite score -5, $Z = -2.1$, $p < 0.01$), **"How many other people have accepted or rejected this recommendation from this system"** (composite score -4, $Z = -1.8$, $p < 0.01$), **"How similar am I to other people who have received this recommendation"** (composite score -4, $Z = -1.58$, $p < 0.01$), and **"What have other people like me done in response to this recommendation"** (composite score -3, $Z = -1.57$, $p < 0.01$) as their least important or valuable questions.

Because of their preference for deep system information, and their reluctance to place any priority on other users' behaviors or decisions, we named this factor group the "interested and independent" group. This descriptive name serves to differentiate factor group one from the other groups, as well as to identify a general information seeking strategy found amongst our data (Table 2).

Table 1. Characteristics of factors after rotation. Factors define clusters of participants whose arrangement of questions were very similar, and were mathematically clustered using factor analysis. We refer to these as "factor groups," or by their given typological names throughout the duration of the paper. Factor 1: Interested & Independent; Factor 2: Cautious & Reluctant; Factor 3: Socially Influenced; Factor 4: Egocentric.

Factor characteristics				
	Factor 1	Factor 2	Factor 3	Factor 4
No. of Defining Participants	24	16	24	17
Avg. Rel. Coef.	0.8	0.8	0.8	0.8
Composite Reliability	0.99	0.985	0.99	0.986
S.E. of Factor Z-scores	0.1	0.122	0.1	0.118
Eigenvalue	20	15.34	8.07	7.16
Explained Variance	14%	11%	12%	9%
Male[a]	67%	94%	67%	76%
Female	33%	6%	33%	24%
Experts	29%	75%	29%	41%
Novices	71%	25%	71%	59%
20–30 yrs old	46%	18.8%	58%	53%
30–40 yrs old	12.5%	43.8%	21%	29%
40–50 yrs old	29%	25%	8.5%	12%
50+ yrs old	12.5%	12.5%	12.5%	6%

[a] All participants identified as either male or female

4.2   Factor Group 2 - "Cautious and Reluctant"

Factor group two was defined by 16 participants and explained 11% of the study variance with an eigenvalue of 15.34. 94% were male, 64% were less than 40 years old, and 3/4 had extensive working knowledge of intelligent systems.

This group was exemplified by a deep concern over a system's past performance and reliability. For example, they most wanted to know **"What is the history of the reliability of this system?"** (Composite score 5, Z = 1.85, p< 0.01), followed by **"Under what circumstances has this system been wrong in the past?"** (Composite score 4, Z = 1.4, p< 0.01) and **"What data does the system depend on in order to work properly, and do we know if those dependencies are functioning properly?"** (Composite score 3, Z = 1.19, p<0.05). This group also appeared very interested in information that could help them gauge how the system considers uncertainty and risk, as exemplified by their high ranking of questions like **"How much uncertainty does the system have?"** (Composite score 3, Z = 1.12, p < 0.01) and **"How does the system consider risk, and what is its level of acceptable risk?"** (Composite score 2, Z = 1, p < 0.01).

Participants in factor group two were least interested in whether **"Is there anyone in my social network that has received a similar recommendation?"** (composite score -5, $Z = -1.69$, $p < 0.05$). They also thought little of questions such as **"What does the system *think* I want to achieve? (How does the system represent my priorities and goals)"** (composite score -4, $Z = -1.59$, $p < 0.01$), **"Can I influence the system by providing feedback? Will it listen and consider my input?"** (composite score -4, $Z = -1.42$, $p < 0.01$), and **"Was this recommendation made specifically for ME?"** (composite score -3, $Z = -1.32$, $p < 0.01$).

Because the nature of questions prioritized by factor group two seemed to revolve around the kinds of information that could aid in validating that a system was operating normally, we named this group "Cautious and Reluctant." This group seemed to be the least willing group to interact with and perhaps most suspicious of intelligent autonomous systems, based on their information priorities. Thus this group would represent a particularly vulnerable user group whose needs would most need to be considered in the design of a system such as D-SAM.

4.3   Factor Group 3 - "Socially Influenced"

Factor group three was defined by 24 participants and explained 12% of the study variance with an eigenvalue of 8.07. 67% were male, 79% were less than 40 years old, and 71% had little to no working knowledge of intelligent systems. Participants in this group most wanted to know **"Why is this recommendation the *best* option?"** (composite score 5, $Z = 1.75$, $p < 0.05$) followed closely by **"What are the pros/cons associated with this option?"** (composite score 4, $Z = 1.25$, $p < 0.01$). They also indicated an interested in learning what others have done by ranking **"What is the degree of satisfaction that others have expressed when taking this recommendation?"** (composite score 3, $Z = 0.9$, $p < 0.01$), and **"How many other people have accepted or rejected this recommendation from this system? (What is the ratio of approve to disapprove?)"** (composite score 1, $Z = 0.29$, $p < 0.01$) higher than any other factor group.

Participants in this group were least interested in knowing anything about the qualities of data used by the system. Questions like **"What is the signal-to-noise ratio of this data?"** (composite score -5, $Z = -2.34$, $p < 0.01$), **"Can I see the data for myself?"** (composite score -4, $Z = -2.22$, $p < 0.01$), **"How much data was used to train this system?"** (composite score -4, $Z = -1.53$, $p < 0.01$), and **"Is the system working with solid data, or is the system inferring or making assumptions on 'fuzzy' information?"** (composite score -3, $Z = -1.43$, $p < 0.01$) were all ranked lowest by this factor group.

Analyzing the priorities of this factor group revealed a pattern of preferences related to the behaviors and decisions of other users. While their highest rated questions revolved around understanding the recommendation itself, this group also highly ranked questions related to what other users have done. Relative to other groups, this group was the only one that considered this kind of information relevant or important. Given the majority of these participants were less than 40 years old, these findings could potentially indicate a user group with a posture towards intelligent systems that incorporates social components of usage, suggesting the increasing importance of utilizing features that provide this information in intelligent system designs.

4.4   Factor Group 4 - "Egocentric"

Factor group four was defined by 17 participants and explained 9% of the study variance with an eigenvalue of 7.16. 76% were male, 82% were less than 40 years old, and expertise was almost evenly split between 59% who had little to no working knowledge of intelligent systems, and 41% who had extensive working knowledge of intelligent systems. Participants in this group appear most interested in understanding how recommendations relate to themselves, and others like them. Their top ranked question was **"Was this recommendation made specifically for ME (based on my profile/interests), or was it made based on something else (based on some other model, such as corporate profit, or my friend's interests, etc.)?"** (composite score 5, $Z = 2.6$, $p < 0.01$), followed by **"Precisely what information about ME does the system know?"** (composite score 4, $Z = 1.25$, $p < 0.01$), **"What have other people like ME done in response to this recommendation?"** (composite score 3, $Z = 1.22$, $p < 0.01$), **"How many other people like ME have received this recommendation from this system?"** (composite score 3, $Z = 1$, $p < 0.01$), and **"Is there anyone in my social network that has received a similar recommendation?"** (composite score 3, $Z = 0.98$, $p < 0.01$).

Participants in this group appeared not to care much for details about other options, or how the system considers the concept of risk. They ranked **"What are the pros/cons associated with this option?"** (composite score -5, $Z = -1.99$, $p < 0.01$), **"How does the system consider risk, and what is its level of acceptable risk?"** (composite score -4, $Z = -1.63$, $p < 0.01$), **"Are there any other options not presented here?"** (composite score -4, $Z = -1.42$, $p < 0.01$), **"How many other options are there?"** (composite score -3, $Z = -1.21$, $p < 0.01$) and **"What does the system think is MY level of acceptable risk?"** (composite score -3, $Z = -1.17$, $p < 0.01$) as least important to them.

Interpretations of this group's information priorities revealed a clear preference for self-referential information. Accordingly, we named this group the "Egocentric" group. While the egocentrics were the smallest of our four factor groups, their unambiguous preferences indicated a clear strategy in decision making. When faced with unusual or unexpected results from an intelligent system, at least in our recommendation scenario, these individuals consider themselves in the equation, and consider answers to these questions most important to help them understand and trust system outputs.

5   Discussion

Thus far, we have demonstrated four distinct differences in user information preferences when interacting with unusual or unexpected recommendations from an intelligent financial planning system. These differences characterize different ways that users might seek to resolve conflicts with intelligent systems, especially when faced with unusual, unexpected, or ambiguous system behaviors. In our study we designed an interaction scenario where our participants were presented with a recommendation from a financial management system, and that recommendation seemed potentially unsafe or inadvisable enough that users would need additional information in order to determine whether or not to accept and act on the recommendation, or to reject it. Our findings validate the argument that users use different reasoning strategies, and that those strategies can be quantified and described in sufficient detail to allow design recommendations to be created.

In this next section, we discuss the implications of these findings in terms of how they might be used to prioritize design elements, guide interface development, and structure communication strategies to promote effective human-machine teams. Our first step is to analyze questions that had near-universal consensus in our sample; that is, questions that nearly all users agreed were either very important, or very unimportant. These questions should be considered as most valuable in terms of design priorities, since they all had near full consensus in our sample. Next, we explore questions that produced the highest disagreement amongst all factor groups. These questions represent design elements that could please some users, while aggravating or confounding others. Since these questions were the source of much contention within our sample, we propose these as a starting point for interfaces. Next we analyze questions by category in order to extract valuable design insights and lessons learned, and show how our findings might be translated to design through illustrative interface mockups. We end with a discussion of the limitations of our study, and plans for future work.

5.1   Consensus Amongst Groups

While each factor group had identifying statements that distinguished it from others, there were some questions that all factor groups found either important or unimportant. These are known as consensus questions, or those that do not distinguish between ANY pair of factor groups.

All four of our factor groups thought the question **"What are all of the factors (or indicators) that were considered in this recommendation, and how are they weighted?"** as highly important (average score 3.75, Z score variance 0.06). That this question was the most agreed upon is not surprising, given that other studies have confirmed most individuals demand at least some degree of explanation and justification for system outputs in reference to automated recommendations [7]. Our participants also moderately valued **"What safeguards are there to protect me from getting an incorrect**

Table 2. All four factor groups, with their identifying questions and relative rankings.

RELATIVE RANKINGS OF QUESTIONS FOR ALL FACTOR GROUPS

Interested & Independent

	Interested & Independent	Cautious & Reluctant	Socially Influenced	Egocentric
	5	2	5	1
Highest Ranked				
Why is this recommendation the best option?				
Questions ranked higher in this array than any other factor array				
Can I influence the system by providing feedback? Will it listen and consider my input?	3	4	-2	2
Can I see the data for myself?	2	0	4	-3
What if I decline? How will that decision be used in future recommendations by this system?	0	-3	-3	-2
Questions ranked lower in this array than any other factor array				
How many other people have received this recommendation from this system?	-3	-2	0	3
What have other people like me done in response to this recommendation?	-3	-1	0	3
How similar am I to other people who have received this recommendation?	-4	-1	2	1
How many other people have accepted or rejected this recommendation from this system?	-4	-3	1	-1
Lowest Ranked				
Is there anyone in my social network that has received a similar recommendation?	-5	-5	1	3

Cautious & Reluctant

	Cautious & Reluctant	Interested & Independent	Socially Influenced	Egocentric
	5	-1	1	-1
Highest Ranked				
What is the history of the reliability of this system?				
Questions ranked higher in this array than any other factor array				
Under what circumstances has this system been wrong in the past?	4	-2	0	0
What data does it depend on, and do we know if those dependencies are functioning properly?	3	1	0	2
How much uncertainty does the system have?	3	0	-1	0
How does the system consider risk, and what is its level of "acceptable risk?"	2	-1	-1	-4
Questions ranked lower in this array than any other factor array				
Was this recommendation made specifically for ME, or on something else?	-3	2	4	5
Can I influence the system by providing feedback?	-4	3	-2	2
What does the system *think* I want to achieve?	-4	3	2	0
Lowest Ranked				
Is there anyone in my social network that has received a similar recommendation?	-5	-5	1	3

Socially Influenced

	Socially Influenced	Interested & Independent	Cautious & Reluctant	Egocentric
	5	5	2	1
Highest Ranked				
Why is this recommendation the best option?				
Questions ranked higher in this array than any other factor array				
What are the pros/cons associated with this option?	4	2	-1	-5
Does the system know and understand my goals?	3	3	-1	1
How satisfied are others who took this recommendation?	3	-2	-2	-1
Questions ranked lower in this array than any other factor array				
Is it working with clean, or 'fuzzy' data?	-3	-2	-2	-2
How much data was used to train this system?	-4	-1	0	-2
Can I see the data for myself?	-4	2	0	-3
Lowest Ranked				
What is the signal to noise ratio of this data?	-5	-3	0	-2

Egocentric

	Egocentric	Interested & Independent	Cautious & Reluctant	Socially Influenced
	5	2	-3	4
Highest Ranked				
Was this recommendation made specifically for ME, or on something else?				
Questions ranked higher in this array than any other factor array				
Precisely what information about me does the system know?	4	0	-2	0
What have other people like me done in response to this recommendation?	3	-3	-1	0
How many other people have received this recommendation from this system?	3	-3	-2	0
Is there anyone in my social network that has received a similar recommendation?	3	-5	-5	1
Questions ranked lower in this array than any other factor array				
How many other options are there?	-3	-1	0	0
Are there any other options not presented here?	-4	0	1	1
How does the system consider risk, and what is its level of "acceptable risk?"	-4	-1	2	2
Lowest Ranked				
What are the pros/cons associated with this option?	-5	2	-1	4

recommendation?" (average score 1.5, Z score variance 0.031) across all factor groups. Despite the wide array of differences in information priorities and decision making heuristics we found amongst our participants, these two questions were agreed upon by all as having at least moderate importance for users of intelligent systems that make recommendations. These questions should therefore be considered highly valuable to answer through an interface, and those design elements should be considered a high priority in intelligent recommender systems such as the one described in our study.

On the other end of the spectrum, none of the factor groups found the questions **"Is my data uniquely different from the data on which the system has been trained?"** (average score −0.75, Z score variance 0.122), and **"Is the system working with solid data, or is the system inferring or making assumptions on fuzzy information?"** (average score −2.25, Z score variance 0.109) as being very important or valuable to them. These questions are likely important to some people, such as programmers who may appreciate this granularity of information about the underlying data, but to end users they are unlikely to be very meaningful or to improve trust or acceptance. In contrast to the questions earlier, answering these questions through an interface would likely add either confusion or become an irritation to users of systems such as D-SAM. Examining what questions produced agreement from across all participants allowed us to quickly narrow down our potential design elements, illustrating a clear benefit of a mixed methods approach to user-centered design.

5.2   Disagreement Amongst Groups

Just as we examined questions that all groups found equally important or unimportant to them, we also examined questions that produced the greatest disagreement between groups. These polarizing questions can help identify potential design elements that may be points of contention to some users. To analyze these questions in a way that is both detailed, yet practical, we arranged all questions into five categories, based on their similarity to one another. The first category was named parameters and logic, and describes aspects of system features that constrain its operations, such as how sensor data is used in deriving system outputs. The next category was named qualities of data, and describes features of relevance about data itself, such as its age, provenance, level of noise, etc. The next category was named user personalization, which describes how systems consider the user in deriving system outputs (this is especially relevant in recommender systems, such as the system we developed for this study). The next category was named justification of options, and describes how options are arranged, how they are prioritized by the system (again, this category of explanation is highly relevant to recommender systems that may generate several potential recommended options, but may only display one to the user). The final category was named social influence, which describes the behaviors and decisions of other users. This is a somewhat unique explanation strategy to recommender systems, commonly seen in music or movie recommendations (i.e., users who watched this show also enjoyed this other show). We will discuss how each factor group valued and prioritized these categories of questions in the sections below.

5.3 System Parameters and Logic

Questions explaining the inner workings of a system, including its reasoning, logic, policies, and limitations, were termed System Parameters & Logic. These questions produced a low degree of disagreement (average Z score variance 0.33) across all groups, with most questions averaging around the mean (score of 0). With the exception of the Cautious and Reluctant group (who were most interested in questions about reliability, uncertainty, and risk), all others found these questions to be of moderate to low importance, indicating them as medium to low priority design elements that are perhaps best delivered through menu options that can be accessed by those most interested. Designing explanations that provide information of this sort, therefore, is advisable, given that most participants, regardless of their factor group, ranked these questions moderately important.

5.4 Qualities of Data

Overall, questions pertaining to the qualities of data, such as age, noise and provenance generated moderate agreement between all factor groups (average Z score variance 0.419). Questions such as **"How current is the data used in making this recommendation?"**, **"How clean or accurate is the data used in making this recommendation?"**, and **How is this data weighted or what data does the system prioritize?** all averaged between 0–1 across all factor groups. It is important to note here that the forced distribution used for this experiment results in a mean score of 0. That these questions were all ranked around the mean indicates they are questions which the majority of stakeholders would like addressed in some form, plausibly in order to better understand and trust intelligent system recommendations.

Other questions related to the qualities of data, however, proved more divisive, and may be too much for some users to appreciate. As discussed in the section on Consensus, none of the factor groups found the questions **"Is my data uniquely different from the data on which the system has been trained?"** or **"Is the system working with solid data, or is the system inferring or making assumptions on fuzzy information?"** very important to them, indicating a potential limit of the usefulness of displaying qualities of data as a means of improving intelligibility. While the Interested and Independent group demonstrated the most willingness and interest in these types of questions, none of the other factor groups was especially interested.

5.5 User Personalization

We termed questions aimed at helping users understand what of *their* data is known, and how that data is used to derive recommendations as User Personalization questions. This category generated a wider range of sentiment than questions about the qualities of data (average Z score 0.744), including the most divisive question **"Was this recommendation made specifically for ME (based on my profile/interests), or was it made based on something else (based on some other model, such as corporate profit, or my friend's interests, etc.)?"** On average, the Socially Influenced and Egocentric groups favored these types of questions more than the more analytical Interested and Independents, and Cautious and Reluctant. Examining user sentiment surrounding these questions helps perhaps to understand why variance was so high. For instance, people in the Cautious and Reluctant group commented things like *"I don't think 'me' is important... I need objective metrics!"*, whereas people in the Socially Influenced group expressed a different sentiment, *I want to know that the system has made the right choice for me and my lifestyle/preferences, and whether it has it really taken all my situations and personal feelings into consideration.*

Yet, the recent increasing concern over potentially inappropriate collection and uses of personal data by social media and others, combined with the moderate rankings of many questions in our sample, such as **Does the system know and understand my goals?** (average score 1.5, Z score variance 0.51), and **Precisely what about me does the system know?** (average score 0.5, Z score variance 0.59), suggests new efforts should be made towards affording users information about and answers to these kinds of questions. Considering the strong prioritization of these questions by the Socially Influenced and Egocentrics, we strongly suggest designers consider making these affordances available wherever possible. To demonstrate one example of how some of these questions can be addressed in order to make systems and algorithms more transparent to users, we have provided a screen shot of a restaurant recommendation app currently under development, which will be used in follow on studies.

Justification of Options. Closely related to explanations, justifications offer assertions about reasons for decisions or choices, examples, alternatives that are eliminated, or counterfactuals [23]. All factor groups in our study agreed that a justification of **Why this recommendation is the BEST option** is important and valuable to them (average score 3.25, Z score variance 0.66). Other questions related to justification of options were also agreed upon as *not* being valuable or useful to our factor groups, such as **Are there any other options not presented here** (average score −0.75, Z score variance 0.5), and **How many options are there?** (average score −1.25, Z score variance 0.25). These questions are likely too in depth for most stakeholders to appreciate, especially given that one of the principal reasons for leveraging decision support tools is to ease the burden of choice [24].

One question: **What are the pros and cons associated with this option?** produced a very high amount of variance between groups (average score 0, Z score variance 1.56). Both Interested and Independents (composite score 2) and Socially Influenced (composite score 4) felt this question was important to them, while the Cautious and Reluctant (composite score −1) and Egocentrics

(−5) did not. Since the Interested and Independents and Socially Influenced were not significantly aligned on any other questions, it is worth exploring why they should both see this question as one they would like answered through an interface.

Understanding the reasoning behind these user priorities is an important component of this research, and if we consider the above question in relation to what other questions these groups found valuable, we may better understand how designs can afford users answers that are meaningful to them.

In this case, while both Interested and Independents, and Socially Influenced want to know the pros and cons associated with a recommendation, precisely *how to answer that* is decidedly different. While the Socially Influenced are more likely to seek answers in the form of what other people report, such as user satisfaction metrics, Interested and Independents would prefer to understand what data was used and how it was weighted. Questions like the above are precisely those that motivate our research, since they have the potential to both confirm and confound user sentiment, depending on a variety of individual factors which are often difficult to measure.

Social Influence. We termed questions that pertained to the actions or opinions of others, or to how users are characterized and grouped with others as Social Influence questions. Questions in this category produced the greatest amount of disagreement between groups (average Z score variance 0.98), suggesting that as design elements they represent potentially polarizing options. Averaging all questions in this category, we see that the Egocentrics (average score 1.33) and Socially Influenced (average score 1.17) both consider this information valuable and useful to their decision making, while the Cautious and Reluctant (average score −2.33) and Interested and Independent (average score −3.5) clearly do not.

Socially-related information, such as how users are characterized and grouped into personas, and what other people like them have done in similar circumstances, is commonly used in current systems that offer recommendations, such as Netflix, Spotify, or Amazon (e.g., others who purchased this also bought XYZ). These features may improve decision making for some, like the Egocentrics, while they may be ignored by others, like the Interested and Independent. What is of potential interest, however, is how this type of information may soon be featured in other applications with greater scope.

There is considerable room for this kind of information to be considered useful, for instance, as crowd sourcing becomes a more common feature in several domains. There are already several notable examples, such as citizen science [25], personal wellness [26], and even app design [27] which make use of a community of distributed participants that collaborate to form something. These projects often feature consensus building activities that leverage the concept of "hive mind" or "wisdom of the crowd" to achieve common goals. While there are certainly limits to the use of crowd sourcing, especially in highly personalized domains such as clinical medicine or personal financial management, these approaches may very well become more commonplace as intelligent systems broaden and consume greater market presence in our everyday lives. Designers that choose to feature socially-related information into their products may well find those features appreciated and valued, especially as a younger techno-centric generation assumes more of the user base.

5.6 Design Implications

Of the 36 questions in our sample bank, most were of value to one factor group or another, and (as we showed in the Consensus section above) very few were totally unimportant. For prospective designers of transparent intelligent systems, this presents something of a quandary. The most obvious solution–to present all data that could be relevant to someone–would result in impractical long lists of information that is not especially relevant to anyone.

Our approach has uncovered a detailed view of the different manner in which users reason about systems, and can help designers better understand how some explanations can have a greater or lesser effect on user trust, engagement, and acceptance. For example, an explanation and justification of options is most important to people like the Interested & Independent group, while users in the Socially Influenced group might respond well to social navigation cues. The Cautious & Reluctant group would be more satisfied with a detailed description of the data that fed the model, and appreciate control over which data are used to make recommendations. Using mixed methods approaches such as Q-methodology can add significant value to traditional user-centered investigations, and offer data that is qualitatively nuanced, while being quantitatively rigorous. An approach such as this could be successfully used early in the design cycle–and indeed we suggest and encourage the early involvement of users, but would also be appropriate in later stages as well. Designers can readily make use of data such as these to help resolve potential conflicts in priorities, and guide their communication strategies.

6 Conclusion

We have explored potential design features for enhancing the intelligibility and transparency of intelligent systems to end users using a novel mixed methods approach. We have described a variety of reasoning and information seeking strategies of potential users of said systems, and have detailed them into a robust user typology. We have explored this typology in detail, and have compared and contrasted user preferences related to understanding system functions and behaviors to demonstrate how they can be used to guide design strategies. We have compared and contrasted potential design features in order to determine which may be more efficient and valuable to end users in the context of interactions with intelligent recommender systems. Our findings support and reinforce the argument that system transparency is a multi-dimensional construct requiring at least some consideration for user preference and individual differences in order to achieve the desired effect of improving trust, usability, and technology acceptance.

References

1. Gregor, S., Benbasat, I.: Explanations from intelligent systems: theoretical foundations and implications for practice. MIS Q. **23**(4), 497 (1999)
2. Sherry, L., Feary, M., Polson, P., Palmer, E.: What's it doing now?: taking the covers off autopilot behavior. Presented at the 11th International Symposium on Aviation Psychology, Dayton, OH, USA (2001)
3. Herlocker, J.L., Konstan, J.A., Riedl, J.: Explaining collaborative filtering recommendations. Presented at the 2000 ACM Conference for Computer Supported Social Work, Philadelphia, PA, USA (2000)
4. Swearingen, K., Sinha, R.: Beyond algorithms: an HCI perspective on recommender systems. In: ACM SIGIR 2001 Workshop on Recommender Systems, New Orleans, LA, USA (2001)
5. Lim, B.Y., Dey, A.K.: Toolkit to support intelligibility in context-aware applications. Presented at the 12th ACM International Conference on Ubiquitous Computing, Copenhagen, Denmark, pp. 13–22 (2010)
6. Ribeiro, M.T., Singh, S., Guestrin, C.: Why should i trust you?. Presented at the the 22nd ACM SIGKDD International Conference, San Francisco, CA, USA, pp. 1135–1144 (2016)
7. Lim, B.Y., Dey, A.K., Avrahami, D.: Why and why not: explanations improve the intelligibility of context-aware intelligent systems. Presented at the ACM CHI Conference on Human Factors in Computing Systems, Boston, MA, USA, pp. 2119–2128 (2009)
8. Doshi-Velez, F., Kim, B.: Towards a rigorous science of interpretable machine learning, AirXiv (2017)
9. Glass, A., McGuinness, D.L., Wolverton, M.: Toward establishing trust in adaptive agents. Presented at the 13th International Conference on Intelligent User Interfaces, New York, New York, USA, p. 227 (2008)
10. Ososky, S., Sanders, T., Jentsch, F., Hancock, P.A., Chen, J.Y.C.: Determinants of system transparency and its influence on trust in and reliance on unmanned robotic systems. Presented at the SPIE Defense + Security, vol. 9084 (2014)
11. Krause, J., Perer, A., Ng, K.: Interacting with predictions. Presented at the the 2016 CHI Conference, New York, New York, USA, pp. 5686–5697 (2016)
12. Cooper, A.: The inmates are running the asylum: why high-tech products drive us crazy and how to restore the sanity, London, UK (2004)
13. Miller, T.: Explanation in artificial intelligence: insights from the social sciences, airXiv, pp. 1–57, June 2017
14. Buchannan, B., Shortliffe, E.: Rule-Based Expert Systems: The MYCIN Experiments of the Stanford Heuristic Programming Project. Addison Wesley, Reading (1984)
15. Swartout, W.R., Moore, J.D.: Explanation in second generation expert systems. In: David, J.M., Krivine, J.P., Simmons, R. (eds.) Second Generation Expert Systems, pp. 543–585. Springer, Heidelberg (1993). https://doi.org/10.1007/978-3-642-77927-5_24

16. Vorm, E.S., Miller, A.D.: Assessing the value of transparency in recommender systems: an end-user perspective. Presented at the RecSys 2018 Proceedings of the 12th ACM conference on Recommender Systems, Vancouver, Canada (2018)
17. Brown, S.R.: A primer on Q methodology. Operant Subj. **16**, 91–138 (1993)
18. O'Leary, K., Wobbrock, J.O., Riskin, E.A.: Q-methodology as a research and design tool for HCI. Presented at the CHI 2013, Paris, France, pp. 1941–1950 (2013)
19. Ram, A.: AQUA: Questions that Drive the Explanation Process. Lawrence Erlbaum, Hillsdale (1993)
20. Silveira, M.S., de Souza, C.S., Barbosa, S.D.J.: Semiotic engineering contributions for designing online help systems. Presented at the 19th Annual International Conference on Computer Documentation, Santa Fe, NM, USA, p. 31 (2001)
21. Ford, J.K., MacCallum, R.C., Tait, M.: The application of exploratory factor analysis in applied psychology: a critical review and analysis. Pers. Psychol. **39**(2), 291–314 (1986)
22. Devore, J.: Probability and Statistics for Engineering and the Sciences, 4th edn. Brooks/Cole, New York (1995)
23. Biran, O., Cotton, C.: Explanation and justification in machine learning: a survey. In: IJCAI-17 Workshop on Explainable Artificial Intelligence (XAI), Melbourne, Australia (2017)
24. Eriksson, A., Stanton, N.: Takeover time in highly automated vehicles: noncritical transitions to and from manual control. Hum. Factors **59**(4), 689–705 (2017)
25. Thakur, G.S., Sparks, K., Li, R., Stewart, R.N., Urban, M.L.: Demonstrating PlanetSense. Presented at the 24th ACM SIGSPATIAL International Conference, New York, New York, USA, pp. 1–4 (2016)
26. Agapie, E., et al.: Crowdsourcing exercise plans aligned with expert guidelines and everyday constraints. Presented at the 2018 CHI Conference, New York, New York, USA, pp. 1–13 (2018)
27. Huang, T.H.K., Chang, J.C., Bigham, J.P.: Evorus. Presented at the 2018 CHI Conference, Montreal, Quebec, Canada, pp. 1–13 (2018)

The Role of Gaze as a Deictic Cue in Human Robot Interaction

Efecan Yılmaz[1], Mehmetcan Fal[1,2], and Cengiz Acartürk[1(✉)]

[1] Institute of Informatics, Department of Cognitive Science, Middle East Technical University, Ankara, Turkey
{efecan,acarturk}@metu.edu.tr
[2] TAI, Turkish Aerospace Industries, Ankara, Turkey
mehmetcan.fal@tai.com.tr

Abstract. Gaze has a major role in social interaction. As a deictic reference, gaze aims at attracting visual attention of a communication partner to a referred entity in the environment. Gaze direction in natural faces is a well-investigated domain of research at behavioral and neurophysiological levels. However, our knowledge about deictic role of gaze in Human Robot Interaction is limited. The present study focuses on a comparative analysis of the deictic role of gaze direction in alternative face morphologies. We report an experimental study that investigated deictic gaze in a virtual reality environment. Human participants identified object locations by utilizing deictic gaze cues provided by avatar faces, as well as natural human faces. Our findings reveal a facilitating role in the accuracy of objection detection in favor of gaze embedded in natural faces compared to gaze embedded in synthetic avatar face morphologies.

Keywords: Deictic gaze · Joint attention · Human-avatar interaction · Human-robot interaction · Synthetic faces

1 Introduction

Gaze has a major role in social interaction. Gaze direction is a well-investigated research domain at behavioral and neurophysiological levels in both humans and primates. Early research on the role of gaze direction and orientation of the face revealed nerve cell selectivity in macaqua monkey [1]. Gaze direction and head orientation have been investigated within the context of Human Computer Interaction [2–4].

In natural communication settings, the role of gaze direction can be studied in accompany with language use, since language and vision are two complementary modalities that employ deixis in communication. Verbal deictic expressions are necessarily accompanied by other modalities, such as pointing gestures or pointing by gaze. The presence of multiple modalities serves various purposes, such as disambiguating spatial reference to objects in the environment.

The present study investigates deictic gaze within the context of natural vs. synthetic faces. We designed a joint action setting to investigate the role of gaze deixis in disambiguating spatial reference to objects in a virtual reality (VR) environment. The use of

© Springer Nature Switzerland AG 2020
D. D. Schmorrow and C. M. Fidopiastis (Eds.): HCII 2020, LNAI 12197, pp. 466–478, 2020.
https://doi.org/10.1007/978-3-030-50439-7_32

VR allowed us to investigate human-avatar interaction in an immersive, social interaction setting, as well as giving the flexibility for studying the differences between natural and synthetic face morphologies. Below, we introduce the concept of deixis within the context of Human Robot Interaction.

1.1 Deixis in Human-Robot Interaction (HRI)

Joint action settings have been frequently used for studying communication in Human Robot Interaction (HRI) by employing specific communication settings, such as joint attention, action observation, task sharing, and action coordination [5, 6]. These studies focus on various aspects of HRI, such as the properties of shared environments [7], and collaboration between humans and robots by means of shared tasks and/or deictic expressions [8–10]. They also address multimodal aspects of communication such as gaze and gestures [11–13] as well as language [8, 9]. For instance, Admoni and Scassellati [11] describe gaze-based interaction between humans and robots within the context of *intuitive interactions*. They state that the communication between the interlocutors in an HRI setting take place in both verbal and non-verbal modalities, such as gaze and gestures. More generally, *social robotics* is a recent domain of research that addresses the investigation of communication modalities, methods, and interaction patterns between interlocutors [14]. The present study employs the concept of deixis to study the role of face morphologies in HRI.

Deixis can be realized in multiple, interdependent modalities in communication, including both non-verbal forms (e.g., gaze or gesture) and/or verbal forms (e.g., deictic referring expressions). Various aspects of deictic interaction have been reported previously. For instance, Devault, et al. [15] investigated collaborative reference from a linguistics point of view. To propose a dialogue framework in a human-human communication setting, they analyzed collaborative reference by a supervised machine learning model that evaluated interlocutor affirmations to verbal descriptions. They aimed to simulate a collaborative communication setting where they conceived collaborative reference as information state and linguistic reference in both utterance planning and understanding. Another relevant study on gaze interaction in multi-agent HRI environments is by [16]. They investigated how gaze cues could be utilized in HRI settings, particularly in information exchange tasks. They investigated the forms of cues that an avatar may provide to human participants to regulate conversational roles.

In the present study, we designed a VR environment, where the participants were immersed in a deictic communication setting. This setting was used for an experiment in which we explored human responses to gaze direction of virtual avatars (avatars' pointing to objects by their gaze direction vectors). The avatars pointed at objects by their gaze, and the human participants reported the gazed object by naming them. For communication, we used *explicit* referring expressions (e.g., "what is the object at the top-left?"), as well as *implicit* referring expressions resolved by avatars' gaze vectors (e.g., "what is there?"). This setting allowed us to investigate HRI in VR from a joint-action perspective with the following research questions: (1) How does face morphology of avatars (vs. natural faces) influence accuracy and response time to identify the intended object? (2) Does the number of avatars in have an influence on those variables? (3) How is the human participant's gaze distributed on the avatars during the joint task?

We used four avatar designs (humanoid avatar, non-humanoid stick figure avatar, human-image avatar, and blinking non-humanoid stick figure avatar). The number of avatars and their deictic gaze vectors on the objects were varied for each new scene (a set of objects on a table) by randomization. We investigated how gaze distribution, as an indicator of gaze dispersion, was influenced by the task and the number of avatars in this setting. We evaluated gaze distribution of the participants by an entropy analysis, which can be interpreted as a measure of confidence in response. The entropy anaylsis has been used as a measurement of uncertainty inherent in a random variable [17], [18, p. 480]. The following sections introduce the experimental investigation in more detail.

2   Methodology

The experiment was conducted in a VR environment with a head-mounted display (HMD) capable of eye-tracking (a 60 Hz SMI tracker embedded in Samsung Gear VR headset). The experiment had a 4 (avatar design) $\times 2$ (explicit vs. implicit referring expression) $\times 3$ (the number of avatars) design. The avatar designs differed in face and body morphology: Humanoid Robot (Fig. 1a), Stick Figure (Fig. 1b), Human Avatar (Fig. 1c, 1d), Blinking Stick Figure (Fig. 1e, 1f).

The first two avatar designs (Fig. 1a, 1b) employed identical, cartoon-like eyes, whereas human-image avatar design (Fig. 1c, 1d) used model humans' eyes in sets of images, approximately angled to match the gaze vectors from the first two designs. For the sake of naturality of the eyes in the robot face, we added a third face morphology that employed eyelids on the same cartoon-like eyes from the first two designs, furthermore, a blinking animation was also added to the eyes with the lids (Fig. 1e, 1f). The four avatar designs were presented to the participants in four distinct experimental sessions (i.e., the avatar design was a between-subject factor).

The participants attended two experiment sessions (i.e., the joint task was a within-subject factor with two conditions, namely *explicit* and *implicit*, described in more detail below). The number of avatars was also a within-subject factor. In each trial, the number of avatars was varied between one avatar, three avatars, and five avatars (Fig. 2).

3   Participants, Materials, and Procedure

Ninety-three participants participated in the experiment. The participants were either students or academic staff (e.g., assistants, instructors). Participants' ages varied between 18 and 50 (M = 24.55, SD = 5.45). Participants were randomly allocated to one of four groups (cf. four face morphology conditions). All participants were native speakers of the experiment language.

Each experiment session consisted of two joint tasks, namely an *explicit* joint task and an *implicit* joint task. In the explicit joint task, the participants were presented questions that included explicit referring expressions (shown in italics in this example), such as "what is the object at *the top left*?". The questions were presented auditorily. The refering expressions were simply descriptions of geometric locations of the objects on a table in the VR environment. Accordingly, the questions in the explicit joint task were "what is the object at the {*left top, left bottom, right top, right bottom*}?". In the

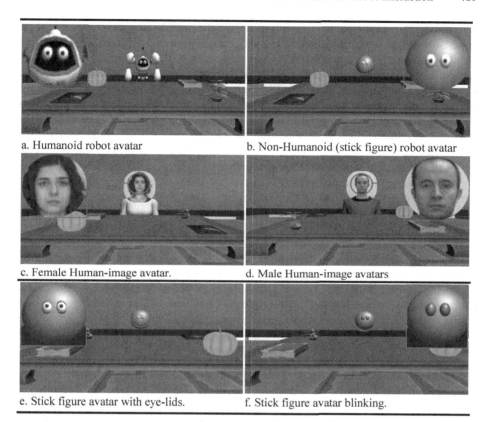

a. Humanoid robot avatar   b. Non-Humanoid (stick figure) robot avatar

c. Female Human-image avatar.   d. Male Human-image avatars

e. Stick figure avatar with eye-lids.   f. Stick figure avatar blinking.

Fig. 1. Alternative face morhologies utilized in the virtual reality environment.

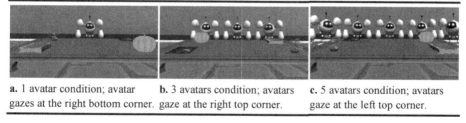

a. 1 avatar condition; avatar   **b.** 3 avatars condition; avatars   **c.** 5 avatars condition; avatars
gazes at the right bottom corner. gaze at the right top corner.   gaze at the left top corner.

Fig. 2. Variation in the number of avatars. Condition repeated for all face morphologies.

implicit joint task, the questions included an implicit deictic reference, i.e. "what is the object there?". The participants had to reply the questions by inspecting the avatar(s) gaze direction(s), which were all consistent in case of multiple avatars. The location gazed by the avatars also was randomized for each of the fourty (thirthysix recorded and four practice) trials. The participants replied to an audio recording verbally in both joint tasks and their answers were transcribed after the experiment.

The accuracy of the participants in identifying the intended object on the table, their response time (the duration between the end of the question and the onset of the answer), and their gaze interaction with the environment (i.e., the distribution and dispersion of participants' gaze on the avatars, as well as their gaze contact ratio with the avatars) were dependent variables. Additionally, the number of avatars were randomized for each trial in both (explicit and implicit) tasks such that 1, 3, or 5, avatars were presented randomly. The explicit task also had a no-avatar condition, which served as a condition for technical verification (i.e. to check whether the environment setting technically worked as expected). In the trials where there was more than one avatar, the gaze vectors of all the avatars consistently pointed at the same object.

Accuracy, response time, gaze distribution on the avatars were analyzed using a repeated-measures ANOVA with the number of avatars (within-subject), the joint task (*explicit, implicit* – within-subject), and the avatar design (between-subject) factors. For gaze disribution, we performed an entropy analysis that used gaze shifts between objects in the VR environment. We assumed that, in the transition matrix, each data point where a participant shifted their gaze from an object to another represented a gaze shift (e.g., from the table to the pumpkin, or from an avatar to the book), whereas each data point where the participant fixated on the same object as before was a gaze fixation. The entropy measure was calculated by the following equation, where R is the normalized matrix with r_i being its values, and p representing probability.

$$H(R) = -\sum p(r_i)\log_2 p(r_i) \text{ for } r_i \in R$$

The term Entropy was originally defined by Shannon [17], and proposed as a measure of dispersion in eye tracking studies by [18, p. 480]. The present procedure resulted in an entropy value for each trial. This allowed us to calculate the uncertainty in gaze distribution as a measure of confidence in response, as well as the randomness in the scanpaths of participants in their gaze interaction with the avatars.

4   Results

This section presents the results of the experiment. In particular, we report the analyses of 4 (avatar designs) ×3 (number of avatars) ×2 (joint tasks) conditions for participants' accuracy in identifying the intended objects, their response time and gaze distribution on the environment.

4.1   Accuracy

Firstly, we report the technical verification condition, where no-avatar existed in the environment. In the *explicit* task, where participants responded to questions, such as "what is the object at the left top?" or similar, virtually no errors were observed ($M = 98.47, SD = 0.09$). This showed that the VR experiment setting worked as intended. We excluded one participant from the results due to the misunderstanding of the task. All the participants were included in the following analyses.

A Repeated Measures ANOVA test was conducted to evaluate the effects of both within-subject (number of avatars, task) factors and the between-subject factor (avatar design). Firstly, a Mauchly's test indicated that the sphericity had not been violated for any of our within-subject effects, $\chi^2_{\text{number-of-avatars}}$ (2) $= .185, p > .5$, $\chi^2_{\text{number-of-avatars * task}}$ (2) $= 1.137$, $p > .05$. Therefore, we report our values without any correction for participants' accuracy. A significant effect was obtained for within-subject factor of task $F(1, 88) = 122.809$, p < 001. This showed that the accuracy of the participants differed significantly between explicit and implicit task conditions. The number of avatars also revealed a significant effect $F(1, 176) = 11.295$, p $< .001$, along with a significant interaction effect for task and number of avatars $F(2, 176) = 8.584$, p $< .001$. Finally, the results for the between-subject factor (avatar designs) show that there was a significant interaction effect for both joint tasks $F(3, 88) = 20.602$, p $< .001$. However, the interaction between the number of avatars and the avatar design was not significant.

To further evaluate the accuracy results, we conducted a series of t-tests. The first was a comparison of each step in the number of avatars for the two tasks. The results showed that the accuracy was significantly different between explicit and implicit tasks for all numbers of avatars (1, 3, and 5 avatars respectively) $t(91) = 6.461, p < .001, t(91) = 5.604, p < .001, t(91) = 7.727, p < .001$. These first t-test results were in line with the significant effects present in the repeated measures analysis. Secondly, we analyzed for the significant effects observed in the number of avatars. In the explicit joint task, the number of avatars did not have a significantly different result in terms of accuracy in any of the steps for the number of avatars for 1 and 3, 1 and 5, and 3 and 5 avatars. In contrast, for the implicit joint task the accuracy results were significantly different between 1 and 5 avatars; $t(92) = 2.406, p < .05$, as well as for 3 and 5 avatars; $t(92) = 4.318, p < .001$. For each pair of values, the descriptive results are shown in Table 1.

Table 1. Descriptive statistics for the accuracy of participants' responses for varying numbers of avatars appearing in the VR environment in the two joint tasks. The results are presented in n/1 (i.e. a result of 1.000 means 100% accuracy).

Number of avatars	Explicit joint task	Implicit joint task
1 avatar	M = .984 (SD = .044)	M = .866 (SD = .164)
3 avatars	M = .989 (SD = .092)	M = .893 (SD = .157)
5 avatars	M = .977 (SD = .088)	M = .825 (SD = .188)

Finally, we conducted several independent-samples analyses to further evaluate the role of the avatar face morphology on the accuracy in the tasks. The results showed that in the explicit task, the accuracy values exhibited a similar pattern among all conditions (Table 2). However, in the implicit task, the non-human avatar faces exhibited a similar pattern, which was different than the accuracy values in human-image avatars. In particular, the use of human-image avatars resulted in much higher accuracy values in the implicit task condition, compared to the accuracy scores in the use of non-human avatars (Table 3).

In summary, when considering all results (4 avatar designs, 2 tasks, and 3 conditions in the number of avatars), the accuracy in the explicit condition did not differ significantly among avatar designs. This result was expected since the participants were able to detect

Table 2. Explicit joint task only descriptive statistics for accuracy by avatar designs and the number of avatars.

Number of avatars	Humanoid avatars	Non-humanoid avatars	Human-Image Avatars	Blinking non-humanoid Avatars
1 avatar	$M = 1.000$ (SD = .000)	$M = .977$ (SD = .059)	$M = .973$ (SD = .051)	$M = .989$ (SD = .036)
3 avatars	$M = 995$ (SD = .024)	$M = .988$ (SD = .039)	$M = .981$ (SD = .043)	$M = .995$ (SD = .024)
5 avatars	$M = .994$ (SD = .027)	$M = .988$ (SD = .037)	$M = .968$ (SD = .096)	$M = .991$ (SD = .029)

Table 3. Implicit joint task only descriptive statistics for accuracy by avatar designs and the number of avatars.

Number of avatars	Humanoid avatars	Non-humanoid avatars	Human-Image avatars	Blinking non-humanoid avatars
1 avatar	$M = .808$ (SD = .147)	$M = .776$ (SD = .152)	$M = .991$ (SD = .029)	$M = .830$ (SD = .202)
3 avatars	$M = .870$ (SD = .142)	$M = .854$ (SD = .177)	$M = .994$ (SD = .014)	$M = .810$ (SD = .190)
5 avatars	$M = .752$ (SD = .195)	$M = .748$ (SD = .185)	$M = .993$ (SD = .019)	$M = .734$ (SD = .166)

the intended object without employing the gaze directions of the avatars. On the other hand, in the implicit joint task, the accuracy results for human-image avatars differed significantly from non-human avatars for all number of avatars. The following section presents the results for response times of the participants.

4.2 Response Times

The response time analysis was conducted as a repeated-measures ANOVA. A Mauchly's test indicated that the sphericity had not been violated for any of our within-subject effects, $\chi^2_{\text{number-of-avatars}}(2) = 2.717$, $p > .05$ and $\chi^2_{\text{number-of-avatars * task}}(2) = 1.420$, $p > .05$. Therefore, we report our values without any correction for participants' response times. The results indicated a statistically insignificant effect on participants' response times between the tasks $F(1, 88) = 2.249$, p $> .05$. However, the number of avatars had a significant effect; $F(2, 176) = 34.680$, p $< .001$. Additionally, the interaction effect between the number of avatars and the tasks was significant; $F(2, 176) = 25.529$, p $< .001$. The results are shown in Table 4.

Table 4. Descriptive statistics for response times of participants in their answering to the explicit verbal expressions (explicit task) or in resolving the deictic gaze references (implicit task).

Number of avatars	Explicit joint task	Implicit joint task
1 avatar	$M = 2.387$ ($SD = .538$)	$M = 2.240$ ($SD = .589$)
3 avatars	$M = 2.424$ ($SD = .552$)	$M = 2.492$ ($SD = .743$)
5 avatars	$M = 2.423$ ($SD = .546$)	$M = 2.668$ ($SD = .905$)

The results also revealed an interaction between the tasks and avatar design, $F(3, 88) = 5.853$, p $< .01$. In a paired samples t-test, we evaluated how the number of avatars interacted with the response times (Table 5 and Table 6). The t-test results revealed significant differences between 1 and 3 avatars, $t(92) = -5.537$, p $< .001$, between 1 and 5 avatars, $t(92) = -7.740$, p $< .001$, and between 3 and 5 avatars, $t(92) = -3.596$, p $< .01$.

Table 5. Explicit joint task only descriptive statistics for response time by avatar designs and the number of avatars.

Number of avatars	Humanoid avatars	Non-humanoid avatars	Human-Image Avatars	Blinking non-humanoid avatars
1 avatar	$M = 2.319$ ($SD = .564$)	$M = 2.171$ ($SD = .502$)	$M = 2.569$ ($SD = .564$)	$M = 2.420$ ($SD = .445$)
3 avatars	$M = 2.421$ ($SD = .610$)	$M = 2.188$ ($SD = .532$)	$M = 2.629$ ($SD = .559$)	$M = 2.379$ ($SD = .414$)
5 avatars	$M = 2.328$ ($SD = .541$)	$M = 2.166$ ($SD = .511$)	$M = 2.651$ ($SD = .550$)	$M = 2.458$ ($SD = .471$)

In summary, the response time analyses showed that the response times did differ between the two tasks only for a certain number of avatars in the environment.

4.3 Gaze Interaction and Distribution

To reiterate the experiment setting, it was only in the implicit task that gaze interaction with the avatars was required, whereas in the explicit task, the participants were able to answer correctly by replying the question. The change in the nature of the task was observed in gaze distributions. We followed a two-step approach for gaze analyses. Firstly, the gaze data were analyzed to find the total gaze contact with the avatars. Secondly, the gaze shifts and gaze fixations were analyzed as a transition matrix in an entropy analysis. We removed data from two participants, one due to the misunderstanding of the task, the other due to technical problems in gaze data collection.

A repeated-measures ANOVA was conducted for analyzing gaze contact with the avatars. A Mauchly's test of sphericity indicated that the sphericity had been violated for

Table 6. Implicit joint task only descriptive statistics for response time by avatar designs and the number of avatars.

Number of avatars	Humanoid avatars	Non-humanoid avatars	Human-Image avatars	Blinking non-humanoid avatars
1 avatar	$M = 2.468$ (SD = .705)	$M = 1.964$ (SD = .537)	$M = 2.262$ (SD = .574)	$M = 2.263$ (SD = .463)
3 avatars	$M = 2.775$ (SD = .948)	$M = 2.253$ (SD = .577)	$M = 2.378$ (SD = .662)	$M = 2.629$ (SD = .707)
5 avatars	$M = 2.999$ (SD = 1.113)	$M = 2.263$ (SD = .519)	$M = 2.553$ (SD = .964)	$M = 2.901$ (SD = .769)

the within-subject factor of number of avatars $\chi^2_{number\text{-}of\text{-}avatars}$ $(2) = 15.121, p < .001$. In contrast, the results for the interaction between the task and the number of avatars indicated that the sphericity had not been violated $\chi^2_{task\,*\,number\text{-}of\text{-}avatars}$ $(2) = 4.556, p > .05$. As a result, we report corrected (Greenhouse-Geisser) results for the number of avatars within-subjects factor and its interaction effects, whereas reporting task factor without corrections.

The results showed that the task condition (*explicit* vs. *implicit*) had a significant effect on gaze contact of the participants with the avatars $F(1, 89) = 1078.0, p < .001$. The number of avatars factor also revealed a significant main effect $F(1.725, 151.784) = 55.675, p < .001$. There was also a significant interaction between the task conditions and the number of avatars $F(1.903, 167.456) = 22.244, p < .001$. Lastly, the avatar design interacted significantly with the task $F(3, 88) = 4.581, p < .01$, but not with the number of avatars $F(5.174, 151.784) = 1.676, p > .05$. To analyze the within factor differences a paired samples t-test was conducted. The results showed that the total gaze on the avatars was significantly different both within and in between the two tasks. The t-test results are that participants gazed at the avatars in differing amounts of time between the two joint tasks for all numbers in which the avatars were present in the VR environment: 1 avatar; $t(91) = -24.473, p < .001$, 3 avatars; $t(91) = -30.245, p < .001$, and 5 avatars; $t(91) = -29.643, p < .001$.

The mean differences in (values in Table 7) were also analyzed in a paired samples t-test for the within-subjects factor of number of avatars. In the explicit joint task, participants gazed at the avatars at rates significantly differently between 1 avatar and 3 avatars; $t(91) = -3.886, p < .001$, as well as between 1 and 5 avatars; $t(91) = -2.658$, $p < .01$. However, between 3 avatars and 5 avatars, there was no significant difference; $t(91) = .950, p > .05$. The same trend continued for the implicit joint task; in that the significant differences were in between 1 and 3 avatars; $t(92) = -9.670, p < .001$, and in between 1 and 5 avatars; $t(92) = -9.383, p < .001$, and not significant in between 3 and 5 avatars; $t(92) = .332, p > .05$.

Finally, we also conducted a repeated-measures ANOVA and follow up paired samples t-test using the entropy values of participants' gaze interaction. A Mauchly's test of sphericity indicated that the sphericity had not been violated for either the factor of

Table 7. Gaze contact ratios in which participants gazed at the avatars in the VR environment for the total number of gaze samples in each trial, for the number of avatars and the two joint tasks.

Number of avatars	Explicit joint task	Implicit joint task
1 avatar	$M = 26.31\%$ $(SD = 16.75\%)$	$M = 68.70\%$ $(SD = 13.27\%)$
3 avatars	$M = 30.00\%$ $(SD = 17.40\%)$	$M = 79.16\%$ $(SD = 10.93\%)$
5 avatars	$M = 29.09\%$ $(SD = 18.05\%)$	$M = 78.95\%$ $(SD = 11.10\%)$

the number of avatars $\chi^2_{number-of-avatars}$ $(2) = 2.366, p > .05$ or for the interaction of the number of avatars and task $\chi^2_{task * number-of-avatars}$ $(2) = 4.077, p > .05$. Therefore, all results are reported without corrections. The results showed that there was not a significant effect of the within-subject factor task (the two joint tasks) on the entropy results of participants' gaze interaction in the VR environment; $F(1, 87) = .281, p > .05$. In contrast, the interaction between task and avatar design was a significant effect; $F(3, 87) = 7.510, p < .001$. Furthermore, the number of avatars $F(2, 174) = 77.071, p < .001$ had a significant effect, whereas the interaction between the number of avatars and the avatar design $F(6, 174) = 1.991, p > .05$ did not have a significant effect. Finally, the interaction between the task and the number of avatars was also significant $F(2, 174) = 28.447, p < .001$.

The paired samples t-test results (mean values are shown at Table 8) showed that the entropy values – except for one case – were significantly different for both the joint tasks and in all variance point for the number of avatars; t-test values reported in Tables 9, 10, 11.

Table 8. Mean entropy values for each variance in the number of avatars for the two joint tasks.

Number of avatars	Explicit joint task	Implicit joint task
1 avatar	$M = .173$ $(SD = .029)$	$M = .161$ $(SD = .031)$
3 avatars	$M = .180$ $(SD = .027)$	$M = .197$ $(SD = .034)$
5 avatars	$M = .187$ $(SD = .033)$	$M = .193$ $(SD = .032)$

Table 9. Tasks comparison results of paired samples t-test for mean differences in entropy of participants' gaze dispersion.

Number of avatars	Explicit task – implicit task
1 avatar	$t(92) = 3.331, p < .01$
3 avatars	$t(92) = -3.922, p < .001$
5 avatars	$t(92) = -1.388, p > .05$

Table 10. Explicit task results of paired samples t-test for mean differences in entropy of participants' gaze dispersion.

Number of avatars	3 avatars	5 avatars
1 avatar	t(92) = −3.116, p < .01	t(92) = −5.704, p < .001
3 avatars	–	t(92) = −2.640, p < .05

Table 11. Implicit task results of paired samples t-test for mean differences in entropy of participants' gaze dispersion.

Number of avatars	3 avatars	5 avatars
1 avatar	t(92) = −9.518, p < .001	t(92) = −10.656, p < .001
3 avatars	–	t(92) = 1.708, p > .05

The entropy analysis show that participants' gaze dispersion is significantly different between the two joint tasks when 1 or 3 avatars are present in the VR environment. Furthermore, the dispersion within each joint task also differ significantly between 1 avatar and both 3 or 5 avatars in the VR environment (Tables 12 and 13).

Table 12. Descriptive statistics for the entropy values for the explicit joint task and the number of avatars.

Number of avatars	Humanoid avatars	Non-humanoid avatars	Human-Image avatars	Blinking non-humanoid avatars
1 avatar	M = .181 (SD = .027)	M = .180 (SD = .026)	M = .165 (SD = .025)	M = .176 (SD = .024)
3 avatars	M = .188 (SD = .026)	M = .188 (SD = .024)	M = .170 (SD = .025)	M = .179 (SD = .028)
5 avatars	M = .202 (SD = .035)	M = .199 (SD = .036)	M = .169 (SD = .026)	M = .186 (SD = .025)

5   Discussion and Conclusion

The results of the experimental study showed that an explicit statement of a referring expression, as in "what is the object at the left bottom?" in a joint, deictic gaze task revealed different patterns in terms of accuracy, response times and gaze allocation of the participants, compared to an implicit statement of a referring expression, as in "what is the object there?". The differences were shaped by both the number of avatars in

Table 13. Descriptive statistics for the entropy values for implicit joint task and the number of avatars.

Number of avatars	Humanoid avatars	Non-humanoid avatars	Human-Imagen avatars	Blinking non-humanoid avatars
1 avatar	$M = .154$ (SD = .029)	$M = .179$ (SD = .034)	$M = .155$ (SD = .029)	$M = .160$ (SD = .026)
3 avatars	$M = .188$ (SD = .031)	$M = .201$ (SD = .034)	$M = .208$ (SD = .037)	$M = .196$ (SD = .027)
5 avatars	$M = .176$ (SD = .030)	$M = .203$ (SD = .032)	$M = .195$ (SD = .032)	$M = .195$ (SD = .030)

the VR environment (1, 3 and 5 avatars) and their design (humanoid, non-humanoid, and human-image avatars). In particular, in the explicit joint task, the accuracy of the responses to human-image avatars was strikingly high, close to 100% compared to the accuracy of the responses to humanoid and non-humanoid avatars. This finding shows that deictic gaze is difficult to represent in non-human-image avatars, may they have either humanoid or non-humanoid designs, possibly due to humans' highly efficient biological endowment in responding to real human faces.

Human-image avatars took longer to respond in the explicit joint task, possibly due to its salience, since it was not required to extract information from the avatar in the explicit joint task. When it was necessary to resolve the gaze direction of the avatars (i.e., in the implicit task), the response time difference was lost. This finding suggests that humans are more efficient in resolving gaze deixis when they look at a human-image avatar, whereas this efficiency is not related to response time. The participants allocated their gaze on the avatar(s), in contrast to gaze allocation to other entities in the environment, i.e. the table and the objects on the table, more frequently when there were multiple avatars in the environment. However, non-significant differences between 3-avatar conditions and 5-avatar conditions show that the number of avatars in the environment may not have a significant impact on deictic gaze resolution, other than attracting more gaze compared to a single-avatar.

In conclusion, the present study provides evidence for the roles of avatar agent design and collaborative reference in disambiguation deixis in VR environments. Further research aims at studying other design factors, such as the movement of the avatars, as well as their spatial allocation and visual design parameters.

Acknowledgments. This project has been supported by TÜBİTAK 117E021 "A gaze-mediated framework for multimodal Human Robot Interaction".

References

1. Perrett, D., Smith, P., Potter, D., et al.: Visual cells in the temporal cortex sensitive to face view and gaze direction. In: Proceedings of the Royal Society of London, Series B, Biological Sciences, vol. 223, no. 1232, pp. 293–317. Royal Society (1985)
2. Stiefelhagen, R., Zhu, J.: Head orientation and gaze direction in meetings. In: CHI 2002 Extended Abstracts on Human Factors in Computing Systems (CHI EA 2002), pp. 858–859. Association for Computing Machinery, New York City (2002)
3. Katzenmaier, M., Stiefelhagen, B., Schultz, T.: Identifying the addressee in human-human-robot interactions based on head pose and speech. In: ICMI 2004 - Sixth International Conference on Multimodal Interfaces, pp. 144–151. Association for Computing Machinery, New York City (2004)
4. Tan, Z., Thomsen, N.B., Duan, X., et al.: iSocioBot: a multimodal interactive social robot. Int. J. Soc. Robot. **10**(1), 5–19 (2018)
5. Galantucci, B.: An experimental study of the emergence of human communication systems. Cogn. Sci. **29**(5), 737–767 (2005)
6. Sebanz, N., Bekkering, H., Knoblich, G.: Joint action: bodies and minds moving together. Trends Cogn. Sci. **10**(2), 70–76 (2006)
7. Haddadin, S., Croft, E.: Erratum to: Physical human–robot interaction. In: Siciliano, B., Khatib, O. (eds.) Springer Handbook of Robotics, p. E1. Springer, Cham (2016). https://doi.org/10.1007/978-3-319-32552-1_81
8. Fang, R., Doering, M., Chai, J.Y.: Embodied collaborative referring expression generation in situated human-robot interaction. In: Proceedings of the 10th Annual ACM/IEEE International Conference on Human-Robot Interaction, pp. 271–278. Association for Computing Machinery, New York City (2015)
9. Lemaignan, S., Warnier, M., Sisbot, E.A., et al.: Artificial cognition for social human–robot interaction: an implementation. Artif. Intell. **247**(1), 45–69 (2017)
10. Piwek, P.: Salience in the generation of multimodal referring acts. In: Proceedings of the 2009 International Conference on Multimodal Interfaces, pp. 207–210 (2009)
11. Admoni, H., Scassellati, B.: Social eye gaze in human-robot Interaction: a review. J. Hum.-Robot Interact. **6**(1), 25–63 (2017)
12. Ruhland, K., Peters, C.E., Andrist, S., et al.: A review of eye gaze in virtual agents, social robotics and HCI: behaviour generation, user interaction and perception. Comput. Graph. Forum **34**(6), 299–326 (2015)
13. Yücel, Z., Salah, A.A., Meriçli, Ç., et al.: Joint attention by gaze interpolation and saliency. IEEE Trans. Cybern. **43**(3), 829–842 (2013)
14. Breazeal, C., Dautenhahn, K., Kanda, T.: Social robots. In: Siciliano, B., Khatib, O. (eds.), Springer Handbook of Robotics. 2nd edn. pp. 1935–1961. Springer Handbooks, Heidelberg (2016).https://doi.org/10.1007/978-3-319-32552-1
15. Devault, D., Kariaeva, N., Kothari, A., et al.: An information-state approach to collaborative reference. In: Proceedings of the ACL Interactive Poster and Demonstration Sessions, pp. 1–4 (2005)
16. Mutlu, B., Shiwa, T., Kanda, T., et al.: Footing in human-robot conversations: how robots might shape participant roles using gaze cues. Hum. Factors **2**(1), 61–68 (2009)
17. Shannon, C.E.: A mathematical theory of communication. Bell Syst. Tech. J. **1**(1), 379–423 (1948)
18. Holmqvist, K., Andersson, R.: Paradigms - visual search. In: Eye Tracking: A Comprehensive Guide to Methods and Measures, pp. 409–413. Oxford University Press, Oxford (2011)

Author Index

Printed in the United States
By Bookmasters